AF389475

Synthetic Aperture Radar (SAR) Meets Deep Learning

Synthetic Aperture Radar (SAR) Meets Deep Learning

Editors

Tianwen Zhang
Tianjiao Zeng
Xiaoling Zhang

MDPI • Basel • Beijing • Wuhan • Barcelona • Belgrade • Manchester • Tokyo • Cluj • Tianjin

Editors
Tianwen Zhang
University of Electronic Science
and Technology of China
China

Tianjiao Zeng
University of Hong Kong
Hong Kong

Xiaoling Zhang
University of Electronic Science
and Technology of China
China

Editorial Office
MDPI
St. Alban-Anlage 66
4052 Basel, Switzerland

This is a reprint of articles from the Special Issue published online in the open access journal *Remote Sensing* (ISSN 2072-4292) (available at: https://www.mdpi.com/journal/remotesensing/special_issues/synthetic_aperture_radar_meets_deep_learning).

For citation purposes, cite each article independently as indicated on the article page online and as indicated below:

LastName, A.A.; LastName, B.B.; LastName, C.C. Article Title. *Journal Name* **Year**, *Volume Number*, Page Range.

ISBN 978-3-0365-6382-4 (Hbk)
ISBN 978-3-0365-6383-1 (PDF)

Contents

Preface to "Synthetic Aperture Radar (SAR) Meets Deep Learning"

A synthetic aperture radar (SAR) is an important active microwave imaging sensor, whose all-day and all-weather working capacity give it an important place in the remote sensing community. Since the United States launched the first SAR satellite, SAR has received much attention in the remote sensing community, e.g., in geological exploration, topographic mapping, disaster forecast, and traffic monitoring. It is valuable and meaningful, therefore, to study SAR-based remote sensing applications.

In recent years, deep learning represented by convolution neural networks has promoted significant progress in the computer vision community, e.g., in face recognition, the driverless field and Internet of things (IoT). Deep learning can enable computational models with multiple processing layers to learn data representations with multiple-level abstractions. This can greatly improve the performance of various applications. Today, scholars are realizing the potential value of deep learning in remote sensing. Many remote sensing application techniques incorporate deep learning, e.g., in target and oil spill detection, traffic surveillance, topographic mapping, AI-based SAR imaging algorithm updating, coastline surveillance, and marine fisheries management.

Interestingly, when SAR meets deep learning, how to use this advanced technology correctly needs to be considered carefully, and how to achieve the best performance of this "black-box" model also needs careful consideration. Notably, deep learning uncritically abandons traditional hand-crafted features and relies excessively on the abstract features of deep networks. Is this reasonable? Can the abstract features of deep networks fully represent a real SAR? Should the traditional hand-crafted features provided with mature theories and elaborate techniques be abandoned completely? These questions are worth pondering when one applies various deep learning techniques to the SAR remote sensing community. In general, deep learning is always proposed for natural optical images whose imaging mechanisms are greatly different from those of SARs.

When SAR meets deep learning, should an SAR adapt itself to deep learning, or should deep learning adapt itself to an SAR? The relationship between the two needs further exploration and research. Furthermore, is deep learning really suitable for SARs? The number of SAR samples is far smaller than that of natural optical images. In this case, could we ensure that deep networks are able to thoroughly learn SAR mechanisms?

This Special Issue provides a platform for researchers to handle the above significant challenges and present their innovative and cutting-edge research results when applying deep learning to SARs in various manuscript types, e.g., articles, letters, reviews and technical reports.

Tianwen Zhang, Tianjiao Zeng, and Xiaoling Zhang

Editors

Editorial

Synthetic Aperture Radar (SAR) Meets Deep Learning

Tianwen Zhang [1], Tianjiao Zeng [2,3] and Xiaoling Zhang [1,*]

[1] School of Information and Communication Engineering, University of Electronic Science and Technology of China, Chengdu 611731, China
[2] Electrical and Electronic Engineering, University of Hong Kong, Hong Kong 999077, China
[3] School of Aeronautics and Astronautics, University of Electronic Science and Technology of China, Chengdu 611731, China
* Correspondence: xlzhang@uestc.edu.cn

Citation: Zhang, T.; Zeng, T.; Zhang, X. Synthetic Aperture Radar (SAR) Meets Deep Learning. *Remote Sens.* **2023**, *15*, 303. https://doi.org/10.3390/rs15020303

Received: 6 December 2022
Accepted: 3 January 2023
Published: 4 January 2023

1. Introduction

Synthetic aperture radar (SAR) is an important active microwave imaging sensor. Its all-day and all-weather working capacity makes it play an important role in the remote sensing community. Since the launch of the first SAR satellite by the United States [1], SAR has received extensive attention in the remote sensing community [2], e.g., geological exploration [3], topographic mapping [4], disaster forecast [5,6], and marine traffic management [7–10]. Therefore, it is valuable and meaningful to study SAR-based remote sensing applications [11].

In recent years, with the rapid development of artificial intelligence, deep learning (DL) [12] has been applied to all walks of life, such as face recognition, automatic driving, search recommendation, internet of things, and so on. The DL represented by convolutional neural network (CNN) is promoting the evolution of many algorithms and the innovation of advanced technologies. At present, scholars are exploring the application value of DL in SAR remote sensing field. Many SAR remote sensing application technologies based on DL have emerged, such as land surface change detection, ocean remote sensing, sea-land segmentation, traffic surveillance and topographic mapping.

Aiming to promote the application of DL in SAR, we initiated this Special Issue and collected a total of 14 papers (including 12 articles, 1 review and 1 technical note) covering various topics, e.g., object detection, classification and tracking, SAR image intelligent processing, data analytics in the SAR remote sensing community and interferometric SAR technology. The overview of contribution is in the following section.

2. Overview of Contribution

On the topic of object detection, classification and tracking, Li et al. [13] summarized the dataset, algorithm, performance, DL framework, country and timeline of DL-based ship detection methods. They analyzed the 177 published papers about DL-based SAR ship detection and attempted to stimulate more research in this field. Xia [14] proposed a visual transformer framework based on contextual joint-representation learning referred to as CRTransSar. CRTransSar combined the global contextual information perception of transformers and the local feature representation capabilities of convolutional neural networks (CNNs). It was found to produce more accurate ship detection results than other most advanced methods. Note that the authors also released a larger-scale SAR multiclass target detection dataset called SMCDD. Feng et al. [15] established a lightweight position-enhanced anchor-free SAR ship detection algorithm called LPEDet. They designed a lightweight multiscale backbone and a position-enhanced attention strategy for balancing detection speed and accuracy. The results showed that their method achieved a higher detection accuracy and a faster detection speed than other state-of-the-art (SOTA) detection methods. Xu et al. [16] presented a unified framework combining triangle distance IoU loss

(TDIoU loss), an attention-weighted feature pyramid network (AW-FPN), and a Rotated-SARShip dataset (RSSD) for arbitrary-oriented SAR ship detection. Their method showed superior performance on both SAR and optical image datasets, significantly outperforming the SOTA methods. Xiao et al. [17] proposed a simple, yet effective, self-supervised representation learning (Lite-SRL) algorithm for the scene classification task. Note that they successfully evaluate the on-board operational capability of Lite-SRL by transplanting Lite-SRL to the low-power computing platform NVIDIA Jetson TX2. Kačan et al. [18] explored object classification on a raw and a reconstructed Ground-based SAR (GBSAR) data. They revealed how processing raw data provides overall better classification accuracy than processing reconstructed data, and revealed the value of this method in industrial GBSAR applications where processing speed is critical. Bao et al. [19] proposed a guided anchor Siamese network (GASN) for arbitrary targets of interest (TOI) tracking in Video-SAR. GASN used a matching function for returning the most similar area, followed by a guided anchor subnetwork to suppress false alarms. GASN realized the TOI tracking with high diversity and arbitrariness, outperforming SOTA methods.

On the topic of SAR image intelligent processing, Tan et al. [20] proposed a feature-preserving heterogeneous remote sensing image transformation model. Through decoupling network design, the method enabled enhancing the detailed information of the generated optical images and reducing its spectral distortion. The results in SEN-2 satellite images revealed that the proposed model has obvious advantages in feature reconstruction and the economical volume of the parameters. Zhang et al. [21] proposed a self-supervised despeckling algorithm with an enhanced U-Net called SSEUNet. Unlike previous self-supervised despeckling works, the noisy-noisy image pairs in SSEUNet were generated from real-world SAR images through a novel generation training pairs module, making it possible to train deep convolutional neural networks using real-world SAR images. Finally, experiments on simulated and real-world SAR images show that SSEUNet notably exceeds SOTA despeckling methods. Habibollahi et al. [22] proposed a DL-based change detection algorithm for bi-temporal polarimetric SAR (PolSAR) imagery called TCD-Net. In particular, this method applied three steps as follows: (1) pre-processing, (2) parallel pseudo-label training sample generation based on a pre-trained model and the fuzzy C-means (FCM) clustering algorithm, and (3) classification. TCD-Net could learn more strong and abstract representations for the spatial information of a certain pixel, and was superior to other well-known methods. Fan et al. [23] proposed a high-precision, rapid, large-size SAR image dense-matching method. The method mainly included four steps: down-sampling image pre-registration, sub-image acquisition, dense matching, and the transformation solution. The experimental results demonstrated that the proposed method is efficient and accurate, which provides a new idea for SAR image registration. Zhang et al. [24] proposed A low-grade road extraction network Based on the fusion of optical and SAR images at the decision level called SDG-DenseNet. Furthermore, they verified that the decision-level fusion of road binary maps from SAR and optical images can significantly improve the accuracy of low-grade road extraction from remote sensing images.

On the topic of data analytics in the SAR remote sensing community, Wangiyana et al. [25] explored the impact of several data augmentation (DA) methods on the performance of building detection on a limited dataset of SAR images. Their results showed that geometric transformations are more effective than pixel transformations and DA methods should be used in moderation to prevent unwanted transformations outside the possible object variations. The study could provide potential guidelines for future research in selecting DA methods for segmentation tasks in radar imagery.

On the topic of interferometric SAR technology, Pu et al. [26] proposed a robust least squares phase unwrapping method called PGENet that works via a phase gradient estimation network based on the encoder–decoder architecture for InSAR. Experiments on simulated and real InSAR data demonstrated that PGENet outperformed the other five well-established phase unwrapping methods and was robust to noise.

3. Conclusions

Recently, as many SAR systems have been put into use, massive SAR data are available, providing important support for exploring how to apply DL to SAR fields. A large number of SAR data coupled with the DL methodology jointly promote the development of SAR fields. The Special Issue shows innovative applications in object detection, classification and tracking, SAR image intelligent processing, data analytics in the SAR remote sensing community and interferometric SAR technology. There is no doubt that applying DL to more SAR fields (such as terrain classification, SAR agriculture monitoring, SAR imaging algorithm updating, SAR forest applications, marine pollution, etc.) is of great significance for earth remote sensing. In addition, we welcome scholars who are interested in applying DL to SAR to contribute to the scientific literature on this subject.

Author Contributions: Writing—original draft preparation, T.Z. (Tianwen Zhang); writing—review and editing, T.Z. (Tiaojiao Zeng) and X.Z. All authors have read and agreed to the published version of the manuscript.

Data Availability Statement: Not applicable.

Acknowledgments: We thank all authors, reviewers and editors for their contributions.

Conflicts of Interest: The authors declare no conflict of interest.

References

1. Born, G.H.; Dunne, J.A.; Lame, D.B. Seasat mission overview. *Science* **1979**, *204*, 1405–1406. [CrossRef] [PubMed]
2. Moreira, A.; Prats-Iraola, P.; Younis, M.; Krieger, G.; Hajnsek, I.; Papathanassiou, K.P. A tutorial on synthetic aperture radar. *IEEE Geosci. Remote Sens. Mag.* **2013**, *1*, 6–43. [CrossRef]
3. De Novellis, V.; Castaldo, R.; Lollino, P.; Manunta, M.; Tizzani, P. Advanced Three-Dimensional Finite Element Modeling of a Slow Landslide through the Exploitation of DInSAR Measurements and in Situ Surveys. *Remote Sens.* **2016**, *8*, 670. [CrossRef]
4. Da Silva, A.D.Q.; Paradella, W.R.; Freitas, C.C.; Oliveira, C.G. Evaluation of Digital Classification of Polarimetric SAR Data for Iron-Mineralized Laterites Mapping in the Amazon Region. *Remote Sens.* **2013**, *5*, 3101–3122. [CrossRef]
5. Khan, S.I.; Hong, Y.; Gourley, J.J.; Khattak, M.U.; De Groeve, T. Multi-Sensor Imaging and Space-Ground Cross-Validation for 2010 Flood along Indus River, Pakistan. *Remote Sens.* **2014**, *6*, 2393–2407. [CrossRef]
6. Martinis, S.; Twele, A.; Strobl, C.; Kersten, J.; Stein, E. A Multi-Scale Flood Monitoring System Based on Fully Automatic MODIS and TerraSAR-X Processing Chains. *Remote Sens.* **2013**, *5*, 5598–5619. [CrossRef]
7. Xu, X.; Zhang, X.; Zhang, T. Lite-YOLOv5: A Lightweight Deep Learning Detector for On-Board Ship Detection in Large-Scene Sentinel-1 SAR Images. *Remote Sens.* **2022**, *14*, 1018. [CrossRef]
8. Zhang, T.; Zhang, X. A Mask Attention Interaction and Scale Enhancement Network for SAR Ship Instance Segmentation. *IEEE Geosci. Remote Sens. Lett.* **2022**, *19*, 4511005. [CrossRef]
9. Xu, X.; Zhang, X.; Shao, Z.; Shi, J.; Wei, S.; Zhang, T.; Zeng, T. A Group-Wise Feature Enhancement-and-Fusion Network with Dual-Polarization Feature Enrichment for SAR Ship Detection. *Remote Sens.* **2022**, *14*, 5276. [CrossRef]
10. Zhang, T.; Zhang, X. HTC+ for SAR Ship Instance Segmentation. *Remote Sens.* **2022**, *14*, 2395. [CrossRef]
11. Zhang, L.; Zhang, L. Artificial Intelligence for Remote Sensing Data Analysis: A review of challenges and opportunities. *IEEE Geosci. Remote Sens. Mag.* **2022**, *10*, 270–294. [CrossRef]
12. LeCun, Y.; Bengio, Y.; Hinton, G. Deep learning. *Nature* **2015**, *521*, 436–444. [CrossRef] [PubMed]
13. Li, J.; Xu, C.; Su, H.; Gao, L.; Wang, T. Deep Learning for SAR Ship Detection: Past, Present and Future. *Remote Sens.* **2022**, *14*, 2712. [CrossRef]
14. Xia, R.; Chen, J.; Huang, Z.; Wan, H.; Wu, B.; Sun, L.; Yao, B.; Xiang, H.; Xing, M. CRTransSar: A Visual Transformer Based on Contextual Joint Representation Learning for SAR Ship Detection. *Remote Sens.* **2022**, *14*, 1488. [CrossRef]
15. Feng, Y.; Chen, J.; Huang, Z.; Wan, H.; Xia, R.; Wu, B.; Sun, L.; Xing, M. A Lightweight Position-Enhanced Anchor-Free Algorithm for SAR Ship Detection. *Remote Sens.* **2022**, *14*, 1908. [CrossRef]
16. Xu, Z.; Gao, R.; Huang, K.; Xu, Q. Triangle Distance IoU Loss, Attention-Weighted Feature Pyramid Network, and Rotated-SARShip Dataset for Arbitrary-Oriented SAR Ship Detection. *Remote Sens.* **2022**, *14*, 4676. [CrossRef]
17. Xiao, X.; Li, C.; Lei, Y. A Lightweight Self-Supervised Representation Learning Algorithm for Scene Classification in Spaceborne SAR and Optical Images. *Remote Sens.* **2022**, *14*, 2956. [CrossRef]
18. Kačan, M.; Turčinović, F.; Bojanjac, D.; Bosiljevac, M. Deep Learning Approach for Object Classification on Raw and Reconstructed GBSAR Data. *Remote Sens.* **2022**, *14*, 5673. [CrossRef]
19. Bao, J.; Zhang, X.; Zhang, T.; Shi, J.; Wei, S. A Novel Guided Anchor Siamese Network for Arbitrary Target-of-Interest Tracking in Video-SAR. *Remote Sens.* **2021**, *13*, 4504. [CrossRef]

20. Tan, D.; Liu, Y.; Li, G.; Yao, L.; Sun, S.; He, Y. Serial GANs: A Feature-Preserving Heterogeneous Remote Sensing Image Transformation Model. *Remote Sens.* **2021**, *13*, 3968. [CrossRef]
21. Zhang, G.; Li, Z.; Li, X.; Liu, S. Self-Supervised Despeckling Algorithm with an Enhanced U-Net for Synthetic Aperture Radar Images. *Remote Sens.* **2021**, *13*, 4383. [CrossRef]
22. Habibollahi, R.; Seydi, S.T.; Hasanlou, M.; Mahdianpari, M. TCD-Net: A Novel Deep Learning Framework for Fully Polarimetric Change Detection Using Transfer Learning. *Remote Sens.* **2022**, *14*, 438. [CrossRef]
23. Fan, Y.; Wang, F.; Wang, H. A Transformer-Based Coarse-to-Fine Wide-Swath SAR Image Registration Method under Weak Texture Conditions. *Remote Sens.* **2022**, *14*, 1175. [CrossRef]
24. Zhang, J.; Li, Y.; Si, Y.; Peng, B.; Xiao, F.; Luo, S.; He, L. A Low-Grade Road Extraction Method Using SDG-DenseNet Based on the Fusion of Optical and SAR Images at Decision Level. *Remote Sens.* **2022**, *14*, 2870. [CrossRef]
25. Wangiyana, S.; Samczyński, P.; Gromek, A. Data Augmentation for Building Footprint Segmentation in SAR Images: An Empirical Study. *Remote Sens.* **2022**, *14*, 2012. [CrossRef]
26. Pu, L.; Zhang, X.; Zhou, Z.; Li, L.; Zhou, L.; Shi, J.; Wei, S. A Robust InSAR Phase Unwrapping Method via Phase Gradient Estimation Network. *Remote Sens.* **2021**, *13*, 4564. [CrossRef]

Technical Note

Data Augmentation for Building Footprint Segmentation in SAR Images: An Empirical Study

Sandhi Wangiyana *, Piotr Samczyński and Artur Gromek

Institute of Electronic Systems, Faculty of Electronics and Information Technology, Warsaw University of Technology, 00-665 Warsaw, Poland; p.samczynski@elka.pw.edu.pl (P.S.); a.gromek@elka.pw.edu.pl (A.G.)
* Correspondence: sandhi.wangiyana.dokt@pw.edu.pl

Abstract: Building footprints provide essential information for mapping, disaster management, and other large-scale studies. Synthetic Aperture Radar (SAR) provides consistent data availability over optical images owing to its unique properties, which consequently makes it more challenging to interpret. Previous studies have demonstrated the success of automated methods using Convolutional Neural Networks to detect buildings in Very High Resolution (VHR) SAR images. However, the scarcity of such datasets that are available to the public can limit research progress in this field. We explored the impact of several data augmentation (DA) methods on the performance of building detection on a limited dataset of SAR images. Our results show that geometric transformations are more effective than pixel transformations. The former improves the detection of objects with different scale and rotation variations. The latter creates textural changes that help differentiate edges better, but amplifies non-object patterns, leading to increased false positive predictions. We experimented with applying DA at different stages and concluded that applying similar DA methods in training and inference showed the best performance compared with DA applied only during training. Some DA can alter key features of a building's representation in radar images. Among them are vertical flips and quarter circle rotations, which yielded the worst performance. DA methods should be used in moderation to prevent unwanted transformations outside the possible object variations. Error analysis, either through statistical methods or manual inspection, is recommended to understand the bias presented in the dataset, which is useful in selecting suitable DAs. The findings from this study can provide potential guidelines for future research in selecting DA methods for segmentation tasks in radar imagery.

Keywords: image augmentation; building extraction; SAR; semantic segmentation

Citation: Wangiyana, S.; Samczyński, P.; Gromek, A. Data Augmentation for Building Footprint Segmentation in SAR Images: An Empirical Study. *Remote Sens.* **2022**, *14*, 2012. https://doi.org/10.3390/rs14092012

Academic Editors: Bahram Salehi, Tianwen Zhang, Tianjiao Zeng and Xiaoling Zhang

Received: 2 March 2022
Accepted: 20 April 2022
Published: 22 April 2022

Publisher's Note: MDPI stays neutral with regard to jurisdictional claims in published maps and institutional affiliations.

1. Introduction

Buildings are the main structures in any urban area. A building's footprint is a polygon surrounding a building's area when viewed from the top. Maintaining this geographic information is vital for city planning, mapping, disaster preparedness, or other large-scale studies. Synthetic Aperture Radar (SAR) imagery provides an advantage over optical sensors by penetrating through clouds and capturing data day-night and in all-weather conditions. This consistently available remote sensing data has attracted researchers to study areas frequently covered by clouds, such as in disastrous situations. Temporal changes are better identified using methods such as Change Detection in large-scale areas [1]. However, its unique properties are difficult for non-experts to analyze. This fact leads to the exploitation of automated methods such as deep learning using the Convolutional Neural Network (CNN).

CNN is known for extracting relevant features automatically by learning the underlying function that maps a pair of input and output examples. Automated building detection in Very High Resolution (VHR) SAR images was demonstrated using CNN in [2,3]. Despite this, detecting buildings in an urban SAR scenery is challenging due to the complex background and multi-scale objects [2]. In an urban area, buildings are visible but difficult to distinguish from each other. The major indicator of a building is the double bounce scattering formed by the grounds and walls that are visible to the sensor. However, in large cities, high-rise buildings can be challenging to detect due to a phenomenon called layover, which projects the building's wall at the ground toward the sensor. A different approach was taken in [4], where a CNN model was trained to predict these layover areas instead of the traditional building footprint. Predicting the visible building region is indeed intuitive in SAR images, but for most GIS applications, the building footprints are still desirable. An extensive search space on various architectures, pre-trained weights, and loss functions for segmenting building footprints from optical and SAR images was performed in [5]. It was found that the diverse building areas and heights in different cities were problematic. Small-area buildings, mostly found in Shanghai, Beijing, and Rio, were undetectable, while high-rise buildings (mostly in San Diego and Hong Kong) degraded the model's performance due to extreme geometric distortions. Those models performed well in cities such as Barcelona and Berlin because most of the buildings were of moderate size and height.

Predicting well on unseen data or the ability to generalize is the main goal of training a deep learning model. It is a generally accepted fact that deep neural networks perform well on computer vision tasks by relying on large datasets to avoid overfitting [6]. Overfitting happens when a model fits its training set too well. This results in low accuracy predictions on novel data. For the task of building footprint extraction, a handful of datasets from optical sensors exist [7,8], but unfortunately, not many datasets with VHR SAR data are available for public usage. Open, high-quality datasets can be used as a standard benchmark to compare different algorithms and methods. The release of the SpaceNet6 dataset [9] was aimed to promote further research on this topic.

For data that are expensive to collect and label, such as radar or medical images, a common technique to boost performance is using data augmentation. Data augmentations (DA) increase the set of possible data points, artificially growing the dataset's size and diversity. It potentially helps the model avoid focusing on features that are too specific to the data used for training, therefore, increasing generalization (the ability to predict well on data not seen during training) without the need to acquire more images [10].

The use of data augmentation for CNN models has proved effective for classifying rare, deadly skin lesions through basic geometric transformations [11] and unique data cleaning methods [12]. In remote sensing, [13] investigated data augmentations for hyperspectral remote sensing images during inference, while [14] performed object augmentation to increase the number of buildings in optical remote sensing images and demonstrated better building extraction performance. In SAR imagery, [15] showed improvements in paddy rice semantic segmentation by applying quarter-circle rotations and random flipping. Random erasing [16] on target ships was performed in [17] to simulate information loss in radar imagery and improve the robustness of object detection. Conventional methods can be used as a form of DA, as demonstrated in [18], which combined handcrafted features from a Histogram of Oriented Gradients (HOG) with automated features from CNN to improve ship classification in SAR imagery.

Generative models are also a popular data augmentation method, by generating synthetic samples that still retain similar characteristics to the original set. It uses an architecture called Generative Adversarial Network (GAN). However, due to the high computation cost, these are generally applied for low-resolution images, such as in target recognition [19,20], or in a limited scene understanding, such as reducing speckle filters [21,22]. Another promising form of adding synthetic data for SAR is by using simulations. In [23], Inverse SAR (ISAR) images were generated using a simulation to augment the limited SAR data for marine floating raft aquaculture detection. Variations in imaging methods and environmental conditions can be consistently created while retaining accurate labels and information about the target. These would be expensive to both collect and analyze using direct measurements. However, the current state-of-the-art simulators are still unable to produce high fidelity images that create a gap between measured and synthetic SAR imagery [24].

To the best of our knowledge, no prior research investigated and compared augmentation methods specifically for building detection from SAR images. In this study, we explored several geometric and pixel transformations and their performance on the SpaceNet6 dataset [9]. The article is organized as follows: Section 2 provides a technical overview of the dataset, CNN model, and training details. All data augmentation methods used in this research are briefly explained in Section 3. The results of the ablation study and the main experiments are presented in Section 4 and discussed in Section 5. Finally, we conclude the article in Section 6.

The main contributions of this paper are:

- Extensive experimentation on data augmentation methods for automated building footprint extraction in SAR images.
- Demonstration and validation showing which methods are effective and their trade-off between performance and resource cost.

2. Materials and Methods

2.1. Dataset Overview

SpaceNet6 was released as a dataset for a competition to extract building footprints from multi-sensor data. The competition data consist of a training set and a testing set. The latter we do not use because there were no labels provided to verify our predictions. The former consists of 3401 tiles of satellite imagery, each from optical RGB, near-infrared (NIR), and SAR. Several works made use of this multi-modal data to boost segmentation performance [5,25,26]. In this study, we are only interested in using the SAR data. A quad polarization X-band sensor mounted on an aerial vehicle was used to take images over the largest port in Europe, the Rotterdam port. The 120 km^2 coverage is split into tiles of 450 m × 450 m, with 0.5 m/pixel spatial resolution in both range and azimuth direction. An example of an image tile is shown in Figure 1.

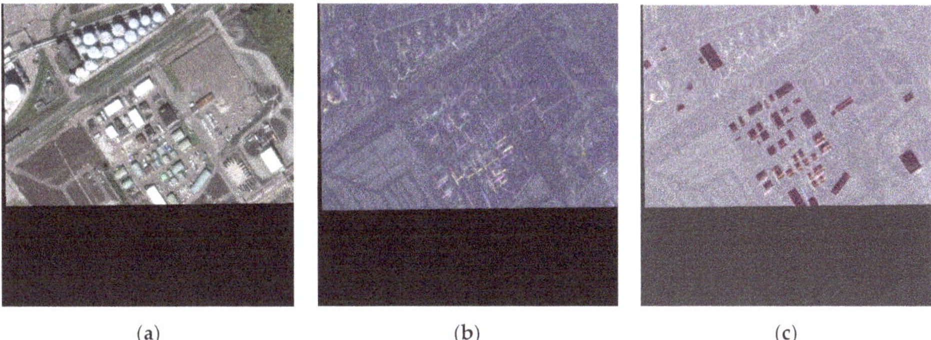

(a)   (b)   (c)

Figure 1. A tile example shows a pair of (**a**) optical RGB and (**b**) SAR shown as false-colored (R = HH, G = VV, B = VH) over a small oil storage area. (**c**) shows the building footprints overlayed on top of the false-colored SAR. One main challenge of this dataset is differentiating patterns of buildings against objects with high backscatter intensity (such as containers, oil silos, and ships) in the port area.

The SAR data has 2 orientations; both were captured using a north and a south facing sensor. Figure 2a shows the tile over a base map of Rotterdam city, marking the position of images from orient1 (or north-facing) in green and orient0 (or south-facing) in red. The direction of flight is indicated by the azimuth (*az*) arrow, while the direction where the sensor is facing is given by the range (*rg*) arrow.

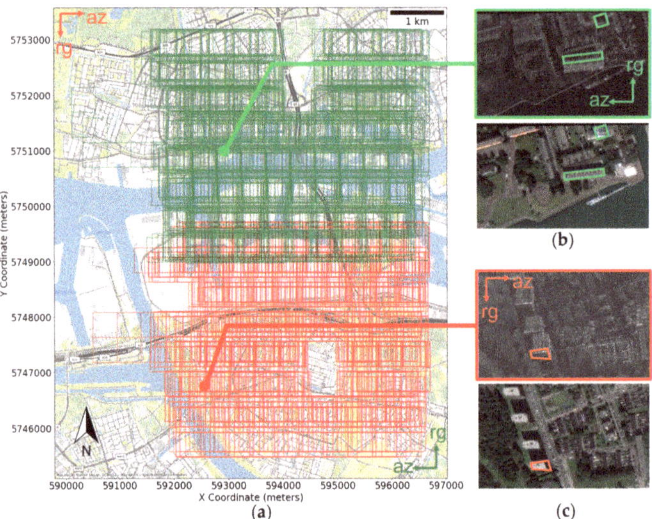

Figure 2. (**a**) The map of Rotterdam Port area (UTM Zone 31N) overlayed with tile boundaries for all 3401 tiles in the SpaceNet6 training set. Green tiles are orient1 while red tiles are orient0. To showcase the data augmentation methods in this study, we only use the green tiles (orient1) for training. Some building footprints are highlighted as an example of layover effects in (**b**) orient1 and (**c**) orient0 tiles. Notice the direction of a layover is always projected towards the sensor's position (near range) while the shadow is cast away from the sensor. The optical RGB image is provided for comparison.

Each orientation creates different characteristics of how a building looks, namely the shadows and layover. The layover creates a projection of the building's body onto the ground. It is caused by the side-looking nature of SAR imaging and the fact that SAR imaging maps pixel values based on the order of received echo. This is especially prominent in tall buildings.

The quad-channel SAR (also known as polarimetric SAR) is excellent for deep learning applications that benefit from the extra data. However, to showcase the effectiveness of data augmentation methods, we only used tiles from orient1 (covering the northern part of the city) and only a single channel HH polarization. This constraint enforces overfitting due to a limited data situation while still maintaining distinct land use cover of the port area and residential buildings. We chose only the HH polarization channel due to it being mostly available in SAR constellations such as GaoFen3, TerraSARX, and Sentinel 1 [27].

2.2. Preparing the Dataset

The purpose of training a Deep Learning model is to predict well on new data, which the model never sees during training. This is called generalization. The proper way to develop a model is to prepare 3 separate datasets called the training set, the validation set, and the testing set. As their names suggest, each is used in a different phase of a model's development. For each iteration of training, called an epoch, the model updates its weights by learning from the training set. At the end of each epoch, the model is tested on the validation set without updating its weights. This is used as a guideline for the Deep Learning practitioner to optimize the model or training parameters by studying the generalization ability over each epoch. Finally, when training is finished, the true performance of a model is determined by its score on a separate testing set.

To train our model, we used roughly half of the SpaceNet6 training set, which was constrained to a single sensor orientation and a single HH channel. We did not use a separate testing set. To evaluate the performance of our model and as a guideline for the impact of augmentation methods, we generated a validation dataset using the expanded version of SpaceNet6. This extra data was released later after the SpaceNet6 competition finished and consisted of unprocessed Single Look Complex (SLC) SAR data with additional polygon labels. We generated tile images from the SLC rasters with the same constraint as our training set: a single orientation and a single HH channel. We matched the exact preprocessing steps explained in [9] using their provided Python library. The steps are:

1. Radiometric calibration using the Capella Space factor provided in each SLC raster's metadata.
2. Converted the complex image into SAR-Intensity.
3. Performed multi-look to reduce speckle noise, using an average convolution operation with a 2×2 kernel.
4. Converted the pixel values to a log scale to create SAR Log-Intensity images.
5. Orthorectified to correct geometric distortions caused by the Earth's complex topography. Furthermore, resampled the raster to a spatial resolution of 0.5 m $\times$ 0.5 m per pixel. An example of a preprocessed SLC raster is shown in Figure 3.
6. Generated non-overlapping tiles with a target resolution of 900×900 pixels within the boundaries of the testing set labels (outside the area in Figure 2a) provided in the expanded dataset. We discard tiles with more than 40% of the no-data region.

Figure 3. One of the SLC stripes after preprocessed and cropped to the boundaries of the testing set labels. These will be further cropped into non-overlapping tiles and later used as our validation dataset.

We used a search algorithm to crop and remove the no-data regions. These are the black regions shown in the left and bottom parts of the image tile in Figure 1. They were produced during the pre-processing stage of the dataset, stemming from the need to have a square non-overlapping tile image. The borders between the raster and the no-data regions created jagged lines due to slight affine rotation during orthorectification. Our experiments have shown these jagged lines to affect the results of pixel transformations, so in order to obtain consistent results, we have cropped them to a clean rectangle shape tile. The code we used in our methods is available as open source in https://github.com/sandhi-artha/sn6_aug (last accessed on 9 April 2022).

2.3. Segmentation Model

Segmentation is the task of partitioning the image into regions based on the similarity (alike characteristic within the same class) and discontinuity (the border or edge between different classes) of pixel intensity. For building footprint extraction, there are only 2 classes: the positive examples, i.e., pixels belonging to a building's region, and negative examples, which are the rest of the pixels (non-building). The deep learning model is given a pair of images and labels for training. The iterative process outputs a similar-sized image classifying each pixel as belonging to one of the two classes.

An encoder-decoder type architecture is commonly used for segmentation models, popularized by the famous UNet [28] and its variations. For aerial imagery, DeepLab v3 [29], PSPNet [30], and Feature Pyramid Network (FPN) [31] are commonly used. Based on a previous study [32], we used the FPN architecture combined with the EfficientNet B4 backbone. EfficientNet is a family of CNN models generated using compound scaling to determine an optimal network size [33]. For a deep learning model, the architecture refers to how each layer in the network is connected, while the backbone refers to the feature extraction part of the model.

Building footprints taken from overhead images have various sizes. To differentiate a building from the background, we need to see enough pixels representing the whole or most of the building. This means a higher spatial resolution is required for detecting buildings with a smaller area. A common method in computer vision to help the model learn these multi-sized objects is to use multi-scale input, i.e., the input image downscaled to different pixel resolutions. This is called an image pyramid (Figure 4).

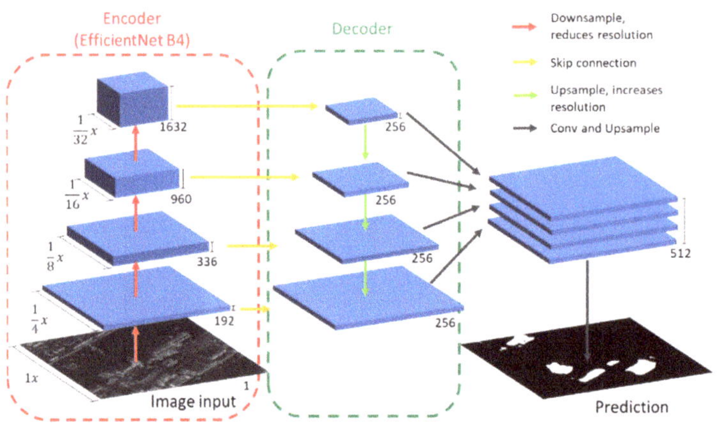

Figure 4. FPN architecture with EfficientNet backbone for segmentation tasks.

FPN uses feature pyramids instead. In the Encoder, the image input is scaled down using a dilated convolution operation which cuts the image dimension in half at each pyramid level. As the data flows up the pyramid, the top layer will have the least width and height (the original input's size divided by 32) but the richest semantic information (1632 feature maps or channels). In a classification task, this is compressed further to output a vector with the same size as the number of classification labels [31]. For a segmentation task, an output with the same spatial size as the input is required. Therefore, the top layer needs to be upscaled.

A 1×1 convolution filter is applied to the final layer in the encoder pyramid to reduce the number of feature maps to 256, without modifying the image dimension. As data flows down the Decoder's pyramid, the width and height increase $2\times$ using nearest neighbors upsampling. In the skip connections (yellow arrow), feature maps from the same pyramid level in the Encoder and Decoder were concatenated. A 1×1 convolution was used to scale the feature maps from the Encoder pyramid to 256. This provides context for better localization as the image gradually recovers in pixel resolution. Afterward, feature pyramids from the Decoder go through a Conv and Upsample operation (black arrow), resulting in modules with 128 feature maps and image dimension 1/4 of the original input. These are then stacked channel-wise, creating a module of 512 feature maps. A final Conv and Upsample operation reduces the number of channels to 1 and restores the image dimension back to the original input [34].

2.4. Training Details

The training was performed in a Kaggle Kernel, a cloud computing environment equipped with a 2-core processor and an Nvidia P100 GPU with 16 GB of video memory (VRAM). The training pipeline was built using the TensorFlow framework. The

Segmentation-Models library [35] was used to combine the FPN architecture with the EfficientNet B4 backbone with no pre-trained weights. Adam [36] was used as the optimizer with default parameters. We used a learning rate scheduler, which configures the learning rate α to gradually decay by a factor of $0.5 * (1 + \cos(n\pi/N))$ where n is the current epoch and N is the total number of epochs.

To evaluate the performance of our model, we use the Intersection over Union (IoU) as the metric. IoU is the ratio of overlapping between the predicted area and the real area (Figure 5). In this case, it is a pixel-based metric. A higher IoU indicates a better predictive accuracy.

$$\text{IoU} = \frac{y_{gt} \cap y_{pred}}{y_{gt} \cup y_{pred}} = \frac{\text{TP}}{\text{TP} + \text{FP} + \text{FN}} \tag{1}$$

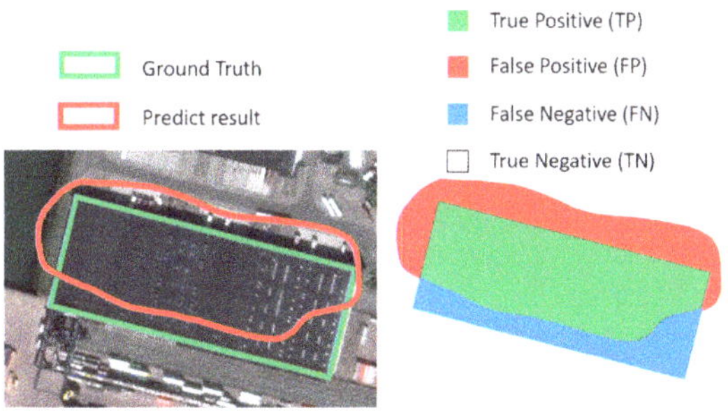

Figure 5. How IoU is calculated over an optical image of a warehouse building.

True Positives (TP) are pixels labeled as building and are correctly predicted as building. True Negatives (TN) are pixels labeled as background and are correctly predicted. False Negatives (FN) are misclassified background pixels, while False Positives (FP) are misclassified pixels of buildings.

Calculating statistics over each image tile in the training set, 20% of tiles have less than 1% positive samples (pixels classified as buildings) (Figure 6a). This indicates that most tiles contain high negative samples (background pixels). One must be cautious in selecting a loss function for training a model on a skewed data distribution such as this because the negative samples will dominate the predictions. For example, using a binary cross-entropy as the loss function, the model will obtain a minor error even if it predicted the whole image as background pixels.

We experimented with several loss functions and concluded that Dice Loss leads to better convergence in this dataset. It is based on the Dice Coefficient, which is used to calculate the similarity between 2 samples based on the degree of overlapping. Dice loss is simply $1 - Dice\ Coefficient$. This results in a loss or error score ranging from 0 to 1, where 0 indicates a perfect and complete overlap.

$$Loss_{Dice} = 1 - 2 * \frac{y_{gt} \cap y_{pred}}{y_{gt} + y_{pred}} \tag{2}$$

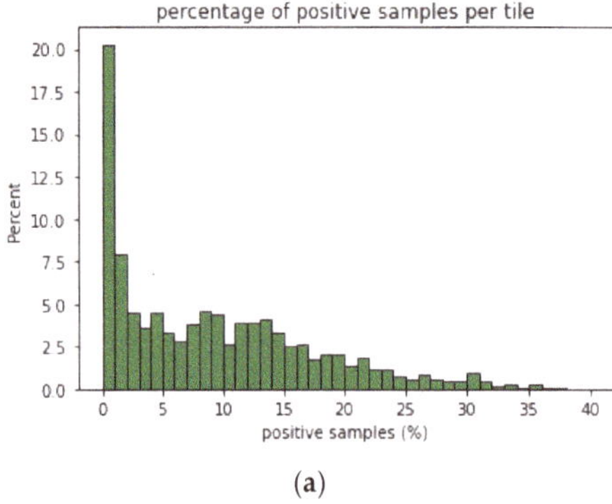 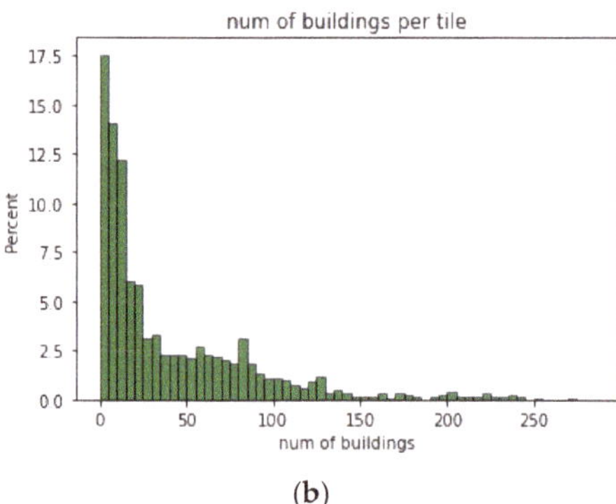

Figure 6. Per image tile statistics for the training set, normalized. (**a**) shows the distribution of positive samples compared to the total number of pixels in an image tile and 20% of tiles have less than 1% of total pixels categorized as buildings. (**b**) shows the number of buildings count for each image tile. Most tiles (17.5%) contain less than 5 buildings.

2.5. Ablation Study

First, we studied the impact of each augmentation method in an ablation study, applying the same model and training configuration but applying different transformations to the dataset. To speed up this process, we trained and validated only a subset of our main training set. We divided it into 5 identical columned regions. The first and second columns were used as the mini-training dataset, while the last column was for the mini-validation dataset. We did not include the middle area in the mini training set to create a buffer (notice the high overlapping of tiles in Figure 2a) which prevents data leakage between both sets. The mini-training set and the mini-validation set contain 37% and 23% images of the main training set, respectively. After concluding which augmentation works well for the mini dataset in the ablation study, we applied combinations of positively impactful transformations to the main dataset.

3. Data Augmentation

This section describes the data augmentations used in this study and how they were implemented during the model's development. In general, the geometric transformations (including reduce transformation) were applied using TensorFlow operations, while pixel transformations were applied using the help of the Albumentation library [37].

3.1. Types of Data Augmentation

3.1.1. Reduce Transformation

Cropping the no-data regions result in a varying aspect ratio for each tile that must be addressed. Square-shaped images are preferred as data input to simplify spatial compression and expansion of the image inside the model. The image size also needs to be reduced to fit into the GPU's memory. We decided on the target resolution of 320 pixels by 320 pixels, which allowed the batch size of eight for the single P100 GPU. All resizing methods used the bilinear interpolation. Two main resize methods were tested:

- **Pad Resize**: no-data regions are added back to create a square aspect ratio, taking the minimum pixel value of 0.0. This is similar to the original tile, but now the raster is centered with no jagged borders. Next, the image is downsampled to the target resolution.
- **Distorted Resize**: rectangle shape tile is resized to the square target resolution, distorting the shape of the image, but no black regions are present.

This downsampling process can be exploited to introduce randomness that further increases the diversity of the training samples. Cropping at random locations gave better details than just resizing the whole image (Figure 7). However, because it introduces randomness, these methods cannot be used as a reduction method for the validation dataset:

- **Random Crop**: a random region with the size of 320 × 320 pixels is cropped out of the rectangle image. This preserves pixel scale since no downsampling is performed.
- **Random Crop and Resize**: crops a random location with a random scaling, then downsample it to the target resolution.

Figure 7. Reduce Transformations comparison.

3.1.2. Geometric Transformations

In computer vision tasks, geometric transformations are cheap and easy to implement. However, it is important to be aware of choosing the transformations' magnitude that preserves the label in the image. For example, in optical character recognition, rotating a number by 180° can result in a different label interpretation in the case of the numbers six and nine.

Flipping an image along the horizontal or vertical centerline is a common data augmentation method. Referring to Figure 2a, the range direction rg for this dataset is on the vertical or y-axis, while the flight direction az is on the horizontal or x-axis. The **Horizontal Flip** does not alter the properties of a radar image. It would be as if the vehicle carrying the sensor was moving in the opposite direction. In contrast, the **Vertical Flip** makes the shadows and layovers appear on the opposite side, creating inconsistency.

Rotation helps the model learn the invariant orientation of a building. We performed **Rotation90** or quarter circle rotations $\{90°, 180°, 270°\}$ and **Fine Rotation** with a randomized angle range, e.g., $[-10°, 10°]$. Similar to Vertical Flip, the quarter circle rotation affects the imaging properties of the radar. The fine rotation exposes an area where image data are unknown. There are several ways to "fill" this blank area, and our experiments showed that leaving it to a dark pixel (0.0) gave the best result.

Shear is a distortion along a specific axis used to modify or correct perception angles (Figure 8). Despite SAR being a side-looking imaging device, the processed SAR image appears flat owing to the orthorectification process that corrects geometric distortions. In **ShearX**, the edges of the image that are parallel to the x-axis stay the same, while the other two edges are displaced depending on the shear angle range. **ShearY** is the exact opposite. We set the shear rotation to be randomized between an angle range of $[-10°, 10°]$.

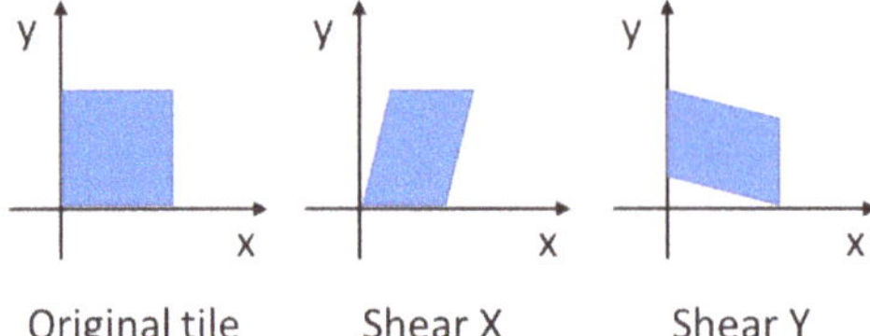

Figure 8. Shear transformation in 2 directions.

Random Erasing is an augmentation method inspired by the drop-out regularization technique [16]. It simply selects a random patch or region in the image and erases the pixels within that region. The goal is to increase robustness to occlusion by forcing the model to learn an alternative way of recognizing the covered objects. We have decided to fill the patch values with a dark pixel (0.0). The random erasing was implemented using CoarseDropout class from the Albumentation library. The region's width and size were randomized from 30 pixels to 40 pixels, and the number of regions created was randomized from two to ten patches. The proposed geometric transformations are shown in Figure 9.

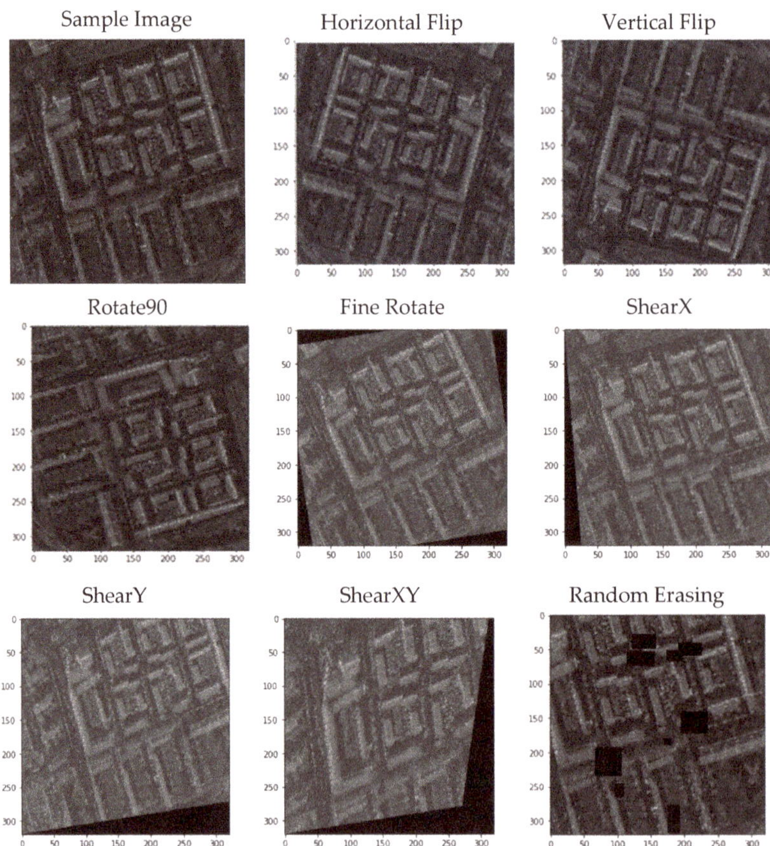

Figure 9. Summary of Geometric Transformations.

3.1.3. Pixel Transformations

In airborne sensors, unknown perturbations of the sensor's position relative to its expected trajectory can cause several defects such as radiometric distortions and image defocusing. To increase the model's robustness when encountering these defects, we used noise injection methods. The common **Gaussian Noise** was generated using the GaussNoise class, while **Speckle Noise** was amplified by multiplying each pixel with random values using the MultiplicativeNoise class, both from the Albumentation library. Some images suffered from defocusing due to an unpredicted change in the flight trajectory, causing fluctuations in the microwave path length between the sensor and the scene [38]. We simulate this defocusing by applying **Motion Blur** with random kernel size using the MotionBlur class from Albumentation.

We used **Sharpening** with a high pass filter to improve edge detection. Consequently, this also increases other high-frequency components such as speckle.

The SAR images in this dataset are SAR-Intensity in the log scale, also referred to as dB images. From samples of the dataset, we observed that the dB images have a pixel distribution close to Rayleigh and mostly populate a narrow area. Figure 10 shows how histogram equalization methods stretch the pixel distribution to span the full range, increasing contrast and improving edge visibility. Contrast Limited Adaptive Histogram Equalization (**CLAHE**) works best when there are large intensity variations in different parts of the image, limiting the contrast amplification to preserve details. We applied it using the CLAHE class from the Albumentation library. The proposed pixel transformations are shown in Figure 11.

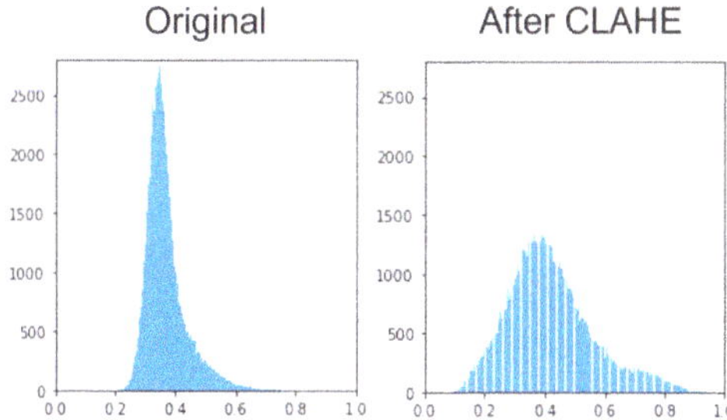

Figure 10. Histogram of a SAR Log Intensity image, before and after applying histogram equalization using CLAHE.

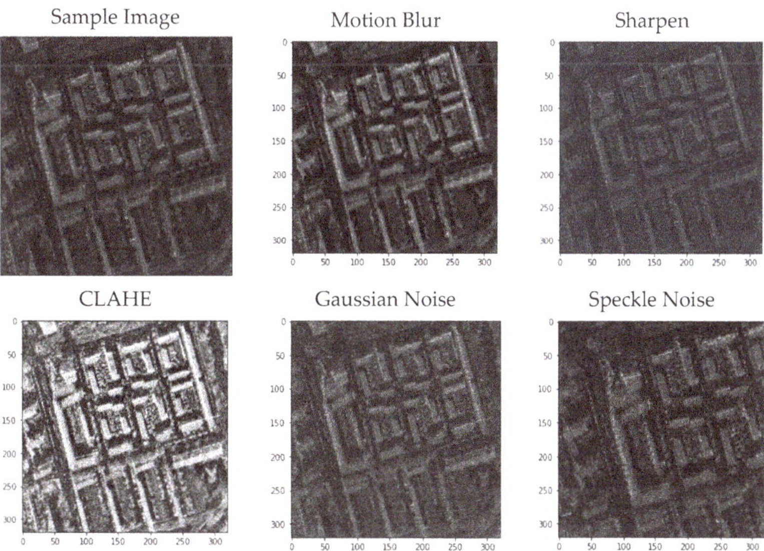

Figure 11. Summary of Pixel Transformations.

3.1.4. Speckle Filters

Although still categorized as Pixel Transformation, this subsubsection was created to provide more explanation for Speckle Filters. Similar to other coherent imaging methods (laser, ultrasound), SAR suffers from noise-like phenomena called speckle. This happens due to the interference of multiple scatterers within a resolution cell. Speckle takes form in granular variations of pixel values that can lower the interpretability of a SAR image for computer vision systems as well as human practitioners.

Speckle reduction filters such as a Box filter can smooth the speckle using a local averaging window. This is effective in homogenous areas, but in applications requiring high-frequency information such as edges, filters that can adapt to local texture can better preserve information in heterogeneous areas [39]. A previous study [32] has shown a slight performance gain by applying Low Pass filters with varying strength on the UNet model. In this research, we inspected the use of well-known adaptive speckle filters as a form of data augmentation, namely Enhanced Lee Filter (**eLee**), **Frost** Filter, and Gamma Maximum A Posteriori (**GMAP**) Filter.

In Figure 12a, two sample crops of size 150×150 pixels are shown for comparing filtration results. A good filter should retain the average mean of an image while reducing speckle [40]. In homogenous areas, the standard deviation should ideally be 0. Speckle filters were applied in MATLAB to the SAR Intensity image (linear scale) and later converted back to the Log-Intensity image (dB scale). Results of filtering are shown in Figure 12c,d, while the quantitative comparison is presented in Table 1. We can observe that the GMAP filter was slightly better at preserving the average value and reducing variance in crops 1 and 2, respectively. The image's Equivalent Number of Looks (ENL) also showed a reduction in speckle variance for all filters.

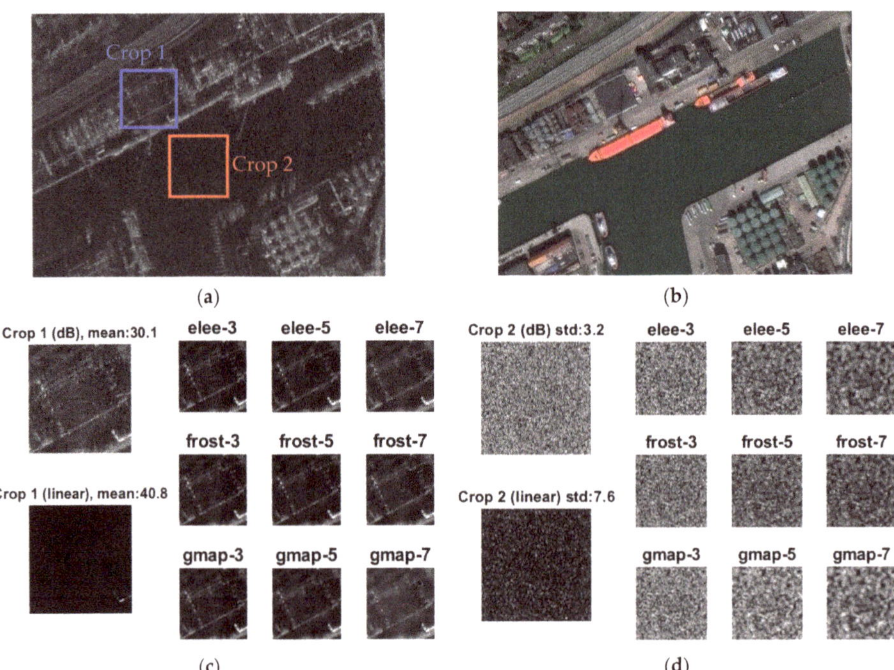

Figure 12. A sample SAR Log Intensity (dB) Image (**a**) with its equivalent optical RGB (**b**). We further analyze two distinct crop regions of a building (crop 1) and a homogenous water area (crop 2). The results of filtering with selected speckle filters and various kernel sizes are given in (**c**,**d**). The image in linear scale appeared dark because of the wide dynamic range.

Table 1. Quantitative comparison of the applied filters. The mean, standard deviation (std), and Equivalent Number of Looks (ENL) are calculated in the linear scale.

Filter	Size	Crop 1—Mean	Crop 2—Std	Image ENL
	3	43.3	4.3	7.0
eLee	5	44.0	2.8	17.2
	7	44.2	2.1	33.6
	3	43.3	4.6	6.1
Frost	5	44.0	3.4	12.8
	7	44.0	2.8	19.2
	3	42.5	4.3	7.1
GMAP	5	41.9	2.8	17.5
	7	41.3	2.1	34.2
Original		40.8	7.6	1.0

3.2. Data Augmentation Design and Strategy

We set a probability of 50% for an image to load either the original version or with data augmentations applied. The magnitude of the transformation was also randomized in a value range, increasing the variation in every iteration of training, except for flipping and quarter circle rotation, which had a limited set of transformations.

The order of transformations is important when multiple augmentations are combined during the main experiment. Pixel transformations are applied first to prevent the presence of no-data pixels from affecting the results. Following it is a reduced transformation, and finally, a geometrical transformation. When using multiple pixel transformations, it is important to combine them into a "One Of" group. By chance, only one of the transformations will be applied, preventing the creation of a disastrous result. In geometric transformation, there is no grouping, so an image has a chance to go through all transformations, which might increase no-data pixels but are generally less harmful than multiple filtering operations.

As categorized in [10], there are three stages of applying Data Augmentation (DA): Online (on-the-fly), Offline, and Test Time Augmentation (TTA).

In Online DA, the input data is manipulated during training. This can lead to a bottleneck if a fast accelerator is used in training, but the augmentation algorithm is slow, leaving the accelerator mostly waiting for data. The advantage is that it does not store the inflated data in storage. On the other hand, Offline DA allows complex manipulation and will not bloat training time. However, since it is applied before training, it takes up storage, and the variations are pre-determined (less randomness). We used Offline DA only for speckle filtered images since they were processed outside the TensorFlow environment, and an image was stored for every applied filter. Other transformations in this study used Online DA, which can have a finer degree of randomness in every iteration.

In TTA, A additional images are generated from each test image x, where A is the number of augmentations applied during the inference or prediction stage. The model will then predict on $A + 1$ samples and the average sum will be taken as the final prediction. This method of predicting multiple transformed versions of the input mimics the theory of ensemble learning, where a group of models using different architectures or trained on different data combines their prediction to increase generalization. This was investigated in [41], concluding that TTA helped reduce overconfident incorrect predictions compared to when using only a single model.

In a classification task, averaging predictions are straightforward since the output is an array with a size equal to the number of classes. In a segmentation task, one must be cautious to perform augmentations that modify the location of labels (in our dataset, the building footprints). If such methods are used, the solution is simply to revert back to the transformation before averaging the predictions.

4. Results

4.1. Ablation Study

In this isolated experiment, we measured the impact of each augmentation method. The model was trained on the mini-training dataset and evaluated on the mini-validation dataset. Results are shown in Table 2. Loss and IoU are the scores for the training set, while Val Loss and Val IoU are scores from the validation dataset. The training lasted for 60 epochs. The four metrics were taken at the best epoch, which is when the model obtained the highest Val IoU. We chose not to take the final score at epoch 60 to showcase the best performance of each augmentation objectively. The score at the last epoch was always worse than the best epoch because the training scores kept getting better while the validation scores stagnated or even deteriorated. This can be visualized in Figure 13. The gap between both training and validation scores was less when an augmentation method was applied, which delayed overfitting, enabling the model to move into what is known as a Local Minima or a temporary performance peak.

Table 2. Results of Ablation study. All scores are in percentage units. For loss, the lower is better. For IoU, the higher the better. Scores are color-coded in comparison to Pad Resize, where green color projects positive gain while red projects negative gain.

Method	Loss	Val Loss	IoU	Val IoU
Reduce Transformations				
Pad Resize	22.04	47.40	64.10	36.33
Distorted Resize	26.35	46.57	58.56	36.95
Random Crop	34.75	47.97	48.92	35.63
Random Crop Resize	27.59	44.98	57.34	38.59
Geometric Transformations				
Horizontal Flip	27.35	46.04	57.58	37.84
Vertical Flip	35.01	53.23	48.52	31.00
Rotation90	45.43	58.10	37.90	27.15
Fine Rotation [−10,10]	24.75	45.73	60.80	37.93
ShearX [−10,10]	22.07	47.42	64.11	36.32
ShearY [−10,10]	23.36	45.76	62.44	37.88
Random Erasing	22.48	47.84	63.56	36.01
Pixel Transformations				
Motion Blur	25.79	48.10	59.45	35.88
Sharpening	20.82	48.20	65.81	36.02
CLAHE	31.20	48.36	52.69	35.53
Gaussian Noise	31.67	46.99	52.67	36.66
Speckle Noise	25.82	46.76	59.43	36.85
Speckle Filter—eLee	23.94	49.43	61.61	34.56
Speckle Filter—Frost	23.39	49.34	62.30	34.78
Speckle Filter—GMAP	20.36	47.39	66.40	36.38

To fit the image resolution to the model's input, we used two basic Reduce methods: Pad Resize and Distorted Resize. A slightly better score obtained using Distorted Resize might indicate that the black paddings were distracting the model because the full scope of the raster's space was not utilized. However, a problem arises when reshaping back the distorted prediction to the original aspect ratio, so we used Pad Resize as the Reduce method for the validation data. Adding randomness to the Reduce method using Random Crop Resize had the biggest performance gain compared to other augmentations. Randomizing the crop size gives the chance to see the image at different scales and details. Random Crop did not perform well due to the small static crop size of a 160 m × 160 m area, increasing the chance of encountering partial parts of a building.

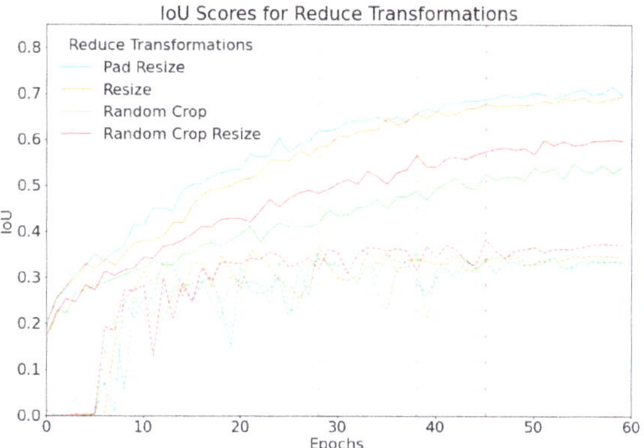

Figure 13. IoU scores comparing the four Reduce Transformations. The solid lines show training IoU scores, while the dashed lines show validation IoU scores. The loosely dotted vertical lines show where the best epoch (highest Val IoU) for a given method. Adding more variations to the input delays the overfitting and is shown by a later best epoch.

Geometric transforms generally increased performance, except for Vertical Flip and Rotation90. Both were detrimental to the performance as they caused the extreme displacement of shadows and layovers' location compared to the actual building footprint. Pixel transforms were not as effective, giving similar or slightly worse scores than the baseline. These augmentations affect the recognition of texture, an important feature when the edges of a building or its shape are unrecognizable due to occlusion or noise. However, this filtering can also be destructive as it also amplifies non-building patterns.

Training scores were generally lower when augmentations were applied, as the model struggles to find the underlying function among the additional variations. A strong increase in training scores for the GMAP speckle filter indicates better recognition of the training data. However, these variations were not shown among the validation data, hence the lower validation scores.

To validate the effects of our proposed data augmentation, we performed the Ablation Study on different mini datasets which we briefly described in the following:

- SAR Orient 0, which had the north facing sensor, opposite Orient 1 (see Figure 2).
- PAN, also from the same SpaceNet6 dataset but uses the single band panchromatic images instead of SAR.
- Inria, a VHR optical RGB aerial imagery at 30 cm spatial resolution. To enforce the limited data configuration, we took only 15 images each from Austin and Chicago as the mini training dataset. As for the mini testing set, 10 images from Vienna were used. Each image was divided into 25 image tiles of 1000 × 1000 pixels. The RGB images were converted into grayscale for a fair comparison with the other single-channel mini datasets.

Figure 14 shows that performance gain and loss on the other datasets mostly agree with results in SAR orient1. Random Crop Resize, Horizontal Flip, and Fine rotation showed consistent gains over all datasets. Meanwhile, Rotation90 showed consistent dips in performance, which are more prominent in datasets from SpaceNet6. The result from PAN highlights the method's impact on a similar geographic region (Rotterdam Port) but a different modality, while the result from Inria highlights the impact when exposed to the different urban settlements of multiple cities. However, due to the stochastic nature of deep neural networks, using an optimized model and training method fitted to one dataset might not translate to an optimal solution on another dataset, which has different distribution [6]. Therefore, these directive insights should be further tweaked when working on a different dataset.

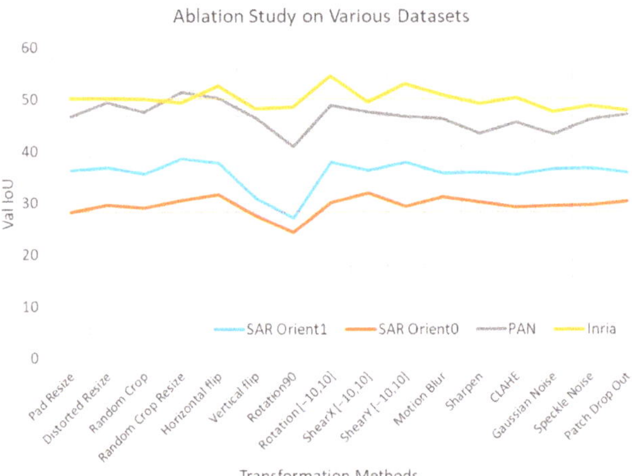

Figure 14. Results of the Ablation Study compared with three other datasets. SAR Orient 0 and PAN are from the same SpaceNet 6 dataset, while Inria is taken from the Inria Aerial Image Labeling dataset [7]. The IoU scores for the baseline method (Pad Resize) are marked with dotted horizontal lines for a straightforward comparison to other augmentation methods.

4.2. Main Experiment

Using similar concepts to those from the Ablation study, we experimented with several combinations of augmentation methods applied to the main training set and evaluated on our prepared validation set. The same model and training parameters were used except for the increased training duration to 90 epochs. This compensates for the additional variation and gives more room for the model to learn. A callback was set to monitor the best Val IoU score and labeled it as the best epoch. The following augmentation schemes were applied:

1. Baseline: No changes after applying a Reduce Transformation
2. Light Pixel: Motion Blur, Sharpen, and Additive Gaussian Noise
3. Light Geometry: Horizontal Flip, Fine Rotation $[-10, 10]$, ShearY $[-10, 10]$
4. Heavy Geometry: Horizontal Flip, Fine Rotation $[-20, 20]$, ShearX $[-10, 10]$, ShearY $[-10,10]$, Random Erasing
5. Combination: Light Pixel + Light Geometry

Only Baseline used Pad Resize as the Reduce method, while the other combinations used Random Crop Resize. For every augmentation scheme, the model's performance was taken at the best epoch and shown in Table 3. Due to different datasets, the scores in Table 2 should not be directly compared to results from this main experiment. The training time t_{train} was measured as the duration of training but shown in the table as a scaling factor compared to Baseline where $t_{train} = \frac{t_{train\_scheme}}{t_{train\_baseline}}$.

In line with results from the Ablation Study, geometric transformations had better scores than pixel transformations. Increasing the magnitude of transformation did not lead to an increase in performance, as shown by the lower scores obtained in Heavy Geometry. Increasing the diversity of transformations in Combination also did not improve performance despite consisting of transformations that showed positive impacts during the Ablation Study.

All models predicted well on medium height elongated residential buildings (Figure 15a). Applying augmentation increases confidence, modeling a more accurate shape characterized by rooftop patterns. However, fine details of a building structure and small buildings remained undetected.

Table 3. Results of combining multiple augmentations. All loss and IoU scores are in percentage. t_{train} is the scale of training time compared to the Baseline. A lower value indicates faster training time. For loss, lower is better. For IoU, higher is better. Scores are color-coded where a darker green indicates a better value.

Augmentation Scheme	Loss	Val Loss	IoU	Val IoU	Best Epoch	t_{train}
Baseline	17.94	44.54	69.82	42.13	47	1.00
Light Pixel	23.47	44.57	62.28	42.72	81	0.98
Light Geometry	28.24	39.85	56.24	47.25	60	1.01
Heavy Geometry	28.90	41.02	55.46	46.12	67	1.09
Combination	29.04	41.27	55.36	46.05	74	1.06

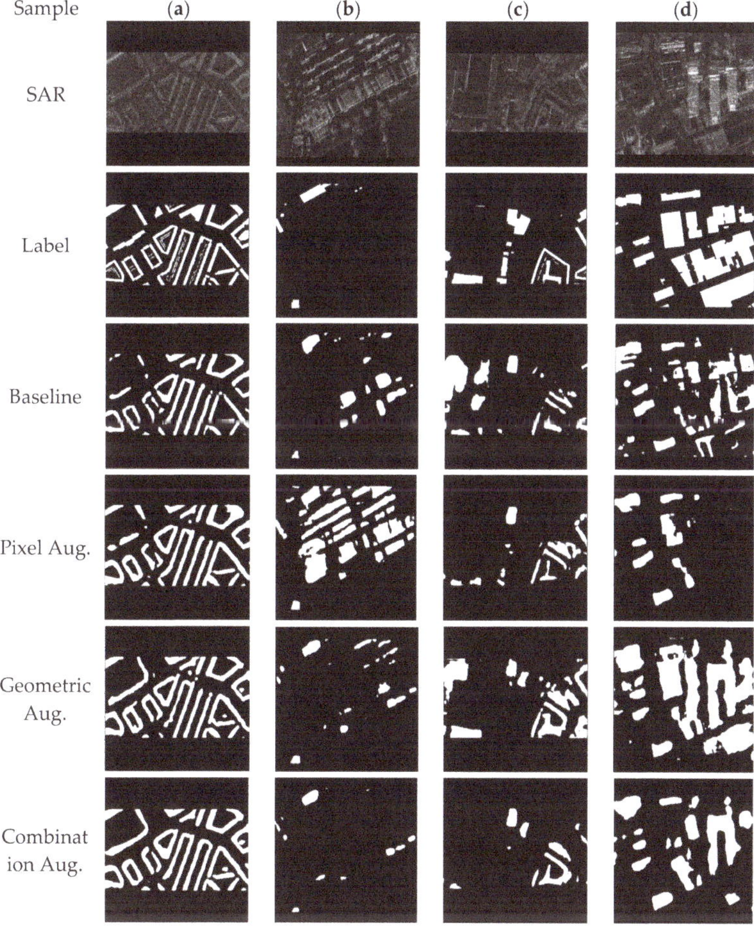

Figure 15. Comparison of predictions from the trained models of different scene objects: (**a**) medium-height residential buildings, (**b**) containers in a shipping port, (**c**) outdoor sports field, and (**d**) high-rise buildings.

For an image tile of large negative samples (pixels belonging to non-building), pixel augmentations drive extra attention to high backscattering objects such as container storage and large shipping/port equipment made of metal (Figure 15b). This leads to an increase in false positives. Geometric augmentations were less prone to this. However, Geometric augmentations overfit non-building objects with building-shaped backscatters, such as the fences surrounding a sports field (Figure 15c). A combination of geometric and pixel augmentations seems to tune down these false positives and correctly recognize them as non-object patterns.

Occlusion was the biggest problem, especially related to high-rise buildings in dense areas (Figure 15d). All models failed to recognize buildings occluded by the overlay of a neighboring high-rise building. Interestingly Geometric augmentations tend to classify the overlayed parts as positives.

Despite not using any augmentations in Baseline, the Light Pixel scheme, applied using the Albumentation library, managed to train slightly faster. This was caused by the callback function, saving the state of the model each time it sees a better-monitored value, requiring additional time on certain epochs. It shows that training duration is not an objective measurement due to the randomness involved. However, we still included it as a comparison. The other schemes show that adding more transformation methods will increase training time.

4.3. Test-Time Augmentation

The state of the models in the Main Experiment was saved at their best epoch, and TTA was applied after the training ended. We experimented with two methods for TTA:

1. TTA_1: Pad Resize was used for reducing image resolution. Transformations include Horizontal Flip, Rotate $\{-10°, 10°\}$ and ShearY $\{-10°, 10°\}$. After predicting each sample variation, an inverse transformation was applied, and the average sum will be used as the final prediction. Total predictions per test sample: six.
2. TTA_2: The rectangle image tile was divided into two square patches with some overlapping in the middle. This slightly increased the detail by utilizing the whole image space and removing the black bars. Afterward, TTA_1 was performed on each tile. Finally, the two prediction patches were combined by averaging the pixel values in the overlapping region. Total predictions per test sample: twelve.

We applied TTA to the Baseline model and the best model from the Main Experiment, which was trained on the Light Geometry scheme. TTA comes at the cost of additional inference time t_{test}, which is a scaling factor compared to the Baseline's inference time. It increases proportionally to the number of augmentations applied. Compared to the time required by re-training a model, the additional inference time to implement TTA was negligible. Results for TTA are shown in Table 4.

Table 4. Results of applying TTA to the Baseline model and Light Geometry. All loss and IoU scores are in percentage. For loss, lower is better. For IoU, higher is better. Scores are color-coded where a darker green indicates a better value.

TTA Scheme	Val Loss	Val IoU	t_{test}
Baseline	44.54	42.13	1.00
Baseline + TTA_1	41.79	44.65	3.35
Baseline + TTA_2	48.01	38.25	6.37
Light Geometry + TTA_1	39.05	47.90	3.39
Light Geometry + TTA_2	37.11	49.69	6.39

The Baseline and Light Geometry model benefit from TTA_1, which consist of simple Geometric transformations. Interestingly, when TTA_2 was applied to the Baseline model, the performance was lower, as it predicted fewer positive samples from the two square patches. The Baseline model, trained on images with a fixed scale, had less confidence in predicting medium-sized buildings compared to the Light Geometry model, which had the chance to see variations of scaling thanks to the Random Crop Resize reduction method.

5. Discussion

Deep learning methods use vast parameter exploration. Results vary with different data, models, and training parameters (also known as hyperparameters). Our goal is to provide intuition based on our experiments of modifying the data while maintaining the model and hyperparameters unchanged. Radar images, including SAR, have unique properties that differ from optical images. Therefore, several transformations can result in poor performance. Selecting augmentation methods requires knowledge of the biases in the training data, either through statistical analysis, in the case of a large dataset, or through the manual inspection of samples. This helps reduce the search space instead of trying every available method.

Tiling is required in remote sensing images as it is impossible to fit a large raster directly to a model. The choice of target resolution will affect the detection of multi-scale objects, such as buildings. Introducing randomness by varying the scale and crop size during dataset loading is a cheap way of boosting performance since there is no need to store extra images, such as in the case of tiling with overlapping regions. However, cropping too much will increase the chance of a large object covering the whole space and hinder performance. No-data regions are inevitable when tiling a large raster, and in our experience, it is better to remove them before feeding the image tile to the model.

Our study shows that pixel transformations are not as effective as geometric transformations. The reason might be that kernel filters, which are the base of most pixel transformations, are already an integral component of the CNN model itself, thus, learnable by an adequately sized model.

Applying transformations at different stages of the model development has different tradeoffs. For instance, Offline DA can utilize the training images with their stored variations, all at once to train the model. We experimented on a dataset transformed with speckle filters and achieved similar scores to Random Crop Resize in Table 2 but at the cost of 3.5× training time. This might be a good option when the data consist of only a few hundred samples and gathering extra data is unfeasible, such as in the case of classifying rare medical images.

We demonstrated that Test Time Augmentations (TTA) is a cheap method to boost test scores. However, applying a set of augmentations during the test did not achieve better scores when compared to applying the same set of augmentations during training. The model predicted the varying test samples better, had it seen these variations during training. Therefore, applying augmentations in both stages will result in better scores. When using Shear and Fine Rotation in TTA, the angle must be kept low because it removes some portion of the image (outside the image boundary) where it will not return when doing the inverse transformation after prediction. This is why quarter-circle rotations and flips are more commonly used as TTA because they retain the full image after inverse transformation.

6. Conclusions

This paper presents several data augmentation methods for semantic segmentation of building footprints in SAR imagery. By artificially increasing the training dataset, we improved the model's generalization on unseen samples from the validation set, thereby reducing overfitting. The results show a 5% increase in Val IoU score when comparing the best augmentation scheme to the baseline model (no augmentation). Data augmentation can be very helpful in situations with limited data, either due to proprietary licenses or an expensive collection process.

For building detection in SAR, geometric transformations were more effective than pixel transformations. However, some transformations (such as vertical flip and quarter circle rotations) that alter key features of a building in SAR images were proven to be detrimental. Therefore, data augmentations must not be overused, especially since it takes more resources to train (either storage or processing time), which does not always lead to a better result. Test Time Augmentations showed an additional performance gain compared to augmentations applied only during training.

We hope this study can be used as a guide for future research to optimize object detection in a limited set of radar imagery or as an inspiration for investigating alternative methods to augment radar images. The search for effective data augmentation methods can be expensive; thus, automated approaches can save time and computing resources. These approaches have not yet been studied in this article. Furthermore, generating new data through generative models or the use of simulations remain interesting avenues to explore.

Author Contributions: Conceptualization, S.W. and P.S.; methodology, S.W.; software, S.W. and A.G.; validation, S.W., P.S. and A.G.; formal analysis, S.W.; investigation, S.W.; resources, S.W.; data curation, S.W.; writing—original draft preparation, S.W.; writing—review and editing, S.W., P.S. and A.G.; visualization, S.W.; supervision, P.S. and A.G.; project administration, S.W. and P.S. All authors have read and agreed to the published version of the manuscript.

Funding: This research received no external funding.

Data Availability Statement: Not applicable.

Conflicts of Interest: The authors declare no conflict of interest.

References

1. Kang, M.; Baek, J. Sar image change detection via multiple-window processing with structural similarity. *Sensors* **2021**, *21*, 6645. [CrossRef] [PubMed]
2. Jing, H.; Sun, X.; Wang, Z.; Chen, K.; Diao, W.; Fu, K. Fine Building Segmentation in High-Resolution SAR Images Via Selective Pyramid Dilated Network. *IEEE J. Sel. Top. Appl. Earth Obs. Remote Sens.* **2021**, *14*, 6608–6623. [CrossRef]
3. Shahzad, M.; Maurer, M.; Fraundorfer, F.; Wang, Y.; Zhu, X.X. Buildings detection in VHR SAR images using fully convolution neural networks. *IEEE Trans. Geosci. Remote Sens.* **2019**, *57*, 1100–1116. [CrossRef]
4. Sun, Y.; Hua, Y.; Mou, L.; Zhu, X.X. CG-Net: Conditional GIS-aware Network for Individual Building Segmentation in VHR SAR Images. *IEEE Trans. Geosci. Remote Sens.* **2022**, *60*, 1–15. [CrossRef]
5. Xia, J.; Yokoya, N.; Adriano, B.; Zhang, L.; Li, G.; Wang, Z. A Benchmark High-Resolution GaoFen-3 SAR Dataset for Building Semantic Segmentation. *IEEE J. Sel. Top. Appl. Earth Obs. Remote Sens.* **2021**, *14*, 5950–5963. [CrossRef]
6. Sun, C.; Shrivastava, A.; Singh, S.; Gupta, A. Revisiting Unreasonable Effectiveness of Data in Deep Learning Era. *Proc. IEEE Int. Conf. Comput. Vis.* **2017**, *2017*, 843–852. [CrossRef]
7. Maggiori, E.; Tarabalka, Y.; Charpiat, G.; Alliez, P. Can semantic labeling methods generalize to any city? The inria aerial image labeling benchmark. *Int. Geosci. Remote Sens. Symp.* **2017**, *2017*, 3226–3229. [CrossRef]
8. Van Etten, A.; Lindenbaum, D.; Bacastow, T.M. SpaceNet: A Remote Sensing Dataset and Challenge Series. *arXiv* **2018**, arXiv:1807.01232.
9. Shermeyer, J.; Hogan, D.; Brown, J.; Van Etten, A.; Weir, N.; Pacifici, F.; Hansch, R.; Bastidas, A.; Soenen, S.; Bacastow, T.; et al. SpaceNet 6: Multi-sensor all weather mapping dataset. *IEEE Comput. Soc. Conf. Comput. Vis. Pattern Recognit. Work.* **2020**, *2020*, 768–777. [CrossRef]
10. Shorten, C.; Khoshgoftaar, T.M. A survey on Image Data Augmentation for Deep Learning. *J. Big Data* **2019**, *6*, 60. [CrossRef]
11. Ronneberger, O.; Fischer, P.; Brox, T. U-Net: Convolutional Networks for Biomedical Image Segmentation. In Proceedings of the International Conference on Medical Image Computing and Computer-Assisted Intervention, Munich, Germany, 5–9 October 2015.
12. Bisla, D.; Choromanska, A.; Berman, R.S.; Stein, J.A.; Polsky, D. Towards automated melanoma detection with deep learning: Data purification and augmentation. *IEEE Comput. Soc. Conf. Comput. Vis. Pattern Recognit. Work.* **2019**, *2019*, 2720–2728. [CrossRef]
13. Nalepa, J.; Myller, M.; Kawulok, M. Hyperspectral Data Augmentation. *arXiv* **2019**, arXiv:1903.05580.
14. Illarionova, S.; Nesteruk, S.; Shadrin, D.; Ignatiev, V.; Pukalchik, M.; Oseledets, I. Object-Based Augmentation for Building Semantic Segmentation: Ventura and Santa Rosa Case Study. In Proceedings of the IEEE/CVF International Conference on Computer Vision, Montreal, BC, Canada, 11–17 October 2021; pp. 1659–1668. [CrossRef]
15. Yang, L.; Huang, R.; Huang, J.; Lin, T.; Wang, L.; Mijiti, R.; Wei, P.; Tang, C.; Shao, J.; Li, Q.; et al. Semantic Segmentation Based on Temporal Features: Learning of Temporal-Spatial Information From Time-Series SAR Images for Paddy Rice Mapping. *IEEE Trans. Geosci. Remote Sens.* **2021**, *60*, 1–16. [CrossRef]
16. Zhong, Z.; Zheng, L.; Kang, G.; Li, S.; Yang, Y. Random erasing data augmentation. In Proceedings of the AAAI Conference on Artificial Intelligence, New York, NY, USA, 7–15 February 2020; pp. 13001–13008. [CrossRef]

17. Song, T.; Kim, S.; Kim, S.; Lee, J.; Sohn, K. Context-Preserving Instance-Level Augmentation and Deformable Convolution Networks for SAR Ship Detection. *arXiv* **2022**, arXiv:2202.06513.

18. Zhang, T.; Zhang, X.; Ke, X.; Liu, C.; Xu, X.; Zhan, X.; Wang, C.; Ahmad, I.; Zhou, Y.; Pan, D.; et al. HOG-ShipCLSNet: A Novel Deep Learning Network with HOG Feature Fusion for SAR Ship Classification. *IEEE Trans. Geosci. Remote Sens.* **2022**, *60*, 1–22. [CrossRef]

19. Ding, J.; Chen, B.; Liu, H.; Huang, M. Convolutional Neural Network with Data Augmentation for SAR Target Recognition. *IEEE Geosci. Remote Sens. Lett.* **2016**, *13*, 364–368. [CrossRef]

20. Cui, Z.; Zhang, M.; Cao, Z.; Cao, C. Image Data Augmentation for SAR Sensor via Generative Adversarial Nets. *IEEE Access* **2019**, *7*, 42255–42268. [CrossRef]

21. Wang, P.; Zhang, H.; Patel, V.M. SAR Image Despeckling Using a Convolutional Neural Network. *IEEE Signal Process. Lett.* **2017**, *24*, 1763–1767. [CrossRef]

22. Zhang, G.; Li, Z.; Li, X.; Xu, Y. Learning synthetic aperture radar image despeckling without clean data. *J. Appl. Remote Sens.* **2020**, *14*, 026518. [CrossRef]

23. Wang, D.; Han, M. SA-U-Net++: SAR marine floating raft aquaculture identification based on semantic segmentation and ISAR augmentation. *J. Appl. Remote Sens.* **2021**, *15*, 016505. [CrossRef]

24. Lewis, B.; Scarnati, T.; Levy, M.; Nehrbass, J.; Zelnio, E.; Sudkamp, E. Machine learning techniques for SAR data augmentation. In *Deep Neural Network Design for Radar Applications*; Scitech Publishing: London, UK, 2021; pp. 163–206.

25. Zheng, Z.; Ma, A.; Zhang, L.; Zhong, Y. Deep multisensor learning for missing-modality all-weather mapping. *ISPRS J. Photogramm. Remote Sens.* **2021**, *174*, 254–264. [CrossRef]

26. Adriano, B.; Yokoya, N.; Xia, J.; Miura, H.; Liu, W.; Matsuoka, M.; Koshimura, S. Learning from multimodal and multitemporal earth observation data for building damage mapping. *ISPRS J. Photogramm. Remote Sens.* **2021**, *175*, 132–143. [CrossRef]

27. Hoeser, T.; Bachofer, F.; Kuenzer, C. Object detection and image segmentation with deep learning on earth observation data: A review—Part II: Applications. *Remote Sens.* **2020**, *12*, 3053. [CrossRef]

28. Russakovsky, O.; Deng, J.; Su, H.; Krause, J.; Satheesh, S.; Ma, S.; Huang, Z.; Karpathy, A.; Khosla, A.; Bernstein, M.; et al. ImageNet Large Scale Visual Recognition Challenge. *Int. J. Comput. Vis.* **2015**, *115*, 211–252. [CrossRef]

29. Chen, L.-C.; Papandreou, G.; Schroff, F.; Adam, H. Rethinking Atrous Convolution for Semantic Image Segmentation. *arXiv* **2017**, arXiv:1706.05587.

30. Zhao, H.; Shi, J.; Qi, X.; Wang, X.; Jia, J. Pyramid scene parsing network. In Proceedings of the IEEE Conference on Computer Vision and Pattern Recognition, Honolulu, HI, USA, 21–26 July 2017; pp. 6230–6239. [CrossRef]

31. Lin, T.Y.; Dollár, P.; Girshick, R.; He, K.; Hariharan, B.; Belongie, S. Feature pyramid networks for object detection. In Proceedings of the IEEE Conference on Computer Vision and Pattern Recognition, Honolulu, HI, USA, 21–26 July 2017; pp. 936–944. [CrossRef]

32. Wangiyana, S.; Samczynski, P.; Gromek, A. Effects of SAR Resolution in Automatic Building Segmentation Using CNN. In Proceedings of the 2021 Signal Processing Symposium (SPSympo), Lodz, Poland, 20–23 September 2021; pp. 289–293.

33. Tan, M.; Le, Q.V. EfficientNet: Rethinking model scaling for convolutional neural networks. In Proceedings of the International Conference on Machine Learning, Long Beach, CA, USA, 10–15 June 2019; pp. 10691–10700.

34. Seferbekov, S.; Iglovikov, V.; Buslaev, A.; Shvets, A. Feature pyramid network for multi class land segmentation. In Proceedings of the IEEE Computer Society Conference on Computer Vision and Pattern Recognition Workshops, Salt Lake City, UT, USA, 18–23 June 2018; Volume 2018, pp. 272–275.

35. Yakubovskiy, P. Segmentation Models. 2019. Available online: https://github.com/qubvel/segmentation_models (accessed on 10 December 2021).

36. Kingma, D.P.; Ba, J.L. Adam: A method for stochastic optimization. In Proceedings of the 3rd International Conference on Learning Representations (ICLR 2015), San Diego, CA, USA, 7–9 May 2015; pp. 1–15.

37. Buslaev, A.; Iglovikov, V.I.; Khvedchenya, E.; Parinov, A.; Druzhinin, M.; Kalinin, A.A. Albumentations: Fast and flexible image augmentations. *Information* **2020**, *11*, 125. [CrossRef]

38. Oliver, C.; Quegan, S. *Understanding Synthetic Aperture Radar Images*; EngineeringPro Collection; SciTech Publ.: Raleigh, NC, USA, 2004; ISBN 9781891121319.

39. Parrilli, S.; Poderico, M.; Angelino, C.V.; Verdoliva, L. A nonlocal SAR image denoising algorithm based on LLMMSE wavelet shrinkage. *IEEE Trans. Geosci. Remote Sens.* **2012**, *50*, 606–616. [CrossRef]

40. Shi, Z.; Fung, K.B. Comparison of digital speckle filters. *Int. Geosci. Remote Sens. Symp.* **1994**, *4*, 2129–2133. [CrossRef]

41. Wang, G.; Li, W.; Aertsen, M.; Deprest, J.; Ourselin, S.; Vercauteren, T. Test-time augmentation with uncertainty estimation for deep learning-based medical image segmentation. In Proceedings of the MIDL 2018 Conference, Amsterdam, The Netherlands, 4–6 July 2018; pp. 1–9.

Review

Deep Learning for SAR Ship Detection: Past, Present and Future

Jianwei Li [1], Congan Xu [1,2,*], Hang Su [1], Long Gao [1] and Taoyang Wang [3]

[1] Information Fusion Institute, Naval Aviation University, Yantai 264000, China; lgm_jw@163.com (J.L.); shpersonal_email@163.com (H.S.); gaolong14@nudt.edu.cn (L.G.)
[2] Advanced Technology Research Institute, Beijing Institute of Technology, Jinan 250300, China
[3] School of Remote Sensing Information Engineering, Wuhan University, Wuhan 430000, China; wangtaoyang@whu.edu.cn
[*] Correspondence: 7520220053@bit.edu.cn

Abstract: After the revival of deep learning in computer vision in 2012, SAR ship detection comes into the deep learning era too. The deep learning-based computer vision algorithms can work in an end-to-end pipeline, without the need of designing features manually, and they have amazing performance. As a result, it is also used to detect ships in SAR images. The beginning of this direction is the paper we published in 2017BIGSARDATA, in which the first dataset SSDD was used and shared with peers. Since then, lots of researchers focus their attention on this field. In this paper, we analyze the past, present, and future of the deep learning-based ship detection algorithms in SAR images. In the past section, we analyze the difference between traditional CFAR (constant false alarm rate) based and deep learning-based detectors through theory and experiment. The traditional method is unsupervised while the deep learning is strongly supervised, and their performance varies several times. In the present part, we analyze the 177 published papers about SAR ship detection. We highlight the dataset, algorithm, performance, deep learning framework, country, timeline, etc. After that, we introduce the use of single-stage, two-stage, anchor-free, train from scratch, oriented bounding box, multi-scale, and real-time detectors in detail in the 177 papers. The advantages and disadvantages of speed and accuracy are also analyzed. In the future part, we list the problem and direction of this field. We can find that, in the past five years, the AP_{50} has boosted from 78.8% in 2017 to 97.8 % in 2022 on SSDD. Additionally, we think that researchers should design algorithms according to the specific characteristics of SAR images. What we should do next is to bridge the gap between SAR ship detection and computer vision by merging the small datasets into a large one and formulating corresponding standards and benchmarks. We expect that this survey of 177 papers can make people better understand these algorithms and stimulate more research in this field.

Keywords: SAR ship detection; SAR dataset; single-stage detector; two-stage detector; anchor free; train from scratch; oriented bounding box; multi-scale detection; deep learning; computer vision

Citation: Li, J.; Xu, C.; Su, H.; Gao, L.; Wang, T. Deep Learning for SAR Ship Detection: Past, Present and Future. *Remote Sens.* **2022**, *14*, 2712. https://doi.org/10.3390/rs14112712

Academic Editor: Dusan Gleich

Received: 6 May 2022
Accepted: 2 June 2022
Published: 5 June 2022

1. Introduction

Synthetic aperture radar (SAR) remote sensing has become one of the important methods for marine monitoring due to its all-day, all-weather advantage. Ship detection in SAR images has broad prospects in both military and civilian fields [1,2].

The traditional detection method includes three steps: sea-land segmentation, CFAR (constant false alarm rate) detection, and discrimination [3,4]. In the sea-land segmentation step, the land pixels are rejected to avoid interference with the CFAR step. The common method is based on GIS (geographic information system) or image features. The gray histogram is the classical feature used for segmentation. In the second step, CFAR is usually used for ship detection. The distribution function is assumed to fit the pixel distribution of the SAR image. K, Weibull, and Rayleigh distribution are usually used in this step. To keep the probability of a false alarm at a constant value, the CFAR algorithm compares the testing pixel with an adaptive threshold that is generated by the local background surrounding the testing pixel. After the pre-screening by CFAR, a discriminator is needed to reject the background. Discriminator includes two procedures: feature designing and classifier designing. According to the feature difference between ship chips and non-ship chips, this step can reduce the number of false alarms. The traditional detection method dominated this field for a long time.

With the development of deep learning-based object detection algorithms in computer vision (CV) [5], SAR researchers also began to seek inspiration from computer vision. There are three reasons that can explain the revival of deep learning. They are the arising of computing power, big data, and corresponding algorithms. As SAR images are not easily accessed, the deep learning-based detection method cannot be used for SAR ship detection at the beginning.

This problem was solved in 2017, as the first dataset SSDD (SAR Ship Detection Dataset) was open to the public. SSDD provides the same data and evaluation criteria for researchers, and it solves the problem that the traditional algorithms lack data and are not comparable in this field. Since then, more and more researchers adopt a deep learning-based method in this area. The deep learning-based algorithms also show great results compared with the traditional CFAR-based method. The active and open characteristics of computer vision also further promote the development of this field. We think that the emergence of SSDD means that this field comes into the deep learning era.

As far as we know, there are 177 papers [6–182] that use deep learning-based algorithms to detect ships in SAR images. However, there are no papers that review them yet. In order to summarize the achievements of the 177 papers and show the way for the future, we specially wrote this paper, hoping to contribute to the development of this field.

The rest of this review is arranged as shown in Figure 1. Section 2 briefly analyzes some work related to our paper. Section 3 summarizes the past of the traditional detection algorithms in SAR images. It mainly includes CFAR, hand-crafted features, and limited shallow representation. Section 4 introduces the present in deep learning-based detectors. We review the 177 papers, divide them into 10 categories and analyze them, respectively. Section 5 shows the future direction of this field. Section 6 is the conclusion of the paper.

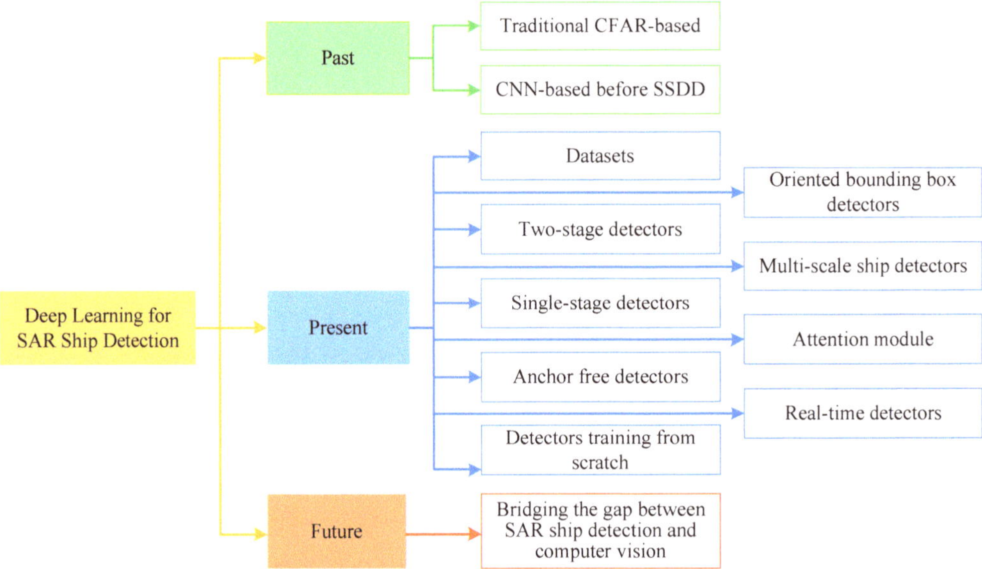

Figure 1. The overall architecture of the paper.

2. Related Work

As far as we know, few researchers have written review papers about this direction. This is partly due to the fact that this direction is new to some extent. At present, only three papers [105,127,170] have performed work related to our work.

Jerzy et al. [105] reviewed the papers from the last 5 years that discuss SAR ship detection. They mainly introduce the development of CFAR methods, CNN (convolutional neural network) based methods, GLRT (generalized likelihood ratio test) based methods, feature extraction-based methods, weighted information entropy-based methods, and variational Bayesian inference-based methods. Compared with paper [105], we mainly focus on the deep learning-based detection methods and do not focus on the traditional methods.

Mao et al. [127] solved the problem of the lack of performance benchmark for state-of-the-art methods on SSDD. Through this work, researchers can compare their work in the same experimental setup. They present 21 advanced detection models, including single-stage, two-stage, train from scratch detection algorithms, and so on. Compared with paper [127], we not only introduce the performance of different public datasets, but also classify all the papers, and summarize the principles and results of the algorithms.

Zhang et al. [170] solved the problem of the coarse annotations and ambiguous standards in SSDD. These improvements are beneficial for a fair comparison. It has played a great role in promoting the healthy development of this field. We suggest that researchers use the standards specified in this paper in the future. Compared with paper [170], our work is not limited to SSDD but introduces other datasets in this field. More importantly, our team has systematically analyzed, classified, and commented on the methods used, and pointed out the future research direction, which is beneficial to the development of this field.

In short, our work is different from the other papers. It is the first comprehensive review of SAR ship detection.

3. Past—The Traditional SAR Ship Detection Algorithms

Traditional detection algorithms in SAR images are based on hand-crafted features and limited shallow-learning representation. It can be divided into three steps: preprocessing, candidate region extraction, and discrimination.

CFAR is a common method for candidate region extraction. It can select potential ship regions. It first statistically models the clutter and then obtains the threshold value according to the false alarm rate. The pixels above the threshold are regarded as ship pixels, and those below the threshold are regarded as background. CFAR is essentially a segmentation-based algorithm, that is, the pixels are classified into two categories (ship or non-ship) according to the gray size, and then the ship pixel region is merged into the ship region. The performance of this method largely depends on the statistical modeling of sea clutter and the parameter estimation of the selected model. According to different SAR image products and practical application requirements, different statistical models such as Gaussian distribution, gamma distribution, log-normal distribution, Weibull distribution, and K distribution are proposed. Gaussian distribution and K distribution are the most commonly used. Generally speaking, when the scene is relatively simple, the CFAR method can achieve better results. However, for small ships and complex offshore scenes, due to the difficulty of modeling, it will have more false positives and poor detection performance.

Discrimination is generally realized by using artificially designed features and training classifiers. In addition to the simple features such as length, width, aspect ratio, and scattering point position, the features introduced from computer vision are also commonly used and have stronger robustness. Such as integral image features, HoG (histogram of oriented gradients), SURF (speeded up robust features), and LBP (local binary pattern). These features improve the performance of the detection algorithm. In classifier designing, decision trees, SVM, gradient boosting, and their improved versions also further improve the performance.

Feature and classifier designing have pushed this field forward in the past few years. However, since the rise of deep learning in 2012 [5], the above ideas are dwarfed in speed and accuracy. The object detection algorithm based on deep learning is an end-to-end processing method, as shown in Figure 2. It does not need to optimize multiple independent steps like the traditional method. It optimizes the whole detection system uniformly. It can adapt to various complex scenes (there is no need for sea–land segmentation in nearshore and port) and has very strong robustness. Therefore, in recent years, deep learning-based SAR ship detection algorithms have become a new research hotspot.

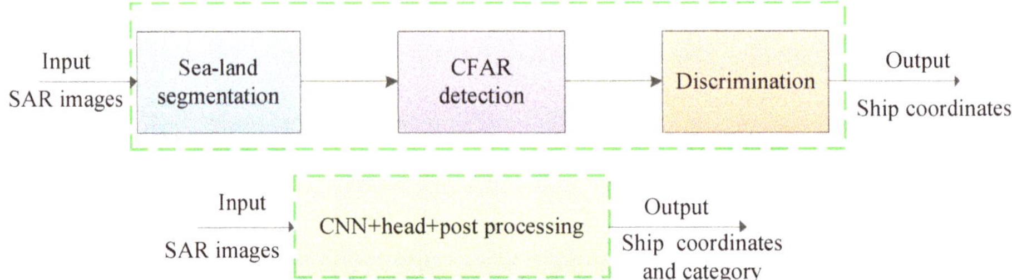

Figure 2. The differences between the CFAR-based detector and the deep learning-based detector.

The advantages and disadvantages of deep learning and traditional-based detection algorithms in SAR images can be proved by qualitative analysis or quantitative experiments.

The detection method based on CFAR has four main shortcomings through qualitative analysis. Firstly, CFAR needs to set the size of the protection window according to the size of the ship. It works well in the case of local uniform clutter in a single ship. If several ships with different sizes are close, the inconsistency between the change of ship size and the fixed protective window will lead to missed detection. Secondly, the CFAR algorithm needs to accurately model SAR images, which is difficult to implement. Thirdly, the essence of CFAR is an unsupervised algorithm, and its performance is essentially worse than the supervised algorithm (Faster R-CNN (region-based convolutional neural network), YOLO (you only look once), SSD (single shot detector), etc.). Fourthly, the CFAR algorithm and discrimination algorithm is a system pieced together after multiple links are debugged separately, and its performance cannot be compared with the end-to-end deep learning algorithm.

Sun Xian carried out a comparative experiment between the classical ship detection algorithm and the deep learning algorithm [65]. In the paper, the classical ship detection algorithms (optimal entropy automatic threshold method and CFAR method based on K distribution) are tested and analyzed on the AIR-SARShip-1.0 dataset. The experiment results are shown in Table 1. We can find that the performance of the deep learning algorithm is significantly better than that of the traditional algorithm.

Table 1. The performance difference between traditional-based detectors and deep learning-based detectors on the same condition [65].

Algorithms	AP (Average Precision)
CFAR method based on K distribution	19.2%
optimal entropy automatic threshold method	28.2%
Faster R-CNN	79.3%
SSD-512	74.3%

Before SSDD, there are six papers that use convolutional neural networks [1–6] to detect ships in SAR images. We think that these six papers are not based on deep learning. The reasons are as follows. Firstly, some algorithms are not end-to-end, they just use CNN as a component in the traditional detection process. Secondly, although some algorithms are end-to-end, the dataset and evaluation criteria are not public, which is difficult for future researchers to reproduce, and the results are also not comparable.

Due to the important role of SSDD, we take the publication date of SSDD paper as the time separation point between the traditional and deep learning-based detection algorithms. Therefore, we believe that ship detection in SAR images entered the era of deep learning on 1 December 2017, as shown in Figure 3. A large number of researchers gradually began to abandon the traditional detection algorithm based on CFAR and adopt the advanced detection method based on deep learning [6–182]. The overview of these deep learning-based detectors in SAR images is the focus of this paper.

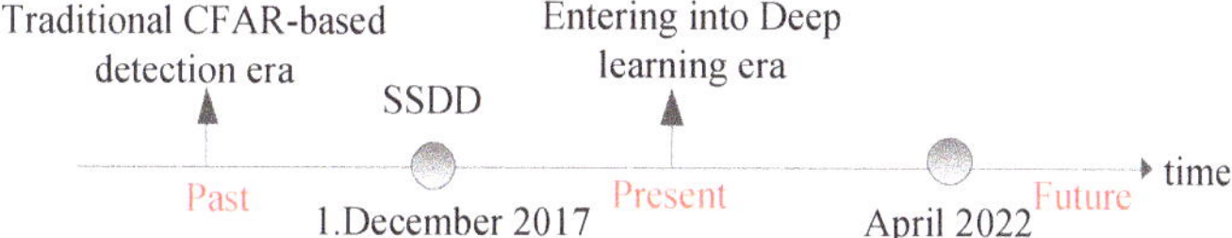

Figure 3. The time divisions of past, present and future.

4. Present—The Deep Learning-Based SAR Detection Algorithms

4.1. The General Overview of the 177 Papers

4.1.1. The Countries

In the country view, we can find that 90% of the papers' authors are Chinese, which is shown in Figure 4. There is no doubt that Chinese researchers have been the mainstream in this direction. Several public datasets are constructed by Chinese researchers, which further prove the above opinion.

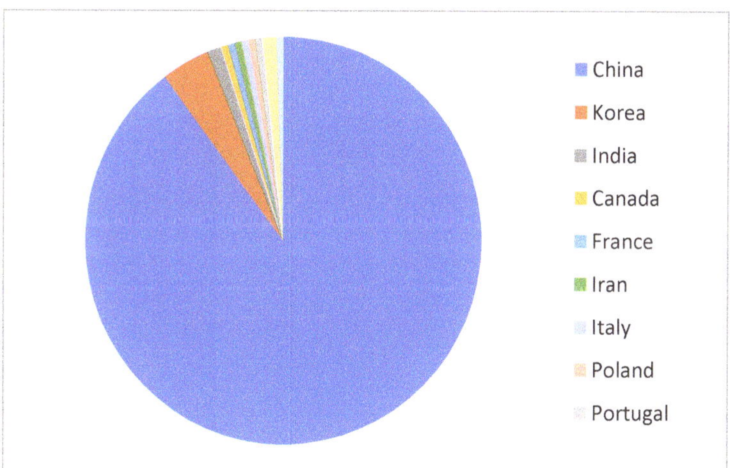

Figure 4. The percentages of the author countries.

4.1.2. Journal or Conference

A total of 63% of the 177 papers are published in journals, and 37% are in conferences. The most common journals and conferences are Remote Sensing and IEEE International Geoscience and Remote Sensing Symposium (IGARSS), respectively.

4.1.3. Timeline of the 177 Papers

The timeline of the deep learning based SAR ship detectors is shown in Figure 5.

Figure 5. The timeline of the 177 papers.

Gray lines and gray circles in the figure represent the time, the purple bar represents the number of papers in the current year, and the red circles represent the public time of the dataset. From the timeline, we can find that in the passing five years, the number of papers about deep learning-based SAR ship detectors becoming more and more. The period 2016–2017 is in the transitional period between traditional and deep learning methods, and there are sporadic papers. Additionally, due to the lack of a unified dataset and the lack of in-depth understanding of deep learning and computer vision algorithms, the application of deep learning algorithms is not thorough. This situation did not change until the emergence of the first dataset (SSDD) paper at the end of 2017. SSDD discloses the data and evaluation criteria it used, which lays the foundation for the rapid development of this field. Since then, the fast lane of research was opened, and a large number of papers were published. The milestones of this field are the several open-access datasets, which are shown in the red circle above. With the increase in available datasets, more and more researchers are paying attention to this field.

4.1.4. The Datasets and Satellites That Are Used

The datasets that those papers used are shown in Table 2. We can see that SSDD is the most frequently used dataset for now. It was used 83 times, 62.4% percent of the total. Additionally, the usage of several public datasets shows a gradual upward trend.

Table 2. The datasets that are used in the past five years.

Datasets	2016	2017	2018	2019	2020	2021	2022	Total
SSDD	0	1	2	19	28	29	4	83
SSDD+	0	0	1	2	1	2	0	6
SAR-Ship-Dataset	0	0	0	1	4	14	1	20
AIR-SARShip1.0/2.0	0	0	0	1	3	5	0	9
HRSID	0	0	0	0	2	6	1	9
LS-SSDD-v1.0	0	0	0	0	1	1	1	3
Official-SSDD	0	0	0	0	0	1	0	1
SRSD-v1.0	0	0	0	0	0	1	0	1
RSDD-SAR	0	0	0	0	0	0	1	1
Total	0	1	3	23	40	58	8	133

Before the first dataset paper was published in 2017, different researchers adopted different SAR images and indicators to test their detectors. Thus, the results of the papers are not comparable. This phenomenon is not beneficial for the development of this field. In order to overcome this, we constructed the first dataset SSDD and it is open to the public. Meanwhile, we provide another dataset called SSDD+, which shares the same images with SSDD but has an oriented bounding box. With the rapid development of deep learning-based computer vision algorithms after 2019, SSDD draws more attention to researchers. Zhang analyzed the usage situation of SSDD in paper [170]. From this paper, we can find that SSDD becomes the most popular dataset, though it has many drawbacks. In addition to this, SAR-Ship-Dataset and AIR-SARShip are showing great potential to be the popular dataset. The other datasets were seldom used, as their public dates are a little late. As deep learning models need more data to prevent overfitting, the future of this field is to merge them into a large dataset and provide the benchmarks on the large dataset with the common detection algorithms in computer vision. We think that, if the dataset is big enough, the benchmark is whole enough, and the maintenance is regular enough, it will be accepted by most researchers. This is the focus of our future work.

Table 3 shows the SAR satellites that papers used besides the public dataset. We can find that SAR images from Sentinel-1 are the most frequently used all the time. This is because the data are easy to acquire, and can be downloaded for free.

Table 3. The satellites that are used in the paper beside the datasets.

Satellites	2016	2017	2018	2019	2020	2021	2022	Total
Sentinel-1	1	4	7	6	6	2	0	26
RadarSat-2	1	0	2	2	2	0	0	7
ALOS PALSAR	0	1	0	0	0	0	0	1
TerraSAR-X	0	1	0	0	3	0	0	4
Gaofen-3	0	1	5	6	5	2	1	20
COSMO_SKYMed	0	0	1	0	2	0	0	3
AISSAR	0	0	1	0	0	0	0	1

However, as China's first C-band multi-polarization SAR satellite Gaofen-3 was officially put into use on 23 January 2017, the policy of obtaining Gaofen-3images has become easier and easier. More and more papers use Gaofen-3 as the source image.

4.1.5. Deep Learning Framework

A deep learning framework can reduce the workload of researchers [183–187]. So, since the emergence of CAFFE (Convolution Architecture For Feature Extraction) [188] in 2017, it gets more and more attention from researchers. Table 4 shows the deep learning framework those 177 papers used. We can find that in the beginning years (2017–2018), CAFFE is the most frequently used framework. It is because CAFFE is the first common deep learning framework that researchers use and most of the detection algorithms in computer vision are based on CAFFE, for example, Faster R-CNN and SSD. In order to improve the efficiency of the deep learning framework, Google provided TensorFlow [189] in 2017. Compared with CAFFE, Tensorflow is more powerful and easier to use. A lot of researchers adopt Tensorflow as their framework. After Tensorflow, PyTorch [190] was promoted by Facebook FAIR in 2017. PyTorch is more suitable for researchers, and the number of users surpasses the Tensorflow gradually. In addition to CAFFE, Tensorflow and Pytorch, Keras, DarkNet and PaddlePaddle are also used by some researchers. Due to the fact that most detection algorithms in computer vision are based on Tensorflow and Pytorch, we recommend researchers in this area use them as the deep learning framework.

Table 4. The deep learning framework those papers used.

Framework	2016	2017	2018	2019	2020	2021	2022	Total
Caffe	0	3	9	3	6	2	0	23
Tensorflow	0	2	3	12	5	7	0	**29**
Pytorch	0	0	0	3	19	18	6	**44**
Keras	0	0	0	1	3	3	0	7
DarketNet	0	0	0	0	1	3	0	4
PaddlePaddle	0	0	0	0	0	1	0	1

4.1.6. Performance Evolution

Tables 5–11 show the performance of several public datasets. In the 'AP' column, the large number represents AP_{50}, and the small one represents AP. AP50 refers to the average precision with IoU (intersection over union) = 50%. AP refers to the value of IoU from 50% to 95% in steps of 5% and then calculates the average value of AP under these IoUs. Normally, AP_{50} is higher than AP. AP_{50} and AP are usually used in PASCAL VOC and MS COCO, respectively. Since the dataset contains only one class of ships, the mAP value is the same as the AP value.

In the table, the italics represent the performance of two-stage detectors, and the non-italics represent the performance of single-stage detectors. The blue, red, green, purple, and golden colors represent anchor-free, train from scratch, oriented bounding box, multi-scale, and attention detectors. The underlines represent real-time detectors.

The number of papers in Tables 5–11 is less than in Table 2. This is because some of the papers in Table 2 did not use the AP or AP_{50} as the evaluation indicator, so we do not show them in Tables 5–11.

From Table 5, we can see that there are 52 papers that are trained and tested on SSDD. Additionally, in the past five years, AP_{50} of detectors on SSDD boosted from 78.8% in 2017 to 97.8 % in 2022. The testing time is also getting faster and faster. What should be noticed is that as the train-test division is ambiguous in the original SSDD, so the AP in Table 5 is not comparable to some extent. That is also why we recommend the following researchers adopt the Improve SSDD [170] as the new standard.

Table 5. The performance evolution of detectors on SSDD (The data come from the 177 papers).

No.	Date	AP	Time	No.	Date	AP	Time
11	*1 December 2017*	*78.8%*	*173 ms*	**104**	**14 October 2020**	**92.6%** **56.5%**	7.39 ms
15	9 March 2019	91.3%	**96 ms**	111	16 November 2020	91.84%	
39	2 April 2019	89.76%	10.938 ms	*115*	*2 December 2020*	*89.79%*	
40	**21 May 2019**	**90.16%**	**21 ms**	116	3 December 2020	90.7%	13.6 ms 74 FPS
43	24 July 2017	79.78%	28.4 ms	117	4 December 2020	88.33%	15 FPS
51	23 September 2019	80.12%	9.28 ms	**118**	**7 December 2020**	**94.6%**	**3.9 ms** **258 FPS**
54	**24 October 2019**	**94.13%**	**9.03 ms** **111 FPS**	121	28 December 2020	95.1%	33 ms
55	24 October 2019	90.04%	87 ms	131	12 February 2021	95.7% 63.4%	
56	14 November 2019	94.7%		**132**	**17 February 2021**	**93.78%**	**202 FPS**
62	*14 November 2019*	*83.4%*		134	27 February 2021	80.45%	
63	*14 November 2019*	*90.44%*	*96.04 ms*	149	17 March 2021	94.41%	31 FPS
68	December 2019	96.93%	8.72 ms	146	23 March 2021	95.52%	
69	*2 January 2020*	*97.9%* *64.6%*	*103 ms*	148	31 March 2021	92.09%	
74	19 March 2020	96.4% 67.4%	106.4 ms	151	13 May 2021	88.08%	12.25 ms
78	*30 March 2020*	*94.2%* *59.5%*	*0.93 M*	154	9 June 2021	98.4%	
81	3 April 2020	94%		**157**	30 June 2021	61.4%	45 FPS
82	16 April 2020	90.08% 68.1%		158	1 July 2021	96.8% 62.7%	438 ms
84	22 April 2020	97.07%	233 FPS	*160*	*13 July 2021*	97.2% 61.5%	
90	25 May 2020	93.96%		161	14 July 2021	95.29%	11 FPS
93	24 June 2020	94.72%	63.2 ms	170	6 December 2021	97.8% 64.9%	
96	**21 July 2020**	**96.08%**	**4.51 ms** **222 FPS**	171	10 December 2021	82.2%	5.2 ms
98	21 August 2020	81.17%	24 ms	173	22 December 2021	97.4%	42.5 FPS
99	21 August 2020	83.4%		174	6 February 2022	95.6% 61.1%	
100	31 August 2020	90.57%	17.2 ms	175	25 February 2022	97.8 %	17.5 FPS
102	**6 October 2020**	**86.3%**		176	25 February 2022	95.03%	47 FPS
103	14 October 2020	95.6% 61.5%		177	19 March 2022	97.0%	

From Table 6, we can see that there are only four papers that are trained and tested on SSDD+, and the AP_{50} performance is increased from 84.2% in 2018 to 94.46% in 2021. The overall performance is a bit lower than that of SSDD. That is because the detectors on SSDD+ should predict an additional parameter (angle). We also find that the SSDD+ is seldom used compared with SSDD. That is, few researchers are interested in oriented bounding box detection in this area.

Table 6. The performance evolution of detectors on SSDD+ (the data come from the 177 papers).

No.	Date	AP	Time
20	29 August 2018	84.2%	40 FPS
41	26 June 2019	81.36%	
83	20 April 2020	90.11%	62.77 ms
124	8 January 2021	94.46%	

From Table 7, we can see that there are 14 papers that are trained and tested on SAR-Ship-Dataset, and the AP_{50} performance is boosted from 89.07% in 2019 to 96.1% in 2021. The running speed is also accelerated to 60.4 FPS with 96.1% AP. The overall performance is a bit lower than that of SSDD. That is because this dataset is relatively larger than SSDD.

Table 7. The performance evolution of detectors on SAR-Ship-Dataset (the data come from the 177 papers).

No.	Date	AP	Time	No.	Date	AP	Time
38	29 March 2019	89.07%		138	17 February 2021	92.4%	
89	20 May 2020	94.7%	18 ms	157	19 May 2021	93.46%	339 FPS
113	30 November 2020	91.89%	12.05 FPS	158	8 June 2021	95.52%	
114	30 November 2020	91.07%		163	1 July 2021	95.8%	
123	5 January 2021	90.25%	22 ms	166	14 July 2021	94.39%	
133	17 February 2021	93.9%		178	22 December 2021	96.1	60.4 FPS
136	17 February 2021	95.1%		179	6 February 2022	95.1	

From Table 8, we can see that there are only four papers are trained and tested on AIR-SARShip, and the AP_{50} performance is boosted from 88.01% in 2019 to 92.49% in 2021. In addition, the running speed becomes 7.98 times faster (from 41.6 ms to 5.22 ms). The overall performance is a bit lower than that of SSDD.

Table 8. The performance evolution of detectors on AIR-SARShip (the data come from the 177 papers).

No.	Date	AP	Time	Version
65	1 December 2019	88.01%	24 FPS	1.0
97	13 August 2020	86.99%		1.0
130	8 February 2021	80.9%		1.0
171	1 December 2021	92.49%	5.22 ms	2.0

From Table 9, we can see that there are only nine papers are trained and tested on HRSID, and the AP_{50} performance is boosted from 89.3% in 2019 to 94.4% in 2021. The overall performance is a bit lower than that on SSDD. That is because this dataset is relatively larger than SSDD.

Table 9. The performance evolution of detectors on HRSID (the data come from the 177 papers).

No.	Date	AP	No.	Date	AP
94	29 June 2020	89.3% 69.4%	168	6 August 2021	89.2% 68%%
110	*10 November 2020*	*not given* *84.4%*	174	14 February 2022	91.4% 66.4%
120	23 December 2020	91.99% 68.5%	175	6 December 2021	94.4% 72%
131	12 February 2021	92.4% 69.5%	178	22 December 2021	88.3%
165	**13 July 2021**	**90.7%** **69.4%**			

From Table 10, we can see that there are only three papers are trained and tested on LS-SSDD-v1.0, and the AP performance is boosted from 72.3% in 2019 to 75.5% in 2022. The overall performance is a bit lower than that on SSDD. LS-SSDD-v1.0 is specially used for large-scale SAR ship detection, which is fit for satellite-based SAR systems. It should be used more in the future.

Table 10. The performance evolution of detectors on LS-SSDD-v1.0 (the data come from the 177 papers).

No.	Date	AP
101	15 September 2020	75.3%
168	6 August 2021	71.7%
180	25 February 2022	75.5%

The above datasets are relatively smaller than the datasets used in computer vision. In order to improve the generalization ability of the detector, researchers should use a large dataset. Some researchers merge several datasets into a large one as shown in Table 11. From Table 11, we can see that there are three papers that are trained and tested on the composite dataset, and the AP performance is 81.13%, 71.4%, and 95.1%, respectively. As deep learning-based detectors are data-hungry, we should merge the public datasets into a large one to prevent over-fitting.

Table 11. The performance evolution of detectors on other datasets (the data come from the 177 papers).

No.	Date	AP$_{50}$	Time	Datasets
108	30 October 2020	81.13%	35.5 ms	SSDD + SAR-Ship-Dataset
125	27 January 2021	71.4%	2920 ms	SAR-Ship-Dataset +AIRSAR-Ship-1.0
167	26 July 2021	95.1%		HRSID + SSDD + IEEE 2020 Gaofen Challenge

4.2. The Algorithm Taxonomy of the 177 Papers

We divide the 177 papers into 10 categories, they are papers about datasets, two-stage detectors, single-stage detectors, anchor-free detectors, train from scratch detectors, detectors with the oriented bounding box, multi-scale detectors, detectors with attention module, real-time detectors, and others. The percentages of each algorithm are shown in Table 12. What should be explained is that the summation of the percentages is larger than 1. This is because many algorithms in the papers have several attributes. For example, it not only belongs to the single-stage detector but is also trained from scratch.

Table 12. The percentage of each algorithm.

Algorithms	Datasets	Two-Stage	Single-Stage	Anchor Free	Scratch
Percentage	5%	26.7%	25.6%	5.1%	4.0%
Algorithms	**Oriented**	**Multi-scale**	**Attention**	**Real-time**	**Others**
Percentage	5.7%	14.2%	5.1%	13.1%	14.2%

From Table 12, we can find the following conclusions. Firstly, there are eight papers that introduce the datasets to the researchers. They make a great contribution to this field. Secondly, two-stage detectors used in this field are slightly more than single-stage detectors. This is partly because the two-stage detectors have higher accuracy than the single-stage detectors in most cases. In addition, accuracy is the first consideration at the moment. Thirdly, anchor-free detectors, detectors trained from scratch, oriented bounding box detectors, and detectors with attention modules almost have a percentage of 5–6% among the 177 papers. This is because, as the above four directions are rare, they are not yet noticed by many researchers. In fact, these directions can overcome the problems of the ship size distribution abnormal and the lack of SAR images. They should be paid more attention in the future. Fourthly, almost 14% of papers are about multi-scale SAR ship detection, which is a little higher than other directions. This is because, compared with objects in computer vision images, ships in SAR are rather small. In order to improve the performance, detectors should pay more attention to multi-scale ships. Fifthly, 14.20% of papers are classified as others, which represents that these papers do not belong to the nine categories. Sixthly, only three papers, which is 1.7% of the 177 papers are reviewed in this field. Considering the active research in this field, it is not enough for now. This is one of the motivations for our work.

4.3. The Public Datasets

4.3.1. Overview

As far as we know, there are 10 public datasets that are used for training and detecting ships in SAR images. They are SSD(SSD+) [11], SAR-Ship-Dataset [38], AIR-SARShip1.0 [65], HRSID [94], LS-SSDD-v1.0 [101], AIR-SARShip2.0 [191], Official-SSDD [170], SRSDD-v1.0 [177] and RSDD-SAR [192]. Table 13 shows the detailed information of the 10 public datasets, in which the annotations of SSDD+, Official-SSDD, SRSDD-v1.0, and RSDD-SAR are the oriented bounding box.

Table 13. Detail information of existing public datasets.

Dataset	Date	Source	Resolution	Image Size	Images/Ships	Annotation
SSDD (SSDD+)	1 December 2017	RadarSat-2 TerraSAR Sentinel-1	1 m–15 m	190–668	1160/2456	vertical oriented
SAR-Ship-Dataset	29 March 2019	Gaofen-3 Sentinel-1	3 m–25 m	256 × 256	43,918/59,535	vertical
AIR-SARShip-1.0 AIR-SARShip-2.0	1 December 2019 25 August 2021	Gaofen-3	1 m, 3 m	3000 × 3000 1000 × 1000	31 300	vertical
HRSID	29 June 2020	Sentinel-1 TerraSAR	0.5 m, 1 m, 3 m	800 × 800	5604/16,951	polygon
LS-SSDD-v1.0	15 September 2020	Sentinel-1	5 m, 20 m	24,000 × 16,000	15/6015	vertical
Official-SSDD	15 September 2021	The same as SSDD				polygon
SRSDD-v1.0	15 December 2021	Gaofen-3	1 m	1024 × 1024	666/2275	oriented recognition
RSDD-SAR	April 2022	Gaofen-3 TerraSAR	2–20 m	512 × 512	7000/10,263	oriented

In addition to these, SMCDD [182] is a good dataset based on China's first commercial SAR satellite HISEA-1. It has 1851 bridges, 39,858 ships, 12,319 oil tanks, and 6368 aircraft. It shows a great advantage in multi-class ship detection.

In the future, it is very necessary to combine the above datasets into a large one to avoid the problem of overfitting.

In the following part, we will introduce the details of the datasets and evaluate their advantages and drawbacks.

4.3.2. SSDD, SSDD+ and Official-SSDD

We made our dataset SSDD publicly available at the conference of 2017BIGSARDATA in Beijing [11]. SSDD is the first open dataset in this community. It can be a benchmark for researchers to train and evaluate their algorithms. In SSDD, there are a total of 1160 images and 2456 ships. The ships in SSDD have rich diversity, including small-size ships, complex backgrounds, and dense arrangements near the wharf. We also give the statistical results of the length, width, and aspect ratio of the ship bounding box in SSDD. The papers that used SSDD and their performance are shown in Table 5.

At the same time, based on 1160 SAR images of SSDD, we use the oriented bounding box to relabel the ship and obtain the dataset SSDD+. SSDD+ is the first dataset for SAR ship detection with an oriented bounding box. The papers that used SSDD+ and their performance are shown in Table 6.

At that time, there were some problems in SSDD due to the lack of understanding of computer vision and deep learning. The drawbacks of SSDD are the coarse annotations and ambiguous standards of use. It hinders fair comparisons and effective academic exchanges in this field.

In September 2021, Zhang [170] systematically analyzed and improved the problem of SSDD; they call it Official-SSDD. Zhang relabeled ships in SSDD and proposed three new datasets; they are bounding box SSDD, rotatable bounding box SSDD, and polygon segmentation SSDD. In addition, they also formulate some standards: the train-test division, the inshore-offshore protocol, the ship-size definition, the determination of the densely distributed small ship ships, and the determination of the densely parallel berthing at ports ship samples. We suggest that follow-up researchers use the Official-SSDD and standards proposed in paper [170] to carry out their relevant research.

4.3.3. SAR-Ship-Dataset

The training of the deep learning model depends on a large amount of data, and the amount of SSDD is relatively small. To solve this problem, Wang Chao [38] constructed a dataset called SAR-Ship-Dataset. SAR-Ship-Dataset contains 43,819 images and 59,535 ships, which are more than SSDD. The sources of SAR-Ship-Dataset are 102 Gaofen-3 images and 108 Sentinel-1 SAR images. These ships have distinct scales and backgrounds. The resolution, incident angle, polarization mode, and imaging mode are also diverse, which are helpful for the deep learning models to fit different conditions. The papers that used SAR-Ship-Dataset and their performance are shown in Table 7.

4.3.4. AIR-SARShip

A dataset containing more diverse scenes and covering various types of ships will help to train a model with better performance, stronger robustness, and higher practicability. In order to achieve the above purpose, Sun Xian constructed a dataset based on the Gaofen-3 satellite, named AIR-SARShip-1.0 [65].

It contains a total of 31 large images. A total of 21 images are training data and the other 10 images are testing data. The image resolutions include 1 m and 3 m. The image size is about 3000×3000 pixels. The information of each image in the dataset includes image number, pixel size, resolution, sea state, scene, and the number of ships. The dataset has the characteristics of a large scene and a small ship.

On the basis of version 1.0, Sun Xian and other researchers added more Gaofen-3 data to build AIR-SARShip-2.0. The dataset contains 300 SAR images. The scene types include ports, islands, reefs, sea surfaces with different levels of sea conditions, etc. The annotation information includes the location of ships, which has been confirmed by professional interpreters.

The papers that used AIR-SARShip and their performance are shown in Table 8.

4.3.5. HRSID

The original SAR image used to construct HRSID [94] includes 99 Sentinel-1B images, 36 TerraSAR-X images, and 1 TanDEM-X image. HRSID has 5604 high-resolution SAR images with 800×800 pixels to meet the needs of actual training for GPU. It is designed for ship detection and segmentation based on CNN, and it only contains one category of ships. It is divided into 65% training set and 35% testing set. It uses polygons to label the ship. In order to reduce the deviation of the ship detection algorithm, the interference derived from the ship is marked as a part of the ship.

According to statistics, the total number of ships marked in HRSID is 16,951, and each SAR image contains an average of three ships. The number of small ships, medium ships, and large ships accounted for 54.5%, 43.5%, and 2% of all ships, respectively. The bounding box areas of small ships, medium ships, and large ships account for more than 0~0.16%, 0.16~1.5%, and 1.5% of SAR images, respectively. Therefore, ships are sparsely distributed in SAR images.

The papers that used HRSID and their performance are shown in Table 9.

4.3.6. LS-SSDD-v1.0

Zhang Xiaoling [101] constructed the SAR ship detection dataset LS-SSDD-v1.0 with a large scene and small ships. The dataset consists of 15 pieces with a size of $24{,}000 \times 16{,}000$ pixels Sentinel-1 SAR images. Each image is directly divided into 600 sub-images with 800×800 pixels. The dataset contains 6015 ships. LS-SSDD-v1 can support researchers to flexibly apply the dataset. The optical information provided in Google Earth software and ship information provided by AIS is used for the annotation of LS-SSDD-v1.0. The coastline of the imaging area in the dataset is relatively complex, the land area is smaller than the ocean area, and the ships in the inland river are densely distributed. The dataset has the following characteristics: contains large scenes, focus on the small ships, rich pure background, etc. It also provides a large number of performance benchmarks of detection algorithms on datasets.

The papers that used LS-SSDD-v1.0 and their performance are shown in Table 10.

4.3.7. SRSDD-v1.0

The original images of SRSDD-v1.0 are from Gaofen-3 [177]. It contains 30 panoramic SAR images of port areas. It is annotated with an oriented bounding box. Optical images (Google Earth or GF-2) are used to assist the annotation. The image size is set to 1024×1024. The annotation format is the same as DOTA. The coordinates of the four corners of the box, the category, and whether it is difficult to identify are given in annotation files.

It contains 666 images. A total of 420 images with 2275 ships include the land cover. A total of 246 images with 609 ships only contain the sea in the background. It has six categories: ore-oil ships (166), bulk cargo ships (2053), fishing boats (288), law enforcement ships (25), dredger ships (263), and container ships (89). The dataset has a certain data imbalance problem.

4.3.8. RSDD-SAR

The RSDD-SAR dataset consists of 84 scenes of Gaofen-3 and 41 scenes of TerraSAR-X. RSDD-SAR has 7000 images, including 10,263 ships, of which 5000 are randomly selected as the training set and the other 2000 as the testing set. By analyzing the distribution of ship angle and aspect ratio in the dataset, it can be found that the angle of ships in the dataset is evenly distributed between $0°$ and $180°$, and the aspect ratio is concentrated between two and six. It indicates that the dataset has the characteristics of arbitrary rotation direction and a large aspect ratio. The dataset has the characteristics of a high proportion of small ships, which can be used to verify the performance of a small ship detection algorithm. The RSDD-SAR dataset contains vast sea areas, ports, docks, waterways, and other scenes with different resolutions, which are suitable for practical applications.

4.4. Two-Stage Detectors

The deep learning-based object detection algorithm can be divided into single-stage detectors and two-stage detectors. The single-stage detectors use a full convolution network to classify and regress these anchor boxes once to obtain the detection results. The two-stage detectors use a CNN to classify and regress these anchor boxes twice to obtain the detection results. The principles of single-stage and two-stage detection algorithms are shown in Figure 6.

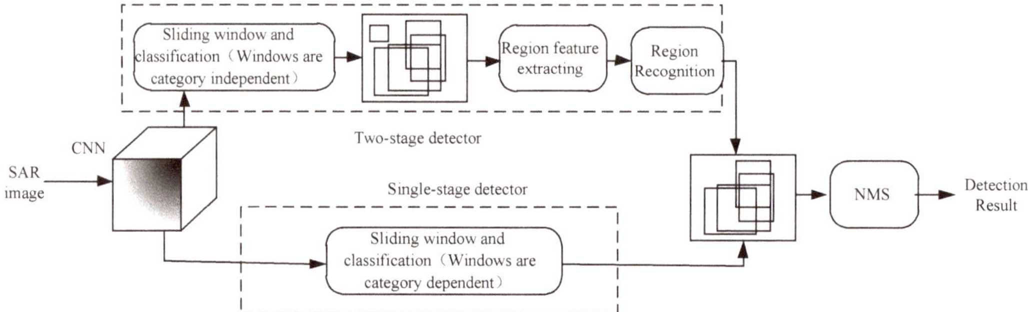

Figure 6. The principle of single-stage and two-stage detectors.

Classical two-stage detectors are Faster R-CNN, R-FCN (fully convolutional network) [193], feature pyramid networks (FPN) [194], Cascade R-CNN [195], Mask R-CNN [196], and so on [197]. Faster R-CNN is the foundation work, and most of the two-stage detectors are improved based on it.

Among the 177 papers, most of the papers are improved from the following aspects: backbone network, region proposal network (RPN), anchor box, loss function, and non-maximum suppressing (NMS). They are shown in Figure 7. Compared with computer vision, the research in this field lags behind, and other more advanced two-stage detection algorithms have not been used here.

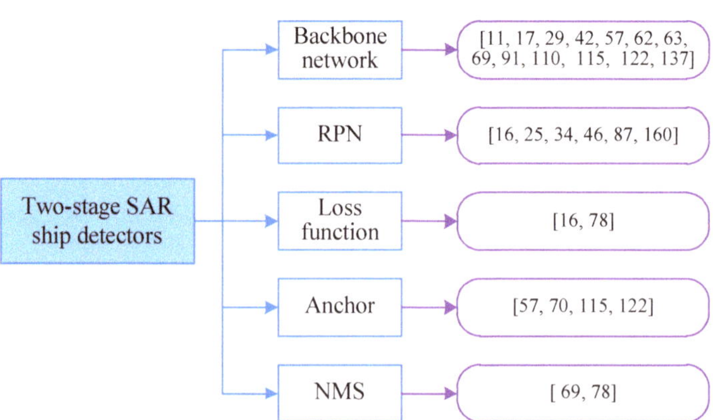

Figure 7. The two-stage SAR ship detectors.

4.4.1. Backbone Network

There are three main directions in the improvement of the backbone network, namely FPN, feature fusion, and attention.

FPN produces a feature pyramid structure that combines low-resolution, which has strong semantic features with high-resolution, which has weak semantic features. It includes a bottom-up channel, a top-down channel, and a skipping connection. It predicts independently at all levels, which only brings minimal additional calculation and storage consumption. It improves the detection result of small-size ships and thus is widely used. A lot of work has been completed to improve FPN in the computer vision field, such as ASFF [198], NAS-FPN [199], and BiFPN [200].

There are six papers [42,62,63,69,91,110] adopted and improved FPN in this field. Cui et al. [42] proposed a DAPN (dense attention pyramid network) structure. It densely connects the convolution block attention module from the top to the bottom of the pyramid network. By this, rich features including resolution and semantic information are extracted for multi-scale ship detection. Li et al. [62] used a convolution block attention module (CBAM) [201] to control the degree of upper- and lower-level feature fusion in FPN. Liu et al. [63] proposed a scale-transferrable pyramid network. It densely connects each feature map from top-to-down using scale-transfer layer. It can expand the resolution of feature maps, which is helpful for detection. Wei et al. [69] adopted a parallel high-resolution feature pyramid network to make full use of the feature mapping of high-resolution and low-resolution convolution for SAR ship detection. Zhao et al. [91] adopted receptive fields block and convolutional block attention module to build a top-down fine-grained feature pyramid. It can capture features of ships with large aspect ratios and enhance local features with their global dependences. Hu et al. [110] used a dense connection to a feature pyramid network, in which the shallow features and deep features are processed differently. It considered the differences between different levels.

There are three papers [11,57,115] that improved the backbone network through feature fusion. Li et al. [11] fused the feature maps from convolutional layer 3 to layer 5. The fusion includes the normalization and 1×1 convolution. Normalizing each RoI (region of interest) pooling tensor can reduce the scale differences between the following layers. It can prevent the 'larger' features from dominating the 'smaller' ones and make the algorithm more robust. This modification stabilizes the system and increases the accuracy. Yue et al. [57] fused the semantically strong features with the low-level high-resolution features, which is helpful for reducing false alarms. Li et al. [115] presented a jump connection structure to extract the features of each scale target in the SAR image. It can improve the ability of recognition and localization.

There are five papers [17,29,62,122,137] that improve the backbone network through the attention module (SENet). It squeezes the feature map along the space and the channel direction, which can explicitly model the interdependence between feature channels, and then automatically obtain the importance of each feature channel through learning. It can improve the useful features and suppress the features that are not useful for the current task according to the importance.

4.4.2. RPN

Another direction is improving the RPN module of Faster R-CNN. Paper [16,25,34,46,87] did not use a single feature map to generate proposals but generated proposals from each fused feature map. Liu et al. [36] designed a scale-independent proposal generation module, which extracts the features such as edge, super-pixel, and strong scattering component from SAR image to obtain ship proposals, and sorts whether the proposals contain ships from the integrity and tightness of the contour. In paper [160], candidate proposals are extracted from the original SAR image and the denoised SAR image, respectively, and then combined to reduce the impact of noise in the SAR image on ship detection. They can improve the performance of multi-size ships to some extent.

4.4.3. Loss Function

Faster R-CNN forces the ratio of positive and negative samples to 1:3 to solve the problem of unbalanced positive and negative proposals. Similar work in the field of computer vision includes focal loss, OHEM (online hard example mining) [202], GHM (gradient harmonizing mechanism) [203], and Libra R-CNN [204]. The paper [16,78], respectively, adopted focal loss to increase the weight of hard negative samples and reduce the weight of simple samples, so as to avoid the problem that a large number of simple samples cover a small number of hard negative samples in the training process.

4.4.4. Anchor and NMS

Faster R-CNN uses three scales and three aspect ratios, producing a total of $60 \times 60 \times 9$ anchor boxes. However, the ship size in the SAR image is extremely small and sparse. There will be a waste in using dense anchor boxes for ship detection in SAR images. Yue et al. [57] and Wang et al. [122] set the parameters of the anchor box based on the analysis of the actual size and distribution of the ship, mainly reducing the size of the anchor box and selecting the appropriate shape. Chen et al. [70] and Li et al. [115] used K-means to obtain the distribution of the ship size, so as to obtain the appropriate anchor box and reduce the difficulty of learning.

Wei et al. [69] and Wang et al. [78] used soft NMS [205] to replace NMS. Soft NMS improves the discrimination process of IoU and threshold in the cycle process and uses weights to attenuate scores to avoid accuracy loss.

4.4.5. Others

ISASDNet (instance segmentation assisted ship detection network) was proposed based on Mask R-CNN in the paper [163]. It has two branches: detection and segmentation. The two branches output interaction to improve the detection results. Gui et al. [34] proposed a lightweight detection head with a large separable convolution kernel and position-sensitive pooling, which improves the detection speed.

4.5. Single-Stage Detectors

The two-stage detectors generate a candidate box first and then identify and regress the candidate box, which is quite different from the principle of human eyes. The single-stage detectors only need to look at the picture once and can predict what the object is and where the object is. It is similar to the human eyes. In addition, they are quite faster than two-stage detectors.

Classical single-stage detectors are YOLO, SSD, RetinaNet [206], and CornerNet [207]. YOLO and SSD are the two most popular single-stage detection algorithms, and most of the subsequent single-stage works are based on them.

The single-stage ship detectors in SAR images are shown in Figure 8.

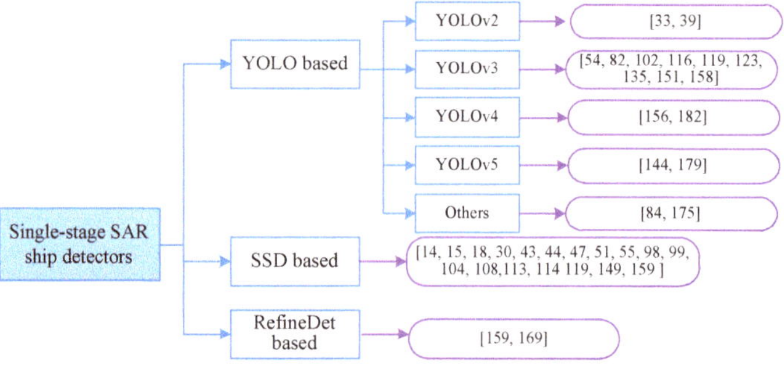

Figure 8. The single-stage SAR ship detectors.

4.5.1. YOLO and SSD Series in Computer Vision

YOLOv1 [186] regards object detection as a regression problem, and it outputs the spatially separated bounding box and related class probability simultaneously. A neural network can predict the bounding box and class probability from the image in one forward calculation. The speed is very fast, but it has an inaccurate location prediction, and the recall is low. YOLOv2 [208] uses the multi-scale training method. It predicts the offset rather than the parameter itself. The offset value is slightly smaller, which can increase the accuracy of prediction. It uses an anchor mechanism to obtain anchor box parameters by clustering the object size in the dataset. The backbone network adopts DarkNet-19. Although the detection head has changed from 7×7 to 13×13, the detection result of a small object is still poor. The YOLOv3 [209] detection head includes three branches: 13×13, 26×26, and 52×52, which can take into account large, medium, and small objects and make the location prediction more accurate. The anchor mechanism of YOLOv3 is the same as that of YOLOv2. YOLOv4 [210] uses two anchors for one ground truth, while YOLOv3 uses only one anchor for one ground truth. With this, the problem of imbalance between positive and negative samples is alleviated. CIoU loss is adopted to solve the problems of MSE (mean squared error) loss, IoU loss, GIoU, and DIoU [211–214]. YOLOv4 also uses several techniques to achieve state-of-the-art results. YOLOv5 adopts adaptive anchors and uses the network to learn anchor parameters. Its detection head is the same as YOLOv3 and YOLOv4. It is slightly weaker than YOLOv4 in performance, but much faster than YOLOv4, and has strong advantages in the rapid deployment of the model.

SSD detection algorithm combines the regression idea with the anchor box (default frame) mechanism. It eliminates the candidate region generation and subsequent pixel or feature resampling stage (RoI pooling) in the two-stage algorithm. It encapsulates all calculations in one network, making it easy to train and very fast. RFBNet [215] and M2Det [216] are two successors of SSD. They use receptive fields and multi-level feature pyramid networks to improve the classical SSD, respectively.

Single-stage SAR ship detection algorithms can be divided into three categories: SAR image ship detection based on the YOLO series, SAR ship detection based on the SSD series, and other algorithms.

4.5.2. SAR Ship Detection Based on YOLO Series

YOLO series are widely used in this field. The improvements mainly focus on lightweight backbone network designing, multi-layer feature fusion, anchor box generation, multi-feature map prediction, loss function, etc.

YOLOv2. Deng et al. [33] and Chang et al. [39] adopted YOLOv2 to detect ships in SAR images. Paper [39] proposed YOLOv2-reduced which reduces some layers of YOLOv2. YOLOv2-reduced has an AP of 89.76% with 10.937 ms and 44.72 BFLOPS compared with YOLOv2, which has an AP of 90.05% with 25.767 ms and 50.17 BFLOPS.

YOLOv3. Zhang et al. [82] accelerated the original YOLOv3 by using DarkNet-19 as the backbone network. Additionally, it reduces the repeated YOLOv3-scale1, YOLOv3-scale2, and YOLOv3-scale3. Zhu et al. [116], Chaudhary et al. [123], and Jiang et al. [135] used the classical YOLOv3 with some techniques to detect ships in SAR images. Wang et al. [119] proposed SSS-YOLO which redesigned the feature extractor network to enhance the spatial and semantic information of small ships. It adopts a PAFN (path argumentation fusion network) to fuse different features in a top-down and bottom-up manner. SSS-YOLO has a better performance for small ships in SAR images. Hong et al. [158] improved the performance of YOLOv3 with some techniques. The improved clustering algorithm K-means++ generates an anchor box, which improves the performance of YOLOv3 for multi-scale ships. The Gaussian parameter for ship detection is introduced to add an uncertainty estimator for the positioning of the bounding box. Four anchor boxes are assigned to each detection scale instead of three in YOLOv3. Zhang et al. [40] used the idea of the YOLO algorithm, the input image meshes, and the depth separable convolution is used to improve the detection speed. MobileNet is used as the feature extractor to detect ships under three scales: 13 × 13, 26 × 26, and 52 × 52. The size of the anchor box can be obtained by the K-means algorithm. D-CNN-13 has a big receptive field with anchor box widths and heights of (9, 11), (11, 22) (14, 26). D-CNN-26 has a medium receptive field with anchor box widths and heights of (16, 40), (17, 12) (27, 57). D-CNN-52 has a small receptive field with anchor box widths and heights of (28, 17), (57, 28) (69, 72). Zhang et al. [54] used depth convolution and point convolution to replace the traditional convolution neural network, and adopt a multi-scale detection mechanism, concatenation mechanism, and anchor box mechanism to improve the detection speed. The detection network is composed of three parts, which means that it can detect an input SAR image under three different scales (5 × 5, 10 × 10, and 20 × 20), and then obtain the final ship detection results. It has nine anchor boxes for three detection scales, so it can detect up to nine ships in the same grid cell. Zhou et al. [102] designed a CNN named LiraNet, which has low complexity, few parameters, and a strong feature representation ability. LiraNet combines the idea of dense connections, residual connections, and group convolution, and it includes stem blocks and extractor modules. The network is the feature extractor of Libra-YOLO. Lira-YOLO has only 2.980 Bflops, and the parameter quantity is only 4.3 MB. It has good accuracy with less memory and computational cost compared with tiny-YOLOv3. In [151], DarkNet-53 with the residual unit is used as the backbone to extract features, and a top-down pyramid structure is added for multi-scale feature fusion. Soft NMS, mix-up, mosaic data augmentation, multi-scale training, and hybrid optimization are used to boost the performance. The 13 × 13, 26 × 26, 52 × 52 feature maps with the large, medium, and small receptive fields are responsible for large, medium, and small ships, respectively. The model is trained from scratch to avoid the learning objective bias of pre-training. The detection speed is fast, about 72 frames per second.

YOLOv4. Ma et al. [156] proposed YOLOv4-light, which is tailored to reduce the model size, detection time, number of computational parameters, and memory consumption. The three-channel images are used for compensating for the loss of accuracy. Liu et al. [181] proposed a detection method based on YOLOv4-Lit [217], whose backbone is MobileNetv2. A receptive field block is used for multi-scale target detection. It has an AP of 95.03% with 47.16 FPS and 49.34 M model size.

YOLOv5. Tang et al. [144] proposed N-YOLO based on YOLOv5. N-YOLO adopts a noise level classifier to classify the noise level of SAR images. SAR ship potential area extraction module is used to extract the complete region of potential ships. Zhou et al. [179] proposed a multi-scale ship detection network based on YOLOv5. It has the cross-stage partial network to improve feature representation capability, and the feature pyramid network with fusion coefficients module to fuse feature maps adaptively. It has a good tradeoff between model size and inference time.

Others. Zhang et al. [84] proposed ShipDeNet-20. It has only 20 convolution layers, and the model size is smaller than 1 MB, which is lighter than the other state-of-the-art detectors. ShipDeNet-20 is based on YOLO and is trained from scratch. Feature fusion module, feature enhance module, and scale share feature pyramid module are proposed to make up the accuracy loss of the raw ShipDeNet-20. It has a good tradeoff between accuracy and speed. Zhu et al. [175] proposed DB-YOLO. It is composed of a feature extraction network, duplicate bilateral feature pyramid network, and detection network. The single-stage network can meet the requirements of real-time detection, and it uses cross-stage partial to reduce redundant parameters. A duplicate bilateral feature pyramid network can enhance the fusion of semantic and spatial information. It alleviates the problem of small ship detection. CIoU loss is used as the loss function, as it has a faster convergence speed and better performance.

4.5.3. SAR Ship Detection Based on SSD Series

Wang et al. [14,18] directly used SSD and do not improve it. Papers [51,98,108] are the detection algorithms trained from scratch based on SSD. Most of the other papers improve the backbone network of SSD to make the model have a stronger feature extraction ability.

Chen et al. [15] adopted a two-stage regression network based on SSD to improve the performance of small ships, namely R2RN (robust two-stage regression network). R2RN connected an anchor modified module and object detection module to inherit the essence of the feature pyramid. Ma et al. [30] proposed an SSD model with multi-resolution input, which can extract richer features. Papers [43,44] applied the attention mechanism to SSD and design a new loss function based on GIoU. Li et al. [47] analyzed the reasons for the low detection accuracy of small and medium-sized ships in SSD and puts forward improvement strategies. Firstly, the anchor box optimization method based on K-means clustering is adopted to improve the matching performance of the anchor box. Secondly, a feature fusion method based on deconvolution is proposed to improve the representation ability of the underlying feature map. Chen et al. [55] adopted the attention mechanism and multi-level features to improve the feature extraction ability of the backbone network. Han et al. [99] used deconvolution to enhance the representation of small ships in the pyramid and improved the detection accuracy of SSD. Zhang et al. [113] token the original SAR image and saliency map as the input and fused the fusion of their features to reduce the computational complexity and network parameters. Chen et al. [114] proposed SSDv2, which adds a deconvolution module and prediction module on the basis of SSD to improve the detection accuracy. Jin et al. [149] improved SSD by feature fusion and squeeze-excitation module.

Sun et al. [162] proposed SANet (semantic attention-based network). It combines semantic attention, focal loss, label, and anchor assigning to improve the performance without increasing computation. Papers [104,159] adopted M2Det to detect ships in SAR images.

4.5.4. Others

RefineDet adopts a two-step cascade regression strategy to predict the position and size of objects. It can make the single-stage detectors obtain the accuracy of the two-stage detector without increasing computation. It is widely used in computer vision. Zhu et al. [159] adopted RefineDet to detect ships in SAR Images, which achieve an AP of 98.4%. In [169], GHM was used as the loss function of RefineDet, so that the network can make full use of all examples, and adaptively increase the weight of difficult cases. A multi-scale feature attention module is added to the network to highlight important information and suppress the interference caused by clutter. It achieves 96.61% precision on AIR-SARShip-1.0.

4.6. Anchor Free Detectors

4.6.1. Development of Anchor Free Detection Algorithm in Computer Vision

The anchor box is the key to the success of Faster R-CNN and SSD. The backbone network extracts features from the input image to obtain the feature map, and each pixel on the map is the anchor point. Taking each anchor point as the central point and artificially setting different scales and aspect ratios, multiple anchor boxes can be obtained. Anchor box has the following two advantages: firstly, it can generate dense candidate boxes, which is convenient for the network to classify and regress the targets. Secondly, it can improve the recall ability and is suitable for small target detection.

However, the anchor box needs to be designed manually by experience, which has the following defects: firstly, hyper-parameters need to be set, such as the number, size, aspect ratio, IoU threshold, etc. Secondly, in order to match the ground box, a large number of anchor boxes need to be generated, which are computationally intensive. Thirdly, most of them are invalid, which will lead to an imbalance between positive and negative samples. Fourthly, it is necessary to adjust the anchor box according to the size and shape distribution of the dataset.

The anchor-free detector opens up another idea by eliminating the predefined anchor box. It can directly predict several key points of the target from the feature map. For example, CornerNet, ExtremeNet [218], CenterNet [219], Objects as Points [220], FCOS (fully convolutional one-stage) [221] and FoveaBox [222].

The anchor-free detectors can avoid various problems and has great application potential in SAR ship detection. For example, due to the small size and sparse distribution of ships, most of the candidate anchor boxes are invalid negative samples. The anchor-free detectors can neglect the invalid anchors and reduce the amount of the predicted boxes, thus improving the accuracy and speed simultaneously. The anchor-free ship detectors in SAR images are shown in Figure 9.

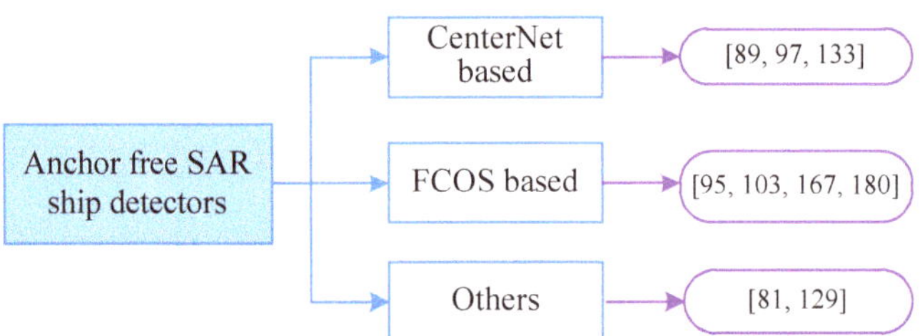

Figure 9. The anchor-free SAR ship detectors.

4.6.2. Development of Anchor-Free SAR Ship Detection Algorithm

Mao et al. [81] proposed a simplified U-Net [223] based anchor-free detector in SAR images. It includes ship bounding boxes regression network and score map regression network. The former is expected to be regressed based on each pixel in the input image. The latter is designed to predict a 2D probability distribution in which each score at each position indicates the likelihood of the current position in the center of any ship. Cui et al. [89] proposed a CenterNet (objects as points) based SAR ship detector. It predicts the center point of the target through key point estimation and uses the image information of the center point to obtain the size and position of the ship. There is no need to set anchors in advance and NMS is not needed, which greatly reduces the number of network parameters and calculations. Anchor mismatching of small ships is also reduced. Spatial shuffle-group enhance attention modules are used to extract features with more semantic information. Fu et al. [95] proposed an attention-guided balanced pyramid based on FCOS to improve the performance of small ships. Zhou et al. [97] proposed an anchor-free detector with dense attention feature aggregation. A lightweight feature extractor and dense attention feature aggregation are used to extract multi-scale features. A center-point-based ship predictor is used to regress the centers and sizes. There is no pre-set anchor and NMS, and thus the computational efficiency is high. Mao et al. [103] proposed a lightweight named ResSARNet with only 0.69 M parameters, and improved FCOS in four aspects: center-ness in bounding box regression branch, not in classification regression branch, center sampling, GIoU loss, and adaptive training sample selection. The network only needs 1.17 M parameters and can achieve 61.5% AP and 70.9% AR. An et al. [129] designed an anchor-free rotatable detector. It designs center point-scale and angle prediction to convert the conventional rotatable prior box mechanism into the center point-scale and angle prediction. The training procedure includes positive sample selection, feature encoding, and loss function designing. Wang et al. [133] proposed a CenterNet-based detector in SAR images. The spatial group-wise enhanced attention module is used to extract more semantic features.

Sun et al. [167] proposed category-position FCOS. The category-position module is used to optimize the position regression branch in the FCOS network. The classification and regression branches are redesigned to alleviate the imbalance between positive and negative samples during training. Zhu et al. [180] adopted FCOS as the base model to reduce the effect of anchors. A new sample definition method is used to replace the IoU threshold according to the differences between SAR images and natural images. The same resolution feature convolution module, multi-resolution feature fusion module, and feature pyramid module are used to extract features. The focal loss and CIoU are used to improve the performance further.

In all, researchers in SAR ship detection realize the benefit of anchor-free detectors. Additionally, more and more papers are appearing in this field. However, there is still a problem: the innovation is relatively weak and some of the existing achievements of computer vision are not used in this field.

4.7. Detectors Trained from Scratch

At present, most SAR image detector backbone needs to pre-train on the classification dataset of natural images, and then fine-tune on the ship detection dataset of SAR images (for example SSDD). This transfer learning can make the detection algorithm initialize better and make up for the problem of insufficient samples. However, there will be the following problems: firstly, there is learning bias. The loss function and category distribution between classification and detection are contradictory in essence. The models trained on classification are not fit for detection. Secondly, most backbone networks will produce a high receptive field through multiple down sampling in the latter layers, which is good for the classification but is harmful for the location. Thirdly, the pre-trained backbone networks are redundant and cannot be modified, which hinders the researchers to design CNN flexibly according to their needs.

In order to solve the problems of transfer learning, algorithms trained from scratch are proposed in computer vision, for example, DSOD (deeply supervised object detectors), DetNet, ScratchDet, and so on [224–227].

The main idea of DSOD and GRP-DSOD to realize training from scratch is by designing the backbone and front-end network elaborately [224]. DetNet [225] retains a large scale in the last few layers, which can have more location information. ScratchDet [226] proposes to adopt the strategy of batch normalization in each layer and increase the learning rate, which can make the detection algorithm more robust and converge faster. Paper [227] replaced the original BN (batch normalization) with group normalization (GN) and asynchronous BN, so as to make the parameters of gradient normalization more accurate. This then made the descending direction of gradient more accurate, so as to accelerate convergence and improve the accuracy.

The model trained from scratch not only has high accuracy but also greatly reduces the size and amount of calculation of the model. Due to the above advantages, it is also used in SAR ship detection.

Most detectors that are trained from scratch in this field have well-designed networks. They are shown in Figure 10.

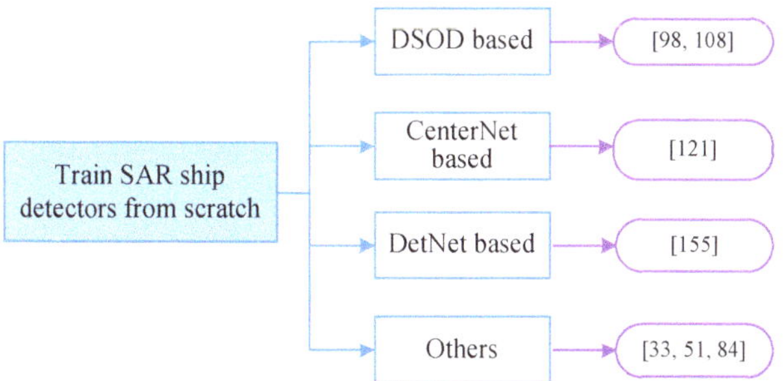

Figure 10. The detectors trained from scratch in SAR images.

Deng et al. [33] designed a dense backbone network composed of multiple dense blocks. The front layer can receive additional supervision from the objective function through dense connections, which makes the training easier, and adopts the feature reuse strategy to make the parameter highly efficient. Zhang et al. [51] designed a lightweight detection algorithm that can be trained from scratch, which can reduce the training and testing time without reducing the accuracy. It adopts the modules of semantic aggregation and feature reusing to improve the performance of multi-scale ships. Zhang et al. [84]

proposed a lightweight detection network ShipDeNet-20. It is designed with fewer layers and convolution kernels and depth separable convolution. It also adopts a feature fusion module, feature enhancement module, and proportionally shared feature pyramid module to improve detection accuracy. Han et al. [98] integrated the lightweight asymmetric square convolution block into SSD to realize training from scratch, and its accuracy and speed are better than the classical DSOD. Han et al. [100] proposed a parallel convolution block of multi-scale kernel and feature reusing convolution module to enhance feature representation and reduce information loss. Han et al. [108] designed two kinds of asymmetric convolution blocks: asymmetric and square convolution feature aggregation block, and asymmetric and square convolution feature fusion block. They replace all 3×3 convolution layers, which are embedded into the classic DSOD to achieve a better result of the training from scratch. Guo et al. [121] proposed an effective and stable single-stage algorithm that is trained from scratch, namely CenterNet++. The model mainly includes three modules: feature matching module, feature pyramid fusion module, and head enhancement module. Zhao et al. [155] used DetNet as the backbone network to realize training from scratch. It uses superposition convolution instead of down sampling to solve the problem of small ship detection and adopts a feature reusing strategy to improve parameter efficiency.

Compared with other directions, fewer researchers in SAR ship detection realize the benefit of training from scratch. Additionally, the papers using training from scratch techniques in this field are not advanced enough. We should adopt more advanced techniques in computer vision in this direction. In all, the detectors trained from scratch are not used to their full extent in this field. Some good conclusions in papers [226,227] should be considered and applied here.

4.8. Detectors with Oriented Bounding Box

The oriented bounding box was originally used in scene text detection. In addition, a large number of achievements have emerged, such as SegLink, RRPN (rotation region proposal network), TextBoxes, TextBoxes++, R2CNN (rotational region convolutional neural network), and so on [228–232]. The ships in remote sensing images also have multi-directional characteristics. The conventional vertical rectangular bounding box often cannot accurately surround the target. With the improvement of ship detection accuracy, the use of oriented bounding boxes to realize multi-directional ship detection has become a research hotspot [233–238]. DOTA (dataset for object detection in aerial images) is a commonly used aerial image target detection dataset in this field, which can be used to develop and evaluate the performance of detection algorithms. Similarly, there are many detection algorithms based on oriented bounding boxes in SAR images, which will be introduced here. At present, the dataset that can be used to train and test the oriented bounding box algorithm are SSDD+, RSDD-SAR, and SRSDD-V1 0, the details have been introduced earlier. The oriented bounding box detectors in SAR images are shown in Figure 11.

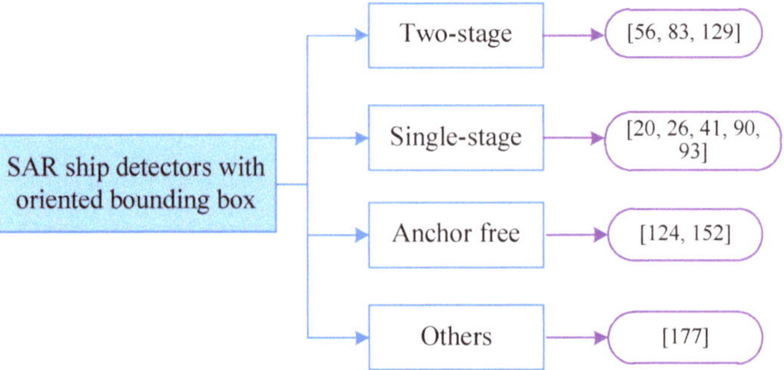

Figure 11. The SAR ship detectors with oriented bounding box.

Two-stage. Chen et al. [56] proposed a multi-scale adaptive recalibration network to detect multi-scale and arbitrarily oriented ships. It can learn the angle information of ships. The anchors, NMS, and loss function are also redesigned to fit the large aspect ratio and arbitrary directionality of ships in SAR images. Pan et al. [83] proposed a multi-stage rotational region-based network (MSR2N) to solve the problem of redundancy regions. MSR2N includes FPN, RRPN, and a multi-stage rotational detection network. It is more suitable and robust for SAR ship detection. An et al. [129] adopted an oriented detector as the based model to solve the problem that conventional CNN models have too many parameters, which increases the difficulty of transfer learning between different tasks.

Single-stage. Wang et al. [20] proposed a SAR ship detector with an oriented bounding box based on SSD. The detector can predict the class, location, and angle information of ships. The semantic aggregation module is used to capture abundant location and semantic information. The attention module is used to adaptively select meaningful features and neglect weak ones. Multi-orientation anchors, angular regression, and the loss function are used to fit the oriented bounding box. Liu et al. [26] adopted DR-Box [239] to detect ships in SAR images. DR-Box is specially designed to detect targets in any direction in remote sensing images. It can effectively reduce the interference of background pixels and locate the target more accurately. An et al. [41] proposed DR-Box-v2 to detect ships in SAR images. A multi-layer anchor box generation strategy for detecting small ships is proposed. A modified encoding scheme is proposed to estimate the position and orientation precisely. Focal loss and hard negative mining are also used to balance the positives and negatives. Yang et al. [90] regarded a rotatable bounding box detector as the base model to solve the problem of negative sample intra-class imbalance in the training stage. Chen et al. [93] proposed a rotated refined feature alignment detector to fit ships with large aspect ratios, arbitrary orientations, and dense distribution properties. A lightweight attention module, modified anchor mechanism, and feature-guided alignment module are proposed to boost the performance of the oriented detector.

Anchor free. Yang et al. [124] proposed R-RetinaNet to beat DRBox-v1, DRBox-v2, and MSR2N (multi-stage rotational region-based network) in this field. R-RetinaNet used a scale calibration method to align the scale distribution. Task-wise attention feature pyramid network is used to alleviate the contradiction of classification and localization. The adaptive IoU threshold training method is used to correct the unbalanced problem. He et al. [152] proposed a method to solve the problem of boundary discontinuity problem in oriented bounding box detectors by learning polar encodings. The encoding scheme uses a group of vectors pointing from the center of the ship to the boundary points to represent an oriented bounding box.

Others. Ding et al. [177] released the SRSDD-v1.0 dataset, which is used for oriented bounding box detectors. The details of the dataset have been described above. They present the performance of several advanced oriented bounding box detection algorithms on the dataset.

Summary. With the emergence of several datasets with oriented bounding boxes, ship detectors in SAR images based on oriented bounding boxes are becoming more and more advanced. However, it is not enough compared with DOTA. Some efforts should be taken in this direction.

4.9. Multi-Scale Ship Detectors

In MS COCO, the proportions of the small, medium, and large size objects are 41.43%, 34.32%, and 24.24%, respectively. However, in SAR images, the proportion of small-size ships is extremely high. For example, the proportion of small, medium, and large ships in the RSDD-SAR dataset are 81.175%, 18.776%, and 0.049%, respectively. In LS-SSDD-v1.0, the proportions are 99.80%, 0.20%, and 0.00%, respectively. Therefore, this field needs to focus on the problem of multi-scale ship detection, especially the small ships.

Although CNN has developed rapidly in computer vision, it has poor performance on small-size object detection. In order to improve the adaptability to multi-scale ships, computer vision often fuses low-level and high-level features (such as FPN), increases the receptive field and improves the anchor box generation and matching strategy.

SAR ship detection also uses the above methods to improve the performance of multi-scale ship detection. The multi-scale ship detectors in SAR images are shown in Figure 12.

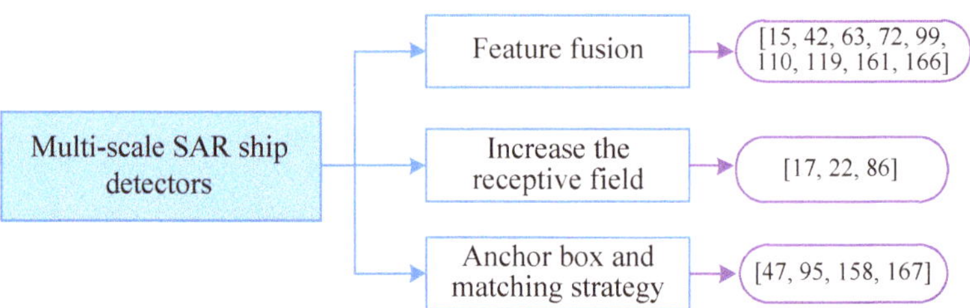

Figure 12. The multi-scale SAR ship detectors.

Feature fusion. Chen et al. [15] proposed a densely connected multi-scale neural network to solve the problem of multi-scale SAR ship detection. It closely connects each feature map with other feature maps from top-to-bottom and generates proposals from each fused feature map. Cui et al. [42] proposed a dense attention pyramid network, which closely connects the convolutional attention module from the top to the bottom of the pyramid network to each feature map, so as to extract rich features containing location information and semantic information for adapting to multi-scale ships. Liu et al. [63] proposed a scale transferable pyramid network to adapt to the detection of multi-scale ships. It constructs a feature pyramid network through horizontal connection and uses a scale transfer layer to closely connect each feature graph from top to bottom. A horizontal connection introduces more semantic information, and a dense scale transfer connection can expand the resolution of the feature map. Jin et al. [72] combined all feature maps from top to bottom to make use of contextual semantic information at all scales and uses extended convolution to increase the receptive field exponentially. Han et al. [99] used deconvolution to enhance the feature representation of small and medium-sized ships in FPN, so as to improve the detection accuracy of SSD. Hu et al. [110] proposed a dense feature pyramid network, which processes shallow features and deep features differently. Compared with traditional FPN, it has stronger adaptability to multi-scale ships. Wang et al. [119] proposed a path argumentation fusion network to fuse different feature maps. It uses bottom-up and top-down methods to fuse more location information and semantic information. Hu et al. [161] proposed a two-way revolution network based on a bidirectional convolution structure, which can effectively process shallow and deep feature information and avoid the loss of small ship information. Zhang et al. [166] proposed a quad feature pyramid network to detect multi-scale ships. It includes deformable convolutional FPN, a content-aware feature reassembly FPN, a path aggregation space attention FPN, and a Balance Scale Global Attention FPN.

Increase the receptive field. Deng et al. [17] designed a feature extractor with multiple receptive fields through ReLU and inception modules. It generates candidate regions in multiple middle layers to match ships of different scales and fuses multiple feature maps so that small-scale ships have a stronger response. Zhao et al. [22] proposed a coupled CNN to detect small-scale ships. It includes a network that generates candidate areas from multiple receptive fields and improves the recognition accuracy by using the context information of each candidate box. Dai et al. [86] did not use a single feature map but fused the feature map in a bottom-up and top-down manner, and generated candidate boxes from each fused feature map.

Anchor box generation and matching strategy. Li et al. [47] first analyzed the reasons for the low detection accuracy of small and medium-sized ships in SSD and made some improvements. The anchor box optimization method based on K-means clustering solves the problem of less positive samples and more negative samples. The feature fusion method based on deconvolution improves the representation ability of the low-level feature map and solves the weak recognition ability of the low-level feature map to small ships. Fu et al. [95] proposed a feature balance and matching network, which uses the anchor-free strategy to eliminate the influence of anchors and uses the attention-guided balance pyramid to balance multiple features at different levels semantically. It has a good performance in the detection of small-scale ships. Hong et al. [158] improved the anchor generation based on an improved K-means++ in YOLOv3. It alleviates the difficulty of multi-scale ship detection in YOLOv3 and changes the number of anchor boxes in the YOLO layer from three to four. Sun et al. [167] show that anchor-free detectors have good adaptability to small ships and have a fast speed.

Summary. Small ship detection is extremely hard but is also extremely important for some applications. That is because people hope to find targets within a long distance, and at this point, the targets must be small in size. SAR ship detection also proves this point. Although the above detection methods for small-size ships have certain effects, they are still far from enough. Innovative work needs to be continued.

4.10. Attention Module

The basic idea of the attention mechanism in computer vision is to make the model ignore irrelevant information and focus on key information. It can be divided into hard attention, soft attention, gaussian attention, spatial transformation, and so on. Attention can be calculated from the spatial domain, channel domain, layer domain, and mixed domain. Representative algorithms include SENet (squeeze and excitation network), SKNet (selective kernel network), CBAM (convolutional block attention module), CCNet (criss-cross attention), OCNet (object context network), DANet (dual attention network), etc. [240–244]. Transformer [245] adopted encoder–decoder architecture, which is the extreme of the attention. It abandons CNN and RNN (recurrent neural network) used in previous deep learning tasks and shows great advantages in the field of NLP (natural language processing) and CV. Swin Transformer [246] makes it compatible with image classification and object detection. It demonstrates the potential of transformer-based models as vision backbones.

Chen et al. [43,44] proposed an attention-based detector. The attention model is mainly composed of the convolution branch and mask branch. Elements in mask maps are similar to the weight of feature maps, which enhance regions of interest and suppress non-target regions. Cui et al. [89] introduced the space shuffle group enhanced attention module to CenterNet. It can extract stronger semantic features and suppress some noise at the same time, so as to reduce false positives caused by inshore and inland interference. Zhao et al. [91] combined the receptive field module and convolution block attention module to construct a top-down fine-grained feature pyramid. Wang et al. [122] designed a feature enhancement module based on a self-attention mechanism. Its spatial attention and channel attention work at the same time to highlight the target and suppress the spot to a certain extent. Wang et al. [131] embedded a soft attention module in the network to suppress the influence of noise and complex background. Zhu et al. [136] proposed a SAR ship detection method based on a hierarchical attention mechanism. The method includes a global attention module and a local attention module. Hierarchical attention strategies are proposed from the image layer and target layer, respectively. Sun et al. [162] introduced a semantic attention mechanism, which highlights the regional characteristics of ships and enhanced the classification ability of the detector. Du et al. [169] embedded the multi-scale feature attention module in the network. By applying the channel and spatial attention mechanism to the multi-scale feature map, it can highlight important information and suppress the interference caused by clutter.

CRTransSar [182] is the first to use a transformer for SAR image ship detection. It is based on Swin Transformer and shows great advantages. CRTransSar combines the global contextual information perception of transformers and the local feature representation capabilities of convolutional neural networks. It innovatively proposes a visual transformer framework based on contextual joint-representation learning. Experiments on SSDD and SMCDD show the effectiveness of the method.

4.11. Real-Time Detectors

At present, deep learning-based detectors need large computation and storage resources, which hinders the application in real-time prediction. In order to solve this problem, there are a lot of acceleration ideas in the evolution of object detection algorithms. Firstly, researchers usually speed up the detection process. This idea is reflected in the evolution process of R-CNN, Fast R-CNN, Faster R-CNN, R-FCN, and Light-Head R-CNN. The above detectors share the features gradually, and the network structures become thinner and faster. Secondly, researchers usually design lightweight detection networks. The backbone network and the detection head can both be accelerated. Thirdly, researchers usually compress and accelerate CNN models. It includes lightweight neural network designing, model pruning, model quantization, and knowledge distillation [247–253].

The exploration of real-time detection algorithms in SAR ship detection can be divided into three directions, which are shown in Figure 13.

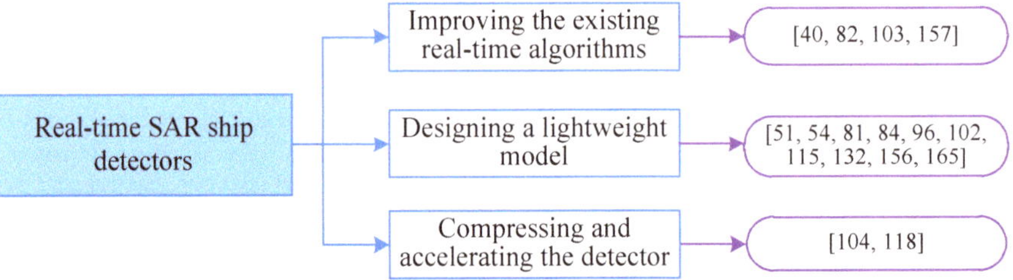

Figure 13. The real-time SAR ship detectors.

4.11.1. Improving the Existing Real-Time Algorithms

Many improvements in this field are based on the YOLO and SSD series, because they have great advantages in running time, especially the YOLO series. Zhang et al. [40] used the idea of the YOLO algorithm and adopted depth separable convolution to accelerate the speed. MobileNet is used as the backbone network to improve the detection speed under the condition of ensuring detection accuracy. Zhang et al. [82] proposed an improved YOLOv3 (using DarkNet-19 and deleting repeated layers). It achieved 90.08% AP_{50} and 68.1% AP on the SSDD dataset. Mao et al. [103] adopted the FCOS detection algorithm with ResSARNet as the backbone network, and center-ness on bounding box regression branch, center sampling, GIoU loss, and adaptive training sample selection were used. It can achieve 61.5% AP with only 1.17 M parameters. Zhong et al. [157] adopted CFAR and YOLOv4 to realize real-time ship detection on China HISEA-1 SAR images.

4.11.2. Designing a Lightweight Model

Zhang et al. [51] designed a lightweight feature optimization network LFO-Net based on SSD. It can be trained from scratch and reduce the training and testing time without reducing the accuracy. The detection performance is further improved by the bidirectional feature fusion module and attention mechanism. It achieved 80.12% AP_{50} with 9.28 ms testing time on SSDD. Zhang et al. [54] used multi-scale detection, cascade, and anchor box mechanism to design a lightweight network for real-time SAR ship detection. It uses depthwise and pointwise to replace the traditional convolution. It achieved 94.13% AP_{50} with 9.03 ms testing time on SSDD. Mao et al. [81] used the simplified U-Net as the feature extraction network, which has only 0.47 million learnable weights, it improves the operation speed and solves the problem caused by the anchor box through the anchor-free method. It has a total of 0.93 million learnable weights, and the AP on the SSDD dataset is 68.1%. Zhang et al. [84] proposed ShipDeNet-20, which has 20 convolution layers and a

0.82 MB model size. It uses fewer layers and kernels, and depthwise separable convolution is also used. It improves the accuracy through the feature fusion module, feature enhancement module, and scale share feature pyramid module. It achieved 97.07% AP_{50} with 233 FPS on SSDD. Zhang et al. [96] proposed HyperLiNet. It realizes high precision through five modules, namely multi receptive field module, divided revolution module, channel and spatial attention module, feature fusion module, and feature pyramid module. It realizes high speed through five modules, namely region-free model, small kernel, narrow channel, separate revolution, and batch normalization fusion. Zhou et al. [102] proposed a lightweight detector Lira YOLO. It combines the idea of dense connections, residual connections, and group convolution, including stem blocks and extractor modules. It achieved 85.46% AP_{50} with a 4.3 MB model size. Li et al. [115] designed a lightweight network of feature relay amplification and multi-scale feature jump connection structure based on Faster R-CNN and improves the selection of anchor boxes and RoI pooling. It achieved 89.8% AP_{50} and the speed increased a lot. Zhang et al. [132] proposed a lightweight detection algorithm ShipDeNet-18, which has fewer layers and fewer convolution kernels. The deep and shallow feature fusion module and a feature pyramid module are adopted to improve the detection accuracy. It achieved 93.78% AP_{50} with 202 FPS. Ma et al. [156] proposed YOLOv4-tiny. It reduces the number of convolutional layers in CSPDarkNet53. It achieves 88.08% AP_{50} with 12.25 ms compared with YOLOv4 with 96.32% AP and 44.21 ms. Sun et al. [165] proposed a lightweight densely connected sparsely activated detector. It can construct a lightweight backbone network, so as to achieve a balance between performance and computational complexity. It achieved 97.2% AP_{50} and 61.5% AP on SSDD.

4.11.3. Compressing and Accelerating the Detector

Mao et al. [104] proposed a knowledge distillation-based network slimming method. YOLOv3 and Darknet-53 are pruned on filter-level to obtain lightweight models. Kullback Leibler Divergence (KLD) knowledge distillation is used to train student and teacher networks (YOLOv3@EfficientNet-B7). The model has only 15.4 M parameters, and the AP decreases by only 1%. Chen et al. [118] proposed the algorithm of Tiny-YOLO-Lite. It designs and prunes the backbone structure, strengthens the channel level sparsity, and uses knowledge partition to make up for the performance degradation caused by pruning. Tiny-YOLO-Lite reduces the size of the model, reduces the number of floating-point operations, and obtains faster accuracy.

4.11.4. Summary

From the above discussion, we can find that real-time ship detection is also a hot topic in SAR images. However, the above works are not enough. It is obvious that the transferred deep learning models from computer vision are abundant in this field. Researchers should do the following work to realize real-time detection. Firstly, the anchor-free and the training from scratch method should be used to design lightweight detection algorithms. Secondly, some model compressing, and accelerating techniques should be used to improve the speed further. Thirdly, the lightweight models should be transplanted to high-performance AI chips (NVIDIA Jetson TX2) to achieve the purpose of running at the edge (satellite, airplane).

4.12. Other Detectors

In this part, we mainly introduce weakly supervised, GAN (generative adversarial network) and data augmentation, which are shown in Figure 14.

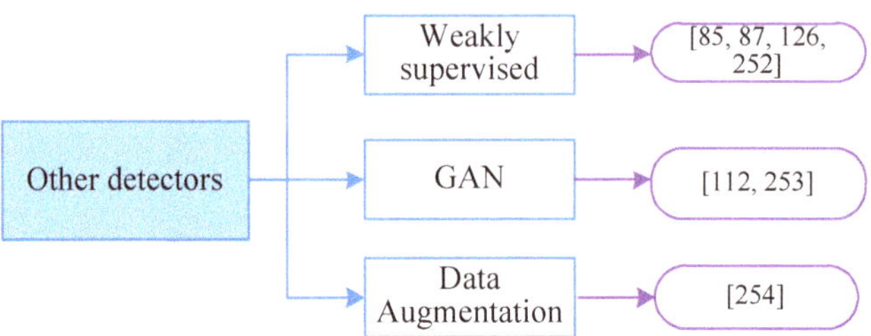

Figure 14. The other SAR ship detectors.

4.12.1. Weakly Supervised

The supervised methods, such as deep learning approaches, need substantial time and manpower to make training samples [254]. Papers [85,87,126] adopted weakly supervised to train ship detection algorithms. The model is trained by two global labels, namely, "ship" and "non-ship," and produces a ship location heatmap, ship bounding box, and pixel-level segmentation product. They can alleviate the problem of annotation partly. However, the accuracy is lower than the supervised method.

4.12.2. GAN

The insufficient SAR samples restrict the performance of the algorithm. Zou et al. [112] used a multi-scale Wasserstein auxiliary classifier generative adversarial network [255] to generate high-resolution SAR ship images. Then, the original dataset and the generated data are combined into a composite dataset to train the YOLOv3 network, so as to solve the problem of low detection accuracy under a small dataset. Based on the idea of generative adversarial networks, an image enhancement module driven by target features is designed. The quality of the ships in the image is improved. The experimental results verify the effectiveness of this method.

4.12.3. Data Augmentation

Data augmentation can expand the size of the dataset several times, so as to improve the detection accuracy [256]. The training method based on a feature mapping mask eliminates the gradient noise introduced by random clipping, so as to improve the detection performance. The SAR images with ships are generated by electromagnetic numerical analysis technology, and the sea clutter model is used to simulate the real SAR image patch containing various SAR slices, so as to improve the performance of SSD.

4.13. Problems

From the 177 papers, we can see that most of the detection algorithms in this field are borrowed from computer vision. Additionally, its development is also behind the detectors in computer vision. Due to the large difference between natural image and SAR image (for example, SAR image is single-channel, ship size is small, and distribution is very sparse), some detection algorithms are not suitable for SAR ship detection. So, we should design detectors according to the real characteristics of ships in SAR images.

The 177 papers mainly use the image essences of SAR images, and the research and application of the scattering mechanism are not enough. This is one problem we should solve in the future.

At present, there are several public small datasets, but we lack a large dataset. The models trained on a small dataset face the problem of over-fitting. What we should do next is merge the small datasets into a large one, and make sure the train-test division standards, evaluation indicators, and benchmarks are clear. These works can promote the development of this field.

5. Future—The Direction of the Deep Learning-Based SAR Ship Detectors

5.1. Anchor Free Detector Deserves Special Attention

The anchor-free detection algorithm has many advantages, which have been introduced in Section 4.6. It should be emphasized that the detection algorithm without an anchor box is especially suitable for SAR images. As SAR images have sparse and small size ships, it can greatly improve the detection speed and avoid various problems in anchor box designing and matching. Therefore, the anchor-free detection algorithm needs to be paid more attention. Fortunately, researchers in this field have realized this and many research results have emerged.

5.2. Train Detector from Scratch Deserves More Attention

At present, there are the following generally accepted conclusions about training from scratch: firstly, pre-training accelerates the convergence speed, especially in the early stage of training. However, the training time of scratch is roughly equivalent to the total time of pre-training and fine-tuning. Secondly, if there are enough target images and computing resources, pre-training is not necessary. Thirdly, if the cost of image collection and image cleaning is considered, a general large-scale classification dataset is not an ideal choice, and collecting images on detection tasks will be a more effective approach. Fourthly, when the target task is to predict spatial positioning (such as ship detection), pre-training does not show any benefits.

Collecting images for detection and training is a solution worth considering, especially when there is a significant gap between the pre-training task and the detection task (such as ImageNet image and SAR image). Therefore, in the field of SAR ship detection, it is very necessary to combine the existing public datasets into a large dataset, so as to ensure training models from scratch.

Due to the difference between natural images and SAR images, it is very necessary to adopt training from scratch detection algorithms in this field, as it can obtain a detection algorithm with stronger adaptability to SAR images and smaller model size. However, the work at this stage is far from enough, so we need to pay more attention to the detection algorithms of training from scratch.

5.3. Many Other Works Need to Be Used for Oriented Bounding Box Detector

The ship in the SAR image has very changeable directionality, and the vertical bounding box cannot adapt to this scene. It is necessary to use an oriented bounding box. In an inshore scenario, a vertical bounding box is susceptible to interference from onshore buildings and other ships, affecting detection performance, while an oriented bounding box can accurately represent the ship target and reduce redundant interference. In addition, for ship targets in an offshore scenario, an oriented bounding box can obtain information such as heading and aspect ratio, which is of great significance for subsequent trajectory prediction and situation estimation tasks. The scene text detection and the aerial remote sensing image dataset DOTA have conducted in-depth research on the oriented bounding box and achieved many results. We should learn from them.

5.4. Small Ship Detection Is an Eternal Topic

The main reasons for the poor detection result of small-size ships are as follows: firstly, the features extracted from small-scale ships are few, and the size and receptive field of the anchor are too large for small ships. Secondly, the size of the anchor is discrete (for example, 16, 32, 64, etc.), while the size of the ship is continuous, which makes the recall rate of small-size ships low. Thirdly, the anchor of a small ship matches less with the ground truth bounding box, resulting in fewer positive samples and too many negative samples.

As the proportion of small ships in SAR images is very high, small object detection is difficult, especially in this field. So, it is an eternal topic to study how to improve the detection effect of small ships.

5.5. Real-Time Detection Is the Key to Application

Real-time SAR ship detection needs to start from many aspects. For example, we can design a lightweight detection network, compress and accelerate the model to improve the speed, and transplant the detection algorithm to high-performance AI chips at the edge (NVIDIA Jetson TX2). At present, most of the work in this field is focused on the first two aspects, and there is less research on the third aspect, which needs to be focused on in the future. Only by realizing this technology can we realize the real-time detection and recognition of ships on satellite or aircraft platforms.

5.6. Transformer Is the Future Trend

In the past two years, transformer shows great advantage in object detection compared with CNN, for example, DETR (detection transformer) [257] and Swin Transformer. DINO [258] can achieve 63.6% AP on the COCO test-dev, which surpasses CNN-based detector by a large margin. Nowadays, the hot topic of computer vision is the transformer. CRTransSar [182] is the first to use a transformer for SAR image ship detection. It shows a great advantage in accuracy (97% AP on SSDD). Although there are still some problems when a transformer is used for SAR ship detection, there is no doubt that the transformer will be the research trend in the future due to its great advantages.

5.7. Bridging the Gap between SAR Ship Detection and Computer Vision

Compared with the field of computer vision, the field of ship detection in SAR images is relatively small and not active enough. Therefore, it is necessary to bring this field to computer vision, and systematically learn from the rich achievements in computer vision.

We should also learn about its openness, standardized evaluation, and easily accessed codes. This work can promote this field to develop rapidly. What we should do is as follows: firstly, the existing public datasets (SSDD, SAR-Ship-Dataset, AIR-SARShip, HRSID, LS-SSDD-v1.0, SRSDD-v1.0, and RSDD-SAR) need to be combined into a large dataset, which can be called LargeSARDataset here. The modes trained on it can avoid over-fitting. Secondly, determine the training samples and testing samples. Thirdly, determine the evaluation indicators. Fourthly, release the benchmark. Fifthly, bring it into the field of computer vision. As shown in Figure 15. Through this work, we can bridge the gap between SAR ship detection and computer vision.

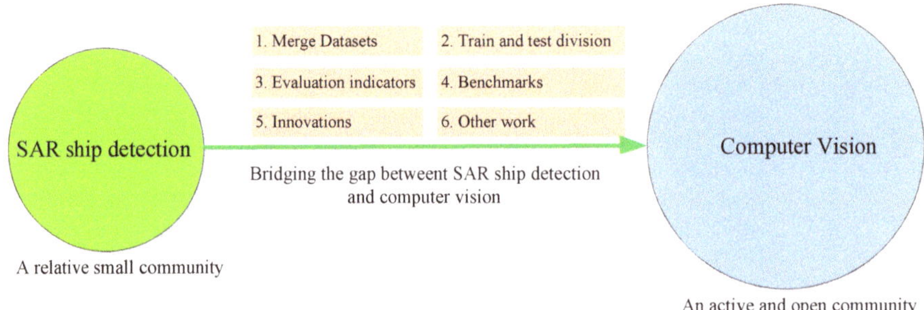

Figure 15. Bridging the gap between SAR ship detection and computer vision.

In addition to detection, classification and segmentation of SAR images also enter into the deep learning era [259–265]. Classification and segmentation algorithms borrowed from computer vision are extensively used in SAR images. We will review them in the future. In the process of detection, only ship and non-ship targets are considered, and the specific content of non-ship targets is not analyzed [266–268]. Some icebergs have great similarities in shape and size with ships, and the algorithms are difficult to distinguish them. So, we will study how to solve this problem in the future.

6. Conclusions

This paper introduces the past, present, and future of deep learning-based ship detection algorithms in SAR images.

Firstly, the history of SAR ship detection is reviewed (before SSDD was public on 1 December 2017). This part mainly introduces the detection algorithm based on CFAR and analyzes the great advantages of deep learning-based algorithms. In addition, they are compared in theory and experiment.

After that, there is a comprehensive overview of the current (from 1 December 2017 to now) ship detection algorithms based on deep learning. This part first analyzes the datasets, country, timeline, deep learning framework, and the performance evolution of the 177 papers. The basic situation of 10 datasets in this field is introduced especially. The 177 papers were classified, and they are two-stage, single-stage, anchor free, train from scratch, oriented bounding box, multi-scale, attention model, real-time detection, and so on. The specific algorithms in those papers are analyzed, including the principle, innovation, performance, and the summary.

Finally, the problems existing in this field and the future development direction are described. The main ideas are to design the detection algorithm according to the specific characteristics of SAR image, focus on the detection algorithm without an anchor box, pay enough attention to the detection algorithm of training from scratch, and learn from the existing achievements of natural scene text detection and DOTA, improve the performance of small ships continuously, pay attention to realize the real-time detection of ships through model acceleration and AI chip. It is emphasized that the future important work is to bridge the gap between SAR ship detection and computer vision by merging the existing small datasets into a larger one and making relevant standards.

This review can provide a reference for researchers in this field or researchers interested in this field so that they can quickly understand the current situation and future development direction of this field.

Author Contributions: Conceptualization, J.L. and C.X.; methodology, H.S., L.G. and T.W.; investigation, J.L.; writing—original draft preparation, J.L.; writing—review and editing, J.L. and C.X.; supervision, C.X.; funding acquisition, C.X. All authors have read and agreed to the published version of the manuscript.

Funding: This research was funded by the National Natural Science Foundation of China, No. 61790550, No. 61790554, No. 61971432, No. 62022092.

Data Availability Statement: No new data were created or analyzed in this study. Data sharing is not applicable to this article.

Conflicts of Interest: The authors declare no conflict of interest.

References

1. Moreira, A.; Prats-Iraola, P.; Younis, M.; Krieger, G.; Hajnsek, I.; Papathanassiou, K.P. A tutorial on synthetic aperture radar. *IEEE Geosci. Remote Sens. Mag.* **2013**, *1*, 6–43. [CrossRef]
2. Reigber, A.; Scheiber, R.; Jager, M.; Prats-Iraola, P.; Hajnsek, I.; Jagdhuber, T.; Papathanassiou, K.P.; Nannini, M.; Aguilera, E.; Baumgartner, S.; et al. Very-high-resolution airborne synthetic aperture radar imaging: Signal processing and applications. *Proc. IEEE* **2013**, *101*, 759–783. [CrossRef]
3. Li, H.; Hong, W.; Wu, Y.; Fan, P. An efficient and flexible statistical model based on generalized Gamma distribution for amplitude SAR images. *IEEE Trans. Geosci. Remote Sens.* **2010**, *48*, 2711–2722. [CrossRef]
4. Achim, A.; Kuruoglu, E.E.; Zerubia, J. SAR image filtering based on the heavy-tailed Rayleigh model. *IEEE Trans. Image Process.* **2006**, *15*, 2686–2693. [CrossRef]
5. Russakovsky, O.; Deng, J.; Su, H.; Krause, J.; Satheesh, S.; Ma, S.; Huang, Z.; Karpathy, A.; Khosla, A.; Bernstein, M.; et al. ImageNet Large Scale Visual Recognition Challenge. *Int. J. Comput. Vis.* **2015**, *115*, 211–252. [CrossRef]
6. Schwegmann, C.P.; Kleynhans, W.; Salmon, B.P.; Mdakane, L.W.; Meyer, R.G.V. Very deep learning for ship discrimination in Synthetic Aperture Radar imagery. In Proceedings of the IEEE International Geoscience and Remote Sensing Symposium (IGARSS), Beijing, China, 10–15 July 2016; pp. 104–107. [CrossRef]
7. Miao, K.; Leng, X.; Zhao, L.; Ji, K. A modified faster R-CNN based on CFAR algorithm for SAR ship detection. In Proceedings of the 2017 International Workshop on Remote Sensing with Intelligent Processing (RSIP), Shanghai, China, 18–21 May 2017. [CrossRef]

8. Liu, Y.; Zhang, M.H.; Xu, P.; Guo, Z. SAR ship detection using sea-land segmentation-based convolutional neural network. In Proceedings of the 2017 International Workshop on Remote Sensing with Intelligent Processing (RSIP), Shanghai, China, 18–21 May 2017. [CrossRef]
9. Kang, M.; Ji, K.; Leng, X.; Lin, Z. Contextual Region-Based Convolutional Neural Network with Multilayer Fusion for SAR Ship Detection. *Remote Sens.* **2017**, *9*, 860. [CrossRef]
10. Wang, Y.; Chao, W.; Hong, Z. Combining single shot multibox detector with transfer learning for ship detection using Sentinel-1 images. In Proceedings of the 2017 SAR in Big Data Era: Models, Methods and Applications (BIGSARDATA), Singapore, 19–22 November 2017.
11. Li, J.; Qu, C.; Shao, J. Ship detection in SAR images based on an improved faster R-CNN. In Proceedings of the Sar in Big Data Era: Models, Methods & Applications, Beijing, China, 13–14 November 2017. [CrossRef]
12. Cozzolino, D.; Martino, G.D.; Poggi, G.; Verdoliva, L. A fully convolutional neural network for low-complexity single-stage ship detection in Sentinel-1 SAR images. In Proceedings of the Geoscience & Remote Sensing Symposium, Fort Worth, TX, USA, 23–28 July 2017. [CrossRef]
13. An, Q.; Pan, Z.; You, H. Ship Detection in Gaofen-3 SAR Images Based on Sea Clutter Distribution Analysis and Deep Convolutional Neural Network. *Sensors* **2018**, *18*, 334. [CrossRef]
14. Wang, Y.; Wang, C.; Zhang, H.; Zhang, C.; Fu, Q. Combing Single Shot Multibox Detector with transfer learning for ship detection using Chinese Gaofen-3 images. In Proceedings of the 2017 Progress in Electromagnetics Research Symposium-Fall (PIERS-FALL), Singapore, 19–22 November 2017. [CrossRef]
15. Chen, S.Q.; Zhan, R.H.; Zhang, J. Robust single stage detector based on two-stage regression for SAR ship detection. In Proceedings of the 2nd International Conference on Innovation in Artificial Intelligence, Shanghai China, 9–12 March 2018; pp. 169–174. [CrossRef]
16. Jiao, J.; Zhang, Y.; Sun, H.; Yang, X.; Gao, X.; Hong, W.; Fu, K.; Sun, X.; Wen, H. A Densely Connected End-to-End Neural Network for Multiscale and Multiscene SAR Ship Detection. *IEEE Access* **2018**, *6*, 20881–20892. [CrossRef]
17. Deng, Z.; Sun, H.; Zhou, S.; Zhao, J.; Lei, L.; Zou, H. Multi-scale object detection in remote sensing imagery with convolutional neural networks. *ISPRS J. Photogramm. Remote Sens.* **2018**, *145*, 3–22. [CrossRef]
18. Wang, Y.; Wang, C.; Zhang, H. Combing a Single Shot Multibox Detector with transfer learning for ship detection using sentinel-1 SAR images. *Remote Sens. Lett.* **2019**, *9*, 780–788. [CrossRef]
19. Wang, R.; Li, J.; Duan, Y.; Cao, H.; Zhao, Y. Study on the Combined Application of CFAR and Deep Learning in Ship Detection. *J. Indian Soc. Remote Sens.* **2018**, *46*, 1413–1421. [CrossRef]
20. Wang, J.; Lu, C.; Jiang, W. Simultaneous Ship Detection and Orientation Estimation in SAR Images Based on Attention Module and Angle Regression. *Sensors* **2018**, *18*, 2851. [CrossRef]
21. Zhao, J.; Zhang, Z.; Yu, W.; Truong, T. A Cascade Coupled Convolutional Neural Network Guided Visual Attention Method for Ship Detection from SAR Images. *IEEE Access* **2018**, *6*, 50693–50708. [CrossRef]
22. Zhao, J.; Guo, W.; Zhang, Z.; Yu, W. A coupled convolutional neural network for small and densely clustered ship detection in SAR images. *Sci. China Inf. Sci.* **2019**, *62*, 42301. [CrossRef]
23. Khan, H.M.; Cai, Y. Ship detection in SAR Image using YOLOv2. In Proceedings of the 2018 37th Chinese Control Conference (CCC), Wuhan, China, 25–27 July 2018. [CrossRef]
24. Sharifzadeh, F.; Akbarizadeh, G.; Kavian, Y.S. Ship Classification in SAR Images Using a New Hybrid CNN–MLP Classifier. *J. Indian Soc. Remote Sens.* **2018**, *47*, 551–562. [CrossRef]
25. Zhou, F.; Fan, W.; Sheng, Q.; Tao, M. Ship Detection Based on Deep Convolutional Neural Networks for Polsar Images. In Proceedings of the 2018 IEEE International Geoscience and Remote Sensing Symposium, Valencia, Spain, 22–27 July 2018. [CrossRef]
26. Lei, L.; Chen, G.; Pan, Z.; Lei, B.; An, Q. Inshore Ship Detection in Sar Images Based on Deep Neural Networks. In Proceedings of the 2018 IEEE International Geoscience and Remote Sensing Symposium, Valencia, Spain, 22–27 July 2018. [CrossRef]
27. Wang, Y.; Chao, W.; Hong, Z. Ship Discrimination with Deep Convolutional Neural Networks in Sar Images. In Proceedings of the 2018 IEEE International Geoscience and Remote Sensing Symposium, Valencia, Spain, 22–27 July 2018. [CrossRef]
28. Schwegma, N.C.P.; Kleynhans, W.; Salmon, B.P.; Mdakane, L.W.; Meyer, R.G.V. Synthetic Aperture Radar Ship Detection Using Capsule Networks. In Proceedings of the 2018 IEEE International Geoscience and Remote Sensing Symposium, Valencia, Spain, 22–27 July 2018. [CrossRef]
29. Lin, Z.; Ji, K.; Leng, X.; Kuang, G. Squeeze and Excitation Rank Faster R-CNN for Ship Detection in SAR Images. *IEEE Geosci. Remote Sens. Lett.* **2019**, *16*, 751–755. [CrossRef]
30. Ma, M.; Chen, J.; Liu, W.; Yang, W. Ship Classification and Detection Based on CNN Using GF-3 SAR Images. *Remote Sens.* **2018**, *10*, 2043. [CrossRef]
31. Chen, S.W.; Tao, C.S.; Wang, X.S.; Xiao, S.P. Polarimetric SAR Targets Detection and Classification with Deep Convolutional Neural Network. In Proceedings of the 2018 Progress in Electromagnetics Research Symposium (PIERS-Toyama), Toyama, Japan, 1–4 August 2018. [CrossRef]
32. Wang, Z.; Yang, T.; Zhang, H. Land contained sea area ship detection using spaceborne image. *Pattern Recognit. Lett.* **2019**, *130*, 125–131. [CrossRef]
33. Deng, Z.; Sun, H.; Zhou, S.; Zhao, J. Learning Deep Ship Detector in SAR Images from Scratch. *IEEE Trans. Geosci. Remote Sens.* **2019**, *57*, 4021–4039. [CrossRef]

34. Gui, Y.; Li, X.; Xue, L. A Multilayer Fusion Light-Head Detector for SAR Ship Detection. *Sensors* **2019**, *19*, 1124. [CrossRef] [PubMed]

35. Wang, Y.; Wang, C.; Zhang, H.; Dong, Y.; Wei, S. Automatic Ship Detection Based on RetinaNet Using Multi-Resolution Gaofen-3 Imagery. *Remote Sens.* **2019**, *11*, 531. [CrossRef]

36. Liu, N.; Cao, Z.; Cui, Z.; Pi, Y.; Dang, S. Multi-Scale Proposal Generation for Ship Detection in SAR Images. *Remote Sens.* **2019**, *11*, 526. [CrossRef]

37. Wang, J.; Zheng, T.; Lei, P.; Bai, X. A Hierarchical Convolution Neural Network (CNN)-Based Ship Target Detection Method in Spaceborne SAR Imagery. *Remote Sens.* **2019**, *11*, 620. [CrossRef]

38. Wang, Y.; Wang, C.; Zhang, H.; Dong, Y.; Wei, S.L. A SAR Dataset of Ship Detection for Deep Learning under Complex Backgrounds. *Remote Sens.* **2019**, *11*, 765. [CrossRef]

39. Chang, Y.L.; Anagaw, A.; Chang, L.; Wang, Y.C.; Hsiao, C.Y.; Lee, W.H. Ship detection based on YOLOv2 for SAR imagery. *Remote Sens.* **2019**, *11*, 786. [CrossRef]

40. Zhang, T.; Zhang, X. High-Speed Ship Detection in SAR Images Based on a Grid Convolutional Neural Network. *Remote Sens.* **2019**, *11*, 1206. [CrossRef]

41. An, Q.; Pan, Z.; Liu, L.; You, H. DRBox-v2: An Improved Detector with Rotatable Boxes for Target Detection in SAR Images. *IEEE Trans. Geosci. Remote Sens.* **2019**, *57*, 8333–8349. [CrossRef]

42. Cui, Z.; Li, Q.; Cao, Z.; Liu, N. Dense Attention Pyramid Networks for Multi-Scale Ship Detection in SAR Images. *IEEE Trans. Geosci. Remote Sens.* **2019**, *57*, 8983–8997. [CrossRef]

43. Chen, C.; Hu, C.; He, C.; Pei, H.; Pang, H.; Zhao, T. SAR ship detection under complex background based on attention mechanism. In *Chinese Conference on Image and Graphics Technologies*; Springer: Singapore, 2019; pp. 565–578.

44. Chen, C.; He, C.; Hu, C.; Pei, H.; Jiao, L. A Deep Neural Network Based on an Attention Mechanism for SAR Ship Detection in Multiscale and Complex Scenarios. *IEEE Access* **2019**, *7*, 104848–104863. [CrossRef]

45. Guo, Q.; Wang, H.; Kang, L.; Li, Z.; Xu, F. Aircraft Target Detection from Spaceborne SAR Image. In Proceedings of the 2019 IEEE International Geoscience and Remote Sensing Symposium, Yokohama, Japan, 28 July–2 August 2019. [CrossRef]

46. Gui, Y.; Li, X.; Xue, L.; Lv, J. A scale transfer convolution network for small ship detection in SAR images. In Proceedings of the 2019 IEEE 8th Joint International Information Technology and Artificial Intelligence Conference (ITAIC), Chongqing, China, 24–26 May 2019. [CrossRef]

47. Li, Y.; Chen, J.; Ke, M.; Li, L.; Ding, Z.; Wang, Y. Small targets recognition in SAR ship image based on improved SSD. In Proceedings of the 2019 IEEE International Conference on Signal, Information and Data Processing (ICSIDP), Chongqing, China, 11–13 December 2019. [CrossRef]

48. Ai, J.; Tian, R.; Luo, Q.; Jin, J.; Tang, B. Multi-Scale Rotation-Invariant Haar-Like Feature Integrated CNN-Based Ship Detection Algorithm of Multiple-Target Environment in SAR Imagery. *IEEE Trans. Geosci. Remote Sens.* **2019**, *57*, 10070–10087. [CrossRef]

49. Fan, Q.; Chen, F.; Cheng, M.; Lou, S.; Xiao, R.; Zhang, B.; Wang, C.; Li, J. Ship Detection Using a Fully Convolutional Network with Compact Polarimetric SAR Images. *Remote Sens.* **2019**, *11*, 2171. [CrossRef]

50. Ayhan, N.; Sen, N. Ship detection in synthetic aperture radar (SAR) images by deep learning. In Proceedings of the Artificial Intelligence and Machine Learning in Defense Applications, Strasbourg, France, 19 September 2019

51. Zhang, X.; Wang, H.; Xu, C.; Lv, Y.; Fu, C.; Xiao, H.; He, Y. Lightweight Feature Optimizing Network for Ship Detection in SAR Image. *IEEE Access* **2019**, *7*, 141662–141678. [CrossRef]

52. Yang, T.; Zhu, J.; Liu, J. SAR Image Target Detection and Recognition based on Deep Network. In Proceedings of the 2019 SAR in Big Data Era (BIGSARDATA), Beijing, China, 5–6 August 2019. [CrossRef]

53. Hong, S.J.; Baek, W.K.; Jung, H.S. Ship Detection from X-Band SAR Images Using M2Det Deep Learning Model. *Appl. Sci.* **2020**, *10*, 7751. [CrossRef]

54. Zhang, T.; Zhang, X.; Shi, J.; Wei, S. Depthwise Separable Convolution Neural Network for High-Speed SAR Ship Detection. *Remote Sens.* **2019**, *11*, 2483. [CrossRef]

55. Chen, S.; Zhan, R.; Zhang, J. Regional attention-based single shot detector for SAR ship detection. *J. Eng.* **2019**, *2019*, 7381–7384. [CrossRef]

56. Chen, C.; He, C.; Hu, C.; Pei, H.; Jiao, L. MSARN: A Deep Neural Network Based on an Adaptive Recalibration Mechanism for Multiscale and Arbitrary-oriented SAR Ship Detection. *IEEE Access* **2019**, *7*, 159262–159283. [CrossRef]

57. Yue, B.; Zhao, W.; Han, S. SAR Ship Detection Method Based on Convolutional Neural Network and Multi-layer Feature Fusion. In *The International Conference on Natural Computation, Fuzzy Systems and Knowledge Discovery*; Springer: Cham, Switzerland, 2019.

58. Wang, Z.; Yang, W.; Chen, J.; Li, C. A Level Set Based Method for Land Masking in Ship Detection Using SAR Images. In Proceedings of the 2019 IEEE International Geoscience and Remote Sensing Symposium, Yokohama, Japan, 28 July–2 August 2019. [CrossRef]

59. Hou, X.; Ao, W.; Xu, F. End-to-end Automatic Ship Detection and Recognition in High-Resolution Gaofen-3 Spaceborne SAR Images. In Proceedings of the 2019 IEEE International Geoscience and Remote Sensing Symposium, Yokohama, Japan, 28 July–2 August 2019. [CrossRef]

60. Wang, R.; Xu, F.; Pei, J.; Wang, C.; Huang, Y.; Yang, J.; Wu, J. An Improved Faster R-CNN Based on MSER Decision Criterion for SAR Image Ship Detection in Harbor. In Proceedings of the 2019 IEEE International Geoscience and Remote Sensing Symposium, Yokohama, Japan, 28 July–2 August 2019. [CrossRef]

61. Li, Y.; Ding, Z.; Zhang, C.; Wang, Y.; Chen, J. SAR Ship Detection Based on Resnet and Transfer Learning. In Proceedings of the 2019 IEEE International Geoscience and Remote Sensing Symposium, Yokohama, Japan, 28 July–2 August 2019. [CrossRef]
62. Li, Q.; Min, R.; Cui, Z.; Pi, Y.; Xu, Z. Multiscale Ship Detection Based on Dense Attention Pyramid Network in Sar Images. In Proceedings of the 2019 IEEE International Geoscience and Remote Sensing Symposium, Yokohama, Japan, 28 July–2 August 2019. [CrossRef]
63. Liu, N.; Cui, Z.; Cao, Z.; Pi, Y.; Lan, H. Scale-Transferrable Pyramid Network for Multi-Scale Ship Detection in Sar Images. In Proceedings of the 2019 IEEE International Geoscience and Remote Sensing Symposium, Yokohama, Japan, 28 July–2 August 2019. [CrossRef]
64. Gao, F.; Shi, W.; Wang, J.; Yang, E.; Zhou, H. Enhanced Feature Extraction for Ship Detection from Multi-Resolution and Multi-Scene Synthetic Aperture Radar (SAR) Images. *Remote Sens.* **2019**, *11*, 2694. [CrossRef]
65. Sun, X.; Wang, Z.; Sun, Y.; Diao, W.; Zhang, Y.; Fu, K. AIR-SARShip-1.0: High-resolution SAR ship detection dataset. *J. Radars* **2019**, *8*, 852–862.
66. Fan, W.; Zhou, F.; Bai, X.; Tao, M.; Tian, T. Ship Detection Using Deep Convolutional Neural Networks for PolSAR Images. *Remote Sens.* **2019**, *11*, 2862. [CrossRef]
67. Dechesne, C.; Lefèvre, S.; Vadaine, R.; Hajduch, G.; Fablet, R. Ship Identification and Characterization in Sentinel-1 SAR Images with Multi-Task Deep Learning. *Remote Sens.* **2019**, *11*, 2997. [CrossRef]
68. Zhang, X.; Zhang, T.; Shi, J.; Wei, S. High-speed and High-accurate SAR ship detection based on a depthwise separable convolution neural network. *J. Radars* **2019**, *8*, 841–851.
69. Wei, S.; Su, H.; Ming, J.; Wang, C.; Yan, M.; Kumar, D.; Shi, J.; Zhang, X. Precise and Robust Ship Detection for High-Resolution SAR Imagery Based on HR-SDNet. *Remote Sens.* **2020**, *12*, 167. [CrossRef]
70. Chen, P.; Li, Y.; Zhou, H.; Liu, B.; Liu, P. Detection of Small Ship Objects Using Anchor Boxes Cluster and Feature Pyramid Network Model for SAR Imagery. *J. Mar. Sci. Eng.* **2020**, *8*, 112. [CrossRef]
71. Milios, A.; Bereta, K.; Chatzikokolakis, K.; Zissisx, D.; Matwin, S. Automatic fusion of satellite imagery and AIS data for vessel detection. In Proceedings of the 2019 22th International Conference on Information Fusion (FUSION), Ottawa, ON, Canada, 2–5 July 2019; pp. 1–5.
72. Jin, K.; Chen, Y.; Xu, B.; Yin, J.; Yang, J. A Patch-to-Pixel Convolutional Neural Network for Small Ship Detection with PolSAR Images. *IEEE Trans. Geosci. Remote Sens.* **2020**, *58*, 6623–6638. [CrossRef]
73. Hou, X.; Ao, W.; Song, Q.; Lai, J.; Wang, H.; Xu, F. FUSAR-Ship: Building a high-resolution SAR-AIS matchup dataset of Gaofen-3 for ship detection and recognition. *Sci. China Inf. Sci.* **2020**, *63*, 140303. [CrossRef]
74. Su, H.; Wei, S.; Liu, S.; Liang, J.; Wang, C.; Shi, J.; Zhang, X. HQ-ISNet: High-Quality Instance Segmentation for Remote Sensing Imagery. *Remote Sens.* **2020**, *12*, 989. [CrossRef]
75. Tanveer, H.; Balz, T.; Mohamdi, B. Using convolutional neural network (CNN) approach for ship detection in Sentinel-1 SAR imagery. In Proceedings of the 2019 6th Asia-Pacific Conference on Synthetic Aperture Radar (APSAR), Xiamen, China, 26–29 November 2019. [CrossRef]
76. Wang, J.; Chen, J.; Wang, P.; Zhao, C.; Pan, X.; Gao, A. An Algorithm for Azimuth Ambiguities Detection in SAR Images Using Faster-RCNN. In Proceedings of the 2019 6th Asia-Pacific Conference on Synthetic Aperture Radar (APSAR), Xiamen, China, 26–29 November 2019. [CrossRef]
77. Zheng, T.; Wang, J.; Lei, P. Deep learning based target detection method with multi-features in SAR imagery. In Proceedings of the 2019 6th Asia-Pacific Conference on Synthetic Aperture Radar (APSAR), Xiamen, China, 26–29 November 2019. [CrossRef]
78. Su, H.; Wei, S.; Wang, M.; Zhou, L.; Shi, J.; Zhang, X. Ship Detection Based on RetinaNet-Plus for High-Resolution SAR Imagery. In Proceedings of the 2019 6th Asia-Pacific Conference on Synthetic Aperture Radar (APSAR), Xiamen, China, 26–29 November 2019. [CrossRef]
79. Wang, C.; Pei, J.; Wang, R.; Huang, Y.; Yang, J. A new ship detection and classification method of spaceborne SAR images under complex scene. In Proceedings of the 2019 6th Asia-Pacific Conference on Synthetic Aperture Radar (APSAR), Xiamen, China, 26–29 November 2019. [CrossRef]
80. Xiao, Q.; Cheng, Y.; Xiao, M.; Zhang, J.; Shi, H.; Niu, L.; Ge, C.; Lang, H. Improved region convolutional neural network for ship detection in multiresolution synthetic aperture radar images. *Concurr. Comput. Pract. Exp.* **2020**, *32*, 5820. [CrossRef]
81. Mao, Y.; Yang, Y.; Ma, Z.; Li, M.; Su, H.; Zhang, J. Efficient Low-Cost Ship Detection for SAR Imagery Based on Simplified U-Net. *IEEE Access* **2020**, *8*, 69742–69753. [CrossRef]
82. Zhang, T.; Zhang, X.; Shi, J.; Wei, S. High-Speed Ship Detection in SAR Images by Improved Yolov3. In Proceedings of the 2019 16th International Computer Conference on Wavelet Active Media Technology and Information Processing, Chengdu, China, 14–15 December 2019; pp. 149–152. [CrossRef]
83. Pan, Z.; Yang, R.; Zhang, A.Z. MSR2N: Multi-Stage Rotational Region Based Network for Arbitrary-Oriented Ship Detection in SAR Images. *Sensors* **2020**, *20*, 2340. [CrossRef]
84. Zhang, T.; Zhang, X. ShipDeNet-20: An Only 20 Convolution Layers and <1-MB Lightweight SAR Ship Detector. *IEEE Geosci. Remote Sens. Lett.* **2020**, *18*, 1234–1238. [CrossRef]
85. Gu, F.; Zhang, H.; Wang, C.; Zhang, B. Weakly supervised ship detection from SAR images based on a three-component CNN-CAM-CRF model. *J. Appl. Remote Sens.* **2020**, *14*, 026506. [CrossRef]

86. Dai, W.; Mao, Y.; Yuan, R.; Liu, Y.; Pu, X.; Li, C. A Novel Detector Based on Convolution Neural Networks for Multiscale SAR Ship Detection in Complex Background. *Sensors* **2020**, *20*, 2547. [CrossRef]

87. Zhou, Y.; Cai, Z.; Zhu, Y.; Yan, J. Automatic ship detection in SAR Image based on Multi-scale Faster R-CNN. *J. Phys. Conf. Ser.* **2020**, *1550*, 042006. [CrossRef]

88. Kang, K.M. Automated Procurement of Training Data for Machine Learning Algorithm on Ship Detection Using AIS Information. *Remote Sens.* **2020**, *12*, 1443. [CrossRef]

89. Cui, Z.; Wang, X.; Liu, N.; Cao, Z.; Yang, J. Ship Detection in Large-Scale SAR Images Via Spatial Shuffle-Group Enhance Attention. *IEEE Trans. Geosci. Remote Sens.* **2020**, *59*, 379–391. [CrossRef]

90. Yang, R.; Wang, G.; Pan, Z.; Lu, H.; Zhang, H.; Jia, X. A Novel False Alarm Suppression Method for CNN-Based SAR Ship Detector. *IEEE Geosci. Remote Sens. Lett.* **2020**, *18*, 1401–1405. [CrossRef]

91. Zhao, Y.; Zhao, L.; Xiong, B.; Kuang, G. Attention Receptive Pyramid Network for Ship Detection in SAR Images. *IEEE J. Sel. Top. Appl. Earth Obs. Remote Sens.* **2020**, *13*, 2738–2756. [CrossRef]

92. Han, L.; Zheng, T.; Ye, W.; Ran, D. Analysis of Detection Preference to CNN Based SAR Ship Detectors. In Proceedings of the 2020 Information Communication Technologies Conference (ICTC), Nanjing, China, 29–31 May 2020. [CrossRef]

93. Chen, S.; Zhang, J.; Zhan, R. R2FA-Det: Delving into High-Quality Rotatable Boxes for Ship Detection in SAR Images. *Remote Sens.* **2020**, *12*, 2031. [CrossRef]

94. Wei, S.; Zeng, X.; Qu, Q.; Wang, M.; Su, H.; Shi, J. HRSID: A High-Resolution SAR Images Dataset for Ship Detection and Instance Segmentation. *IEEE Access* **2020**, *8*, 2031. [CrossRef]

95. Fu, J.; Sun, X.; Wang, Z.; Fu, K. An Anchor-Free Method Based on Feature Balancing and Refinement Network for Multiscale Ship Detection in SAR Images. *IEEE Trans. Geosci. Remote Sens.* **2020**, *59*, 1331–1344. [CrossRef]

96. Zhang, T.; Zhang, X.; Shi, J.; Wei, S. HyperLi-Net: A hyper-light deep learning network for high-accurate and high-speed ship detection from synthetic aperture radar imagery. *ISPRS J. Photogramm. Remote Sens.* **2020**, *167*, 123–153. [CrossRef]

97. Zhou, H. Anchor-free Convolutional Network with Dense Attention Feature Aggregation for Ship Detection in SAR Images. *Remote Sens.* **2020**, *12*, 2649.

98. Han, L.; Zhao, X.; Ye, W.; Ran, D. Asymmetric and square convolutional neural network for SAR ship detection from scratch. In Proceedings of the 2020 5th International Conference on Biomedical Signal and Image Processing, Suzhou, China, 21–23 August 2020; pp. 80–85.

99. Han, L.; Ye, W.; Li, J.; Ran, D. Small ship detection in SAR images based on modified SSD. In Proceedings of the 2019 IEEE International Conference on Signal, Information and Data Processing (ICSIDP), Chongqing, China, 11–13 December 2019. [CrossRef]

100. Han, L.; Ran, D.; Ye, W.; Yang, W.; Wu, X. Multi-size Convolution and Learning Deep Network for SAR Ship Detection from Scratch. *IEEE Access* **2020**, *8*, 158996–159016. [CrossRef]

101. Zhang, T.; Zhang, X.; Ke, X.; Zhan, X.; Shi, J.; Wei, S.; Pan, D.; Li, J.; Su, H.; Zhou, Y. LS-SSDD-v1.0: A Deep Learning Dataset Dedicated to Small Ship Detection from Large-Scale Sentinel-1 SAR Images. *Remote Sens.* **2020**, *12*, 2997. [CrossRef]

102. Zhou, L.; Wei, S.; Cui, Z.; Fang, J.; Yang, X.; Ding, W. Lira-YOLO: A lightweight model for ship detection in radar images. *J. Syst. Eng. Electron.* **2020**, *31*, 950–956. [CrossRef]

103. Mao, Y.; Li, X.; Li, Z.; Li, M.; Chen, S. An Anchor-free SAR Ship Detector with Only 1.17M Parameters. In Proceedings of the ICASIT 2020: 2020 International Conference on Aviation Safety and Information Technology, Weihai, China, 14–16 October 2020.

104. Mao, Y.; Li, X.; Li, Z.; Li, M.; Chen, S. Network slimming method for SAR ship detection based on knowlegde distillation. In Proceedings of the 2020 International Conference on Aviation Safety and Information Technology, Weihai, China, 14–16 October 2020; pp. 177–181.

105. Stefanowicz, J.; Ali, I.; Andersson, S. Current trends in ship detection in single polarization synthetic aperture radar imagery. In Proceedings of the Proc. SPIE 11581, Photonics Applications in Astronomy, Communications, Industry, and High Energy Physics Experiments, Wilga, Poland, 14 October 2020; p. 1158109.

106. Li, K.; Luan, S.; Zhou, D. An Optical-to-SAR Transformation Method for SAR Ship Image Augmentation. In Proceedings of the 2020 IEEE 3rd International Conference on Information Communication and Signal Processing (ICICSP), Shanghai, China, 12–15 September 2020. [CrossRef]

107. Xu, C.; Yin, C.; Wang, D.; Han, W. Fast ship detection combining visual saliency and a cascade CNN in SAR images. *IET Radar Sonar Navig.* **2020**, *14*, 1879–1887. [CrossRef]

108. Han, L.; Ran, D.; Ye, W.; Wu, X. Asymmetric convolution-based neural network for SAR ship detection from scratch. In Proceedings of the 2020 9th International Conference on Computing and Pattern Recognition, Xiamen China, 30 October–1 November 2020; pp. 90–95.

109. Idicula, S.M.; Paul, B. Real time SAR Ship Detection using novel SarNeDe method. In Proceedings of the 2021 IEEE International Geoscience and Remote Sensing Symposium IGARSS, Brussels, Belgium, 11–16 July 2021; pp. 2198–2201. [CrossRef]

110. Hu, W.; Tian, Z.; Chen, S.; Zhan, R.; Zhang, J. Dense feature pyramid network for ship detection in SAR images. In Proceedings of the Third International Conference on Image, Video Processing and Artificial Intelligence, Shanghai, China, 23–24 October 2020. [CrossRef]

111. Zhang, T.; Zhang, X.; Shi, J.; Wei, S.; Wang, J.; Li, J.; Su, H.; Zhou, Y. Balance Scene Learning Mechanism for Offshore and Inshore Ship Detection in SAR Images. *IEEE Geosci. Remote Sens. Lett.* **2020**, *19*, 4004905. [CrossRef]

112. Zou, L.; Zhang, H.; Wang, C.; Wu, F. MW-ACGAN: Generating Multiscale High-Resolution SAR Images for Ship Detection. *Sensors* **2020**, *20*, 6673. [CrossRef]

113. Zhang, G.; Li, Z.; Li, X.; Yin, C.; Shi, Z. A Novel Salient Feature Fusion Method for Ship Detection in Synthetic Aperture Radar Images. *IEEE Access* **2020**, *8*, 215904–215914. [CrossRef]

114. Chen, Y.; Yu, J.; Xu, Y. SAR Ship Target Detection for SSDv2 under Complex Backgrounds. In Proceedings of the 2020 International Conference on Computer Vision, Image and Deep Learning (CVIDL), Chongqing, China, 10–12 July 2020. [CrossRef]

115. Li, Y.; Zhang, S.; Wang, W.Q. A Lightweight Faster R-CNN for Ship Detection in SAR Images. *IEEE Geosci. Remote Sens. Lett.* **2020**, *19*, 4006105. [CrossRef]

116. Zhu, M.; Hu, G.; Zhou, H.; Lu, C. Rapid Ship Detection in SAR Images Based on YOLOv3. In Proceedings of the 2020 5th International Conference on Communication, Image and Signal Processing (CCISP), Chengdu, China, 13–15 November 2020. [CrossRef]

117. Zhang, T.; Zhang, X.; Shi, J.; Wei, S.; Wang, J.; Li, J. Balanced Feature Pyramid Network for Ship Detection in Synthetic Aperture Radar Images. In Proceedings of the 2020 IEEE Radar Conference (RadarConf20), Florence, Italy, 21–25 September 2020. [CrossRef]

118. Chen, S.; Zhan, R.; Wang, W.; Zhang, J. Learning Slimming SAR Ship Object Detector Through Network Pruning and Knowledge Distillation. *IEEE J. Sel. Top. Appl. Earth Obs. Remote Sens.* **2020**, *14*, 1267–1282. [CrossRef]

119. Wang, J.; Lin, Y.; Guo, J.; Zhuang, L. SSS-YOLO: Towards more accurate detection for small ships in SAR image. *Remote Sens. Lett.* **2021**, *12*, 93–102. [CrossRef]

120. Yang, R.; Wang, R.; Deng, Y.; Jia, X.; Zhang, H. Rethinking the Random Cropping Data Augmentation Method Used in the Training of CNN-based SAR Image Ship Detector. *Remote Sens.* **2020**, *13*, 34. [CrossRef]

121. Guo, H.; Yang, X.; Wang, N.; Gao, X. A CenterNet++ model for ship detection in SAR images. *Pattern Recognit.* **2021**, *112*, 107787. [CrossRef]

122. Wang, C.; Su, W.; Gu, H. Two-stage ship detection in synthetic aperture radar images based on attention mechanism and extended pooling. *J. Appl. Remote Sens.* **2020**, *14*, 044522. [CrossRef]

123. Chaudhary, Y.; Mehta, M.; Goel, N.; Bhardwaj, P.; Gupta, D.; Khanna, A. YOLOv3 Remote Sensing SAR Ship Image DetectionM. In *Data Analytics and Management*; Springer: Singapore, 2021; pp. 519–531.

124. Yang, R.; Pan, Z.; Jia, X.; Zhang, L.; Deng, Y. A Novel CNN-Based Detector for Ship Detection Based on Rotatable Bounding Box in SAR Images. *IEEE J. Sel. Top. Appl. Earth Obs. Remote Sens.* **2021**, *14*, 1938–1958. [CrossRef]

125. Liu, C.; Zhu, W. An improved algorithm for ship detection in SAR images based on CNN. In Proceedings of the Twelfth International Conference on Graphics and Image Processing, Xi'an, China, 13–15 November 2021.

126. Wang, J.; Wen, Z.; Lu, Y.; Wang, X.; Pan, Q. Weakly Supervised SAR Ship Segmentation Based on Variational Gaussian G (A) (0) Mixture Model A Learning. In Proceedings of the 2020 Chinese Automation Congress (CAC), Shanghai, China, 6–8 November 2020.

127. Mao, Y.; Li, X.; Su, H.; Zhou, Y.; Li, J. Ship Detection for SAR Imagery Based on Deep Learning: A Benchmark. In Proceedings of the 2020 IEEE 9th Joint International Information Technology and Artificial Intelligence Conference (ITAIC), Chongqing, China, 11–13 December 2020. [CrossRef]

128. Zhao, K.; Zhou, Y.; Chen, X. A Dense Connection Based SAR Ship Detection network. In Proceedings of the 2020 IEEE 9th Joint International Information Technology and Artificial Intelligence Conference (ITAIC), Chongqing, China, 11–13 December 2020. [CrossRef]

129. An, Q.; Pan, Z.; You, H.; Hu, Y. Transitive Transfer Learning Based Anchor Free Rotatable Detector for SAR Target Detection With Few Samples. *IEEE Access* **2021**, *9*, 24011–24025. [CrossRef]

130. Zhang, P.; Luo, H.; Ju, M.; He, M.; Chang, Z.; Hui, B. Brain-Inspired Fast Saliency-Based Filtering Algorithm for Ship Detection in High-Resolution SAR Images. *IEEE Trans. Geosci. Remote Sens.* **2021**, *60*, 5201709. [CrossRef]

131. Wang, R.; Shao, S.; An, M.; Li, J.; Wang, S.; Xu, X. Soft Thresholding Attention Network for Adaptive Feature Denoising in SAR Ship Detection. *IEEE Access* **2021**, *9*, 29090–29105. [CrossRef]

132. Zhang, T.; Zhang, X.; Shi, J.; Wei, S. Shipdenet-18: An Only 1 Mb With Only 18 Convolution Layers Light-Weight Deep Learning Network for Sar Ship Detection. In Proceedings of the IGARSS 2020–2020 IEEE International Geoscience and Remote Sensing Symposium, Waikoloa, HI, USA, 26 September–2 October 2020. [CrossRef]

133. Wang, X.; Cui, Z.; Cao, Z.; Dang, S. Dense Docked Ship Detection via Spatial Group-Wise Enhance Attention in SAR Images. In Proceedings of the IGARSS 2020—2020 IEEE International Geoscience and Remote Sensing Symposium, Waikoloa, HI, USA, 26 September–2 October 2020. [CrossRef]

134. Hou, S.; Ma, X.; Wang, X.; Fu, Z.; Wang, J.; Wang, H. SAR Image Ship Detection Based on Scene Interpretation. In Proceedings of the IGARSS 2020—2020 IEEE International Geoscience and Remote Sensing Symposium, Waikoloa, HI, USA, 26 September–2 October 2020. [CrossRef]

135. Jiang, S.; Zhu, M.; He, Y.; Zheng, Z.; Zhou, F.; Zhou, G. Ship Detection with Sar Based on Yolo. In Proceedings of the IGARSS 2020—2020 IEEE International Geoscience and Remote Sensing Symposium, Waikoloa, HI, USA, 26 September–2 October 2020. [CrossRef]

136. Zhu, C.; Zhao, D.; Liu, Z.; Mao, Y. Hierarchical Attention for Ship Detection in SAR Images. In Proceedings of the IGARSS 2020—2020 IEEE International Geoscience and Remote Sensing Symposium, Waikoloa, HI, USA, 26 September–2 October 2020. [CrossRef]

137. Hou, Z.; Cui, Z.; Cao, Z.; Liu, N. An Integrated Method of Ship Detection and Recognition in Sar Images based on Deep Learning. In Proceedings of the IGARSS 2020—2020 IEEE International Geoscience and Remote Sensing Symposium, Waikoloa, HI, USA, 26 September–2 October 2020. [CrossRef]
138. Wang, X.; Cui, Z.; Cao, Z.; Tian, Y. Ship Detection in Large Scale Sar Images Based on Bias Classification. In Proceedings of the IGARSS 2020—2020 IEEE International Geoscience and Remote Sensing Symposium, Waikoloa, HI, USA, 26 September–2 October 2020. [CrossRef]
139. Shao, P.; Lu, X.; Huang, P.; Xu, W.; Dong, Y. Impact Analysis of Radio Frequency Interference on SAR Image Ship Detection Based on Deep Learning. In Proceedings of the IGARSS 2020—2020 IEEE International Geoscience and Remote Sensing Symposium, Waikoloa, HI, USA, 26 September–2 October 2020. [CrossRef]
140. Li, J.; Guo, C.; Gou, S.; Wang, M.; Chen, J. Ship Segmentation on High-Resolution Sar Image by a 3D Dilated Multiscale U-Net. In Proceedings of the IGARSS 2020—2020 IEEE International Geoscience and Remote Sensing Symposium, Waikoloa, HI, USA, 26 September–2 October 2020. [CrossRef]
141. Ferreira, N.; Silveira, M. Ship Detection in SAR Images Using Convolutional Variational Autoencoders. In Proceedings of the IGARSS 2020—2020 IEEE International Geoscience and Remote Sensing Symposium, Waikoloa, HI, USA, 26 September–2 October 2020.
142. Lee, S.J.; Chang, J.Y.; Lee, K.J.; Oh, K.Y. Data Augmentation for Ship Detection using Kompsat-5 Images and Deep Learning Model. In Proceedings of the IGARSS 2020—2020 IEEE International Geoscience and Remote Sensing Symposium, Waikoloa, HI, USA, 26 September–2 October 2020. [CrossRef]
143. Song, J.; Kim, D.J. Fine Acquisition of Vessel Training Data for Machine Learning from Sentinel-1 SAR Images Accompanied by AIS Imformation. In Proceedings of the IGARSS 2020—2020 IEEE International Geoscience and Remote Sensing Symposium, Waikoloa, HI, USA, 26 September–2 October 2020. [CrossRef]
144. Tang, G.; Zhuge, Y.; Claramunt, C.; Men, S. N-YOLO: A SAR Ship Detection Using Noise-Classifying and Complete-Target Extraction. *Remote Sens.* **2021**, *13*, 871. [CrossRef]
145. Raj, J.A.; Idicula, S.M.; Paul, B. A novel Ship detection method from SAR image with reduced false alarm. *J. Phys. Conf. Ser.* **2021**, *1817*, 012010. [CrossRef]
146. Jiang, K.; Cao, Y. SAR Image Ship Detection Based on Deep Learning. In Proceedings of the 2020 International Conference on Computer Engineering and Intelligent Control (ICCEIC), Chongqing, China, 6–8 November 2020. [CrossRef]
147. Li, D.; Liang, Q.; Liu, H.; Liu, Q.; Liu, H.; Liao, G. A Novel Multidimensional Domain Deep Learning Network for SAR Ship Detection. *IEEE Trans. Geosci. Remote Sens.* **2021**, *60*, 5203213. [CrossRef]
148. Geng, X.; Shi, L.; Yang, J.; Li, P.; Zhao, L.; Sun, W.; Zhao, J. Ship Detection and Feature Visualization Analysis Based on Lightweight CNN in VH and VV Polarization Images. *Remote Sens.* **2021**, *13*, 1184. [CrossRef]
149. Jin, L.; Liu, G. An Approach on Image Processing of Deep Learning Based on Improved SSD. *Symmetry* **2021**, *13*, 495. [CrossRef]
150. Ren, Y.; Li, X.; Xu, H. A Deep Learning Model to Extract Ship Size from Sentinel-1 SAR Images. *IEEE Trans. Geosci. Remote Sens.* **2021**, *60*, 5203414. [CrossRef]
151. Chen, Y.; Duan, T.; Wang, C.; Zhang, Y.; Huang, M. End-to-End Ship Detection in SAR Images for Complex Scenes Based on Deep CNNs. *J. Sens.* **2021**, *2021*, 8893182. [CrossRef]
152. He, Y.; Gao, F.; Wang, J.; Hussain, A.; Yang, E.; Zhou, H. Learning Polar Encodings for Arbitrary-Oriented Ship Detection in SAR Images. *IEEE J. Sel. Top. Appl. Earth Obs. Remote Sens.* **2021**, *14*, 3846–3859. [CrossRef]
153. Tian, L.; Cao, Y.; He, B.; Zhang, Y.; He, C.; Li, D. Image Enhancement Driven by Object Characteristics and Dense Feature Reuse Network for Ship Target Detection in Remote Sensing Imagery. *Remote Sens.* **2021**, *13*, 1327. [CrossRef]
154. Li, Y.; Zhu, W.; Zhu, B. SAR image nearshore ship target detection in complex environment. In Proceedings of the 2021 IEEE 5th Advanced Information Technology, Electronic and Automation Control Conference (IAEAC), Chongqing, China, 12–14 March 2021; Volume 5, pp. 1964–1968. [CrossRef]
155. Zhao, K.; Zhou, Y.; Chen, X.; Wang, B.; Zhang, Y. Ship detection from scratch in Synthetic Aperture Radar (SAR) images. *Int. J. Remote Sens.* **2021**, *42*, 5010–5024. [CrossRef]
156. Ma, Z. High-Speed Lightweight Ship Detection Algorithm Based on YOLO-V4 for Three-Channels RGB SAR Image. *Remote Sens.* **2021**, *13*, 1909. [CrossRef]
157. Zhong, R. On-Board Real-Time Ship Detection in HISEA-1 SAR Images Based on CFAR and Lightweight Deep Learning. *Remote Sens.* **2021**, *13*, 1995. [CrossRef]
158. Hong, Z.; Yang, T.; Tong, X.; Zhang, Y.; Jiang, S.; Zhou, R.; Han, Y.; Wang, J.; Yang, S.; Liu, S. Multi-Scale Ship Detection from SAR and Optical Imagery via A More Accurate YOLOv3. *IEEE J. Sel. Top. Appl. Earth Obs. Remote Sens.* **2021**, *14*, 6083–6101. [CrossRef]
159. Zhu, M.; Hu, G.; Li, S.; Liu, S.; Wang, S. An Effective Ship Detection Method Based on RefineDet in SAR Images. In Proceedings of the 2021 International Conference on Communications, Information System and Computer Engineering (CISCE), Beijing, China, 14–16 May 2021. [CrossRef]
160. Shin, S.; Kim, Y.; Hwang, I.; Kim, J.; Kim, S. Coupling Denoising to Detection for SAR Imagery. *Appl. Sci.* **2021**, *11*, 5569. [CrossRef]
161. Hu, H. TWC-Net: A SAR Ship Detection Using Two-Way Convolution and Multiscale Feature Mapping. *Remote Sens.* **2021**, *13*, 2558. [CrossRef]
162. Sun, W.; Huang, X. Semantic attention-based network for inshore SAR ship detection. In Proceedings of the SPIE 11878, Thirteenth International Conference on Digital Image Processing (ICDIP 2021), Singapore, 30 June 2021; Volume 118782A.

163. Wu, Z.; Hou, B.; Ren, B.; Ren, Z.; Wang, S.; Jiao, L. A Deep Detection Network Based on Interaction of Instance Segmentation and Object Detection for SAR Images. *Remote Sens.* **2021**, *13*, 2582. [CrossRef]

164. Dong, Y.; Zhang, H.; Wang, C.; Zhang, B.; Li, L. Ship Detection based on M2Det for SAR images under Heavy Sea State. In Proceedings of the EUSAR 2021 13th European Conference on Synthetic Aperture Radar. VDE, Online, 29 March–1 April 2021; pp. 1–4.

165. Sun, K.; Liang, Y.; Ma, X.; Huai, Y.; Xing, M. DSDet: A Lightweight Densely Connected Sparsely Activated Detector for Ship Target Detection in High-Resolution SAR Images. *Remote Sens.* **2021**, *13*, 2743. [CrossRef]

166. Zhang, T.; Zhang, X.; Ke, X. Quad-FPN: A Novel Quad Feature Pyramid Network for SAR Ship Detection. *Remote Sens.* **2021**, *13*, 2771. [CrossRef]

167. Sun, Z.; Dai, M.; Leng, X.; Lei, Y.; Xiong, B.; Ji, K.; Kuang, G. An Anchor-free Detection Method for Ship Targets in High-Resolution SAR Images. *IEEE J. Sel. Top. Appl. Earth Obs. Remote Sens.* **2021**, *14*, 7799–7816. [CrossRef]

168. Zhang, X.; Huo, C.; Xu, N.; Jiang, H.; Cao, Y.; Ni, L.; Pan, C. Multitask Learning for Ship Detection from Synthetic Aperture Radar Images. *IEEE J. Sel. Top. Appl. Earth Obs. Remote Sens.* **2021**, *14*, 8048–8062. [CrossRef]

169. Du, Y.; Du, L.; Li, L. An SAR Target Detector Based on Gradient Harmonized Mechanism and Attention Mechanism. *IEEE Geosci. Remote Sens. Lett.* **2021**, *19*, 4017005. [CrossRef]

170. Zhang, T.; Zhang, X.; Li, J.; Xu, X.; Wang, B.; Zhan, X.; Xu, Y.; Ke, X.; Zeng, T.; Su, H. SAR Ship Detection Dataset (SSDD): Official Release and Comprehensive Data Analysis. *Remote Sens.* **2021**, *13*, 3690. [CrossRef]

171. Liu, F.; Li, Y. SAR remote sensing image ship detection method NanoDet based on visual saliency. *J. Radars* **2021**, *10*, 885–894.

172. Zhao, Y.; Zhao, L.; Liu, Z.; Hu, D.; Kuang, G.; Liu, L. Attentional Feature Refinement and Alignment Network for Aircraft Detection in SAR Imagery. *arXiv* **2022**, arXiv:2201.07124. [CrossRef]

173. Li, S.; Xiao, Y.; Zhang, Y.; Chu, L.; Qiu, R.C. Learning Efficient Representations for Enhanced Object Detection on Large-scene SAR Images. *arXiv* **2022**, arXiv:2201.08958.

174. Song, T.; Kim, S.; Kim, S.T.; Lee, J.; Sohn, K. Context-Preserving Instance-Level Augmentation and Deformable Convolution Networks for SAR Ship Detection. *arXiv* **2022**, arXiv:2202.06513.

175. Zhu, H.; Xie, Y.; Huang, H.; Jing, C.; Rong, Y.; Wang, C. DB-YOLO: A Duplicate Bilateral YOLO Network for Multi-Scale Ship Detection in SAR Images. *Sensors* **2021**, *21*, 8146. [CrossRef] [PubMed]

176. Lin, S. A Lightweight Detection Model for SAR Aircraft in a Complex Environment. *Remote Sens.* **2021**, *13*, 5020. [CrossRef]

177. Ding, C. SRSDD-v1.0: A High-Resolution SAR Rotation Ship Detection Dataset. *Remote Sens.* **2021**, *13*, 5104. [CrossRef]

178. Qin, M. A Fast and Lightweight Detection Network for Multi-Scale SAR Ship Detection under Complex Backgrounds. *Remote Sens.* **2021**, *14*, 31. [CrossRef]

179. Zhou, K.; Zhang, M.; Wang, H.; Tan, J. Ship Detection in SAR Images Based on Multi-Scale Feature Extraction and Adaptive Feature Fusion. *Remote Sens.* **2022**, *14*, 755. [CrossRef]

180. Zhu, M.; Hu, G.; Zhou, H.; Wang, S.; Feng, Z.; Yue, S. A Ship Detection Method via Redesigned FCOS in Large-Scale SAR Images. *Remote Sens.* **2022**, *14*, 1153. [CrossRef]

181. Liu, S.; Kong, W.; Chen, X.; Xu, M.; Yasir, M.; Zhao, L.; Li, J. Multi-Scale Ship Detection Algorithm Based on a Lightweight Neural Network for Spaceborne SAR Images. *Remote Sens.* **2022**, *14*, 1149. [CrossRef]

182. Xia, R.; Chen, J.; Huang, Z.; Wan, H.; Wu, B.; Sun, L.; Yao, B.; Xiang, H.; Xing, M. CRTransSar: A Visual Transformer Based on Contextual Joint Representation Learning for SAR Ship Detection. *Remote Sens.* **2022**, *14*, 1488. [CrossRef]

183. Everingham, M.; Eslami, S.; Gool, L.V.; Williams, C.; Winn, J.; Zisserman, A. The pascal visual object classes challenge: A retrospective. *Int. J. Comput. Vis.* **2015**, *111*, 98–136. [CrossRef]

184. Lin, T.Y.; Maire, M.; Belongie, S.; Hays, J.; Perona, P.; Ramanan, P.; Dollár, P.; Zitnick, C.L. Microsoft Coco: Common Objects in Context. In *European Conference on Computer Vision*; Springer: Cham, Switzerland, 2014; pp. 740–755. [CrossRef]

185. Ren, S.Q.; He, K.M.; Girshick, R.; Sun, J. Faster R-CNN: Towards real-time object detection with region proposal networks. *IEEE Trans. Pattern Anal. Mach. Intell.* **2017**, *39*, 1137–1149. [CrossRef]

186. Redmon, J.; Divvala, S.; Girshick, R.; Farhadi, A. You only look once: Unified, real-time object detection. In Proceedings of the 2016 IEEE Conference on Computer Vision and Pattern Recognition, Las Vegas, NV, USA, 26 June–1 July 2016; IEEE Press: Piscataway, NJ, USA, 2016; pp. 779–788. [CrossRef]

187. Liu, W.; Anguelov, D.; Erhan, D.; Szegedy, C.; Reed, S.; Fu, C.Y.; Berg, A.C. *SSD: Single Shot MultiBox detectorM. Computer Vision—ECCV 2016*; Springer International Publishing: Cham, Switzerland, 2016; pp. 21–37. [CrossRef]

188. Jia, Y.Q.; Shelhamer, E.; Donahue, J.; Karayev, S.; Long, J.; Girshick, R.; Guadarrama, S.; Darrell, T. Caffe: Convolutional architecture for fast feature embedding. In Proceedings of the 22nd ACM International Conference on Multimedia; ACM Press: New York, NY, USA, 2014; pp. 675–678.

189. Abadi, M.; Agarwal, A.; Barham, P.; Brevdo, E.; Chen, Z.; Citro, C.; Corrado, G.S.; Davis, A.; Dean, J.; Devin, M.; et al. Tensorflow: Large-scale machine learning on heterogeneous distributed systems. *arXiv* **2016**, arXiv:1603.04467. [CrossRef]

190. Paszke, A.; Gross, S.; Massa, F.; Lerer, A.; Bradbury, J.; Chanan, G.; Killeen, T.; Lin, Z.M.; Gimelshein, N.; Antiga, L.; et al. Pytorch: An imperative style, high-performance deep learning library. *Adv. Neural Inf. Process. Syst.* **2019**, *32*, 8024–8035.

191. Available online: http://radars.ie.ac.cn/web/data/getData?newsColumnId=74fe223a-0b01-4830-8d99-1ba276e67ad8&pageType=en (accessed on 23 April 2021).

192. Xu, C.; Su, H.; Li, J.; Li, Y.; Yao, L.; Gao, L.; Yan, W. RSDD-SAR: Rotated Ship Detection Dataset in SAR Images. *J. Radars*. in press.

193. Dai, J.; Li, Y.; He, K.; Sun, J. R-fcn: Object detection via region-based fully convolutional networks. *Adv. Neural Inf. Process. Syst.* **2016**, *29*. [CrossRef]
194. Lin, T.Y.; Dollár, P.; Girshick, R.; He, K.; Hariharan, B.; Belongie, S. Feature pyramid networks for object detection. In Proceedings of the 2017 IEEE Conference on Computer Vision and Pattern Recognition, Honolulu, HI, USA, 21–26 July 2017; IEEE Press: Piscataway, NJ, USA, 2017; pp. 936–944.
195. Cai, Z.; Vasconcelos, N. Cascade r-cnn: Delving into high quality object detection. In Proceedings of the IEEE Conference on Computer Vision and Pattern Recognition, Seattle, WA, USA, 18–23 June 2018; pp. 6154–6162. [CrossRef]
196. He, K.M.; Gkioxari, G.; Dollár, P.; Girshick, R.B. Mask R-CNN. *IEEE Trans. Pattern Anal. Mach. Intell.* **2020**, *42*, 386–397. [CrossRef]
197. Li, Z.; Peng, C.; Yu, G.; Zhang, X.Y.; Deng, Y.D.; Sun, J. Light-head r-cnn: In defense of two-stage object detector. *arXiv* **2017**, arXiv:1711.07264. [CrossRef]
198. Liu, S.; Huang, D.; Wang, Y. Learning spatial fusion for single-shot object detection. *arXiv* **2019**, arXiv:1911.09516. [CrossRef]
199. Golnaz, G.; Lin, T.; Le, Q.V. Nas-fpn: Learning scalable feature pyramid architecture for object detection. In Proceedings of the IEEE Conference on Computer Vision and Pattern Recognition, Long Beach, CA, USA, 16–17 June 2019.
200. Tan, M.; Pang, R.; Le, Q.V. EfficientDet: Scalable and Efficient Object Detection. In Proceedings of the 2020 IEEE/CVF Conference on Computer Vision and Pattern Recognition (CVPR), Seattle, WA, USA, 13–19 June 2020. [CrossRef]
201. Woo, S.; Park, J.; Lee, J.Y.; Kweon, I.S. Cbam: Convolutional block attention module. In Proceedings of the European conference on computer vision (ECCV), Glasgow, UK, 23–28 August 2018; pp. 3–19. [CrossRef]
202. Shrivastava, A.; Gupta, A.; Girshick, R. Training Region-based Object Detectors with Online Hard Example Mining. In Proceedings of the IEEE Conference on Computer Vision and Pattern Recognition, Las Vegas, Nevada, USA, 27–30 June 2016; pp. 761–769. [CrossRef]
203. Li, B.; Liu, Y.; Wang, X. Gradient Harmonized Single-Stage Detector. In Proceedings of the AAAI Conference on Artificial Intelligence, New York, NY, USA, 7–12 February 2019; Volume 33, pp. 8577–8584. [CrossRef]
204. Pang, J.; Chen, K.; Shi, J.; Feng, H.J.; Ouyang, W.L.; Lin, D.H. Libra R-CNN: Towards Balanced Learning for Object Detection. In Proceedings of the 2019 IEEE/CVF Conference on Computer Vision and Pattern Recognition (CVPR), Long Beach, CA, USA, 15–20 July 2020. [CrossRef]
205. Bodla, N.; Singh, B.; Chellappa, R.; Davis, L.S. Soft-NMS—improving object detection with one line of code. In Proceedings of the IEEE International Conference on Computer Vision, Montreal, BC, Canada, 11–17 October 2017; pp. 5561–5569. [CrossRef]
206. Lin, T.Y.; Goyal, P.; Girshick, R.; He, K.; Dollár, P. Focal Loss for Dense Object Detection. *IEEE Trans. Pattern Anal. Mach. Intell.* **2017**, *99*, 2999–3007. [CrossRef]
207. Law, H.; Deng, J. CornerNet: Detecting Objects as Paired Keypoints. *Int. J. Comput. Vis.* **2018**, 734–750. [CrossRef]
208. Redmon, J.; Farhadi, A. YOLO9000: Better, faster, stronger. In Proceedings of the 2017 IEEE Conference on Computer Vision and Pattern Recognition, Honolulu, HI, USA, 21–26 July 2017; IEEE Press: Piscataway, NJ, USA, 2017; pp. 6517–6525. [CrossRef]
209. Redmon, J.; Farhadi, A. YOLOv3: An incremental improvement. *arXiv* **2018**, arXiv:1804.02767. [CrossRef]
210. Bochkovskiy, A.; Wang, C.Y.; Liao, H.Y.M. Yolov4: Optimal speed and accuracy of object detection. *arXiv* **2020**, arXiv:2004.10934. [CrossRef]
211. Zheng, Z.; Wang, P.; Ren, D.; Liu, W.; Ye, R.; Hu, Q.; Zuo, W. Enhancing Geometric Factors in Model Learning and Inference for Object Detection and Instance Segmentation. *IEEE Trans. Cybern.* **2020**, *7*, 1–13. [CrossRef] [PubMed]
212. Yu, J.; Jiang, Y.; Wang, Z.; Cao, Z.; Huang, T. UnitBox: An Advanced Object Detection Network. In Proceedings of the 24th ACM International Conference on Multimedia, Amsterdam, The Netherlands, 15–19 October 2016. [CrossRef]
213. Rezatofighi, H.; Tsoi, N.; Gwak, J.Y.; Sadeghian, A.; Reid, I.; Savarese, S. Generalized Intersection Over Union: A Metric and a Loss for Bounding Box Regression. In Proceedings of the 2019 IEEE/CVF Conference on Computer Vision and Pattern Recognition (CVPR), Long Beach, CA, USA, 15–20 July 2019. [CrossRef]
214. Zheng, Z.; Wang, P.; Liu, W.; Li, J.; Ye, R.; Ren, D. Distance-IoU loss: Faster and better learning for bounding box regression. In Proceedings of the AAAI Conference on Artificial Intelligence, New York, NY, USA, 7–12 February 2020; Volume 34, pp. 12993–13000. [CrossRef]
215. Liu, S.; Huang, D.; Wang, Y. 2018. Receptive field block net for accurate and fast object detection. *arXiv* **2017**, arXiv:1711.07767. [CrossRef]
216. Zhao, Q.; Sheng, T.; Wang, Y.; Tang, Z.; Chen, Y.; Cai, L.; Ling, H. M2Det: A Single-Shot Object Detector Based on Multi-Level Feature Pyramid Network. In Proceedings of the AAAI Conference on Artificial Intelligence, New York, NY, USA, 7–12 February 2019; Volume 33, pp. 9259–9266. [CrossRef]
217. Huang, R.; Pedoeem, J.; Chen, C. YOLO-LITE: A Real-Time Object Detection Algorithm Optimized for Non-GPU Computers. In Proceedings of the 2018 IEEE International Conference on Big Data (Big Data), Seattle, WA, USA, 10–13 December 2018. [CrossRef]
218. Zhou, X.; Zhuo, J.; Krhenbühl, P. Bottom-Up Object Detection by Grouping Extreme and Center Points. In Proceedings of the 2019 IEEE/CVF Conference on Computer Vision and Pattern Recognition (CVPR), Long Beach, CA, USA, 15–20 July 2020. [CrossRef]
219. Duan, K.; Bai, S.; Xie, L.; Qi, H.; Huang, Q.; Tian, Q. Centernet: Keypoint triplets for object detection. In Proceedings of the IEEE/CVF International Conference on Computer Vision, Montreal, BC, Canada, 11–17 October 2019; pp. 6569–6578. [CrossRef]
220. Zhou, X.; Wang, D.; Krhenbühl, P. Objects as Points. *arXiv* **2019**, arXiv:1904.07850. [CrossRef]
221. Tian, Z.; Shen, C.; Chen, H.; He, T. FCOS: Fully Convolutional One-Stage Object Detection. In Proceedings of the 2019 IEEE/CVF International Conference on Computer Vision (ICCV), Seoul, Korea, 27 October–2 November 2020. [CrossRef]

222. Kong, T.; Sun, F.; Liu, H.; Jiang, Y.; Shi, J. FoveaBox: Beyond anchor-based object detector. *IEEE Trans. Image Process.* **2020**, *29*, 7389–7398. [CrossRef]

223. Ronneberger, O.; Fischer, P.; Brox, T. *U-Net: Convolutional Networks for Biomedical Image Segmentation*; Springer International Publishing: Berlin/Heidelberg, Germany, 2015. [CrossRef]

224. Shen, Z.; Liu, Z.; Li, J.; Jiang, Y.; Chen, Y.; Xue, X. Object detection from scratch with deep supervision. *IEEE Trans. Pattern Anal. Mach. Intell.* **2019**, *42*, 398–412. [CrossRef] [PubMed]

225. Li, Z.; Peng, C.; Yu, G.; Zhang, X.; Deng, Y.; Sun, J. Detnet: A backbone network for object detection. *arXiv* **2018**, arXiv:1804.06215. [CrossRef]

226. Zhu, R.; Zhang, S.; Wang, X.; Wen, L.; Shi, H.; Bo, L.; Mei, T. ScratchDet: Training Single-Shot Object Detectors from Scratch. In Proceedings of the IEEE Conference on Computer Vision and Pattern Recognition (CVPR), Long Beach, CA, USA, 15–20 July 2019. [CrossRef]

227. He, K.; Girshick, R.; Dollár, P. Rethinking imagenet pre-training. In Proceedings of the IEEE International Conference on Computer Vision, Montreal, BC, Canada, 11–17 October 2019; pp. 4918–4927. [CrossRef]

228. Shi, B.; Bai, X.; Belongie, S. Detecting Oriented Text in Natural Images by Linking Segments. *IEEE Comput. Soc.* **2017**, 2550–2558. [CrossRef]

229. Ma, J.; Shao, W.; Ye, H.; Wang, L.; Wang, H.; Zheng, Y.; Xue, X. Arbitrary-Oriented Scene Text Detection via Rotation Proposals. *IEEE Trans. Multimed.* **2018**, *20*, 3111–3122. [CrossRef]

230. Liao, M.; Shi, B.; Bai, X.; Wang, X.; Liu, W. Textboxes: A fast text detector with a single deep neural network. In Proceedings of the Thirty-first AAAI Conference on Artificial Intelligence, New York, NY, USA, 7–12 February 2017.

231. Liao, M.; Shi, B.; Bai, X. TextBoxes++: A Single-Shot Oriented Scene Text Detector. *IEEE Trans. Image Process.* **2018**, *27*, 3676–3690. [CrossRef] [PubMed]

232. Jiang, Y.; Zhu, X.; Wang, X.; Yang, S.; Li, W.; Wang, H.; Fu, P.; Luo, Z. R2CNN: Rotational region CNN for orientation robust scene text detection. *arXiv* **2017**, arXiv:1706.09579. [CrossRef]

233. Liu, W.; Ma, L.; Chen, H. Arbitrary-oriented ship detection framework in optical remote sensing images. *IEEE Geosci. Remote Sens. Lett.* **2017**, *15*, 937–941. [CrossRef]

234. Xue, Y.; Sun, H.; Fu, K.; Yang, J.; Sun, X.; Yan, M.; Guo, Z. Automatic ship detection in remote sensing images from google earth of complex scenes based on multiscale rotation dense feature pyramid networks. *Remote Sens.* **2018**, *10*, 32.

235. Yang, X.; Sun, H.; Sun, X.; Yan, M.; Guo, Z.; Fu, K. Position detection and direction prediction for arbitrary-oriented ships via multitask rotation region convolutional neural network. *IEEE Access* **2018**, *6*, 50839–50849. [CrossRef]

236. Li, K.; Cheng, G.; Bu, S.; You, X. Rotation-insensitive and context-augmented object detection in remote sensing images. *IEEE Trans. Geosci. Remote. Sens.* **2017**, *56*, 2337–2348. [CrossRef]

237. Ding, J.; Xue, N.; Long, Y.; Xia, G.; Lu, Q. Learning RoI Transformer for Oriented Object Detection in Aerial Images. In Proceedings of the IEEE Conference on Computer Vision and Pattern Recognition(CVPR), Long Beach, CA, USA, 15–20 July 2019; pp. 2849–2858. [CrossRef]

238. Gong, W.; Shi, Z.; Wu, Z.; Luo, J. Arbitrary-oriented ship detection via feature fusion and visual attention for high-resolution optical remote sensing imagery. *Int. J. Remote Sens.* **2021**, *42*, 2622–2640. [CrossRef]

239. Liu, L.; Pan, Z.; Lei, B. Learning a rotation invariant detector with rotatable bounding box. *arXiv* **2017**, arXiv:1711.09405. [CrossRef]

240. Li, X.; Wang, W.; Hu, X.; Yang, J. Selective kernel networks. In Proceedings of the IEEE/CVF Conference on Computer Vision and Pattern Recognition, Seattle, WA, USA, 13–19 June 2019; pp. 510–519. [CrossRef]

241. Huang, Z.; Wang, X.; Wei, Y.; Huang, C.; Wei, Y.; Liu, W. CCNet: Criss-Cross Attention for Semantic Segmentation. *IEEE Trans. Pattern Anal. Mach. Intell.* **2020**. [CrossRef]

242. Yuan, Y.; Wang, J. OCNet: Object Context Network for Scene Parsing. *arXiv* **2018**, arXiv:1809.00916. [CrossRef]

243. Lin, X.; Guo, Y.; Wang, J. Global Correlation Network: End-to-End Joint Multi-Object Detection and Tracking. *arXiv* **2021**, arXiv:2103.12511. [CrossRef]

244. Fu, J.; Liu, J.; Tian, H.; Li, Y.; Bao, Y.; Fang, Z.; Lu, H. Dual Attention Network for Scene Segmentation. In Proceedings of the 2019 IEEE/CVF Conference on Computer Vision and Pattern Recognition (CVPR), Long Beach, CA, USA, 15–20 July 2020. [CrossRef]

245. Vaswani, A.; Shazeer, N.; Parmar, N.; Uszko-reit, J.; Jones, L.; Gomez, A.N.; Kaiser, Ł.; Polosukhin, I. Attention is all you need. *Adv. Neural Inf. Process. Syst.* **2017**, *30*, 5998–6008.

246. Liu, Z.; Lin, Y.; Cao, Y.; Hu, H.; Wei, Y.; Zhang, Z.; Lin, S.; Guo, B. Swin transformer: Hierarchical vision transformer using shifted windows. In Proceedings of the IEEE/CVF International Conference on Computer Vision, Montreal, BC, Canada, 11–17 October 2021; pp. 10012–10022. [CrossRef]

247. Wu, J.; Leng, C.; Wang, Y.; Hu, Q.; Cheng, J. Quantized convolutional neural networks for mobile devices. In Proceedings of the IEEE Conference on Computer Vision and Pattern Recognition (CVPR), Long Beach, CA, USA, 15–20 July 2016. [CrossRef]

248. He, Y.; Zhang, X.; Sun, J. Channel pruning for accelerating very deep neural networks. In Proceedings of the IEEE International Conference on Computer Vision, Montreal, BC, Canada, 11–17 October 2017; pp. 1389–1397. [CrossRef]

249. Gong, Y.; Liu, L.; Yang, M.; Bourdev, L.D. Compressing deep convolutional networks using vector quantization. *arXiv* **2014**, arXiv:1412.6115.

250. Molchanov, P.; Tyree, S.; Karras, T.; Aila, T.; Kautz, J. Pruning convolutional neural networks for resource efficient inference. *arXiv* **2016**, arXiv:1611.06440. [CrossRef]

251. Han, S.; Mao, H.; Dally, W.J. Deep Compression: Compressing Deep Neural Networks with Pruning, Trained Quantization and Huffman Coding. *arXiv* **2016**, arXiv:1510.00149.
252. Han, S.; Pool, J.; Tran, J.; Dally, W. Learning both weights and connections for efficient neural network. *Adv. Neural Inf. Process. Syst.* **2015**, *28*. [CrossRef]
253. Molchanov, P.; Tyree, S.; Karras, T.; Aila, T.; Kautz, J. Pruning convolutional neural networks for resource efficient transfer learning. *arXiv* **2017**, arXiv:1611.06440.
254. Shao, F.; Chen, L.; Shao, J.; Ji, W.; Xiao, S.; Ye, L.; Zhuang, Y.; Xiao, J. Deep Learning for Weakly-Supervised Object Detection and Object Localization: A Survey. *Neurocomputing* **2022**, 192–207. [CrossRef]
255. Creswell, A.; White, T.; Dumoulin, V.; Arulkumaran, K.; Sengupta, B.; Bharath, A. Generative adversarial networks: An overview. *IEEE Signal Process. Mag.* **2018**, *35*, 53–65. [CrossRef]
256. Zoph, B.; Cubuk, E.D.; Ghiasi, G.; Lin, T.; Shlens, J.; Le, Q.V. Learning data augmentation strategies for object detection. In *European Conference on Computer Vision*; Springer: Cham, Switzerland, 2020; pp. 566–583. [CrossRef]
257. Carion, N.; Massa, F.; Synnaeve, G.; Usunier, N.; Kirillov, A.; Zagoruyko, S. End-to-end object detection with transformers. In *European Conference on Computer Vision*; Springer: Cham, Switzerland, 2020; pp. 213–229. [CrossRef]
258. Zhang, H.; Li, F.; Liu, S.; Zhang, L.; Su, H.; Zhu, J.; Ni, L.M.; Shum, H. DINO: DETR with Improved DeNoising Anchor Boxes for End-to-End Object Detection. *arXiv* **2022**. e-prints.
259. Zhang, T.; Zhang, X. Injection of Traditional Hand-Crafted Features into Modern CNN-Based Models for SAR Ship Classification: What, Why, Where, and How. *Remote Sens.* **2021**, *13*, 2091. [CrossRef]
260. Zhang, T.; Zhang, X. HTC+ for SAR Ship Instance Segmentation. *Remote Sens.* **2022**, *14*, 2395. [CrossRef]
261. Zhang, T.; Zhang, X.; Ke, X.; Liu, C.; Xu, X.; Zhan, X.; Wang, C.; Ahmad, I.; Zhou, Y.; Pan, D.; et al. HOG-ShipCLSNet: A novel deep learning network with hog feature fusion for SAR ship classification. *IEEE Trans. Geosci. Remote Sens.* **2021**, *60*, 1–22. [CrossRef]
262. Zhang, T.; Zhang, X. Squeeze-and-excitation Laplacian pyramid network with dual-polarization feature fusion for ship classification in sar images. *IEEE Geosci. Remote Sens. Lett.* **2021**, *19*, 1–5. [CrossRef]
263. Zhang, T.; Zhang, X. A Full-Level Context Squeeze-and-Excitation ROI Extractor for SAR Ship Instance Segmentation. *IEEE Geosci. Remote Sens. Lett.* **2022**, *19*, 1–5. [CrossRef]
264. Zhang, T.; Zhang, X.; Liu, C.; Shi, J.; Wei, S.; Ahmad, I.; Zhan, X.; Zhou, Y.; Pan, D.; Li, J.; et al. Balance learning for ship detection from synthetic aperture radar remote sensing imagery. *ISPRS J. Photogramm. Remote Sens.* **2021**, *182*, 190–207. [CrossRef]
265. Zhang, T.; Zhang, X. A polarization fusion network with geometric feature embedding for SAR ship classification. *Pattern Recognit.* **2022**, *123*, 108365. [CrossRef]
266. Heiselberg, H. Ship-iceberg classification in SAR and multispectral satellite images with neural networks. *Remote Sens.* **2020**, *12*, 2353. [CrossRef]
267. Heiselberg, P.; Sørensen, K.A.; Heiselberg, H.; Andersen, O.B. SAR Ship–Iceberg Discrimination in Arctic Conditions Using Deep Learning. *Remote Sens.* **2022**, *14*, 2236. [CrossRef]
268. Heiselberg, H.; Stateczny, A. Remote sensing in vessel detection and navigation. *Sensors* **2020**, *20*, 5841. [CrossRef] [PubMed]

Article

Serial GANs: A Feature-Preserving Heterogeneous Remote Sensing Image Transformation Model

Daning Tan [1], Yu Liu [1,2,*], Gang Li [2], Libo Yao [1], Shun Sun [1] and You He [1]

[1]   Institute of Information Fusion, Naval Aviation University, Yantai 264001, China; tandaning@yeah.net (D.T.); ylb_rs@126.com (L.Y.); shun2021@yeah.net (S.S.); heyou_f@126.com (Y.H.)

[2]   Department of Electronic Engineering, Tsinghua University, Beijing 100084, China; gangli@tsinghua.edu.cn

*   Correspondence: liuyu77360132@126.com

Abstract: In recent years, the interpretation of SAR images has been significantly improved with the development of deep learning technology, and using conditional generative adversarial nets (CGANs) for SAR-to-optical transformation, also known as image translation, has become popular. Most of the existing image translation methods based on conditional generative adversarial nets are modified based on CycleGAN and pix2pix, focusing on style transformation in practice. In addition, SAR images and optical images are characterized by heterogeneous features and large spectral differences, leading to problems such as incomplete image details and spectral distortion in the heterogeneous transformation of SAR images in urban or semiurban areas and with complex terrain. Aiming to solve the problems of SAR-to-optical transformation, Serial GANs, a feature-preserving heterogeneous remote sensing image transformation model, is proposed in this paper for the first time. This model uses the Serial Despeckling GAN and Colorization GAN to complete the SAR-to-optical transformation. Despeckling GAN transforms the SAR images into optical gray images, retaining the texture details and semantic information. Colorization GAN transforms the optical gray images obtained in the first step into optical color images and keeps the structural features unchanged. The model proposed in this paper provides a new idea for heterogeneous image transformation. Through decoupling network design, structural detail information and spectral information are relatively independent in the process of heterogeneous transformation, thereby enhancing the detail information of the generated optical images and reducing its spectral distortion. Using SEN-2 satellite images as the reference, this paper compares the degree of similarity between the images generated by different models and the reference, and the results revealed that the proposed model has obvious advantages in feature reconstruction and the economical volume of the parameters. It also showed that Serial GANs have great potential in decoupling image transformation.

Keywords: heterogeneous transformation; SAR image; optical image; conditional generative adversarial nets (CGANs)

Citation: Tan, D.; Liu, Y.; Li, G.; Yao, L.; Sun, S.; He, Y. Serial GANs: A Feature-Preserving Heterogeneous Remote Sensing Image Transformation Model. *Remote Sens.* **2021**, *13*, 3968. https://doi.org/10.3390/rs13193968

Academic Editors: Tianwen Zhang, Tianjiao Zeng and Xiaoling Zhang

Received: 24 August 2021
Accepted: 29 September 2021
Published: 3 October 2021

Publisher's Note: MDPI stays neutral with regard to jurisdictional claims in published maps and institutional affiliations.

1. Introduction

In recent years, there have been more and more applications of remote sensing images in environmental monitoring, disaster prevention, intensive farming, and homeland security. In practice, optical images are widely used due to their high spectral resolution and easy interpretation. The disadvantage is that they are sensitive to meteorological conditions, especially clouds and haze, which severely limits their use for observation and monitoring of ground targets [1]. In contrast, synthetic aperture radar (SAR) sensors can overcome adverse meteorological conditions by creating images using a longer wavelength of radio waves to obtain all-day and all-weather continuous observations. Although SAR images have significant advantages over optical images, their application is still limited by the difficulty of SAR image interpretation. First, because synthetic aperture radar is a side range-measuring instrument, the imaging effect is affected by the distance between the target and the antenna, which can lead to geometric distortion in SAR images [2]. Therefore, compared with optical images, it is more difficult for human eyes to understand the details of SAR images. Secondly, synthetic aperture imaging is a coherent imaging method in which the radio waves in the radar beam are aligned in space and time. While this consistency provides many advantages (required by the synthetic aperture process to work), it also leads to a phenomenon called speckle, which reduces the quality of SAR images and makes image interpretation more challenging [3]. Therefore, it is difficult to distinguish structural information directly from SAR images, which may not necessarily become easier with the increase in spatial resolution [4]. Considering the above two points, how to effectively use and interpret the target and scene information in SAR images has become an important issue that users of SAR data need to pay attention to. Under the condition of reasonable use of SAR image amplitude information, if the SAR image can be converted into a near-optical representation that is easy to recognize by human eyes, this will create new opportunities for SAR image interpretation.

Deep learning is a powerful tool for the interpretation of SAR images. Some scholars have reconstructed clear images by learning hidden nonlinear relations [5–10]. This type of method uses a residual learning strategy to overcome speckle noise by learning the mapping between the speckle image and the corresponding speckle-free reconstruction so that it can be further analyzed and explained. Although this mapping learning may be an ill-posed problem, it also provides a useful reference for SAR image interpretation.

In addition to convolutional neural networks, image translation methods in the field of natural images and human images provide other ideas for SAR-to-optical image transformation, such as through conditional generative adversarial networks (CGANs) [11–14]. This type of method separates the style and semantic information in image transformation, so it can transform from the SAR image domain to the optical image domain, and also ensures the transformed images have the prior structural information of the SAR images and the spectral information of optical images. In previous studies, CGANs were first applied to the translation tasks of text to text [15], text to image [16], and image to image [17,18], and are suitable for generating unknown sequences (text/image/video frames) from known conditional sequences (text/image/video frames). In recent literature, the applications of CGANs in image processing were mostly in image modification. This includes single image super-resolution [17], interactive image generation [18], image editing [19], image-to-image translation [11], etc. CGANs have been used in SAR-to-optical transformation in recent years. In the literature [20–22], different improved SAR-to-optical transformation models based on CycleGAN and pix2pix have been proposed. The general idea of these models is to improve the model structure and loss function, but they are not designed specifically for the differences of imaging principle between SAR images and optical images, so they do not have universal applicability.

In order to solve the problem of heterogeneous image transformation in principle, as shown in Figure 1a, we decomposed the SAR-to-optical transformation task into two steps: the first step was to implement the transformation from the SAR image domain to optical grayscale image domain through the Despeckling GAN. In this step, we aimed to suppress the speckle effect of SAR images and reconstruct the semantic structural information and texture details of SAR images. In the second step, we transformed the optical grayscale images obtained in the first step into optical color images through the Colorization GAN. The two subtasks are relatively independent and have low coupling, which can reduce the semantic distortion and spectral distortion in the process of direct SAR-to-optical transformation.

The main contributions of this paper are as follows.

1. Unlike the existing methods of direct image translation, this paper proposes a feature-preserving SAR-to-optical transformation model, which decouples the SAR-to-optical transformation task into SAR-to-gray transformation and gray-to-color transformation. This design effectively reduces the difficulty of the original task, enhancing the feature details of the generated optical color images and reducing spectral distortion.

2. In this paper, Despeckling GAN is proposed to transform SAR images into optical grayscale images, and its generator is improved on the basis of the U-net [11]. In the processing, Despeckling GAN guides SAR images to generate optical grayscale images based on the texture details of SAR images by gradient maps, thus enhancing the semantic and feature information of transformed images [23].
3. In this paper, Colorization GAN is proposed for despeckled grayscale image colorization. Its generator adopts a convolutional self-coding structure. We establish short-skip connections in different levels and long-skip connections between the same level of encoding and decoding. This structure design enables different levels of image information to flow in the network structure, to generate more realistic images with hue information.

The rest of this paper is structured as follows. Section 2 introduces the materials involved in this paper. Section 3 introduces the method in detail, including the network structure and the loss function. In Section 4, the experimental results are given, which are discussed and evaluated based on indexes. Section 5 shows the discussion of this paper. The last part of the paper (Section 6) gives the conclusions and prospects for future work.

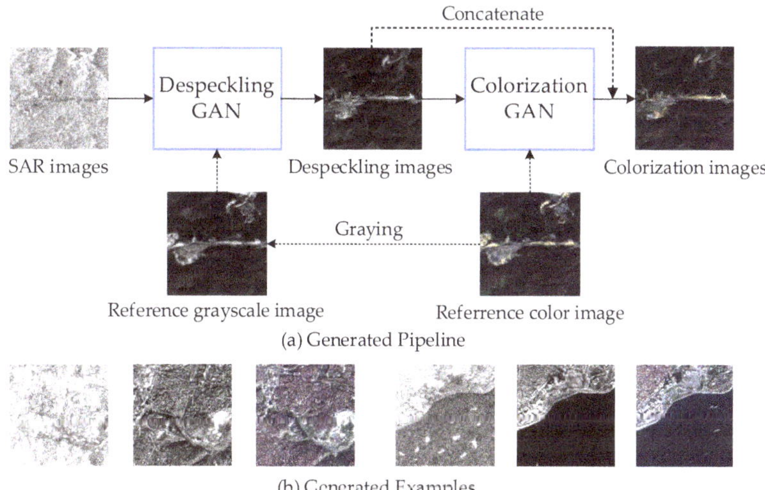

(a) Generated Pipeline

(b) Generated Examples

Figure 1. (**a**) Overview of our method: the SAR image affected by speckling is the input, and the Despeckling GAN generates a corresponding optical grayscale image as output. The optical grayscale image is then sent as input to the second generator network Colorization GAN, and the output is an optical color image. (**b**) Examples of generating optical grayscale images and optical color images through the Serial GANs.

2. Materials

Due to the lack of a large number of paired SAR and optical image datasets, deep learning-based SAR-to-optical translation research has mainly followed the idea of the CycleGAN [12] model; that is, unpaired image transformation. With the decrease in the cost of remote sensing images, a new idea has been presented to solve the cross-modal transformation, by using an image transformation method based on the Generative Adversarial Network. In the literature [24], Schmitt et al. published the SEN1-2 dataset to promote SAR and optical image fusion in deep learning research. The SEN1-2 dataset is a traditional remote sensing image dataset obtained by the SAR and optical sensors of the Sentinel-1 and Sentinel-2 satellites. As part of the Copernicus Project of the European Space Agency (ESA), Sentinel satellites are used for remote sensing tasks in the fields of climate, ocean, and land detection. The mission is being carried out jointly by six satellites with different observation applications. Sentinel-1 and Sentinel-2 provide the two most conventional SAR and optical images respectively, so they have been widely studied in the field of remote sensing image processing. Sentinel-1 is equipped with a C-band SAR sensor, which enables it to obtain high-positioning-accuracy SAR images regardless of weather conditions [25]. In its unique SAR imaging mode, the nominal resolution of Sentinel-1 is not less than 5 m, while providing dual-polarization capability and a very short equatorial access time (about 1 week) [26]. In the SEN1-2 dataset, Sentinel-1 images were collected in the interference wide (IW) swath mode, and the result obtained is the ground-range-detected (GRD) products. These images contain the backscatter coefficient in dB scale for every pixel spacing of 5 m in azimuth and 20 m in range. In order to simplify the operation, the dataset pays more attention to the VV polarization data and ignores the data of VH polarization. Sentinel-2 consists of two polar-orbiting satellites in the same orbit, with a phase difference of 180 degrees [27]. For the Sentinel-2 part of the dataset SEN1-2, the researchers used red, green, and blue channels (i.e., Bands 4, 3, and 2) to generate realistic RGB grid images. Because cloud occlusion will affect the final effect, the cloud coverage of the Sentinel-2 image in the dataset is less than or equal to 1%. SEN1-2 is composed of 282,384 pairs of related image patches, which come from all over the world and all weathers and seasons. It is the first large, open dataset of this kind and has significant advantages for learning a cross-modal mapping from SAR images to optical images. With the aid of the SEN1-2 dataset, we were able to build a new model that is different from the previous methods, the Serial GANs model proposed in this paper. Figure 2 shows some examples of image pairs in SEN1-2.

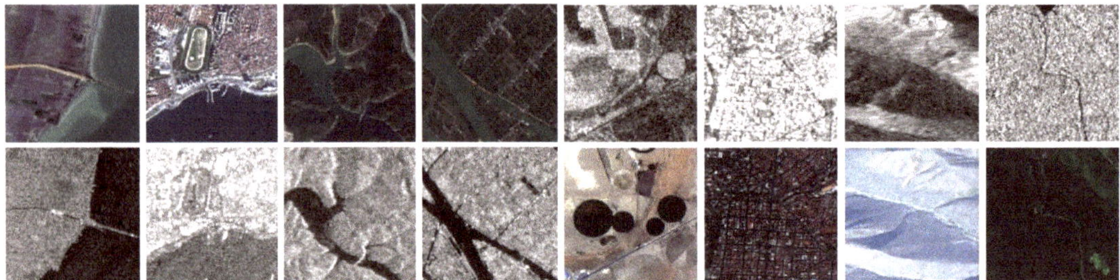

Figure 2. Some example patch pairs from the SEN1-2 dataset. Top row: Sentinel-1 SAR image patches; bottom row: Sentinel-2 RGB image patches.

3. Method

The heterogeneous transformation from SAR images to optical images is an ill-posed problem. The transformation results are often not ideal due to speckle noise, SAR image resolution, and other factors. Inspired by the ideas of pix2pix, CycleGAN and pix2pixHD, as shown in Figure 3a–d, this paper attempted to introduce optical grayscale images as the intermediate transformation domain Y. The transformation task from the SAR image domain X to the optical color image domain Z was completed in two steps by two generators (P and Q) and two discriminators (D_Y and D_Z) as shown in Figure 3e. First, the generator P completes the mapping: $X \to Y$, in which the SAR image is transformed into the optical grayscale image, and the corresponding discriminator D_Y is used to promote the transformation of the SAR image in the source domain X to the optical grayscale image in the domain Y, which is difficult to distinguish from the real optical grayscale image. Then, the generator Q completes the mapping: $Y \to Z$, in which the optical grayscale image is transformed to the optical color image, and the corresponding discriminator D_Z is used to promote the transformation of the optical grayscale image in the intermediate domain Y to the optical color image in the domain Z, which is difficult to distinguish from the optical color image. In this way, the original transformation process from the SAR image to the optical color image is divided into two steps, reducing the semantic distortion and feature loss in the process of direct transformation from the SAR image to the optical color image.

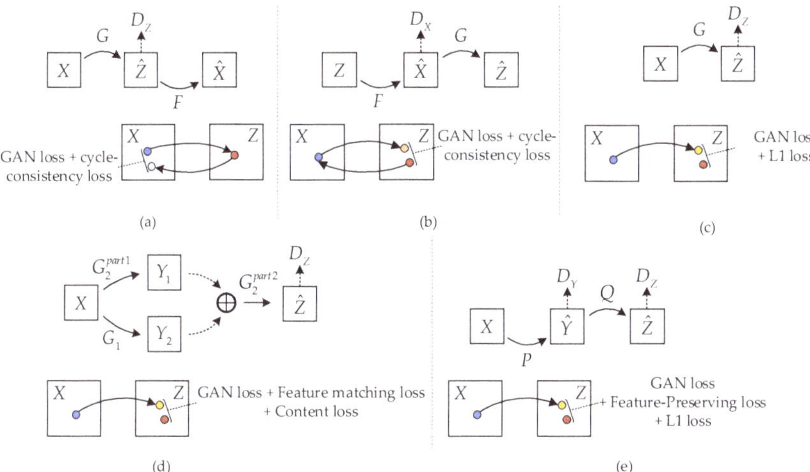

Figure 3. Overview of different methods. (**a**,**b**) CycleGAN. It is essentially two mirror-symmetric GANs, which share two generators G and F with discriminators D_Y and D_X respectively, and it uses GAN loss and cycle-consistency loss; (**c**) pix2pix, which directly transforms the image from the X domain to the Z domain, using GAN loss and L1 loss; (**d**) pix2pixHD. Different from pix2pix, it has two generators, G_1 and G_2, and its loss functions are GAN loss, Feature-matching loss, and Content loss; (**e**) the method proposed in this paper. It uses the intermediate state y as the transition, and its loss functions are GAN loss, Feature-preserving loss, and L1 loss.

As shown in Figure 4, the transformation from SAR images to optical images can be defined as the mapping transformation $T = PQ \ (P : X \to Y, \ Q : Y \to Z)$, from the source domain X to the target domain Y. Suppose that $x^{(i)}$ is a random sample taken from the SAR image domain X, and its distribution function is $\mathbb{P}^{(i)}(x)$, and the random sample $x^{(i)}$ mapped to the optical grayscale image domain is $y^{(i)}$. The final task of the network proposed in this paper is $T : X \to Z$, in which the final distribution function is $\mathbb{P}\left\{ Tx^{(i)} = z^{(i)} \big| x^{(i)} \right\}$ generated from our network.

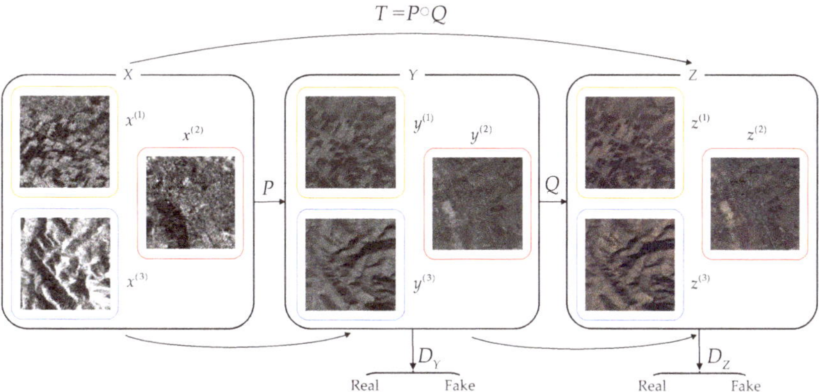

Figure 4. A feature-preserving heterogeneous remote sensing image transformation model is proposed in this paper. Let X, Y, and Z denote the SAR image domain, intermediate optical grayscale image domain, and optical color image domain, respectively, and $x^{(i)} \in X$, $y^{(i)} \in Y$ and $z^{(i)} \in Z$ denote the dataset samples of the corresponding image domain ($i = 1, 2, \cdots, N$, N denotes the total sample number of the data set).

3.1. Despeckling GAN

Generator P: As shown in Figure 5a, this paper used an improved U-net as the generator of Despeckling GAN. The input SAR image was encoded and decoded to output the optical grayscale image. A structure similar to the convolutional self-encoding network enables the generation network to better predict the optical grayscale image corresponding to the SAR image. The encoding and decoding process of the generator works on multiple levels to ensure that the overall contour and local details of the original SAR image are extracted on multi-scales. In the decoding process, the network upsamples the feature map of the previous level to the next level through deconvolution and adds the feature map of the same level in the encoding process through a long-skip connection to get an average merge (Merge). In U-net, this process is completed by concatenation. At the same time, skip connections are also used in each residual block, which has the advantage of overcoming the gradient disappearance problem of the network during training.

Discriminator D_Y: As shown in Figure 5b, PatchGAN, which is commonly used in GAN, was used as the discriminator. The process of heterogeneous image transformation includes the transformation of the content part and feature detail part. The content part refers to the similarity in content between the generated image and the original image, and the feature detail part refers to the similarity in features between the generated image and the target image. With PatchGAN, feature details can be maintained [11].

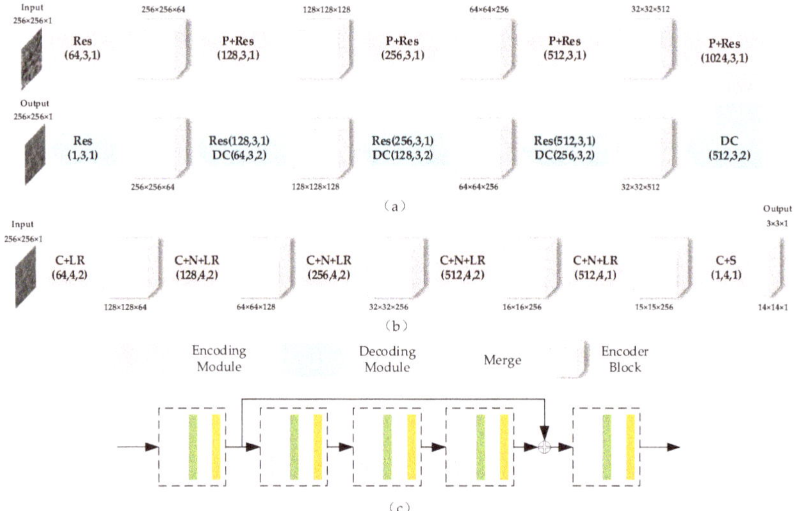

Figure 5. Architecture of the Despeckling GAN. (**a**) Generator (top). (**b**) Discriminator (bottom). (**c**) The detail of the Res. (The purple, green, and orange blocks in the dotted box correspond to the convolutional layer, the batch normalization layer, and the ReLU or Leaky ReLU layer, respectively). The numbers in brackets refer to the number of filters, filter size, and stride, respectively. The numbers above or below the encoder blocks and images indicate the input and output size of each module. Acronyms in the encoding and decoding modules are as follows: Res: Residual block with three convolutional layers and one skip connection, P: Maxpooling, DC: Deconvolution, C: Convolution, N: Batch Normalization, LR: Leaky ReLU, S: Sigmoid, Merge: Sum to average.

The loss function of the Despeckling GAN generator includes CGAN loss, L_1 loss, and feature-preserving loss. Based on the premise of the existing paired training data, this paper used the CGAN loss function to improve the performance of the generator. Through supervised training, the generator P learns the mapping from X to Y, and this makes the discriminator D_Y judge true. The network structure of the discriminator has the function of distinguishing fake images from real images. Therefore, the CGAN loss from X to Y is:

$$\mathcal{L}_{\text{GAN}}(P, D_Y) = \mathbb{E}_{x^{(i)}, y^{(i)}}\left[\log D_Y\left(x^{(i)}, y^{(i)}\right)\right] + \mathbb{E}_{x^{(i)}}\left[\log\left(1 - D_Y\left(P\left(x^{(i)}\right), x^{(i)}\right)\right)\right]. \quad (1)$$

In the reconstruction loss design, the L_1 loss is used to minimize the difference between the optical gray image and the generated image.

$$\mathcal{L}_{\text{Recon}}\left(P, x^{(i)}\right) = \mathbb{E}\left[\left\|P\left(x^{(i)}\right) - x^{(i)}\right\|_1\right]. \quad (2)$$

In the best state T^*, the output of the network $T^*\left(x^{(i)}\right)$ should be similar to the optical gray image $y^{(i)}$. In order to preserve the feature details of SAR images, this paper proposed a gradient-guided feature-preserving loss [28]. If $M(\cdot)$ denotes the operation to calculate the image gradient map, the loss of feature-preserving is:

$$\mathcal{L}_{\text{FP}}\left(P, x^{(i)}\right) = \mathbb{E}\left[M\left(P\left(x^i\right)\right), M\left(y^{(i)}\right)\right]. \quad (3)$$

For images I, $M(\cdot)$ is as follows:

$$\begin{aligned}
I_x(x, y) &= I(x+1, y) - I(x-1, y), \\
I_y(x, y) &= I(x, y+1) - I(x, y-1), \\
\nabla I(x, y) &= \left(I_x(x, y), I_y(x, y)\right), \\
M(I) &= \|\nabla I\|_2.
\end{aligned} \quad (4)$$

Specifically, the operation $M(\cdot)$ can be easily implemented by convolution with a fixed convolution kernel.

Therefore, the total training loss of the Despeckling GAN is:

$$\mathcal{L}_{\text{GAN}^1} = \underset{G}{\arg\min}\underset{D}{\max}\mathcal{L}_{GAN}(P, D_Y) + \beta_1 \mathcal{L}_{\text{Recon}}\left(P, x^{(i)}\right) + \gamma_1 \mathcal{L}_{\text{FP}}\left(P, y^{(i)}\right). \quad (5)$$

where β_1 and γ_1 are weighted values.

3.2. Colorization GAN

Colorization GAN completed the transformation from optical gray images to optical color images. Its principle comes from [29], which proved that, compared with Figure 6a, the colorization result of Figure 6b was better, so the latter was adopted in this paper. When a single channel gray image $\hat{y}^{(i)} \in \mathbb{R}^{H \times W \times 1}$ is input, the model learns the mapping $\hat{z}_{ab}^{(i)} = Q\left(\hat{y}^{(i)}\right)$ from the input gray channel to the corresponding *Lab* space color channels $\hat{z}_{ab}^{(i)} \in \mathbb{R}^{H \times W \times 2}$, where, H and W represent the height and width respectively. Then, the RGB image $\hat{z}^{(i)}$ is obtained by synthesizing $\hat{z}_{ab}^{(i)}$ and $\hat{y}^{(i)}$. The advantage of this method is that it can reduce the ill-posed problem, such that the colorization result is closer to the real image.

As shown in Figure 7, the generator of the Colorization GAN uses a convolutional self-coding structure, which establishes short-skip connections within different levels and long connections between the same levels of encoding and decoding. This kind of structure design enables different levels of image information to flow in the network so that the hue information of the generated image is more real and full. The discriminator of the Colorization GAN is PatchGAN [11]. Recent studies have shown that adversarial loss helps to make colorization more vivid [29–31], and this paper also followed this idea. During training, we input the reference optical color image and the generated image one by one into the discriminator; the discriminator output was 0 (fake) or 1 (real). According to the previous methods, the loss of the discriminator is the sigmoid cross-entropy.

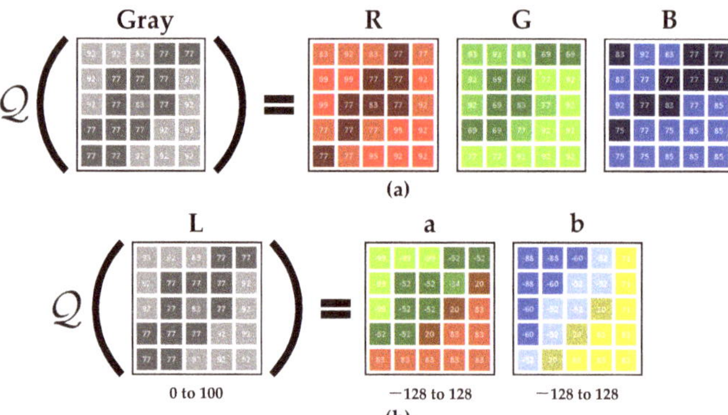

Figure 6. The principle of image colorization. (**a**) The direct mapping from the gray space to the RGB color space; (**b**) the hue Lab mapped from the gray space to the Lab color space.

Among them, the adversarial loss is expressed as follows:

$$\mathcal{L}_{\text{GAN}}(Q, D_Z) = \mathbb{E}_{y^{(i)}, z^{(i)}}\left[\log D_Z\left(y^{(i)}, z^{(i)}\right)\right] + \mathbb{E}_{y^{(i)}}\left[\log\left(1 - D_Z\left(y^{(i)}, Q\left(y^{(i)}\right)\right)\right)\right]. \quad (6)$$

In order to make the generated color distribution closer to the color distribution of the reference image, we defined the $\mathcal{L}_1$ loss in the *Lab* space, which is expressed as follows:

$$\mathcal{L}_1(Q) = \mathbb{E}_{y^{(i)}, z^{(i)}}\left[\left\|Q\left(y^{(i)}\right) - z^{(i)}\right\|_1\right]. \quad (7)$$

Therefore, the total loss function of the Colorization GAN model is as follows:

$$\mathcal{L}_{\text{GAN}^2} = \underset{G}{\arg\min}\underset{D}{\max}\mathcal{L}_{GAN}\left(Q\left(y^{(i)}\right), z^{(i)}\right) + \beta_2\mathcal{L}_1\left(Q, y^{(i)}\right). \quad (8)$$

where β_2 is a weighted value.

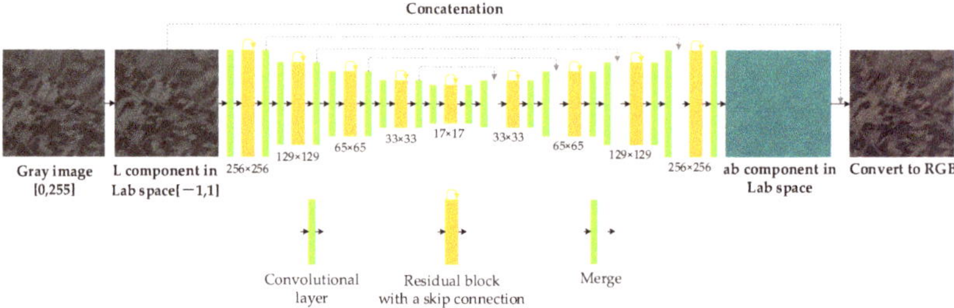

Figure 7. The network structure of the Colorization GAN generator. The gray image of the input model is first transformed into the L channel in the Lab color space and then trained to map to the AB channels through the network. The obtained hue is spliced with the gray image to get the Lab color image. Finally, the Lab image is transformed into an RGB image. The green block represents the convolutional layer, the yellow block represents the residual block, and the green and light-green blocks represent the average by merge.

4. Experiments and Results

As the SEN1-2 dataset covers the whole world and contains 282,384 pairs of SAR and optical color images across four seasons, some of which are overlapped, in order to facilitate the training, the original dataset was randomly sampled according to the stratified sampling method. The dataset was divided into the training dataset, validation dataset, and test dataset, and their respective proportions were about 6:2:2. The experiment of the proposed method was carried out on the computing platform of two 11G GPU GeForce RTX 2080Ti and i9900k CPUs using PyTorch. The input size of the images was 256 × 256, and the batch size was set to 10. In the experimental simulation, 200 epochs were set in the GAN training and optimized by the Adam optimizer. The sum of parameters was set to 0.5 and 0.999, respectively. The initial learning rate of the experiment was set to 0.0002. The first 100 epochs remained unchanged and then decreased to 0 according to the linear decreasing strategy.

Considering that season and landscape will affect the training results of the model, we selected image pairs of different seasons and landscapes and followed the principle of equilibrium [32]. As shown in Table 1, the number of SAR and optical image pairs in four seasons is approximately the same, and the number of image pairs of different landscapes in each season is also approximately the same.

Table 1. Number of different types of images selected in our dataset.

Season	Landscape	Training	Validation	Test	Total	
	River valley	279	90	74	443	
	Mountains and hills	280	91	73	444	
Spring	Urban residential area	275	97	70	442	2218
	Coastal city	278	87	77	442	
	Desert	277	96	74	447	
	River valley	278	92	77	447	
	Mountains and hills	276	96	74	446	
Summer	Urban residential area	279	93	70	442	2219
	Coastal city	276	93	72	441	
	Desert	275	95	73	443	
	River valley	278	96	70	443	
	Mountains and hills	275	89	74	441	
Fall	Urban residential area	278	95	72	445	2215
	Coastal city	279	91	75	445	
	Desert	277	89	77	442	
	River valley	277	91	78	446	
	Mountains and hills	275	94	73	442	
Winter	Urban residential area	278	95	71	444	2218
	Coastal city	278	92	75	445	
	Desert	277	93	71	441	
Total	-	5545	1855	1470	8870	

4.1. Experiment 1

In order to verify the effectiveness of the proposed method, four groups of experiments were designed using the same dataset and different conditions. The four groups of experiments were carried out according to the single variable principle. In Group 1, the unimproved generators P and Q were used, and the loss function included GAN loss and reconstruction loss. In Group 2, the improved generators P and Q were used, and the loss function included GAN loss and reconstruction loss. In Group 3, the unimproved generators P and Q were used, and the loss function included GAN loss, reconstruction loss, and feature-preserving loss. In Group 4, the improved generators P and Q were used, and the loss function included reconstruction loss, reconstruction loss, and feature-preserving loss. The relationship between the four groups of experiments is shown in Table 2.

Table 2. Grouping experiments under different conditions.

	Original Loss	Improved Loss	Original Networks	Improved Networks
Group 1	√		√	
Group 2	√			√
Group 3		√	√	
Group 4		√		√

As shown in Figure 8, the first column shows the SAR images collected by the SEN-1 satellite. The second column shows the SAR images collected by the SEN-2 satellite. The third, fourth, fifth, and sixth columns show the experimental results of Group 1, Group 2, Group 3, and Group 4, respectively. Through visual comparative analysis, it can be seen that improving the network structure and loss function can improve the quality of SAR-to-optical transformation, especially by enhancing the feature detail information of the generated image. It can map the SAR image to the optical color image to the maximum extent and help the interpretation of the SAR image.

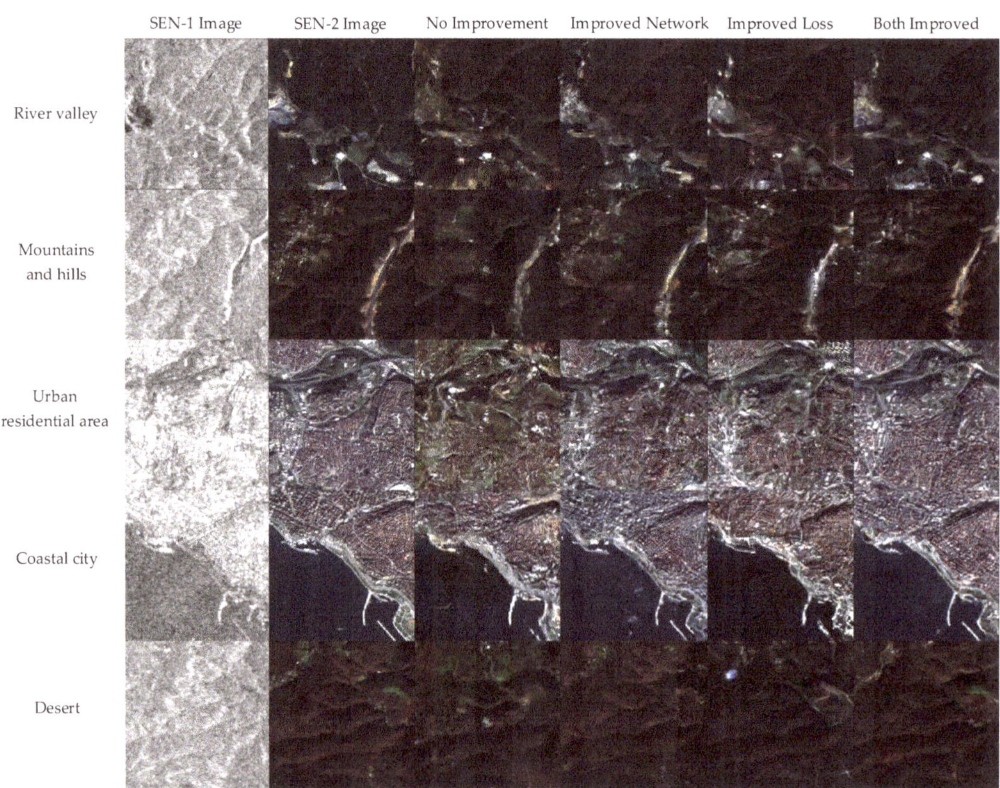

Figure 8. Results produced under different conditions. From top to bottom, the images are remote sensing images of five kinds of landscape: river valley, mountains and hills, urban residential area, seashore, and desert. From left to right: SEN-1 images, SEN-2 images, images generated by Group 1, images generated by Group 2, images generated by Group 3, images generated by Group 4.

In order to compare the detailed information of the generated images, Figure 9 shows the detailed comparison between the SEN-2 images and the four groups of experimental results. According to the subjective evaluation criteria, the results of improving the model and loss function at the same time are closer to the SEN-2 images. Only improving the loss function can improve the details of the generated images, but its effect is inferior to that of improving the model. The detailed comparison of the four groups of experimental results once again proves that the improvement measures proposed in this paper are effective. By comparing the two situations of improving model and improving loss function, it can be found that improving model contributes more to the results.

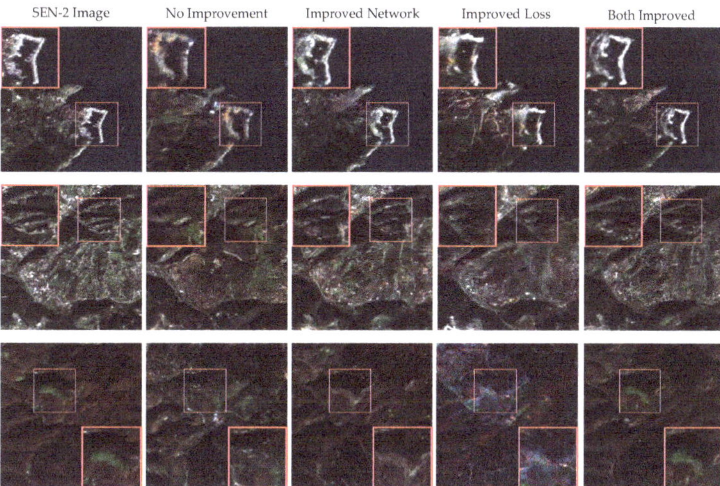

Figure 9. Detailed comparison of Experiment 1. We selected the generated results in three scenarios for detailed comparison with the SEN-2 reference image. The improvement measures proposed in this paper had an obvious effect on improving the quality of the generated images. From left to right: SEN-2 images, images generated by Group 1, images generated by Group 2, images generated by Group 3, and images generated by Group 4.

In order to quantify the effectiveness of the method, the final transformation effect (IQA) was measured by calculating the structural similarity (SSIM) [33,34], and the feature similarity (FSIM) [35]. Both indexes were calculated between the generated image $\hat{z}$ and the corresponding SEN-2 image z. Assuming that the generated image is $\hat{z}$, and the corresponding SEN-2 image is z, the SSIM calculation formula is as follows:

$$SSIM(\hat{z}, z) = [l(\hat{z}, z)]^{\alpha}[c(\hat{z}, z)]^{\beta}[s(\hat{z}, z)]^{\gamma}, \ \alpha, \beta, \gamma > 0 \tag{9}$$

Among them:

$$l(\hat{z}, z) = \frac{2\mu_{\hat{z}}\mu_z + c_1}{\mu_{\hat{z}}^2 + \mu_z^2 + c_1} \tag{10}$$

$$c(\hat{z}, z) = \frac{2\sigma_{\hat{z}z} + c_2}{\sigma_{\hat{z}}^2 + \sigma_z^2 + c_2} \tag{11}$$

$$s(\hat{z}, z) = \frac{\sigma_{\hat{z}z} + c_3}{\sigma_{\hat{z}}\sigma_z + c_3} \tag{12}$$

$l(\hat{z}, z)$, $c(\hat{z}, z)$, and $s(\hat{z}, z)$ in the equation represent the brightness comparison, contrast comparison, and structural comparison, respectively. $\mu_{\hat{z}}$ and μ_z represent the mean of $\hat{z}$ and z, $\sigma_{\hat{z}}$ and σ_z represent the standard deviation of $\hat{z}$ and z, $\sigma_{\hat{z}z}$ represents the covariance of $\hat{z}$ and z, and c_1, c_2 and c_3 are constant constants (so that the parent of the equation is not zero). In actually, $\alpha = \beta = \gamma = 1$, $c_3 = c_2/2$, SSIM is represented as:

$$SSIM(\hat{z}, z) = \frac{(2\mu_{\hat{z}}\mu_z + c_1)(\sigma_{\hat{z}z} + c_2)}{(\mu_{\hat{z}}^2 + \mu_z^2 + c_1)(\sigma_{\hat{z}}^2 + \sigma_z^2 + c_2)} \tag{13}$$

Another index, the FSIM, is a feature similarity evaluation index, which uses phase consistency (phase consistency (PC)) and gradient features (gradient magnitude (GM)), as follows:

$$FSIM = \frac{\sum_{\mathbf{x}\in\Omega} S_L(\mathbf{x}) \cdot PC_m(\mathbf{x})}{\sum_{\mathbf{x}\in\Omega} PC_m(\mathbf{x})}. \tag{14}$$

Which:

$$S_{PC}(\mathbf{x}) = \frac{2PC_1(\mathbf{x}) \cdot PC_2(\mathbf{x}) + T_1}{PC_1^2(\mathbf{x}) + PC_2^2(\mathbf{x}) + T_1} \tag{15}$$

$$S_G(\mathbf{x}) = \frac{2G_1(\mathbf{x}) \cdot G_2(\mathbf{x}) + T_2}{G_1^2(\mathbf{x}) + G_2^2(\mathbf{x}) + T_2} \tag{16}$$

$$S_L(\mathbf{x}) = [S_{PC}(\mathbf{x})]^\alpha \cdot [S_G(\mathbf{x})]^\beta \tag{17}$$

$S_{PC}(\mathbf{x})$, $S_G(\mathbf{x})$, and $S_L(\mathbf{x})$ represent the phase consistent (PC) similarity, gradient feature (GM) similarity, and PC-GM fusion similarity, respectively.

The similarity indicators of the four experimental schemes were calculated as Table 3. By comparing the results of the second, third, and first rows of the table, it can be seen that after improving the generator structure and loss function, both SSIM- and FSIM-generating images had been significantly improved, and the combined use of improved generators and loss functions obtained better results than improving the generator structure or loss functions alone.

Table 3. The model generated result indicators under different improvement measures. The number in bold indicates the optimal value under the corresponding index.

Scheme	SSIM	FSIM
No Improvement	0.2428	0.9000
Improved Network	0.2432	0.9023
Improved Loss	0.2435	0.9015
Both Improvements	**0.2442**	**0.9042**

4.2. Experiment 2

In order to verify the performance of the proposed method in preserving the SAR image features, the proposed algorithm was compared with pix2pix, CycleGAN, and pix2pixHD, respectively. During training, the Serial GANs train the generator P and the discriminator D_Y first, and then the training generator Q and the discriminator D_Z, respectively, with 200 epochs. In Figure 10, the first column shows the SAR images collected by the SEN-1 satellite, the second column shows the optical color images collected by the SEN-2 satellite, and the third, fourth, and fifth columns show the experimental results of pix2pix, CycleGAN, and pix2pixHD, respectively. According to the results, the proposed method can significantly preserve the details of SAR images in the process of heterogeneous transformation, with results as good as pix2pixHD. What is more, the volume of parameters of the model proposed in this paper was significantly lower than in the pix2pixHD model.

In order to compare the details of the images generated by different models, Figure 11 shows the details of the results generated by the proposed method compared with the four methods of pix2pix, CycleGAN and pix2pixHD. According to the subjective evaluation criteria, the results of the proposed method and pix2pixHD are closer to the Sentinel satellite image. The generation results of pix2pix and CycleGAN are inferior to the first two methods. Although the results of the proposed method and pix2pixHD are not significantly different, the subsequent comparison will show that the proposed method is superior to pix2pixHD.

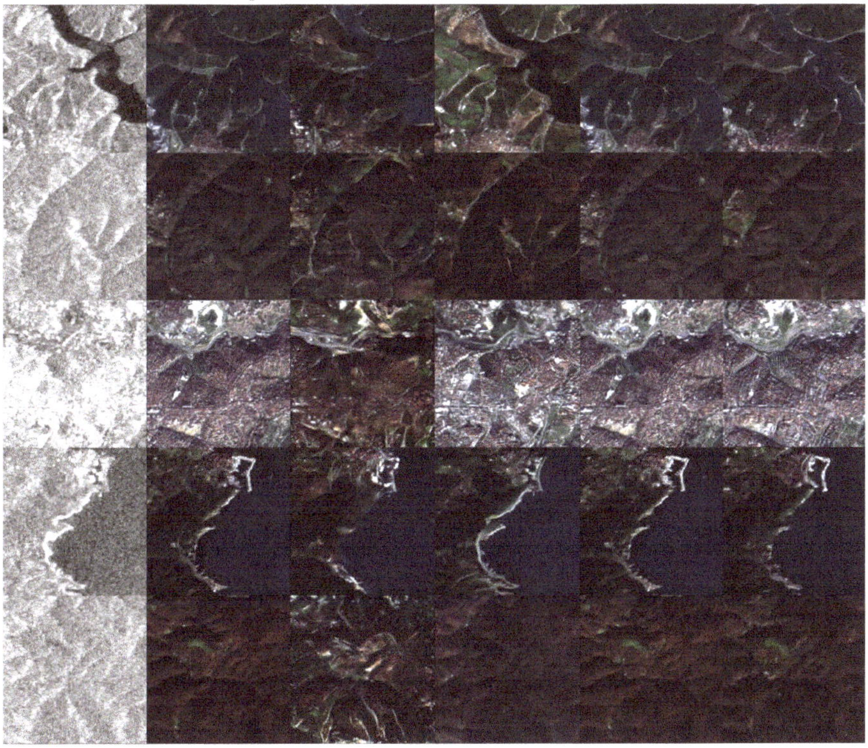

SEN-1 Image SEN-2 Image pix2pix Cycle GAN pix2pix HD Ours

River valley

Mountains and hills

Urban residential area

Coastal city

Desert

Figure 10. Comparison of the results generated by four different heterogeneous transformation models. From the top to the bottom: the remote sensing images of the river valley, mountains and hills, urban residential area, coastal city and desert. From left to right: SEN-1 images, SEN-2 images, images generated by pix2pix, images generated by CycleGAN, images generated by pix2pixHD, and the images generated by our model.

In order to quantitatively measure the advantages of the method, four image quality evaluation indexes (IQA), including PSNR, SSIM, FSIM, and MSE, were selected to quantitatively evaluate the method. As shown in Table 4, it can be seen from the data that the proposed model achieved the best in PSNR, SSIM, and MSE, and the second-best in FSIM.

Table 4. Comparison of the indexes between the images generated by four methods and the SEN-2 images. The number in bold indicates the optimal value under the corresponding index.

	PSNR	SSIM	MSE	FSIM
pix2pix [11]	13.8041	0.2431	0.0673	0.8987
Cycle GAN [21]	13.5052	0.2314	0.0749	0.9039
pix2pix HD [13]	13.4112	0.2347	0.0780	**0.9046**
Ours	**13.9267**	**0.2442**	**0.0669**	0.9042

The above experimental results show the effectiveness of the proposed method and the superiority to pix2pix and CycleGAN from both qualitative and quantitative aspects. In order to further illustrate that our method is better than pix2pixHD, we draw a performance comparison diagram reflecting the model size and FSIM value. As shown in Figure 12, although the FSIM value of our method is 0.0004 lower than that of pix2pixHD, the model size of our method is about half of that of pix2pixHD, so the advantage of our method is more obvious.

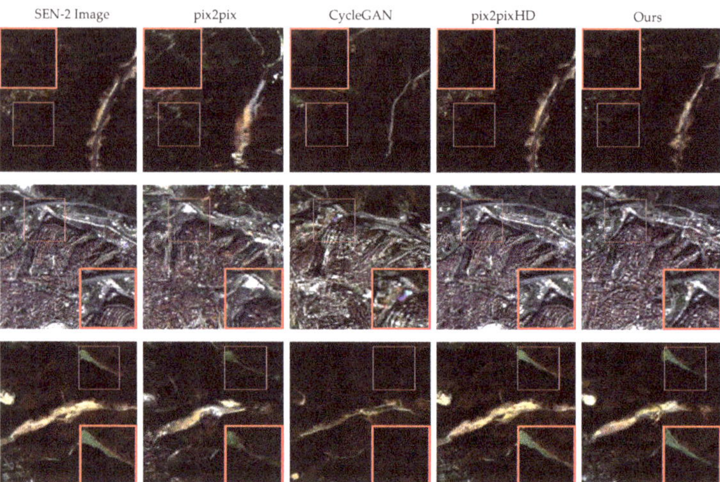

Figure 11. Detailed comparison of Experiment 2. We selected the generated results of different models in three scenarios to compare the details with the SEN-2 reference images. Compared with other image translation models, the proposed model has obvious advantages in improving the generation performance. From left to right: SEN-2 images, images generated by pix2pix, images generated by CycleGAN, images generated by pix2pixHD, and the images generated by our model.

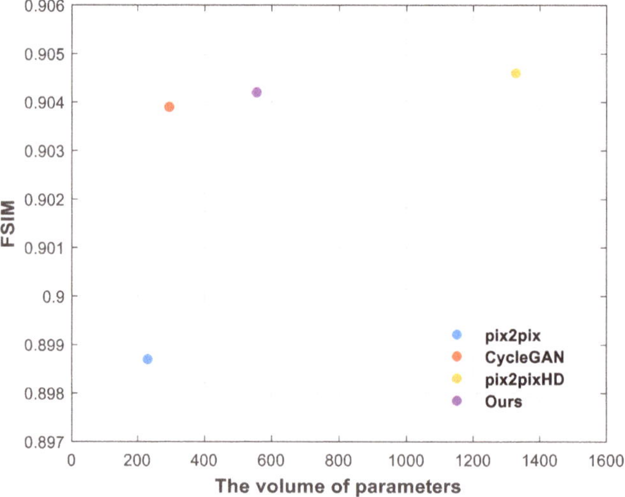

Figure 12. Comparison of the results generated by our method and SOTA methods. The ordinate of the graph represents the normalized FSIM value, and the abscissa represents the parameter size (Mbyte) of the model. The comparison of the four methods is represented by a scatter diagram. The closer the scatter points are to the y-axis $+\infty$, the better the overall cost performance of the model.

5. Discussion

The existing SAR-to-optical method is a one-step transformation method; that is, it directly transforms SAR images into optical RGB images. However, spectral and texture distortions inevitably occur, reducing the accuracy and reliability of the final transformation result. Moreover, the direct use of CycleGAN and pix2pix in SAR-to-optical transformation only reconstructs the original image at the pixel level, without restoring the spectrum and texture. Such results may not be suitable for further image interpretation. Inspired by image restoration and enhancement technology, a Serial GAN image transformation method is proposed here and used for SAR-to-optical tasks.

Based on SEN 1-2 SAR and optical image datasets, the effectiveness of the proposed method was verified through ablation experiments. Through qualitative and quantitative analysis with several SOTA image transformation methods, the superiority of the proposed method was verified. The image transformation method we proposed uses SAR images as prior information to restore and reconstruct SAR images based on the gradient contour and spectrum. The advantage of this is avoiding the mixing distortion caused by directly transforming the SAR image into an optical image, and the final transformation result has better texture detail and an improved spectral appearance. At the same time, our method does not simply involve learning the SAR-optical mapping but restores and reconstructs the SAR image from both the texture information and the spectral information so that it has an interpretation advantage similar to that of the optical image. Note that our proposed method was better than CycleGAN and pix2pix in the index of the transformation results, and some indexes were better than pix2pixHD. From an indicator point of view, this difference was small. However, from intuitive observation, the method proposed in this paper was significantly better than CycleGAN and pix2pix. The reason for this is that our method is not a simple transformation but the reconstruction of SAR images, which restores SAR images from the perspective of image theory. In comparison with the SOTA model pix2pixHD, the proposed method has no obvious advantage in the test value, but the parameter size of the model is about half that of pix2pixHD, which means that our method has more advantages in application. However, the proposed method also has some potential limitations. First, although we considered different seasons and different land types (urban, rural, semi-urban, and coastal areas) in the training data, supervised learning inevitably depends on the data. For different SAR image resolutions and speckle conditions, the results of the transformations will be different. In addition, because supervised learning requires a large number of training samples, the training effect of the model may not be ideal for a dataset with a small sample size. Therefore, problems arising from transfer learning, weakly supervised learning, and cross-modal technology will need to be solved in the future.

6. Conclusions and Prospects

To address the problem of feature loss and distortion in SAR-to-optical tasks, this paper proposed a feature-preserving heterogeneous image transformation model using Serial GANs to maintain the consistency of heterogeneous features, and reduce the distortion caused by heterogeneous transformation. An improved U-net structure was adopted in the model, which was used for SAR image Despeckling GAN, and then the image was colored by Colorization GAN to complete the transformation from a SAR image to an optical color image, which effectively alleviated the uncertainty of transformation results caused by information asymmetry between heterogeneous images. In addition, the end-to-end model architecture also enabled the trained model to be directly used for SAR-to-optical image transformation. At the same time, this paper introduced the feature-preserving loss, which enhanced the feature details of the generated image by constraining the gradient map. Through intuitive and objective comparison, the improved model effectively enhanced the detail of the generated image. In our view, Serial GANs have great potential in other heterogeneous image transformations.

Furthermore, they can provide a common framework for SAR image and photoelectric image transformation. In the future, we will consider incorporating multisource heterogeneous images into a Multiple GANs hybrid model to provide support for the cross-modal interpretation of multisource heterogeneous remote sensing images.

Author Contributions: Conceptualization, Y.L.; Data curation, D.T. and G.L.; Formal analysis, D.T.; Funding acquisition, Y.L. and Y.H.; Investigation, Y.L. and G.L.; Methodology, D.T., G.L. and S.S.; Project administration, L.Y. and Y.H.; Resources, L.Y.; Software, L.Y. and S.S.; Validation, S.S.; Visualization, S.S.; Writing—original draft, D.T.; Writing—review & editing, Y.H. All authors have read and agreed to the published version of the manuscript.

Funding: This research was supported by the National Natural Science Foundation of China, Grant Numbers 91538201, 62022092, and 62171453.

Institutional Review Board Statement: Not applicable.

Informed Consent Statement: Not applicable.

Data Availability Statement: The SEN1-2 dataset was used in this study (accessed 19 July 2021), which is accessible from https://mediatum.ub.tum.de/1436631. It is a dataset consisting of 282,384 pairs of corresponding synthetic aperture radar and optical image patches, acquired by the Sentinel-1 and Sentinel-2 remote sensing satellites, respectively. It is shared under the open access license CC- BY.

Acknowledgments: The authors sincerely appreciate that academic editors and reviewers give their helpful comments and constructive suggestions.

Conflicts of Interest: The authors declare that there are no conflicts of interest regarding the publication of this paper.

References

1. Scarpa, G.; Gargiulo, M.; Mazza, A.; Gaetano, R. A CNN-based fusion method for feature extraction from sentinel data. *Remote Sens.* **2018**, *10*, 236. [CrossRef]
2. Auer, S.; Hinz, S.; Bamler, R. Ray-Tracing Simulation Techniques for Understanding High-Resolution SAR Images. *IEEE Trans. Geosci. Remote Sens.* **2010**, *48*, 1445–1456. [CrossRef]
3. Argenti, F.; Lapini, A.; Bianchi, T.; Alparone, L. A Tutorial on Speckle Reduction in Synthetic Aperture Radar Images. *IEEE Geosci. Remote Sens. Mag.* **2013**, *1*, 6–35. [CrossRef]
4. Reyes, M.F.; Auer, S.; Merkle, N.; Henry, C.; Schmitt, M. SAR-to-optical image translation based on conditional generative adversarial networks-optimization, opportunities and limits. *Remote Sens.* **2019**, *11*, 2067. [CrossRef]
5. Wang, P.; Zhang, H.; Patel, V.M. SAR Image Despeckling Using a Convolutional Neural Network. *IEEE Signal Process. Lett.* **2017**, *24*, 1763–1767. [CrossRef]
6. Lattari, F.; Leon, B.G.; Asaro, F.; Rucci, A.; Prati, C.; Matteucci, M. Deep learning for SAR image despeckling. *Remote Sens.* **2019**, *11*, 1532. [CrossRef]
7. Dalsasso, E.; Yang, X.; Denis, L.; Tupin, F.; Yang, W. SAR Image Despeckling by Deep Neural Networks: From a pre-trained model to an end-to-end training strategy. *Remote Sens.* **2020**, *12*, 2636. [CrossRef]
8. Zhang, K.; Zuo, W.; Chen, Y.; Meng, D.; Zhang, L. Beyond a gaussian denoiser: Residual learning of deep cnn for image denoising. *IEEE Trans. Image Process.* **2017**, *26*, 3142–3155. [CrossRef]
9. Chierchia, G.; Cozzolino, D.; Poggi, G.; Verdoliva, L. SAR image despeckling through convolutional neural networks. In Proceedings of the 2017 IEEE International Geoscience and Remote Sensing Symposium (IGARSS), Fort Worth, TX, USA, 23–28 July 2017; pp. 5438–5441.
10. Qiang, Z.; Yuan, Q.; Jie, L.; Zhen, Y.; Zhang, L. Learning a Dilated Residual Network for SAR Image Despeckling. *Remote Sens.* **2018**, *10*, 196.
11. Isola, P.; Zhu, J.Y.; Zhou, T.; Efros, A.A. Image-to-image translation with conditional adversarial networks. In Proceedings of the 2017 IEEE Conference on Computer Vision and Pattern Recognition (CVPR), Honolulu, HI, USA, 21–26 July 2017; pp. 5967–5976. [CrossRef]
12. Zhu, J.-Y.; Park, T.; Isola, P.; Efros, A.A. Unpaired Image-to-Image Translation using Cycle-Consistent Adversarial Networks In Proceedings of the 2017 IEEE International Conference on Computer Vision (ICCV), Venice, Italy, 22–29 October 2017; Volume 2242–2251.
13. Wang, T.-C.; Liu, M.-Y.; Zhu, J.-Y.; Tao, A.; Kautz, J.; Catanzaro, B. High-Resolution Image Synthesis and Semantic Manipulation with Conditional GANs. In Proceedings of the IEEE Conference on Computer Vision and Pattern Recognition, Salt Lake City, UT, USA, 18–23 June 2018; pp. 8798–8807.

14. Choi, Y.; Choi, M.; Kim, M.; Ha, J.W.; Kim, S.; Choo, J. StarGAN: Unified Generative Adversarial Networks for Multi-domain Image-to-Image Translation. In Proceedings of the IEEE Computer Society Conference on Computer Vision and Pattern Recognition, Salt Lake City, UT, USA, 18–23 June 2018.
15. Li, J.; Monroe, W.; Shi, T.; Jean, S.; Ritter, A.; Jurafsky, D. Adversarial learning for neural dialogue generation. In Proceedings of the EMNLP 2017—Conference on Empirical Methods in Natural Language Processing, Copenhagen, Denmark, 7–11 September 2017.
16. Reed, S.; Akata, Z.; Yan, X.; Logeswaran, L.; Schiele, B.; Lee, H. Generative adversarial text to image synthesis. In Proceedings of the 33rd International Conference on Machine Learning (ICML-2016), New York, NY, USA, 19–24 June 2016; pp. 1681–1690.
17. Denton, E.; Szlam, A.; Fergus, R. Deep Generative Image Models using a Laplacian Pyramid of Adversarial Networks. *arXiv* **2015**, arXiv:1506.05751.
18. Mathieu, M.; Couprie, C.; LeCun, Y. Deep multi-scale video prediction beyond mean square error. In Proceedings of the 4th International Conference on Learning Representations, ICLR 2016, San Juan, Puerto Rico, 2–4 May 2016; pp. 1–14.
19. Ledig, C.; Theis, L.; Huszár, F.; Caballero, J.; Cunningham, A.; Acosta, A.; Aitken, A.; Tejani, A.; Totz, J.; Wang, Z.; et al. Photo-realistic single image super-resolution using a generative adversarial network. In Proceedings of the 30th IEEE Conference on Computer Vision and Pattern Recognition (CVPR), Honolulu, HI, USA, 21–26 July 2016; p. 2017.
20. Zhu, J.-Y.; Krähenbühl, P.; Shechtman, E.; Efros, A.A. Generative Visual Manipulation on the Natural Image Manifold. In Proceedings of the European Conference on Computer Vision (ECCV), Amsterdam, The Netherlands, 11–14 October 2016.
21. Brock, A.; Lim, T.; Ritchie, J.M.; Weston, N. Neural photo editing with introspective adversarial networks. In Proceedings of the 5th International Conference on Learning Representations, Toulon, France, 24–26 April 2017.
22. Fu, S.; Xu, F.; Jin, Y.Q. Translating sar to optical images for assisted interpretation. *arXiv* **2019**, arXiv:1901.03749.
23. Wang, L.; Xu, X.; Yu, Y.; Yang, R.; Gui, R.; Xu, Z.; Pu, F. SAR-to-optical image translation using supervised cycle-consistent adversarial networks. *IEEE Access* **2019**, *7*, 129136–129149. [CrossRef]
24. Zhang, J.; Zhou, J.; Lu, X. Feature-guided SAR-to-optical image translation. *IEEE Access* **2020**, *8*, 70925–70937. [CrossRef]
25. Ma, C.; Rao, Y.; Cheng, Y.; Chen, C.; Lu, J.; Zhou, J. Structure-Preserving Super Resolution With Gradient Guidance. In Proceedings of the 2020 IEEE/CVF Conference on Computer Vision and Pattern Recognition (CVPR) 2020, Seattle, WA, USA, 16–18 June 2020; pp. 7766–7775. [CrossRef]
26. Schmitt, M.; Hughes, L.H.; Zhu, X.X. The SEN1-2 dataset for deep learning in SAR-optical data fusion. *ISPRS Ann. Photogramm. Remote Sens. Spat. Inf. Sci.* **2018**, *4*, 141–146. [CrossRef]
27. Torres, R.; Snoeij, P.; Geudtner, D.; Bibby, D.; Davidson, M.; Attema, E.; Potin, P.; Rommen, B.; Floury, N.; Brown, M.; et al. GMES Sentinel-1 mission. *Remote Sens. Environ.* **2012**, *120*, 9–24. [CrossRef]
28. Zhang, T.; Zhang, X. Injection of Traditional Hand-Crafted Features into Modern CNN-Based Models for SAR Ship Classification: What, Why, Where, and How. *Remote Sens.* **2021**, *13*, 2091. [CrossRef]
29. Schubert, A.; Small, D.; Miranda, N.; Geudtner, D.; Meier, E. Sentinel-1A product geolocation accuracy: Commissioning phase results. *Remote Sens.* **2015**, *7*, 9431–9449. [CrossRef]
30. Drusch, M.; Del Bello, U.; Carlier, S.; Colin, O.; Fernandez, V.; Gascon, F.; Hoersch, B.; Isola, C.; Laberinti, P.; Martimort, P.; et al. Sentinel-2: ESA's optical high-resolution mission for GMES operational services. *Remote Sens. Environ.* **2012**, *120*, 25–36. [CrossRef]
31. Zhang, R.; Isola, P.; Efros, A.A. Colorful image colorization. In Proceedings of the European Conference on Computer Vision (ECCV), Amsterdam, The Netherlands, 11–14 October 2016; pp. 649–666.
32. Zhang, T.; Zhang, X.; Shi, J.; Wei, S.; Wang, J.; Li, J.; Su, H.; Zhou, Y. Balance scene learning mechanism for offshore and inshore ship detection in SAR images. *IEEE Geosci. Remote Sens. Lett.* **2020**. [CrossRef]
33. Sangkloy, P.; Lu, J.; Fang, C.; Yu, F.; Hays, J. Scribbler: Controlling Deep Image Synthesis with Sketch and Color. In Proceedings of the IEEE Conference on Computer Vision and Pattern Recognition (CVPR), Honolulu, HI, USA, 21–26 July 2017.
34. Wang, Z.; Bovik, A.C.; Sheikh, H.R.; Simoncelli, E.P. Image quality assessment: From error visibility to structural similarity. *IEEE Trans. Image Process.* **2004**, *13*, 600–612. [CrossRef]
35. Zhang, L.; Zhang, L.; Mou, X.; Zhang, D. FSIM: A Feature Similarity Index for Image Quality Assessment. *IEEE Trans. Image Process.* **2011**, *20*, 2378–2386. [CrossRef] [PubMed]

 *remote sensing*

Article

Self-Supervised Despeckling Algorithm with an Enhanced U-Net for Synthetic Aperture Radar Images

Gang Zhang [1,*], Zhi Li [1], Xuewei Li [2] and Sitong Liu [1]

[1] Department of Aerospace Science and Technology, Space Engineering University, Bayi Road, Huairou District, Beijing 101416, China; lizhipublic@163.com (Z.L.); liusitong1114@163.com (S.L.)
[2] Institute of Software, Chinese Academy of Sciences, No. 4, South Fourth Street, Zhongguancun, Haidian District, Beijing 100190, China; lixuewei@iscas.ac.cn
* Correspondence: gangzhang1989@126.com; Tel.: +86-186-0056-6200

Abstract: Self-supervised method has proven to be a suitable approach for despeckling on synthetic aperture radar (SAR) images. However, most self-supervised despeckling methods are trained by noisy-noisy image pairs, which are constructed by using natural images with simulated speckle noise, time-series real-world SAR images or generative adversarial network, limiting the practicability of these methods in real-world SAR images. Therefore, in this paper, a novel self-supervised despeckling algorithm with an enhanced U-Net is proposed for real-world SAR images. Firstly, unlike previous self-supervised despeckling works, the noisy-noisy image pairs are generated from real-word SAR images through a novel generation training pairs module, which makes it possible to train deep convolutional neural networks using real-world SAR images. Secondly, an enhanced U-Net is designed to improve the feature extraction and fusion capabilities of the network. Thirdly, a self-supervised training loss function with a regularization loss is proposed to address the difference of target pixel values between neighbors on the original SAR images. Finally, visual and quantitative experiments on simulated and real-world SAR images show that the proposed algorithm notably removes speckle noise with better preserving features, which exceed several state-of-the-art despeckling methods.

Keywords: self-supervised; synthetic aperture radar (SAR); despeckling; enhanced U-Net

Citation: Zhang, G.; Li, Z.; Li, X.; Liu, S. Self-Supervised Despeckling Algorithm with an Enhanced U-Net for Synthetic Aperture Radar Images. *Remote Sens.* **2021**, *13*, 4383. https://doi.org/10.3390/rs13214383

Academic Editors: Tianwen Zhang, Tianjiao Zeng and Xiaoling Zhang

Received: 17 September 2021
Accepted: 28 October 2021
Published: 31 October 2021

Publisher's Note: MDPI stays neutral with regard to jurisdictional claims in published maps and institutional affiliations.

1. Introduction

Synthetic aperture radar (SAR) [1] is an active remote sensing imaging sensor that transmits electromagnetic signals to target in a slant distance manner. Compared with optical imaging sensors, SAR has the imaging ability of all-time and all-weather. Therefore, SAR has become one of the remote sensors used for disaster assessment [2], resource exploration [3], ocean surveillance [4,5] and statistical analysis [6]. Nevertheless, due to the imaging mechanism, the quality of SAR images is inherently affected by speckle noise [7,8]. Speckle noise is a granular disturbance, usually modeled as a multiplicative noise, that affects SAR images, as well as all coherent images [8]. The speckle noise may severely diminish the performances of detection accuracy [9–12] and information extraction [13]. Therefore, the reduction of speckle noise is a key and essential processing step for a number of applications.

In the past few decades, numerous researchers have attempted to reduce the speckle noise in SAR images. Generally, the existing despeckling methods can be roughly summarized as local window methods, non-local mean (NLM) methods and deep learning (DL) methods. In the first group, local window methods are widely used, such as Lee [14], Frost [15] and Kuan [16]. The despeckling performance of local window methods is very dependent on the window size. The larger the size, the smoother the despeckled image and the better the despeckling performance. However, the despeckled image will lose point targets, linear features and textures.

To overcome the disadvantage of local window methods, NLM methods are applied to process SAR images, such as PPB [17], SAR-BM3D [18] and NL-SAR [19]. The NLM methods define the similar pixel and pixel weight by measuring the similarity between a local patch centered on the reference pixel and another local patch centered on a selected non-local neighborhood pixel. The greater the similarity, the larger the weight. However, the NLM methods need to use the equivalent number of looks (ENL) as a prior. In practical applications, it is impossible to obtain the accurate ENL of SAR images. In addition, the time complexity of the local window methods and NLM methods is very high.

Recently, benefiting from new breakthroughs of deep learning [20,21], more and more researchers began to explore DL methods [22–33]. The essence of the DL despeckling methods is to learn a relationship from noisy SAR images to noise-free SAR images. It can be described as follows. Firstly, the input of the DL despeckling methods is a noisy SAR image. Then, the noisy SAR images are encoded and decoded through convolutional layers, pooling layers, batch normalization layers and activation function layers. Finally, a noise-free SAR image is obtained. According to whether there are clean images as targets, the DL despeckling methods can be distinguished into three broad categories: supervised methods, semi-supervised methods and self-supervised methods.

The supervised despeckling methods [22–24,26] use noisy-clean image pairs to train convolutional neural networks (CNNs) and the despeckling models are applied to reduce speckle noise in real-world SAR images. Since there is no noise-free SAR images, the training image pairs of the supervised methods are generated by combining regularization RGB photos (natural images) with simulated speckle noise. The natural images contain camera images [23] and aerial images [24]. The advantage of the generated method is that a large number of noisy-clean image pairs can be easily obtained. A deeper CNN with numerous parameters can be trained. But the disadvantage of the generated method is that it ignores the peculiar characteristics of SAR images. For example, the noise distribution of SAR images is not the same as in natural images, as well as the content, the texture, or the physical meaning of a pixel value [27]. Compared with real-world SAR images, the generated simulated SAR images have quite difference in content and geometry. There are strong scattering points in real-world SAR images, but not in simulated SAR images. Therefore, the despeckling CNNs trained with simulated images will change point targets, linear features and textures in the real-world SAR images. Benefiting from the noise2noise [34] method, semi-supervised methods and self-supervised methods [28–31] were proposed. The method of noise2noise is to directly use noisy-noisy image pairs to train deep CNNs. The semi-supervised despeckling model [27] used a small number of noisy-clean image pairs to train CNNs. Then, the obtained despeckling model was fine-tuned on the time series real-world SAR images. Fine-tune refers to the despeckling model obtained in the first step for training again using the time series real-world SAR images. Compared with the supervised despeckling methods, the semi-supervised despeckling methods can better reduce the speckle noise from real-world SAR images. Nevertheless, the time series SAR images will have differences at different times, which will limit the despeckling performance. The self-supervised despeckling methods directly use the extensive noisy-noisy image pairs to train CNNs. The noisy-noisy image pairs are generated by using natural images with simulated speckle noises [28], time series SAR images [31] or generative adversarial network (GAN) [35], which still limit the practicability of these methods in real-world SAR images.

Through the above analysis, a simple summary can be made. Firstly, the DL despeckling method is raising great interest. However, most of DL methods are focused on the new networks [22–26,32,33,36], while ignoring the most essential problem. In our opinion, the lack of truely noise-free SAR images is the most essential problem. Simulated SAR images can not really solve this deficiency. Secondly, it can be found that the despeckling CNNs are becoming more and more deeper. Meanwhile, the number of trainable parameters is increasing. The simulated SAR images can be easily generated through the pixel-wise product of clean natural images with simulated speckle noise [30]. When processing real-world SAR images, the despeckling models obtained will bring new challenges. The despeckling CNNs trained with simulated images pairs will change point targets, linear features and textures in the real-world SAR images. Thirdly, although noise2noise method can directly use noisy images as targets, the despeckling performance is still affected by natural images [29] and the performance of GAN [30]. In the previous works [27–31], they can not really use real-world SAR data to train the despeckling CNNs. Therefore, in this paper, inspired by noise2noise [34], a novel self-supervised despeckling algorithm with an enhanced U-Net is proposed. This algorithm is called SSEUNet. Compared with previous despeckling CNNs [27–31], SSEUNet can directly use real-world SAR images for training deep CNNs. The SSEUNet is composed of generation training pairs (GTP) module, enhanced U-Net (EUNet) and a self-supervised training loss function with a regularization loss. The main contributions and innovations of the proposed algorithm are as follows:

- Since the noisy-noisy image pairs generated by [28,30,31] can not be used well in the despeckling task of real-world SAR images, we propose GTP module. GTP module can generate training image pairs from noisy images. It can make it possible to train deep CNNs using real-world SAR images.
- Due to the poor feature extraction and fusion capabilities of U-Net [37], we design a novel deep CNN by enhancing U-Net. The novel deep CNN is called enhanced U-Net (EUNet).
- In order to address the difference of target pixel values between neighbors on the original noisy image, a self-supervised training loss function with a regularization loss is put forward.
- Visual and quantitative experiments conducted on simulated and real-world SAR images show that the proposed algorithm notably reduces speckle noise with better preserving features, which outputform several state-of-art despeckling methods.

The rest of this paper in organized as follows. The related work, which includes noisy-clean despeckling methods and noisy-noisy despeckling methods, are analyzed in Section 2. In Section 3, we detailed describe the proposed methods. Section 4 illustrates the results of visual effect and parameter evaluation metrics. Finally, the conclusions are drawn in Section 5.

2. Related Work

2.1. Noisy Clean Despeckling Methods

In recent years, benefiting from deep learning, the noisy-clean despeckling methods have been studied in depth. Inspired by denoising CNN [38], Chierchia et al. [22] proposed the first despeckling CNN, which composed of 17 convolutional layers. The training image pairs of SAR-CNN were generated from time series real-world SAR images. The targets were 25-look SAR images. Zhang et al. [24] directly used dilated convolution layers and skip connections to reduce speckle noise from SAR images. The dilated convolution layer allows CNN to have a lightweight structure and small filter size, but the receptive field of the CNN will not be reduced. In addition, skipping connections reduce the problem of vanishing gradients. Similar to SAR-DRN, Gui et al. [25] proposed a dilated densely connections CNN. After considering the bright distribution of the speckle noise, Shen et al. [26] proposed a recursive deep convolutional neural prior model. The model included a data fitting block and a deep CNN prior block. The gradient descent algorithm was used for the data fitting block and the pre-trained dilated residual channel attention network was applied in the deep CNN prior block. Pan et al. [32] combined multi-channel logarithm gaussian denoising (MuLoG) algorithm with a fast and flexible denoising CNN to deal with the multiplicative noise of SAR images. Li et al. [33] designed a CNN with convolutional block attention module to improve representation power and despeckling performance. In order to help the network retain image details, Zhang et al. [39] proposed a multi-connection CNN with wavelet features. Because the NLM method is one of the most promising algorithm, Cozzolino et al. [23] combined NLM method with deep CNN to design a non-local mean CNN (NLM-CNN). The NLM-CNN used deep CNN to provide interpretable results for target pixel and predicted pixel. Mullissa et al. [36] proposed a two-stage despeckling CNN, called deSpeckleNet. The deSpeckleNet sequentially estimated the speckle noise distribution and the noise-free SAR image. Vitale et al. [40] designed a weighted loss function by considering the contents in the SAR images. Except SAR-CNN, the noisy-clean despeckling methods use synthetic training on the simulated SAR images. Due to the differences in imaging mechanisms and image features between SAR and natural images, i.e., grayscale distribution and spatial correlation [41], training on simulated SAR images is not the best solution [30]. Compared with the most advanced traditional despeckling methods, the deep despeckling CNNs have obvious despeckling advantages, but the lack of truly noise-free SAR images is a major limiting factor in despeckling performance.

2.2. Noisy-Noisy Despeckling Methods

In the real-world, it is impossible to obtain noise-free SAR images. Inspired by noise2noise [34], the noisy-noisy despeckling methods [27–31] were studied in depth. Yuan et al. [28] designed a self-supervised densely dilated CNN (BDSS-CNN) for blind despeckling. In the BDSS-CNN, the noisy-noisy image pairs were generated by adding simulated speckle noise with random ENL to natural images. Then, the generated noisy-noisy image pairs were used to train BDSS-CNN. Finally, the obtained despeckling model was applied on the real-world SAR images. Inspired by blind-spot denoising networks [42], Molini et al. [29] reported a self-supervised bayesian despeckling CNN. The training image pairs were constructed from natural images. Yuan et al. [30] designed a practical SAR images despeckling method to reduce the impact of natural images. The method contained two sub-networks. The first sub-network was a GAN, which used to generate the speckle-speckle training paris. The second network was an enhanced nested-UNet, which was trained by using speckle-speckle training paris. It can be found that the quality of the generated speckle images by GAN will directly affect the despeckling performance of the enhanced nested-UNet. Dalsasso et al. [27] and Ma et al. [31] abandoned the methods of using natural images and GAN, they proposed to use time series real-world SAR images. Whilst, the natural landscape often changes significantly at a short period in the time series real-world SAR images. Therefore, the despeckling performance of [27,31] is still limited.

Although noise2noise methods were used to solve the lack of truly noise-free SAR images, the despeckling performance is still affected by nature images [29] and the performance of GAN [30]. In the previous works [27–31], they can not really use real-world SAR images to train the despeckling CNNs. Therefore, we designed a novel self-supervised despeckling algorithm with an enhance U-Net (SSEUNet). The SSEUNet includes a GTP module, an EUNet and a loss function. The GTP module can directly generate training image pairs from real-world noisy images. The generated image pairs are applied to train EUNet through the proposed loss function. When processing real-world SAR images, the proposed SSEUNet can eliminate the influence of natural images, GAN performance and time series images.

3. Proposed Method

In order to train the despeckling CNN directly using real-world SAR images, we propose a novel self-supervised despeckling algorithm, which is called SSEUNet. The SSEUNet is mainly composed of two GTP module, an enhanced U-Net (EUNet) and a loss function. The GTP module is designed to generate noisy-noisy image pairs for training the proposed EUNet. The EUNet is an enhanced version of U-Net, which has stronger feature extraction and fusion capabilities. The loss function is a self-supervised training loss function with a regularization loss, which is applied to optimal EUNet. In this section, Section 3.1 gives an overview of the proposed SSEUNet framework in detail. The detailed implementation of the proposed GTP module is introduced in Section 3.2. The proposed EUNet will be illustrated in Section 3.3. Section 3.4 introduces the loss function of SSEUNet.

3.1. Overview of Proposed SSEUNet

The proposed SSEUNet is composed of generation training pairs (GTP) module, enhanced U-Net (EUNet) and a self-supervised training loss function with a regularization loss. Figure 1 shows the overview of proposed SSEUNet framework, where y is the input image of proposed SSEUNet. z_1 and z_2 are noisy-noisy image pair generated by proposed GTP module. F_ϕ is the EUNet and ϕ is the parameters of the EUNet. z_1' and z_2' are the despeckled image pair generated by GTP module. $\mathcal{L}$ is the proposed loss function. The proposed SSEUNet are divided into training phase (Figure 1a) and inference phase (Figure 1b). In the training phase, a pair of noisy-noisy images ($\{z_1, z_2\}$) is generated from a noisy image y. The generated masks are recorded as $\mathcal{G}_1$ and $\mathcal{G}_2$. The generated masks come from GTP module. The EUNet takes z_1 and z_2 as input and target, respectively. The input image y is fed into EUNet and the despeckled image y' is obtained in each training epoch, where $y' = F_\phi(y)$. The depseckled-despeckled image pair ($\{z_1', z_2'\}$) of the y' are obtained by GTP module. The loss function is computed by using $F_\phi(z_1)$, z_2, z_1' and z_2'. In the inference phase, the despeckled SAR images are obtained by directly using the trained EUNet.

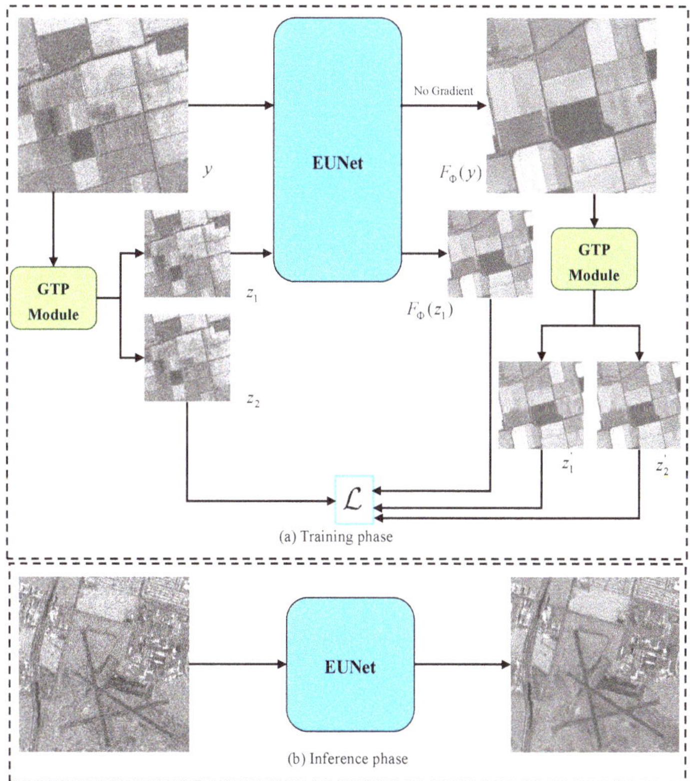

Figure 1. Overview of the proposed SSEUNet framework. (**a**) Complete view of the training phase. (**b**) Inference phase using the trained EUNet.

3.2. Proposed GTP Module

Benefiting from noise2noise despeckling methods, we have conducted further research on the construction of noisy-noisy image pairs. According to Goodman's theory, the fully developed speckle noise in SAR images is completely random and independently distributed noise. Therefore, we attempt to generate noisy-noisy image pairs from the real-world SAR images $y \in \{y_i\}_{i=1}^{N}$, where N is the total number of real-world SAR images.

The height and width of y are H and W, respectively. The noisy-noisy image pair is the $\{z_1, z_2\}$. The generation process of the proposed GTP module is divided into three steps. Firstly, the y is divided into $\lfloor H/k \rfloor \times \lfloor W/k \rfloor$ patches, where the $\lfloor \cdot \rfloor$ is a floor operation and k is the patch size. Secondly, we select a patch at the position (l, m), the two pixels of the patch are randomly extracted. The extracted pixels are used as the pixel of z_1 and z_2 at the position (l, m), respectively. Finally, for $\lfloor H/K \rfloor \times \lfloor W/k \rfloor$ patches, the noisy-noisy image pair $\{z_1, z_2\}$ is obtained by repeating the second step. The size of z_1 and z_2 is $\lfloor H/k \rfloor \times \lfloor W/k \rfloor$. Compared with the methods of using natural images and GAN, the GTP module can directly generate noisy-noisy image pairs from real-world noisy images. Figure 2 shows three noisy-noisy image pairs generation methods, where y is the noisy image, and I is the clean natural image. SN_1 and SN_2 are two independent simulated speckle noise. I_1 and I_2 are simulated SAR images. y_G is the speckled SAR image generated by GAN [35].

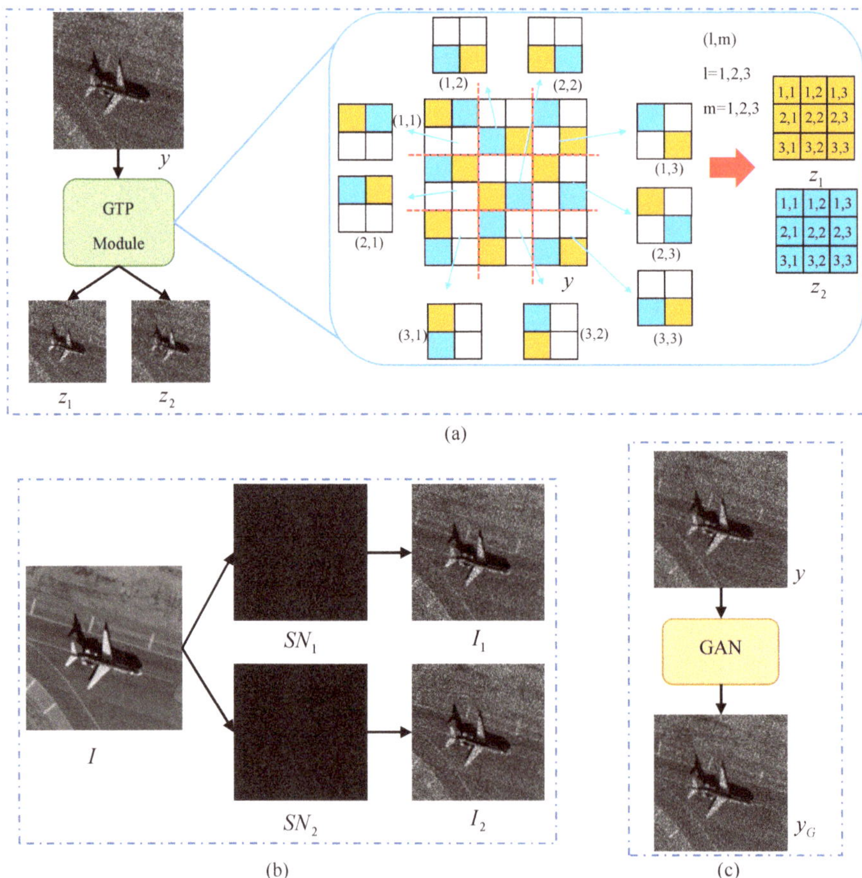

Figure 2. Comparison of different methods for generating noisy-noisy image pairs. (**a**) The method of proposed GTP module. (**b**) The method of using natural images. (**c**) The method of using GAN.

In order to better explain the difference between the three generation methods, k is set to 2. When using GTP module, the y is divided into $\lfloor H/2 \rfloor \times \lfloor W/2 \rfloor$ patches and the size of z_1 and z_2 is a quarter of y. In Figure 2a, the example of generating an image pair is listed right. The original image size is 6×6. The original image is divided into 9 ($\lfloor 6/2 \rfloor \times \lfloor 6/2 \rfloor$) patches. Two pixels are randomly chosen and fill them with orange and blue respectively. The pixel filled with orange is taken as a pixel of a noisy image z_1 and the other pixel filled with blue is taken as a pixel of another noisy image z_2. The noisy-noisy image pair $\{z_1, z_2\}$ is displayed as the orange image and the blue image on the right. In Figure 2b, the clean image I is element-wise multiplied with SN_1 and SN_2 to obtain the noisy-noisy image pair $\{I_1, I_2\}$. The image size of I_1 and I_2 is $H \times W$. In Figure 2c, the image y_G is obtained by using GAN. y and y_G are combined together to construct noisy-noisy image pair $\{y, y_G\}$. The size of y and y_G is $H \times W$. By comparing the three generation methods, it can be seen that the proposed GTP module directly generates the noisy-noisy image pairs from noisy images. Meanwhile, it will not be affected by natural images or GAN performance. The size of the noisy-noisy image pairs do not affect the despeckling performance of CNNs.

In order to verify that the noisy-noisy image pairs generated by GTP module will not change the distribution of the original noisy images, Figure 3 shows the examples of noisy-noisy image pairs generated by GTP module. In Figure 3, μ and σ are the mean and standard variance, respectively. The size of the original images is 256×256. The size of the noisy-noisy image pairs is 128×128. It can be seen that the histograms, μ, σ and visual effects are very similar.

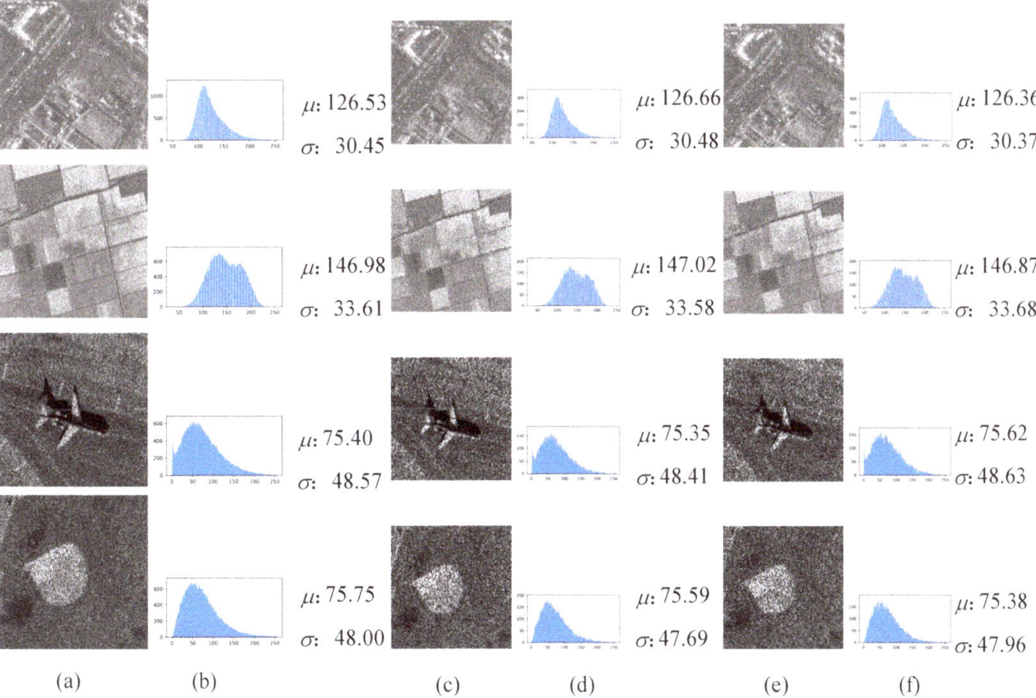

Figure 3. Examples of noisy-noisy image pairs using GTP module. (**a**) Original noisy images y. (**b**) Histograms of y. (**c**) Noisy images z_1. (**d**) Histograms of z_1. (**e**) Noisy images z_2. (**f**) Histograms of z_2.

3.3. Enhanced U-Net

The U-Net [37] was originally proposed for medical image segmentation. It is a fully convolutional network and can be trained with very few images by using data augmentation. The U-Net is widely used in image denoising [43] and super-resolution [44]. At present, there are many variants of U-Net, which are used for various tasks. For example, the nested U-Net [45] and hybrid densely connected U-Net [46] are used for medical image segmentation. Multi-class attention-based U-Net [47] is designed for earthquake detection and seismic phase-picking.

As the speckle noise is modeled as multiplicative noise, the non-linear relationship of between noise-free SAR images and speckle noise can be obtained by a deep CNN. Therefore, an enhanced U-Net (EUNet) is designed to enhance the feature extraction and fusion capabilities of U-Net [37]. The detailed architecture of EUNet is displayed in Figure 4, where *BN-RRDC* represents the residual in residual densely connection with batch normalization (BN-RRDC) block. SSAM1-SSAM4 are sub-space attention module (SSAM) [48], respectively. ISC1-ISC4 denote the proposed improved skip connections, respectively. The *Conv* and *s* mean the convolutional layer and stride, respectively. The *Cat* and *TConv*1-*TConv*4 represent concat layer and transpose convolution layers, respectively. Compared with U-Net [37], EUnet mainly has four enhancements. Firstly, we replace the convolutional layer with the BN-RRDC blocks. The BN-RRDC block is composed of residual in residual densely connections [49] and batch normalization layer, which can significantly improve the feature extraction and representation capabilities of U-Net. Meanwhile, because of the residual structure in the BN-RRDC block, the training difficulty of the EUNet will not increase.

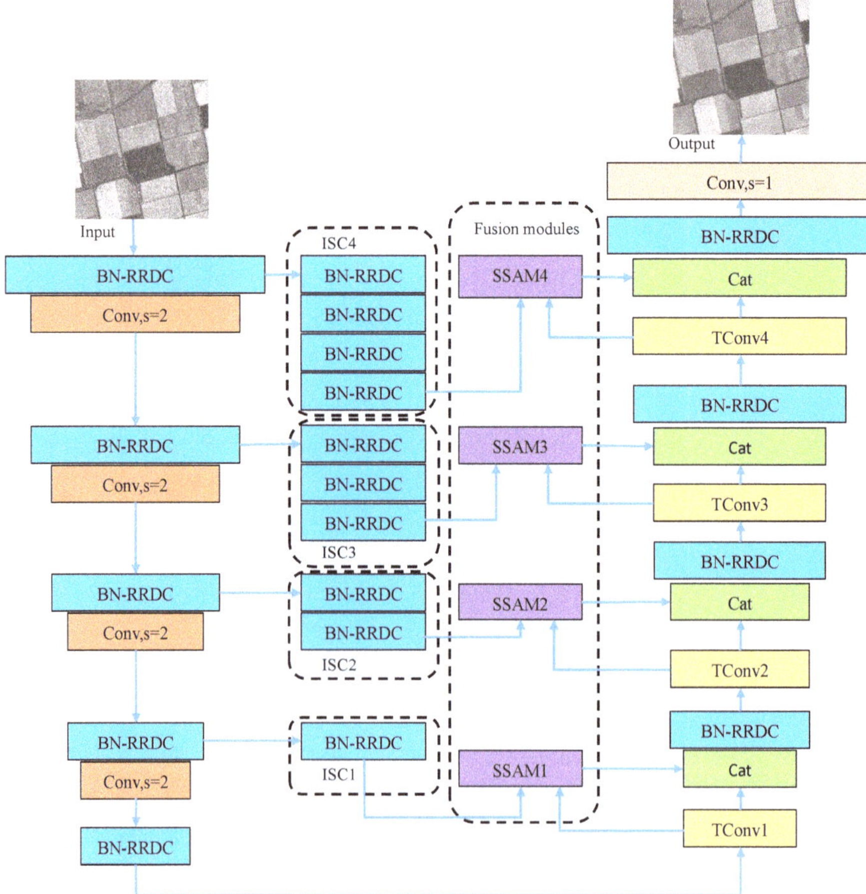

Figure 4. The detailed architecture of EUNet.

Secondly, for reducing features loss, the pooling layers are replaced by convolutional layers with $s = 2$. Thirdly, we design improved skip connections (ISC1-ISC4) to narrow the gap between encoder features and decoder features. In the architecture of U-Net [37], the skip connections directly send the encoder features to the decoder. In the framework of EUNet, the improved skip connections have different numbers of BN-RRDC blocks. In the structures of ISC1-ISC4, the numbers of BN-RRDC blocks are 1, 2, 3 and 4, respectively. Finally, SSAM is introduced into the feature fusion to restore weak texture and high-frequency information of the image. SSAM is a non-local sub-space attention module, which uses the non-local information to generate basis vectors through projection. The reconstructed SAR image can retain most of the original information and suppress speckle noise which is irrelevant to the basis vectors. The SSAM has been achieved state-of-the-art denoise performance for removing noise in the natural images. The detailed structure of BN-RRDC block and SSAM are displayed in Figure 5, where BN is the batch-normalization layer.

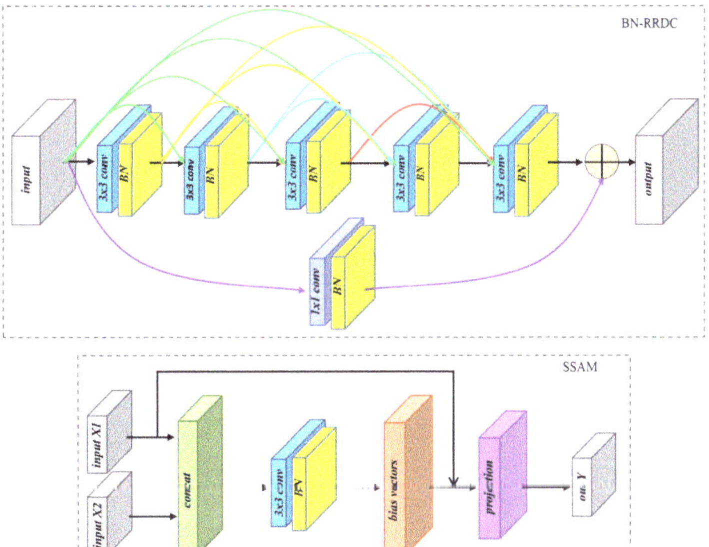

Figure 5. The detailed structure of BN-RRDC block and SSAM.

3.4. Loss Function of SSEUNet

The self-supervised training method is the noise2noise method [34], which does not require clean images as targets. The noise2noise method only requires two noisy image pairs of the same image. Assuming the noisy-noisy image pair is the $\{y_1, y_2\}$ from the clean image x, the noise2noise tries to minimize the following loss function:

$$\mathcal{L}_{mse} = ||F_\phi(y_1) - y_2||_2^2, \tag{1}$$

where F_ϕ is the denoising network and ϕ is the parameters of the denoising network. Equation (1) is the pixel-level loss function. Its optimization is the same as the supervised learning CNNs. In previous noise2noise despeckling methods, Equation (1) is commonly used loss function. Assume that the noisy SAR image is y, and the generation masks of GTP module are $\mathcal{G}_1$ and $\mathcal{G}_1$. The noisy-noisy image pair is $\{z_1, z_2\}$. z_1 and z_2 can be written as:

$$z_1 = \mathcal{G}_1 \odot y, \quad z_2 = \mathcal{G}_2 \odot y, \tag{2}$$

where $\odot$ is the operation of GTP module. When training EUNet, Equation (1) is rewritten as:

$$\mathcal{L}_{mse} = ||F_\phi(z_1) - z_2||_2^2. \tag{3}$$

The despeckled image pair $\{z_1^{'}, z_2^{'}\}$ of $y^{'}$ can be obtained by Equation (4).

$$z_1^{'} = \mathcal{G}_1 \odot y^{'}, \ z_2^{'} = \mathcal{G}_2 \odot y^{'}, \tag{4}$$

where $y^{'} = F_\phi(y)$. Thus, the regularization loss can be defined as:

$$\mathcal{L}_{reg} = ||F_\phi(z_1) - z_2 + z_2^{'} - z_1^{'}||_2^2, \tag{5}$$

Meanwhile, $z_1^{'}$ and $z_2^{'}$ also can be obtained by the despeckling network F_ϕ, which can be written as:

$$z_1^{'} = F_\phi(z_1), \ z_2^{'} = F_\phi(z_2). \tag{6}$$

In an ideal state, $F_\phi^*(z_1)$ and z_2 are exactly the same. F_ϕ^* is the optimal despeckling network. Meanwhile, $z_1^{'}$ and $z_2^{'}$ are also exactly the same. Therefore, $\mathcal{L}_{reg}$ should be equal to 0. Equation (5) provides a regularization loss that is satisfied when a despeckling network F_ϕ is the optimal network F_ϕ^*. The regularization loss can narrow the gap of target pixel values between neighbors on the original noisy image. In order to exploit the regularization loss, we do not directly optimize Equation (3), but add Equation (5) to the Equation (3). Finally, the loss function of SSEUNet is defined as:

$$\mathcal{L} = \mathcal{L}_{mse} + \lambda \mathcal{L}_{reg}, \tag{7}$$

where λ is the hyper-parameter.

4. Experiments and Analysis

We evaluate the effectiveness of the proposed SSEUNet on simulated and real-world SAR datasets. Firstly, we report the implementation details. Secondly, the training and inference datasets, a simulated SAR dataset and a real-world SAR dataset, are introduced in detail. Thirdly, according to the fact of simulated and real-world SAR images, different evaluation metrics are selected. Finally, we present the experimental results and analysis of the simulated and real-world SAR datasets.

4.1. Implementation Details

The proposed SSEUNet does not require a pre-trained model, and it can be trained end-to-end from scratch. In the training phase, the initialization method of the network parameters is He [50]. The optimizer is Adam algorithm [51] and its momentum terms are $\beta_1 = 0.9$ and $\beta_2 = 0.999$. The initial learning rate of SSEUNet is set to 0.00001 for the real-world experiments and 0.0001 for the simulated experiments. The learning rate reduces by reduce-on-plateau strategy and the decay ratio is 0.5. 64×64-sized images are used and 10 instances stack a mini-batch. The training epoch is set to 100. The hyper-parameter of λ in the loss function can control the strength of the regularization term. Empirically, the λ is set to 2 in the simulated SAR experiments and the λ is set to 1 in the real-world SAR experiments. In the inference phase, the input size of the trained ENUet is 256×256. All experiments are conducted on a workstation with Ubuntu 18.04. The hardware is an Intel Xeon(R) CPU E5-2620v3, a NVIDIA Quadro M6000 24GB GPU and 48 GB of RAM. The deep learning framework is PyTorch 1.4.0.

4.2. Datasets

We use UC Merced Land-use (UCML) dataset [52] as a simulated dataset and it is widely used for land-use classification. The UCML dataset is a optical remote sensing dataset. The UCML dataset contains 2100 images, which are extracted from United States

Geological Survey (USGS) National Map of the US regions. The resolution is 0.3 m and the size of each image is 256 × 256 × 3. To generate the simulated SAR images, we convert each image to grayscale image. Then, similar with [53], we generate simulated SAR images by multiplying simulated speckle noise to clean grayscale images. The simulated speckle noise follows Gamma distribution. In our experiments, we only considered single-look SAR images. Because of the high intensity of speckle noise in single-look SAR images, processing single-look SAR images is a very challenging case. The mean and variance of simulated speckle noise are 1. We randomly divide 2100 images into training set (1470), validation set (210) and testing set (420). In the training phase, data augmentation is used to train SSEUNet. The method of data augmentation is cropping and the crop size is 64 × 64. Finally, the augmented training set contains 213,175 patches. The size of training image pairs generated by GTP module is 32 × 32. The validation set and testing set do not use data augmentation. In Figure 6, we displays the examples of the generated simulated SAR images.

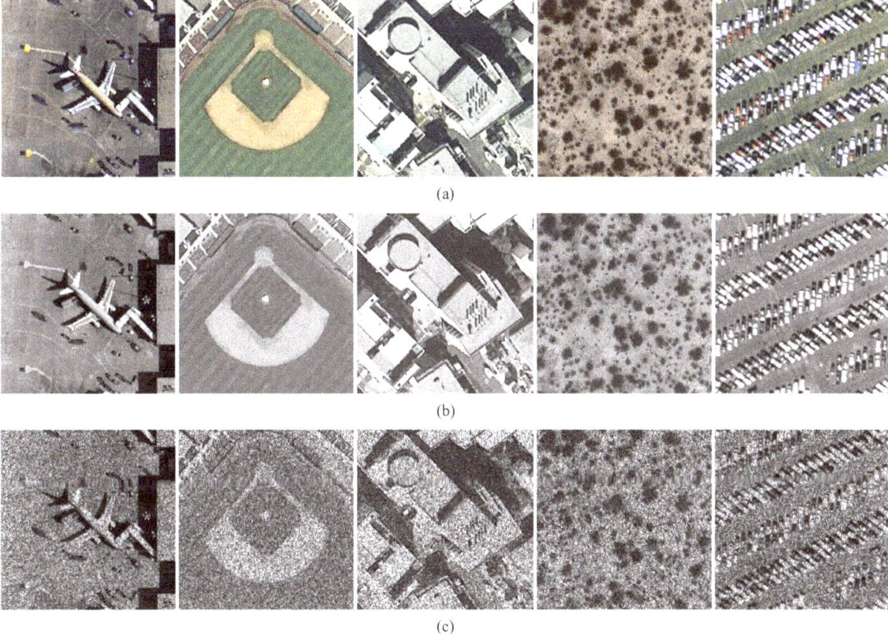

Figure 6. Examples of simulated SAR images. (**a**) RGB images. (**b**) Grayscale images. (**c**) Simulated SAR images.

In order to validate the practicability of SSEUNet, seven large-scale single-look SAR images are used. They were acquired by the ICEYE SAR sensors [54]. The details of each SAR image are listed in Table 1, where the Pol., Level, Mode, Angle and Loc. are polarization type, image format, imaging mode, look angle and imaging area, respectively. The SLC, VV, SL, SM, Asc. and Des. represent the single look complex image, vertical vertical polarization, spotlight, stripmap, ascending orbit and descending orbit, respectively. We convert SLC SAR data (I_{SLC}) into amplitude SAR images (y). The generation method are as follows. Firstly, the amplitude SAR data (I_A) of SLC SAR data is obtained through Equation (8).

$$I_A = \sqrt{I_{Real}^2 + I_{Imag}^2},$$

(8)

where I_{Real} and I_{Imag} are the real data and imaginary data of I_{SLC}, respectively. Secondly, the obtained amplitude SAR data (I_A) is normalized. The normalized method is defined as:

$$I_A^{norm} = 10log_{10}\left(\frac{I_A}{Max(I_A)}\right),\tag{9}$$

where $Max(\cdot)$ means the maximum function. Finally, the normalized SAR data (I_A^{norm}) is encoded for obtaining an amplitude SAR image (y). The encoding method is defined as:

$$y = \frac{I_A^{norm} - Min(I_A^{norm})}{Max(I_A^{norm}) - Min(I_A^{norm})} \times (2^B - 1),\tag{10}$$

where $Min(\cdot)$ represents the minimum function and B is the encoded bit. In our experiments, B is set to 8.

Table 1. Acquisition parameters for the ICEYE-SAR sensors.

Images	Sensor	Pol.	Level	Mode	Angle	Oribit	Pixels	Loc.	Data
Image1	X7	VV	SLC	SL	19.87°	Asc.	4096 × 3840	Airport	23 December 2021
Image2	X7	VV	SLC	SL	27.22°	Asc.	4096 × 3840	Oman	23 December 2021
Image3	X4	VV	SLC	SM	23.61°	Des.	4096 × 2560	Brawley	22 December 2021
Image4	X4	VV	SLC	SM	25.46°	Des.	4096 × 2560	Copeland	2 January 2021
Image5	X4	VV	SLC	SM	27.00°	Des.	4096 × 2560	Corpus	9 August 2020
Image6	X4	VV	SLC	SM	25.50°	Des.	4096 × 4096	Mississippi	3 September 2020
Image7	X4	VV	SLC	SM	21.66°	Asc.	3072 × 4096	Strait	19 February 2021

Each amplitude SAR image (y) is cropped to 256 × 256 with a stride equal to 64. The total number of cropped images is 19,440. They are randomly divided into training set (12,800), validation set (1520) and testing set (5120). The training set uses the same data augmentation as the simulated image. The final training data has 473,600 patches. The examples of testing images are displayed in Figure 7, where the white boxes are selected regions, which are used to compute the no-reference evaluation metrics. R1–R4 are the homogeneous regions. R5–R8 are the heterogeneous regions. R9 and R10 are the whole images. These regions are excluded from training set.

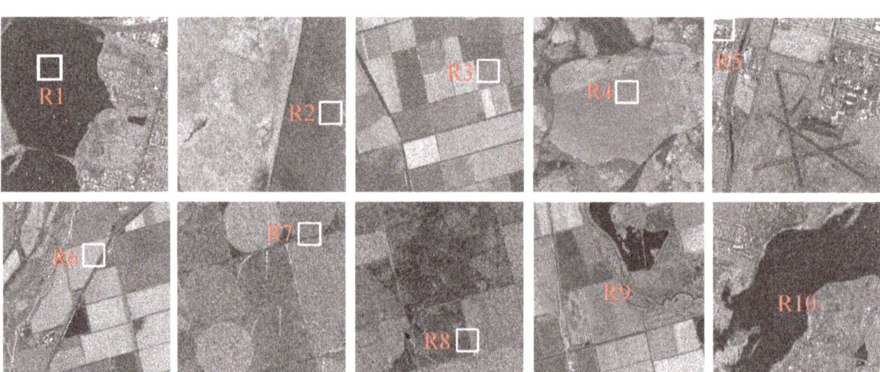

Figure 7. Real-world SAR images for evaluating.

4.3. Evaluation Metrics

For the simulated SAR experiments, the classic evaluation metrics, peak signal-to-noise ratio (PSNR, as high as possible) [55] and structural similarity index (SSIM, as closer to 1 as possible) [56], are used. The PSNR and SSIM are defined as follows:

$$PSNR = 10 \log_{10} \frac{Max_g^2}{MSE(p, g)},$$ (11)

$$SSIM = \frac{(2\mu_p \mu_g + c_1)(2\sigma_{pg} + c_2)}{(\mu_p^2 + \mu_g^2 + c_1)(\sigma_p^2 + \sigma_g^2 + c_2)},$$ (12)

where the p and g are the despeckled image and the clean reference image, respectively. Max_g is the maximum signal power, i.e., 255 for grayscale images. MSE is computed between the clean reference image and its despeckled image. The μ_p, σ_p, μ_g and σ_g represent the mean and standard variance of p and g. The σ_{pg} is the co-variance between p and g. The c_1 and c_2 are constants.

For the real-world SAR experiments, the equivalent number of looks (ENL, as high as possible) [57], and the edge-preservation degree based on ratio of average (ER, as closer to 1 as possible) [58] are used. The ENL and ER are defined as:

$$ENL = \frac{\mu_d^2}{\sigma_d^2},$$ (13)

$$ER = \frac{\sum_i^m |I_{d1}(i) / I_{d2}(i)|}{\sum_i^m |I_{o1}(i) / I_{o2}(i)|},$$ (14)

where the μ_d and σ_d are the mean and standard variance of despeckled image. The i is the index set of the SAR image. The m is the total pixels of the SAR image. The $I_{d1}(i)$ asn $I_{d2}(i)$ represent the adjacent pixel values in the horizontal or vertical directions of the despeckled image, respectively. The $I_{o1}(i)$ and $I_{o2}(i)$ are the adjacent pixel values in the horizontal or vertical directions of the noisy image, respectively.

4.4. Results and Discussion

In order to demonstrate the effectiveness of SSEUNet, we compare it qualitatively and quantitatively with Lee [14], Frost [15], SAR-BM3D [18], PPB [17], MuLoG [59], SAR-CNN [22], SAR-DRN [24] and SSUNet. Lee and Frost are the local window filters. The windows are set to 5×5. SAR-BM3D, PPB and MuLoG belong to the class of NLM methods. The publicly available Matlab codes of SAR-BM3D, PPB and MuLoG are used and the parameters are set as suggested in the original papers. SAR-CNN and SAR-DRN are machine learning methods. We implement and train SAR-CNN and SAR-DRN from scratch on noisy-noisy dataset, following the specifics given by the authors in the original papers. The noisy-noisy dataset is generated by using natural images. Therefore, SAR-CNN and SAR-DRN are unsupervised (noise2noise) methods. SSUNet is a self-unsupervised method and the training dataa is generated by GTP module. The network of SSUNet is the U-Net [37].

4.4.1. Experiments on Simulated SAR Images

For the simulated SAR image despeckling experiments, a combination of quantitative and visual comparisons is used to analyze the effects of the different methods.

In Table 2, we show the average quantitative evaluation results obtained on simulated single-look SAR images with the best performance marked in bold and the second-best marked in underlined. *Times* and *Param.* are the inference average speed and the number of parameters, respectively. *GFLOPS* represents gigabit floating-point operations per second. *Param.* and *GFLOPS* are not marked the best and the second-best. The reason is that the values of *Param.* and *GFLOPS* can not reflect the performance of despeckling methods. The results are computed as the average over the 420 testing images. The local window methods and NLM methods are grounded on the middle part of the table, while noise2noise methods are listed in the lower part. The difference between SSUNet and SSEUNet is the network. The network of SSUNet is the U-Net [37]. The network of SSEUNet is the proposed EUNet. The results of SSUNet and SSEUNet are to verify that EUNet has better despeckling performance than U-Net.

Table 2. Numerical results on simulated single-look images.

Method	MSE	PSNR	SSIM	Times(s)	Param.	GFLOPS
Noisy	0.0512	13.1933	0.1961	-	-	-
Lee	0.0135	19.2929	0.4892	0.0237	1	-
Frost	0.0119	19.7729	0.5402	0.0862	1	-
PPB	0.0117	19.8064	0.5445	9.0554	5	-
SAR-BM3D	0.0118	20.2733	0.6156	15.5699	1	-
MuLoG	0.0138	20.1007	0.6119	7.8607	1	-
SAR-CNN	0.0104	20.4939	0.6170	0.0165	557,057	36.51
SAR-DRN	0.0153	19.9947	0.6238	**0.0150**	185,857	12.18
SSUNet (Ours)	0.0097	20.7164	0.6369	0.0166	698,017	4.62
SSEUNet (Ours)	**0.0087**	**21.1912**	**0.6692**	0.0996	84,390,753	156.28

As can be seen from Table 2, the proposed SSEUNet outperform other methods on the MSE, PSNR and SSIM. Looking at the PSNR and SSIM metrics, noise2noise methods appear to have the potential to provide a clear performance gain over conventional ones. Indeed, although the performance of SAR-CNN is similar to that of the advanced NLM method, SSEUNet is about 1 dB and 0.0536 higher than the best method (SAR-BM3D). But the time it takes to process a image with 256×256 is approximately 157 times faster. By comparing the results of SAR-CNN and SAR-DRN, the more layers of the network, the better the despeckling performance. However, the network with dilated convolution layer (SAR-DRN) can improve the details of the despeckled image. In addition, compared with the second-best results (SSUNet), the proposed SSEUNet achieved a 0.5 dB and 0.0323 improvement, respectively. From the results of SSEUNet and SSUNet, we can find that our improvement to U-Net is effective and the increase of parameters does not significantly reduce the inference speed.

Visual evaluation is another way to qualitatively evaluate the despeckling performance of different methods. To visualize the despeckling results of proposed SSEUNet and other methods, Figure 8 shows the despeckled visual results of a simulated SAR image. The *Ours* is the result of using SSEUNet. It can be seen from the visual results that all despeckling methods can remove speckle noise to a certain extent. But the best despeckling results is achieved by the based on noise2noise methods. The results of the local window methods are too smooth, which causes the image to be blurred, so that the despeckled images lose the boundary information of the image. Compared with the local window methods, the effect of the NLM methods is significantly improved. However, looking at the results of PPB, there is a ringing effect at the boundary, which distorts the image boundary. In the noise2noise methods, as the network depth deepens, the better the despeckling effect, the clearer the image. It can be seen that proposed SSEUNet (*Ours*) shows the best ability to reduce speckle noise and retain texture structure.

4.4.2. Experiments on Real-World SAR Images

In order to further illustrate the practicability of the proposed method, the real-world SAR images are used for the experiments. The real-world SAR images are detailed introduced in Section 4.2. Table 3 lists the quantitative evaluation results of ENL over the selected regions. These selected regions are shown in Figure 7, where R1–R4 are homogeneous regions.

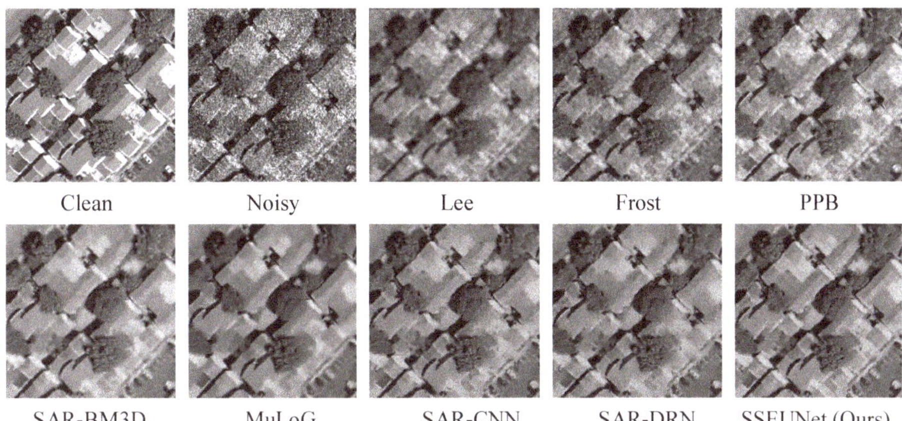

Clean	Noisy	Lee	Frost	PPB

SAR-BM3D	MuLoG	SAR-CNN	SAR-DRN	SSEUNet (Ours)

Figure 8. Visual results of simulated images.

Table 3. ENL results on real-world SAR images.

Method	R1	R2	R3	R4	Average
Noisy	16.82	48.74	78.66	225.81	92.51
Lee	159.19	169.99	245.66	1542.82	529.42
Frost	145.10	169.42	246.17	1539.16	524.96
PPB	136.27	171.74	253.36	1515.63	519.25
SAR-BM3D	155.51	190.93	238.98	1995.89	645.33
MuLoG	160.39	_202.61_	236.39	1561.18	540.14
SAR-CNN	165.13	175.43	232.72	1820.80	598.52
SAR-DRN	163.46	167.19	233.23	1597.36	540.31
SSUNet (Ours)	_170.60_	198.84	_254.92_	_1999.08_	_655.86_
SSEUNet (Ours)	**180.70**	**209.07**	**261.34**	**2070.20**	**680.33**

Clearly, as listed by the ENL values in Table 3, the noise2noise methods show better despeckling performance in homogeneous regions. In the local window methods, the Lee and Frost algorithms show similar despeckling performance. In the NLM methods, the PPB algorithm has the worst despeckling ability. The despeckling ability of SAR-BM3D and MuLoG is similar. Although SAR-DRN and SA-CNN show subtle advantages over NLM methods, the proposed SSEUNet has improved 20.31, 6.46, 24.95 and 509.02 in the four regions over the MuLoG method, respectively. Since the SAR-DRN and SAR-CNN methods use training image pairs generated from natural images, they cannot learn the relationship between speckle noise and noise-free SAR images. Compared with SAR-CNN and SAR-DRN, the despeckling performance of SSUNet is significantly improved. The results SSUNet is obtained through using GTP module to generate noisy-noisy image pairs. In addition, comparing the results of SSUNet and SSEUNet, the EUNet has a stronger despeckling ability than U-Net. In general, the proposed SSEUNet can better process real-world SAR images.

Generally, ENL can reflect the effectiveness of the algorithm, to some extent, but perfectly homogeneous regions are rare in the real-world SAR images. Therefore, an no-reference estimated approach, which is called the *ENL map*, is used to demonstrate the effectiveness of the proposed SSEUNet. The *ENL map* involves calculating small ENLs by using a sliding window (set to 3×3) until the whole SAR image is covered [60]. Figures 9 and 10 show the ENL maps of R9 and R10, which are listed in Figure 7. We only show the ENL maps of the NLM and noise2noise despeckling methods. The reason is that the results of local window methods are too smooth, resulting in a very average ENL value of the despeckled images. The ENL value should have a small change in the heterogeneous region, or even zero. However, it should have a greater improvement in the homogeneous region. This point is proven and shown in Figures 9 and 10. The ENL maps also show that the ability of details losses. From the ENL maps, the most details losses are PPB and MuLoG. Compared with SAR-BM3D, the noise2noise methods can not only better protect the image details, but also better remove the speckle noise. In the noise2noise methods, the proposed SSEUNet has the best despeckling and detail protection ability.

Table 4 displays the numerical results of ER metric on the R5–R8 regions, where V and H are the vertical and horizontal results of the ER metric. It can be seen that the proposed SSEUNet have obvious advantages in protecting horizontal and vertical structures. In the noise2noise methods, the proposed GTP module and EUNet can better process real-word SAR images. The proposed GTP module can use real-world SAR images to construct noisy-noisy image pairs to train deep CNNs, so that the network can better learn the relationship of speckle noise and noise-free SAR images.

Table 4. ER metric on real-world SAR images.

Method	R5		R6		R7		R8	
	V	H	V	H	V	H	V	H
Lee	0.9430	0.9478	0.9876	0.9490	0.9487	0.9563	0.9051	0.9038
Frost	0.9501	0.9544	0.9877	0.9490	0.9587	0.9563	0.9052	0.9117
PPB	0.9565	0.9612	0.9874	0.9688	0.9590	0.9563	0.9054	0.9139
SAR-BM3D	0.9586	0.9613	0.9879	0.9893	0.9579	0.9560	0.9062	0.9043
MuLoG	0.9585	0.9603	0.9879	0.9881	0.9585	0.9557	0.9048	0.9041
SAR-CNN	0.9707	0.9618	0.9706	0.9611	0.9549	0.9578	0.9061	0.9132
SAR-DRN	0.9627	0.9631	0.9707	0.9611	0.9531	0.9575	0.9062	0.9133
SSUNet (Ours)	0.9567	0.9709	0.9876	0.9883	0.9584	**0.9592**	0.9043	0.9136
SSEUNet (Ours)	**0.9923**	**0.9941**	**0.9910**	**0.9912**	**0.9609**	0.9588	**0.9086**	**0.9176**

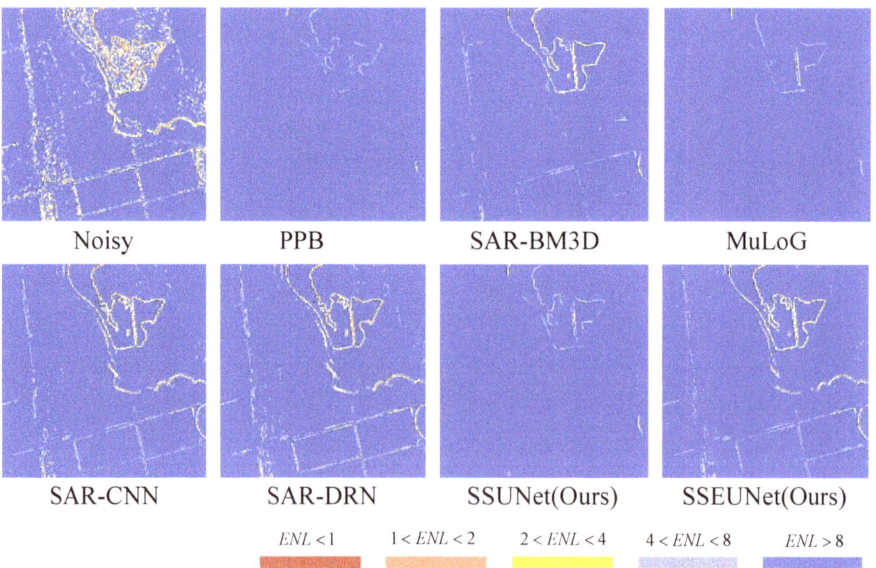

Figure 9. ENL maps of different despeckling methods on R9.

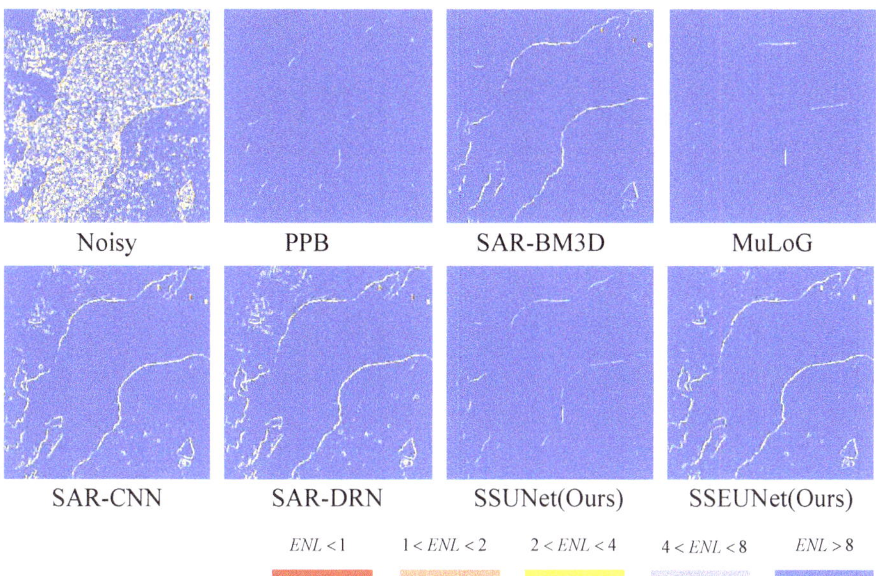

Figure 10. ENL maps of different despeckling methods on R10.

It is worth carefully studying different methods of despeckled SAR images. Figures 11 and 12 shows the details of homogeneous and heterogeneous regions in the real-world SAR images, respectively. It can be seen from the results that the local window methods are too smooth to lose the contrast. Loss the contrast means that the despeckled images structure using local window filters becomes smooth, and the structural contrast between edges and non-edges becomes blurred. In addition, many strong points and linear structures are lost in the NLM despeckling methods. The result of PPB can also cause a ringing effect at the boundary. Compared with other traditional methods, the result of SAR-BM3D performs best and provides an acceptable balance between smoothing and details preservation. Among SAR-DRN, SAR-CNN and SSEUNet methods, the result of SSEUNet is the best. Figure 13 shows the visualization despeckling results of different scenes. It can be seen from the despeckled results that the proposed SSEUNet can deal with real-world SAR images in different scenes.

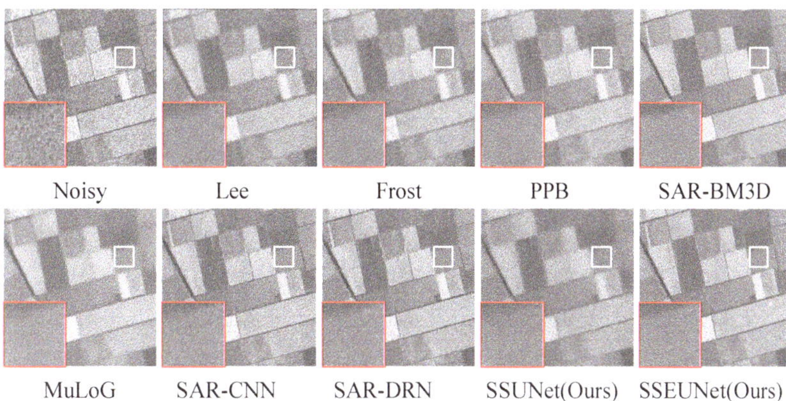

Figure 11. Visual results of homogeneous regions on real-word SAR images.

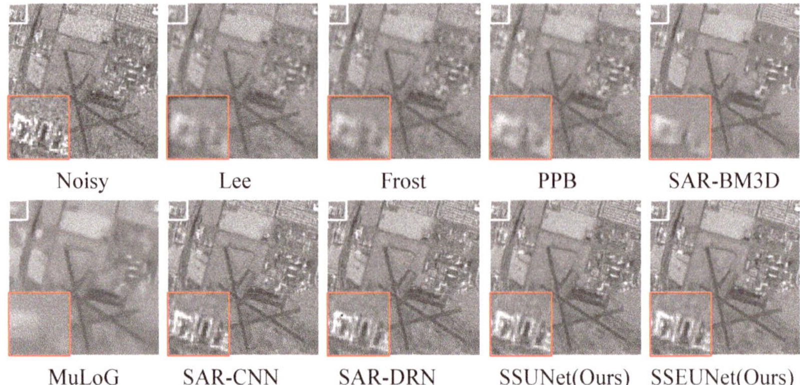

Figure 12. Visual results of heterogeneous regions on real-word SAR images.

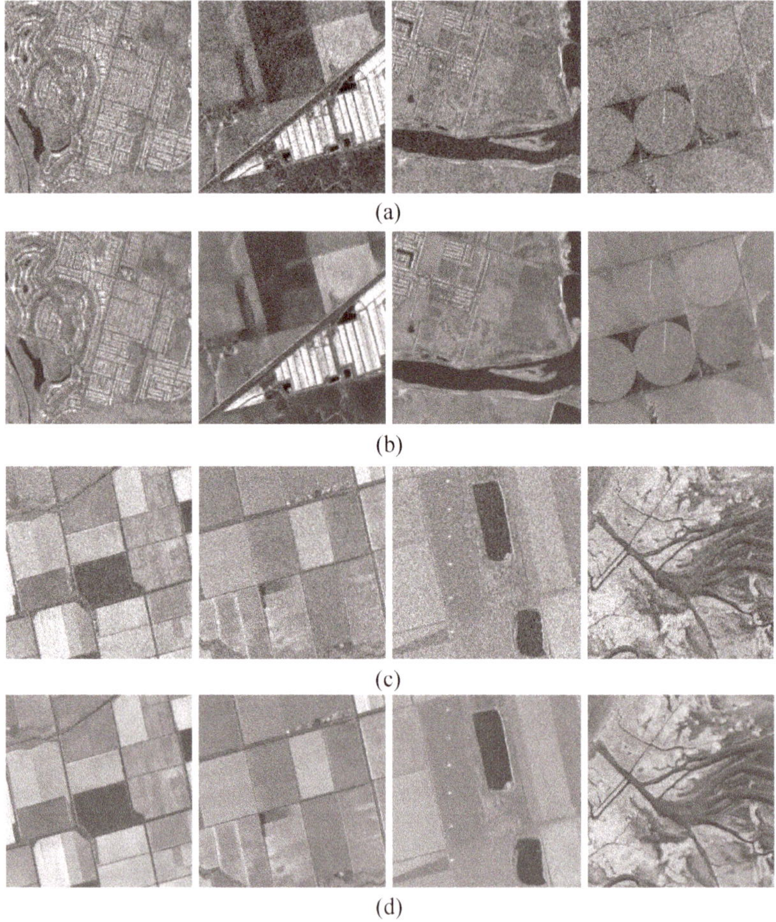

Figure 13. The despeckled results of the SSEUNet in real-world SAR images. (**a**,**c**) Real-world SAR images. (**b**,**d**) Despeckled results.

To gain insight about how the learned SSAM on the real-world SAR images, we pick 8 sample images and inspect the output features of SSAM. As can be seen from Figure 4, the input features of SSAM1 are the output features of ISC1 and TConv1. The inputs of SSAM2 are the outputs of ISC2 and TConv2. The outputs of ISC3 and TConv3 are fed into SSAM3. The input features of SSAM4 are the outputs of ISC4 and TConv4. The output features of SSAM1-SSAM4 are extracted from the *projection* layer in the SSAM1-SSAM4. The structure of SSAM1-SSAM4 are exactly the same. The detailed structure of SSAM is shown in Figure 5. The output feature sizes of SSAM1-SSAM4 are $256 \times 32 \times 32$, $128 \times 64 \times 64$, $64 \times 128 \times 128$ and $32 \times 256 \times 256$, respectively. In the process of visualization, we perform averaging operations on the output features of SSAM1-SSAM4. Each visualization feature is the mean features of all channels. The sizes of visualization features are 32×32, 64×64, 128×128 and 256×256 in the SSAM1-SSAM4, respectively. Figure 14 shows the visualization features of SSAM1-SSAM4. It can be seen that the details of weak texture and structure are gradually restored from noisy SAR images.

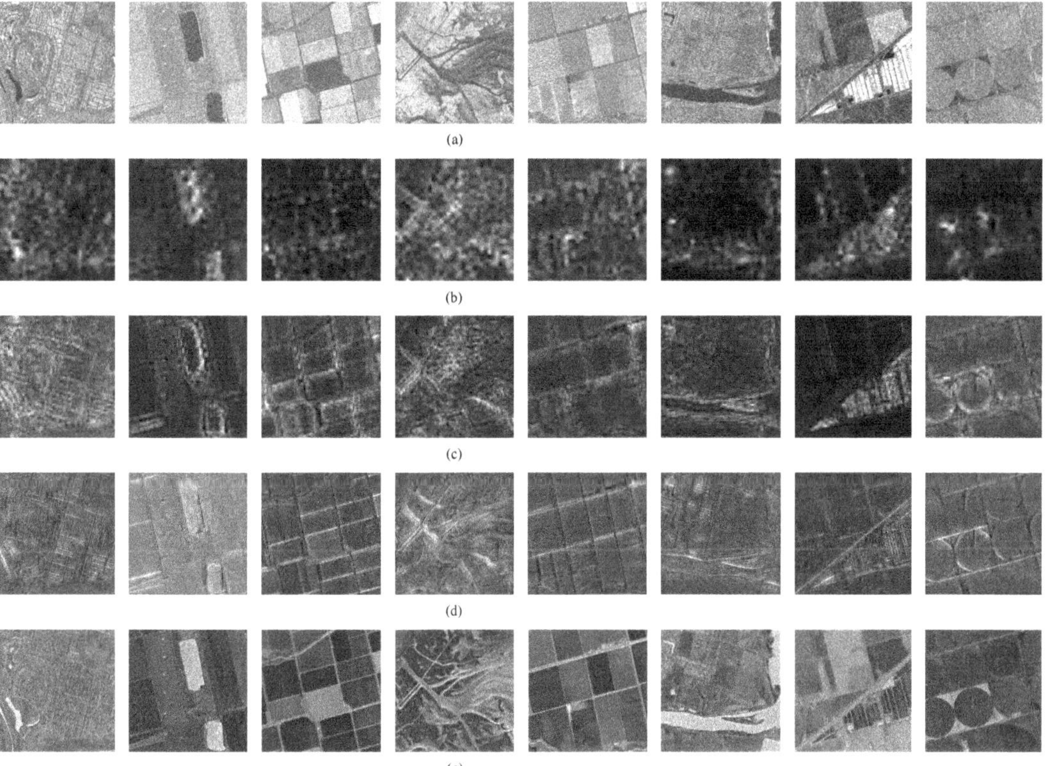

Figure 14. Visualization features of SSAM1-SSAM4. (**a**) Original SAR images. (**b–e**) The visualization features of SSAM1-SSAM4.

5. Conclusions

In this paper, we propose a novel self-supervised despeckling algorithm with an enhanced U-Net (SSEUNet). The proposed SSEUNet is composed of generation training pairs (GTP) module, enhanced U-Net (EUNet) and a self-supervised training loss function with a regularization loss. The proposed SSEUNet has the following advantages. Firstly, unlike previous self-supervised despeckling works, the noisy-noisy image pairs are generated from real-word SAR images through a novel generation training pairs module, which makes it possible to train deep convolutional neural networks using real-world

SAR images. The GTP module can eliminate the effects of natural images, time series images and the performance of GAN. Secondly, the EUNet is designed to improve the features extraction and fusion capabilities of the U-Net. Compared with U-Net, we introduce BN-RRDC blocks, convolutional layers with $s = 2$, improved skip connections and SSAM. Although the EUNet has more parameters and more complex structure, the training difficulty will not increase. The reason is that there is the residual structures in the BN-RRDC block. Thirdly, a self-supervised training loss function is designed to address the difference of target pixel values between neighbors on the original noisy image. The loss function includes a reconstruction loss (MSE) and a regularization loss. Finally, visual and quantitative experiments on simulated and real-world SAR images show that the proposed SSEUNet notably reduces speckle noise with better preserving features, which exceed several state-of-the-art despeckling methods.

However, the inference speed of proposed SSEUNet needs to be improved. The proposed SSEUNet uses complex feature extraction block and sub-space attention modules, which leads to a longer time for evaluating an image. At the same time, the despeckling results of different data augmentation methods also need to be verified on the SAR images. In the future, we plan to explore two works. Firstly, we will explore a lightweight network to replace EUNet for improving inference speed. Secondly, the despeckling effects of different data augmentation methods will be verified.

Author Contributions: Conceptualization, G.Z.; Data curation, G.Z. and S.L.; Formal analysis, G.Z.; Investigation, G.Z. and X.L.; Methodology, G.Z. and X.L.; Resources, Z.L.; Software, G.Z.; Supervision, Z.L.; Validation, G.Z. and X.L.; Visualization, G.Z.; Writing—original draft, G.Z. and X.L. All authors have read and agreed to the published version of the manuscript.

Funding: This research was supported by the National Science Foundation for Distinguished Young Scholars, Grant Numbers 61906213.

Institutional Review Board Statement: Not applicable.

Informed Consent Statement: Not applicable.

Data Availability Statement: UC Merced land-use dataset is available are http://vision.ucmerced.edu/datasets, accessed on 27 October 2021. ICEYE SAR dataset is available at https://www.iceye.com/sar-data, accessed on 27 October 2021.

Conflicts of Interest: The authors declare no conflict of interest.

References

1. Moreira, A.; Prats-Iraola, P.; Younis, M.; Krieger, G.; Hajnsek, I.; Papathanassiou, K.P. A tutorial on synthetic aperture radar. *IEEE Geosci. Remote Sens. Mag.* **2013**, *1*, 6–43. [CrossRef]
2. Yang, Z.; Wei, J.; Deng, J.; Gao, Y.; Zhao, S.; He, Z. Mapping Outburst Floods Using a Collaborative Learning Method Based on Temporally Dense Optical and SAR Data: A Case Study with the Baige Landslide Dam on the Jinsha River, Tibet. *Remote Sens.* **2021**, *13*, 2205. [CrossRef]
3. Qin, Y.; Xiao, X.; Wang, J.; Dong, J.; Ewing, K.; Hoagland, B.; Hough, D.J.; Fagin, T.D.; Zou, Z.; Geissler, G.L.; et al. Mapping Annual Forest Cover in Sub-Humid and Semi-Arid Regions through Analysis of Landsat and PALSAR Imagery. *Remote Sens.* **2016**, *8*, 933. [CrossRef]
4. Zhang, T.; Zhang, X.; Ke, X.; Zhan, X.; Shi, J.; Wei, S.; Pan, D.; Li, J.; Su, H.; Zhou, Y.; et al. LS-SSDD-v1.0: A Deep Learning Dataset Dedicated to Small Ship Detection from Large-Scale Sentinel-1 SAR Images. *Remote Sens.* **2020**, *12*, 2997. [CrossRef]
5. Zhang, T.; Zhang, X.; Ke, X.; Liu, C.; Xu, X.; Zhan, X.; Wang, C.; Ahmad, I.; Zhou, Y.; Pan, D.; et al. HOG-ShipCLSNet: A Novel Deep Learning Network with HOG Feature Fusion for SAR Ship Classification. *IEEE Trans. Geosci. Remote Sens.* **2021**. [CrossRef]
6. Herzfeld, U.C.; Williams, S.; Heinrichs, J.; Maslanik, J.; Sucht, S. Geostatistical and Statistical Classification of Sea-Ice Properties and Provinces from SAR Data. *Remote Sens.* **2016**, *8*, 616. [CrossRef]
7. Singh, P.; Diwakar, M.; Shankar, A.; Shree, R.; Kumar, M. A Review on SAR Image and its Despeckling. *Arch. Comput. Methods Eng.* **2021**. [CrossRef]
8. Argenti, F.; Lapini, A.; Bianchi, T.; Alparone, L. A Tutorial on Speckle Reduction in Synthetic Aperture Radar Images. *IEEE Geosci. Remote Sens. Mag.* **2013**, *1*, 6–35. [CrossRef]
9. Zhang, T.; Zhang, X.; Shi, J.; Wei, S.; Wang, J.; Li, J.; Su, H.; Zhou, Y. Balance Scene Learning Mechanism for Offshore and Inshore Ship Detection in SAR Images. *IEEE Geosci. Remote Sens. Lett.* **2020**, 1–5. [CrossRef]

10. Zhang, T.; Zhang, X. ShipDeNet-20: An Only 20 Convolution Layers and <1-MB Lightweight SAR Ship Detector. *IEEE Geosci. Remote Sens. Lett.* **2021**, *18*, 1234–1238.

11. Zhang, T.; Zhang, X. A Polarization Fusion Network with Geometric Feature Embedding for SAR Ship Classification. *Pattern Recognit.* **2021**, *123*, 108365. [CrossRef]

12. Zhang, T.; Zhang, X. Squeeze-and-Excitation Laplacian Pyramid Network with Dual-Polarization Feature Fusion for Ship Classification in SAR Images. *IEEE Geosci. Remote Sens. Lett.* **2021**, 1. [CrossRef]

13. Wei, Y.; Li, Y.; Ding, Z.; Wang, Y.; Zeng, T.; Long, T. SAR Parametric Super-Resolution Image Reconstruction Methods Based on ADMM and Deep Neural Network. *IEEE Trans. Geosci. Remote Sens.* **2021**. [CrossRef]

14. Lee, J.S. Speckle analysis and smoothing of synthetic aperture radar images. *Comput. Graph. Image Process.* **1981**, *17*, 24–32. [CrossRef]

15. Frost, V.S.; Stiles, J.A.; Shanmugan, K.S.; Holtzman, J.C. A Model for Radar Images and Its Application to Adaptive Digital Filtering of Multiplicative Noise. *IEEE Trans. Pattern Anal. Mach. Intell.* **1982**, *4*, 157–166. [CrossRef]

16. Kuan, D.T.; Sawchuk, A.A.; Strand, T.C.; Chavel, P. Adaptive Noise Smoothing Filter for Images with Signal-Dependent Noise. *IEEE Trans. Pattern Anal. Mach. Intell.* **1985**, *7*, 165–177. [CrossRef]

17. Deledalle, C.; Denis, L.; Tupin, F. Iterative Weighted Maximum Likelihood Denoising with Probabilistic Patch-Based Weights. *IEEE Signal Process.* **2009**, *18*, 2661–2672. [CrossRef] [PubMed]

18. Parrilli, S.; Poderico, M.; Angelino, C.V.; Verdoliva, L. A Nonlocal SAR Image Denoising Algorithm Based on LLMMSE Wavelet Shrinkage. *IEEE Geosci. Remote Sens.* **2012**, *50*, 606–616. [CrossRef]

19. Deledalle, C.A.; Denis, L.; Tupin, F.; Reigber, A.; Jager, M. NL-SAR: A unified nonlocal framework for resolution-preserving (Pol)(In)SAR denoising. *IEEE Trans. Geosci. Remote Sens.* **2015**, *53*, 2021–2038. [CrossRef]

20. Zhang, T.; Zhang, X.; Li, J.; Xu, X.; Wang, B.; Zhan, X.; Xu, Y.; Ke, X.; Zeng, T.; Su, H.; et al. SAR Ship Detection Dataset (SSDD): Official Release and Comprehensive Data Analysis. *Remote Sens.* **2021**, *13*, 3690. [CrossRef]

21. Zhang, T.; Zhang, X.; Shi, J.; Wei, S. HyperLi-Net: A hyper-light deep learning network for high-accurate and high-speed ship detection from synthetic aperture radar imagery. *ISPRS J. Photogramm. Remote Sens.* **2020**, *167*, 123–153. [CrossRef]

22. Chierchia, G.; Cozzolino, D.; Poggi, G.; Verdoliva, L. SAR image despeckling through convolutional neural networks. In Proceedings of the IEEE International Geoscience and Remote Sensing Symposium, Fort Worth, TX, USA, 23–28 July 2017; pp. 5438–5441.

23. Cozzolino, D.; Verdoliva, L.; Scarpa, G.; Poggi, G. Nonlocal CNN SAR Image Despeckling. *Remote Sens.* **2020**, *12*, 1006. [CrossRef]

24. Zhang, Q.; Yuan, Q.; Li, J.; Zhen, Y.; Zhang, L. Learning a Dilated Residual Network for SAR Image Despeckling. *Remote Sens.* **2018**, *10*, 196. [CrossRef]

25. Gui, Y.; Xue, L.; Li, X. SAR image despeckling using a dilated densely connected network. *Remote Sens. Lett.* **2018**, *9*, 857–866. [CrossRef]

26. Shen, H.; Zhou, C.; Li, J.; Yuan, Q. SAR Image Despeckling Employing a Recursive Deep CNN Prior. *IEEE Geosci. Remote Sens.* **2021**, *59*, 273–286. [CrossRef]

27. Dalsasso, E.; Denis, L.; Tupin, F. SAR2SAR: A Semi-Supervised Despeckling Algorithm for SAR Images. *IEEE J. Sel. Top. Appl. Earth Obs. Remote Sens.* **2021**, *14*, 4321–4329. [CrossRef]

28. Yuan, Y.; Sun, J.; Guan, J. Blind SAR Image Despeckling Using Self-Supervised Dense Dilated Convolutional Neural Network. *arXiv* **2019**, arXiv:1908.01608.

29. Molini, A.B.; Valsesia, D.; Fracastoro, G.; Magli, E. Speckle2Void: Deep Self Supervised SAR Despeckling with Blind-Spot Convolutional Neural Networks. *IEEE Geosci. Remote Sens.* **2021**. [CrossRef]

30. Yuan, Y.; Guan, J.; Feng, P.; Wu, Y. A Practical Solution for SAR Despeckling with Adversarial Learning Generated Speckled-to-Speckled Images. *IEEE Geosci. Remote Sens. Lett.* **2020**. [CrossRef]

31. Ma, X.; Wang, C.; Yin, Z.; Wu, P. SAR Image Despeckling by Noisy Reference-Based Deep Learning Method. *IEEE Geosci. Remote Sens.* **2020**, *58*, 8807–8818. [CrossRef]

32. Pan, T.; Peng, D.; Yang, W.; Li, H.C. A Filter for SAR Image Despeckling Using Pre-Trained Convolutional Neural Network Model. *Remote Sens.* **2019**, *11*, 2379. [CrossRef]

33. Li, J.; Li, Y.; Xiao, Y.; Bai, Y. HDRANet: Hybrid Dilated Residual Attention Network for SAR Image Despeckling. *Remote Sens.* **2019**, *11*, 2921. [CrossRef]

34. Lehtinen, J.; Munkberg, J.; Hasselgren, J.; Laine, S.; Karras, T.; Aittala, M.; Aila, T. Noise2Noise: Learning Image Restoration without Clean Data. *arXiv* **2018**, arXiv:1803.04189.

35. Goodfellow, I.J.; Pouget-Abadie, J.; Mirza, M.; Xu, B.; Warde-Farley, D.; Ozair, S.; Courville, A.C.; Bengio, Y. Generative Adversarial Networks. *arXiv* **2014**, arXiv:1406.2661.

36. Mullissa, A.G.; Marcos, D.; Tuia, D.; Herold, M.; Reiche, J. deSpeckNet: Generalizing Deep Learning-Based SAR Image Despeckling. *IEEE Geosci. Remote Sens.* **2020**. [CrossRef]

37. Ronneberger, O.; Fischer, P.; Brox, T. U-Net: Convolutional Networks for Biomedical Image Segmentation. In Proceedings of the Medical Image Computing and Computer-Assisted Intervention, Munich, Germany, 5–9 October 2015; Volume 9351.

38. Zhang, K.; Zuo, W.; Chen, Y.; Meng, D.; Zhang, L. Beyond a Gaussian Denoiser: Residual Learning of Deep CNN for Image Denoising. *IEEE Signal Process.* **2017**, *26*, 3142–3155. [CrossRef] [PubMed]

39. Zhang, J.; Li, W.; Li, Y. SAR Image Despeckling Using Multiconnection Network Incorporating Wavelet Features. *IEEE Geosci. Remote Sens. Lett.* **2020**, *17*, 1363–1367. [CrossRef]
40. Vitale, S.; Ferraioli, G.; Pascazio, V. Multi-Objective CNN-Based Algorithm for SAR Despeckling. *IEEE Geosci. Remote Sens.* **2020**. [CrossRef]
41. Martino, G.D.; Poderico, M.; Poggi, G.; Riccio, D.; Verdoliva, L. Benchmarking framework for SAR despeckling. *IEEE Trans. Geosci. Remote Sens.* **2014**, *52*, 1596–1615. [CrossRef]
42. Laine, S.; Karras, T.; Lehtinen, J.; Aila, T. High-quality self-supervised deep image denoising. In Proceedings of the 33rd Conference on Neural Information Processing Systems (NeurIPS 2019), 8–14 December 2019; pp. 6968–6978.
43. Heinrich, M.; Stille, M.; Buzug, T. Residual U-Net Convolutional Neural Network Architecture for Low-Dose CT Denoising. *Curr. Dir. Biomed. Eng.* **2018**, *4*, 297–300 [CrossRef]
44. Guo, Y.; Chen, J.; Wang, J.; Chen, Q.; Cao, J.; Deng, Z.; Xu, Y.; Tan, M. Closed-Loop Matters: Dual Regression Networks for Single Image Super-Resolution. In Proceedings of the IEEE/CVF Conference on Computer Vision and Pattern Recognition, Virtual, 14–19 June 2020; pp. 5406–5415.
45. Zhou, Z.; Rahman Siddiquee, M.M.; Tajbakhsh, N.; Liang, J. UNet++: A Nested U-Net Architecture for Medical Image Segmentation. In *Deep Learning in Medical Image Analysis and Multimodal Learning for Clinical Decision Support*; Springer: Cham, Switzerland, 2018; pp. 3–11.
46. Li, X.; Chen, H.; Qi, X.; Dou, Q.; Fu, C.W.; Heng, P.A. H-DenseUNet: Hybrid Densely Connected UNet for Liver and Tumor Segmentation From CT Volumes. *IEEE Trans. Med Imaging* **2018**, *37*, 2663–2674. [CrossRef]
47. Li, W.; Chakraborty, M.; Fenner, D.; Faber, J.; Zhou, K.; Ruempker, G.; Stoecker, H.; Srivastava, N. EPick: Multi-Class Attention-based U-shaped Neural Network for Earthquake Detection and Seismic Phase Picking. *arXiv* **2021**, arXiv:2109.02567.
48. Cheng, S.; Wang, Y.; Huang, H.; Liu, D.; Fan, H.; Liu, S. NBNet: Noise Basis Learning for Image Denoising with Subspace Projection. *arXiv* **2020**, arXiv:2012.15028.
49. Wang, X.; Yu, K.; Wu, S.; Gu, J.; Liu, Y.; Dong, C.; Loy, C.C.; Qiao, Y.; Tang, X. ESRGAN: Enhanced Super-Resolution Generative Adversarial Networks. *Eur. Conf. Comput. Vis.* **2018**, *11133*, 63–79.
50. He, K.; Zhang, X.; Ren, S.; Sun, J. Delving deep into rectifiers: Surpassing-human level performance on imagenet classification. In Proceedings of the IEEE International Conference on Computer Vision, Santiago, Chile, 11–18 December 2015; pp. 1026–1034.
51. Diederik, P.K.; Jimmy, L.B. Adam: A method for stochastic optimization. *arXiv* **2014**, arXiv:1412.6980.
52. Yi, Y.; Shawn, N. Bag-Of-Visual-Words and Spatial Extensions for Land-Use Classification. In Proceedings of the 18th SIGSPATIAL International Conference on Advances in Geographic Information Systems, San Jose, CA, USA, 2–5 November 2010; pp. 270–279.
53. Liu, C.; Tupin, F.; Gousseau, Y. Training CNNs on speckled optical dataset for edge detection in SAR images. *ISPRS J. Photogramm. Remote Sens.* **2020**, *170*, 88–102. [CrossRef]
54. Stringham, C.; Farquharson, G.; Castelletti, D.; Quist, E.; Riggi, L.; Eddy, D.; Soenen, S. The Capella X-band SAR Constellation for Rapid Imaging. In Proceedings of the IEEE International Geoscience and Remote Sensing Symposium, Yokohama, Japan, 28 July–2 August 2019; pp. 9248–9251.
55. Lattari, F.; Gonzalez Leon, B.; Asaro, F.; Rucci, A.; Prati, C.; Matteucci, M. Deep Learning for SAR Image Despeckling. *Remote Sens.* **2019**, *11*, 1532. [CrossRef]
56. Wang, Z.; Bovik, A.C.; Sheikh, H.R.; Simoncelli, E.P. Image quality assessment: From error visibility to structural similarity. *IEEE Trans. Image Process.* **2004**, *13*, 600–612. [CrossRef]
57. Lee, J.S.; Pottier, E. *Polarimetric Radar Imaging: From Basics to Applications*; CRC Press: Boca Raton, FL, USA, 2009.
58. Feng, H.; Hou, B.; Gong, M. SAR image despeckling based on local homogeneous-region segmentation by using pixel-relativity measurement. *IEEE Trans. Geosci. Remote Sens.* **2011**, *49*, 2724–2737. [CrossRef]
59. Deledalle, C.A.; Denis, L.; Tabti, S.; Tupin, F. MuLoG, or How to apply Gaussian denoisers to multi-channel SAR speckle reduction. *IEEE Trans. Image Process.* **2017**, *26*, 4389–4403. [CrossRef]
60. Anfinsen, S.N.; Doulgeris, A.P.; Eltoft, T. Estimation of the equivalent number of looks in polarimetric synthetic aperture radar imagery. *IEEE Trans. Geosci. Remote Sens.* **2009**, *47*, 3795–3809. [CrossRef]

remote sensing

Article

A Novel Guided Anchor Siamese Network for Arbitrary Target-of-Interest Tracking in Video-SAR

Jinyu Bao, Xiaoling Zhang *, Tianwen Zhang, Jun Shi and Shunjun Wei

School of Information and Communication Engineering, University of Electronic Science and Technology of China, Chengdu 611731, China; 201811011909@std.uestc.edu.cn (J.B.); twzhang@std.uestc.edu.cn (T.Z.); shijun@uestc.edu.cn (J.S.); weishunjun@uestc.edu.cn (S.W.)
* Correspondence: xlzhang@uestc.edu.cn

Abstract: Video synthetic aperture radar (Video-SAR) allows continuous and intuitive observation and is widely used for radar moving target tracking. The shadow of a moving target has the characteristics of stable scattering and no location shift, making moving target tracking using shadows a hot topic. However, the existing techniques mainly rely on the appearance of targets, which is impractical and costly, especially for tracking targets of interest (TOIs) with high diversity and arbitrariness. Therefore, to solve this problem, we propose a novel guided anchor Siamese network (GASN) dedicated to arbitrary TOI tracking in Video-SAR. First, GASN searches for matching areas in the subsequent frames with the initial area of the TOI in the first frame are conducted, returning the most similar area using a matching function, which is learned from general training without TOI-related data. With the learned matching function, GASN can be used to track arbitrary TOIs. Moreover, we also constructed a guided anchor subnetwork, referred to as GA-SubNet, which employs the prior information of the first frame and generates sparse anchors of the same shape as the TOIs. The number of unnecessary anchors is therefore reduced to suppress false alarms. Our method was evaluated on simulated and real Video-SAR data. The experimental results demonstrated that GASN outperforms state-of-the-art methods, including two types of traditional tracking methods (MOSSE and KCF) and two types of modern deep learning techniques (Siamese-FC and Siamese-RPN). We also conducted an ablation experiment to demonstrate the effectiveness of GA-SubNet.

Keywords: video synthetic aperture radar (Video-SAR); moving target tracking; guided anchor Siamese network (GASN)

Citation: Bao, J.; Zhang, X.; Zhang, T.; Shi, J.; Wei, S. A Novel Guided Anchor Siamese Network for Arbitrary Target-of-Interest Tracking in Video-SAR. *Remote Sens.* **2021**, *13*, 4504. https://doi.org/10.3390/rs13224504

Academic Editor: Fabio Rocca

Received: 27 August 2021
Accepted: 1 November 2021
Published: 9 November 2021

Publisher's Note: MDPI stays neutral with regard to jurisdictional claims in published maps and institutional affiliations.

1. Introduction

Video synthetic aperture radar (Video-SAR) provides high-resolution SAR images at a faster frame rate, which is conducive to the continuous and intuitive observation of ground moving targets. Due to this advantage, Video-SAR brings about important applications in SAR moving target tracking [1]. Since the Sandia National Laboratory (SNL) of the United States first obtained high-resolution SAR images in 2003 [2], many scholars have investigated the problem of moving target tracking in Video-SAR [3–7]. However, due to different angles of illumination, the scattering characteristics of moving targets change with the movement of the platform. Worse still, it is difficult to track a moving target directly because the imaging results of the moving target usually shift from their true position.

Fortunately, shadow is caused by the ground being blocked by the moving target. Due to the absence of energy reflection, shadows appear at the real position of the moving target in the SAR image, with the advantage of a constant grayscale [8]. Therefore, shadow-aided moving target tracking has become a hot topic in Video-SAR. In recent years, many scholars have worked on shadow-aided moving target tracking in Video-SAR [9–11]. Wang et al. [9] fully considered the constant grayscale of shadows and used data multiplexing to achieve moving target tracking. Zhao et al. [10] applied the saliency-based detection mechanism and used spatial–temporal information to achieve moving target tracking in Video-SAR. Tian et al. [11] utilized the dynamic programming-based particle filter to achieve the track-before-detect algorithm in Video-SAR. However, the features used by these traditional methods are usually simple, which leads to the problem of the background being similar to the shadow, meaning it cannot be easily distinguished. Deep learning methods then emerged to solve shadow tracking due to their high accuracy and fast speed advantages [12–16]. Ding et al. [12] presented a framework for shadow-aided moving target detection using deep neural networks, which applied a faster region-based convolutional neural network (Faster-RCNN) [13] to detect shadows in a single frame and used a bi-directional long short-term memory (Bi-LSTM) [14] network to track the shadows. Zhou et al. [15] proposed a framework by combining a modified real-time recurrent regression network and a newly designed trajectory smoothing long short-term memory network to track shadows. Wen et al. [16] proposed a moving target tracking method based on the dual Faster-RCNN, which combined the shadow detection results in SAR images and the range-Doppler (RD) spectrum to suppress false alarms for moving target tracking in Video-SAR.

However, arbitrary target-of-interest (TOI) tracking is a challenge for the above methods. In this paper, we define TOI as a specific target in a video that one wants to track. TOI refers to the shadow to be tracked in Video-SAR. The reasons why arbitrary TOI tracking is a challenge are as follows: First, these methods are all based on appearance features, such as shape and texture. These methods need to train a large number of labeled training samples to extract appearance features, and the training samples must include the TOI. However, when we track an arbitrary TOI, it is impractical to collect samples of all categories for training because of the targets' diversity and arbitrariness. Moreover, it takes extensive work and material resources to label a large number of SAR images. Therefore, these methods are both impractical and costly when tracking an arbitrary TOI in Video-SAR.

Thus, we propose a novel guided anchor Siamese network (GASN) for arbitrary TOI tracking in Video-SAR. First, the key of GASN lies in the idea of similarity learning, which learns a matching function to estimate the degree of similarity between two images. After training using a large number of paired images, the learned matching function in GASN, given an unseen pair of inputs (TOI in the first frame as the template, and the subsequent frame as the search image), is used to locate the area that best matches the template. As GASN only relies on the template information, which is independent of the training data, it is suitable for tracking arbitrary TOIs in Video-SAR. Additionally, a guided anchor subnetwork (GA-SubNet) in GASN is proposed to suppress false alarms and to improve the tracking accuracy. GA-SubNet uses the location information of the template to obtain the location probability in the search image, and then it selects the location with a probability greater than the threshold to generate sparse anchors, which can exclude false alarms. To improve the tracking accuracy, the anchor that more closely matches the shape of the TOI is obtained by GA-SubNet through adaptive prediction processing.

The main contributions of our method are as follows:

1. We established a new network GASN, which trains a large number of paired images to build a matching function to judge the degree of similarity between two inputs. After similarity learning, GASN matches the subsequent frame with the initial area of the TOI in the first frame and returns the most similar area as the tracking result.
2. We constructed a GA-SubNet embedded in GASN to suppress false alarms, as well as to improve the tracking accuracy. By incorporating the prior information of the template, our proposed GA-SubNet can generate sparse anchors that match the shape of the TOI the most.

To verify the validity of the proposed method, we performed experiments on simulated and real Video-SAR data. The results showed that the tracking accuracy of the proposed network is 60.16% on simulated Video-SAR data, 4.55% and 16.49% higher than the two deep learning methods Siamese-RPN [17] and Siamese-FC [18], as well as 18.36% and 28.95% higher than the two traditional methods MOSSE [19] and KCF [20], respectively. Meanwhile, the tracking accuracy is 54.68% on real Video-SAR data, which is higher than the other four methods by 1.93%, 13.08%, 14.70%, and 25.04%, respectively. This demonstrates that our method can achieve accurate arbitrary TOI tracking in Video-SAR.

The rest of this paper is organized as follows: Section 2 introduces the methodology, including the network architecture, preprocessing, and tracking processes. Section 3 introduces the experiments, including the simulated and real data, the implementation details, the loss function, and the evaluation indicators. Section 4 introduces the simulated and real Video-SAR data tracking results. Section 5 discusses the research on pre-training and robustness and the ablation experiment. Section 6 provides the conclusion.

2. Methodology

2.1. Network Architecture

Figure 1 shows the architecture of GASN for arbitrary TOI tracking in Video-SAR, including the Siamese subnetwork, GA-SubNet, and the similarity learning subnetwork. GASN is based on the idea of similarity learning, which compares a template image z to a search image x and returns a high score if the two images depict the same target.

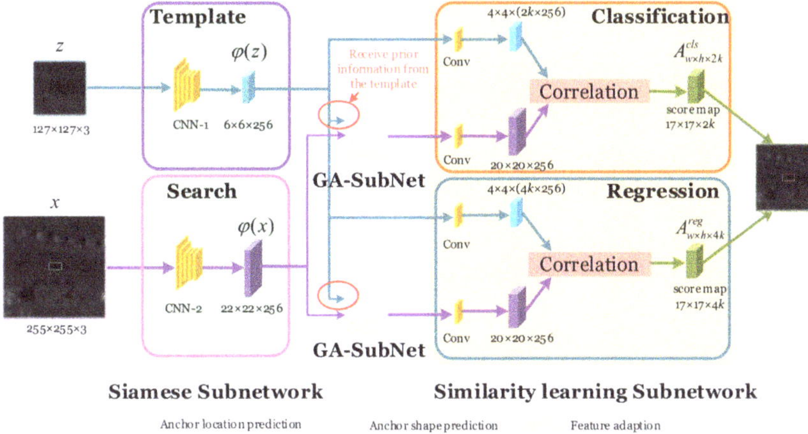

Figure 1. The architecture of GASN.

To prepare for similarity learning, the Siamese subnetwork consists of a template branch and a search branch. The two branches apply identical transformation φ to each input, and the transformation φ can be considered as feature embedding. Then, similarity learning can be expressed as $f(z, x) = g(\varphi(z), \varphi(x))$, where the function g is a similarity metric. To suppress false alarms, GA-SubNet receives the prior information from the template to pre-determine the general location and shape of the TOI in the search image using anchors. When tracking an arbitrary TOI that is different from the training sample, we can use the ability of similarity learning to find the TOI in the next frame by providing the template information of said TOI, such as the position and shape. The similarity learning subnetwork is divided into two branches, one for the classification of the shadow and background, and the other for the regression of the shadow's location and shape. In both branches, the similarity between the shadow template and the search area is calculated, and then the target with the maximum similarity to the template of the TOI is chosen as the tracking result.

GASN always uses the previous frame as the template image and the current frame as the search image. After testing the whole SAR image sequence in such a way, GASN can achieve arbitrary TOI tracking in Video-SAR. In the following, we introduce the three subnetworks of GASN in detail in the order of implementation.

2.1.1. Siamese Subnetwork

The Siamese subnetwork (marked as the green region in Figure 1) [21,22] uses CNN for feature embedding. CNN uses different convolutional kernels for multi-level feature embedding of the image. Therefore, compared to the traditional manual features, the features embedded by the Siamese subnetwork are more representative and can describe the TOI better. To obtain the common features of the previous and current frames, the Siamese subnetwork is divided into a template branch (marked with a purple box) and a search branch (marked with a pink box), and the parameters of CNN-1 and CNN-2 in both branches are shared to ensure the consistency of features. The input of the template branch is the TOI area in the previous frame (denoted as z), and the input of the search branch is the search area in the current frame (denoted as x). See Section 2.2 for details about the preprocessing of the input images. For convenience, we denote the output feature maps of the template and search branches as $\varphi(z)$ and $\varphi(x)$.

2.1.2. GA-SubNet

After obtaining the feature maps, we established a GA-SubNet to suppress false alarms and improve the tracking accuracy. The specific architecture of GA-SubNet is shown in Figure 2, including anchor location prediction, anchor shape prediction, and feature adaptation. In the following, we introduce the three modules of GA-SubNet in detail in the order of implementation.

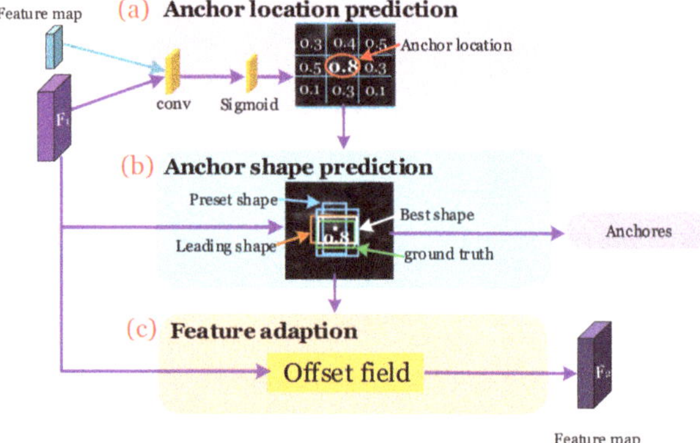

Figure 2. The architecture of GA-SubNet: (**a**) anchor location prediction module generates the sparse location of anchors; (**b**) anchor shape prediction module generates the anchor shape that better conforms to the shape of the shadow; (**c**) feature adaptation module generates a new feature map for the best anchor shape.

The purple region in Figure 2a is the anchor location prediction, i.e., the prediction of the location of the anchor containing the center point of a shadow. First, the input to GA-SubNet is two feature maps, one for the template (marked with a blue cube) and the other for the search area (marked with a purple cube). To obtain the prior information of the template that is independent of the training data, the feature map of the template is used as the kernel to convolute the feature map of search area F_1, so that the score of each location of the output represents the probability that the corresponding location is predicted to be the shadow. Then, the sigmoid function is used to obtain the probability map as shown in the blue box in Figure 2a. After this, the position whose probability exceeds the preset threshold is chosen to be the location of the predicted anchor (marked with a red circle). To learn more information about the shadow, similar to [17], the empirical threshold was chosen as 0.7.

The blue region in Figure 2b is the anchor shape prediction, i.e., the prediction of the anchor shape that better conforms to the shape of a shadow. First, the uniform arbitrary preset anchor shapes are generated (marked with blue boxes) at each location obtained from the anchor location prediction; i.e., several anchor shapes are arbitrarily set at each location, but the anchor shape setting in sparse locations is uniform. The preset anchor shape with the largest IoU with the shadow's ground truth (marked with a green box) is predicted as the leading shape (marked with an orange box). IoU is defined by Equation (1), where P denotes the preset anchor shapes, and G denotes the shadow's ground truth.

$$\text{IoU} = \frac{\text{area}(P \cap G)}{\text{area}(P \cup G)} \tag{1}$$

The leading shape of the anchor is still set arbitrarily and may differ significantly from the shadow's ground truth. To make the IoU larger, the offset between the leading shape and the shadow's ground truth at each location is calculated. After continuously optimizing the offsets using the loss function (described in Section 3.3), the best anchor shape can be obtained (marked with a white box), which better conforms to the shape of the shadow.

The orange region in Figure 2c is the feature adaptation, i.e., the adaptation of the feature map and the SAR image. Because the feature map is obtained by multi-layer convolution of the SAR image, there is a certain correspondence between the feature map and the SAR image; i.e., the leading shape of the anchor in the SAR image corresponds to a specific region in the feature map. However, the leading shape of the anchor at each location is optimized adaptively in the anchor shape prediction, resulting in areas with the same shape in the feature map, corresponding to the areas with different shapes in the SAR image. Therefore, feature adaptation is necessary to satisfy the correspondence between the feature map and the SAR image to ensure the accuracy of tracking. First, 1×1 convolution is used to calculate the offset between the leading shape and the best shape. Then, 3×3 deformed convolution is applied [23,24] based on this offset to the original feature map F_1 of the search area. Finally, the feature map F_2 is obtained for adaptation to the SAR image for the best anchor shape.

2.1.3. Similarity Learning Subnetwork

After obtaining the sparse anchors that better conform to the shadows' shape, the similarity learning subnetwork (marked with a yellow region in Figure 1) is used for classification and regression. The similarity learning subnetwork consists of a classification branch (marked with an orange box in Figure 1) for distinguishing the shadow from the background and a regression branch (marked with a blue box in Figure 1) for predicting the location and shape of the shadow. First, in both branches, to reduce the calculation complexity for subsequent similarity learning, a feature map 6×6 of $\varphi(z)$ is reduced to 4×4 and a feature map 22×22 of $\varphi(x)$ is reduced to 20×20 by using the convolutions (marked with yellow cubes in Figure 1). In addition, the channel of $\varphi(z)$ is adjusted to $2k \times 256$ for the foreground and background classification in the classification branch. The channel of $\varphi(z)$ is adjusted to $4k \times 256$ for determining the location and shape of the shadow in the regression branch. k is the number of anchors, $2k$ represents the probability of the foreground and background for each anchor, and $4k$ represents the location (x, y) and shape (w, h) of the shadow.

$$\begin{aligned} A^{cls}_{w \times h \times 2k} &= [\varphi(x)]_{\text{cls}} \otimes [\varphi(z)]_{\text{cls}} \\ A^{reg}_{w \times h \times 4k} &= [\varphi(x)]_{\text{reg}} \otimes [\varphi(z)]_{\text{reg}} \end{aligned} \tag{2}$$

As shown in Equation (2), the similarity learning subnetwork applies pairwise correlations (marked with red rectangles in Figure 1) to calculate the similarity metric, in which the similarity map $A^{cls}_{w \times h \times 2k}$ is for classification and $A^{reg}_{w \times h \times 4k}$ is for regression. $[\cdot]_{\text{cls}}$ and $[\cdot]_{\text{reg}}$ represent the classification and regression, respectively, and $\otimes$ denotes the convolution operation. We show the feature composition of $A^{cls}_{w \times h \times 2k}$ and $A^{reg}_{w \times h \times 4k}$ in Figure 3. $A^{cls}_{w \times h \times 2k}$ is divided into k groups, and each group contains two feature maps, which indicate the foreground and background probabilities of the corresponding anchors. The anchor is the foreground if the probability of the foreground is higher; otherwise, it is the background. Similarly, $A^{reg}_{w \times h \times 4k}$ is divided into k groups, and each group contains four feature maps (x, y, w, and h), which indicate the similarity metric between the corresponding anchor and the template. According to the highest similarity, the optimal location and the shape of the shadow are obtained.

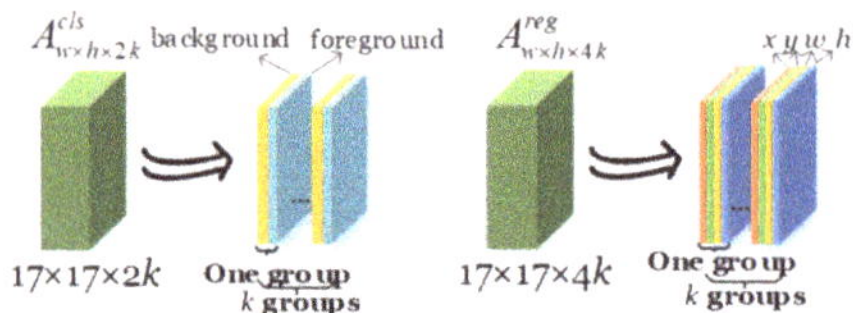

Figure 3. The feature composition of $A^{cls}_{w \times h \times 2k}$ and $A^{reg}_{w \times h \times 4k}$.

2.2. Preprocessing

For all images of Video-SAR to have the same feature dimensions, preprocessing is required before entering GASN. As shown in Figure 4, the input of GASN is a pair of adjacent images in the SAR image sequence. The shadow template is a 127×127 area centered on the center (x, y) of the shadow in frame t-1. Similar to the image preprocessing in [17], we cropped an $((w + h) \times 0.5 + w, (w + h) \times 0.5 + h)$ area in frame t-1 centered on (x, y) and then resized it to 127×127, where (w, h) is the boundary of the shadow. Here, $(x, y, w,$ and $h)$ are known in the training stage, while in the testing stage, the parameters represent the prediction results of the previous frame. Because the template size of all existing methods is 127×127 [17,18], to ensure the rationality of the comparison, we chose 127×127 as the template size. The search area is centered on the center of the shadow in frame t, and we cropped an $(((w + h) \times 0.5 + w) \times 255/127, ((w + h) \times 0.5) + h) \times 255/127)$ area and then resized it to 255×255. This area is larger than the shadow's template to ensure that the shadow is always included in the search area.

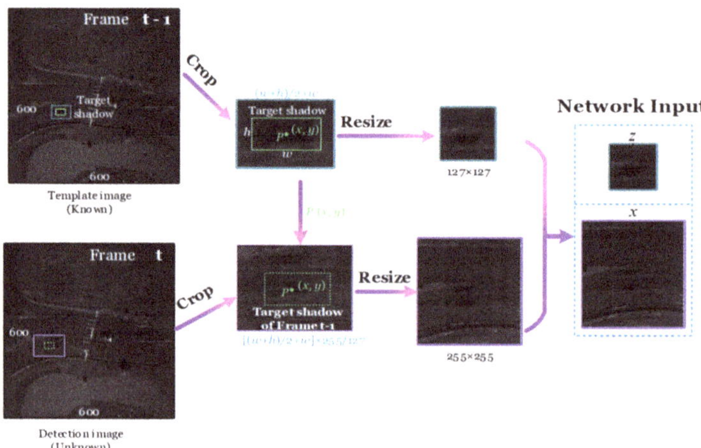

Figure 4. The input preprocessing of GASN.

2.3. Tracking Process

The whole process of TOI tracking based on GASN is shown in Figure 5. The details are as follows.

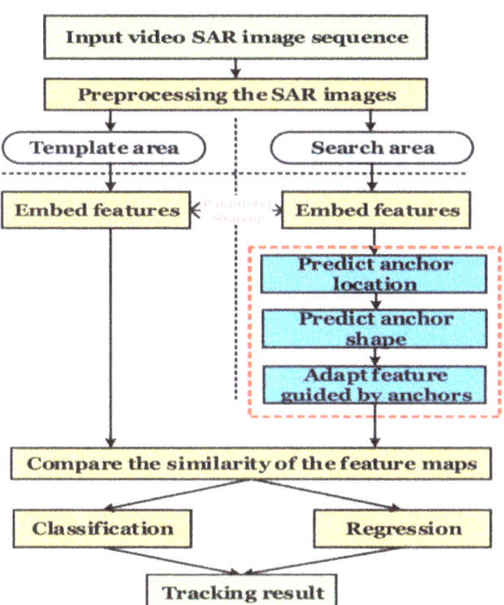

Figure 5. The whole process of arbitrary TOI tracking based on GASN.

Step 1: Input Video-SAR image sequence.

As shown in Figure 6a, N is the number of frames of the input video. For easy observation, we marked the shadow to be tracked with a green box.

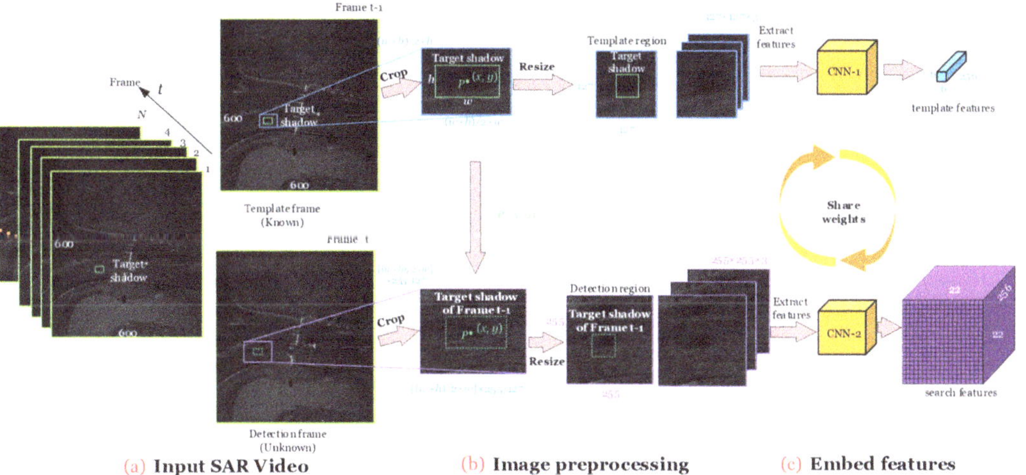

Figure 6. Image preprocessing and feature embedding. Input SAR Video (**a**), preprocess image (**b**), embed features (**c**).

Step 2: Preprocessing SAR images.

For all images of Video-SAR to have the same feature dimensions, we need to crop and resize them. As described in Section 2.2, the shadow in frame t-1 is resized to 127×127 as the template, and frame t is resized to 255×255 as the search area, as shown in Figure 6b. x, y, w, and h represent the center and boundary of the prediction results in the previous frame. Unlike the RGB three-channel optical images, the SAR images are gray; therefore, all three channels are assigned to the same gray value to use the pre-trained weights. Applying models trained on three-channel RGB images to one-channel radar images has been carried out in several published literatures [10,12,15], and the results in Section 5.3 show that it is reasonable to do so.

Step 3: Embed features by the Siamese subnetwork.

After obtaining the template and search areas, the Siamese subnetwork embeds features to better describe the TOI. The Siamese subnetwork is divided into a template branch and a search branch, and the parameters of CNN-1 and CNN-2 in the two branches are shared to ensure the consistency of the features. The template branch outputs $6 \times 6 \times 256$ as the feature map of the template, and the search branch outputs $22 \times 22 \times 256$ as the feature map of the search area, which are shown in Figure 6c.

Step 4: Predict anchor location.

After obtaining the feature maps of the template and the search area, the predict anchor location module pre-determines the general location of the TOI in the search area to suppress false alarms. To only locate the anchors containing the center point of the shadow, the feature map of the template is used to convolute the feature map of the search area to obtain the prior information of the template, so that the score of each location of the output feature map represents the probability that the corresponding location is predicted to be the shadow. Then, the locations whose probability exceeds the preset threshold are used as the locations of the sparse anchors. As shown in Figure 7, the blue regions correspond to the locations of the anchors.

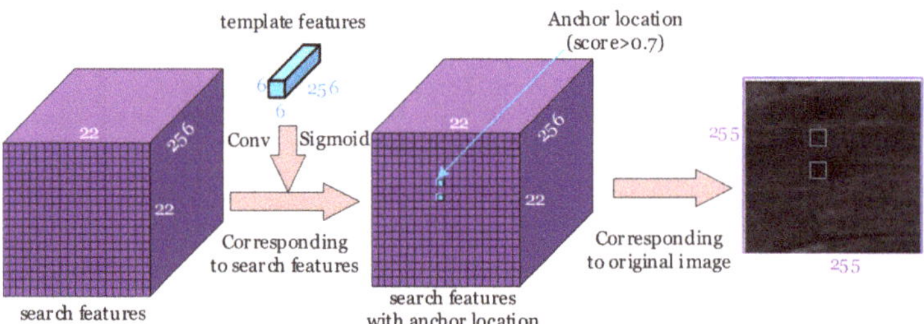

Figure 7. Predicting the anchor location.

Step 5: Predict anchor shape.

To generate the anchor that conforms to the shadow's shape, the anchor shape prediction module generates an anchor shape with the highest coverage of the real shadow's shape by adaptive prediction processing in the sparse locations. First, after anchor generation, the preset anchor shapes (marked with blue boxes in Figure 8) of the anchor are obtained. Among them, the shape with the largest IoU with the shadow's ground truth (marked with a green box) is predicted as the leading shape (marked with an orange box). After this, the leading shape of the anchor is regressed to obtain the best anchor shape (marked with a white box) that better conforms to the shadow's shape.

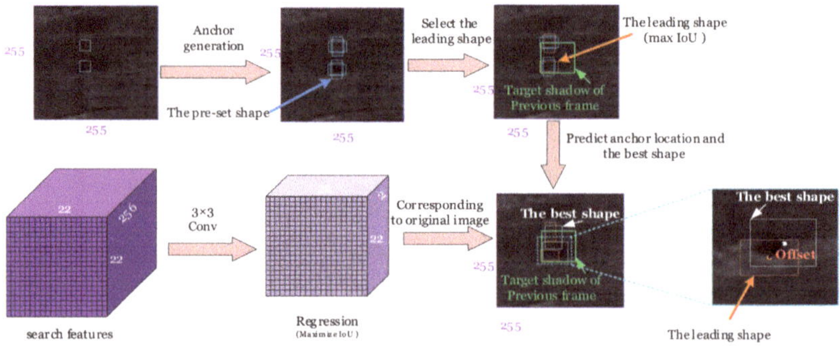

Figure 8. Predict anchor shape.

Step 6: Adapt the feature map guided by anchors.

After the anchor shape prediction, the anchor shape changes, and the feature map needs to be adapted to guarantee the correct corresponding relationship between the feature map and the SAR images. As described in Section 2.1.2, the adapted feature map can be generated by compensating the offset obtained from 1×1 convolution using the 3×3 deformable convolution. Based on the adapted feature map shown in Figure 9 (marked with a dark purple), the higher quality anchors can be used for shadow tracking.

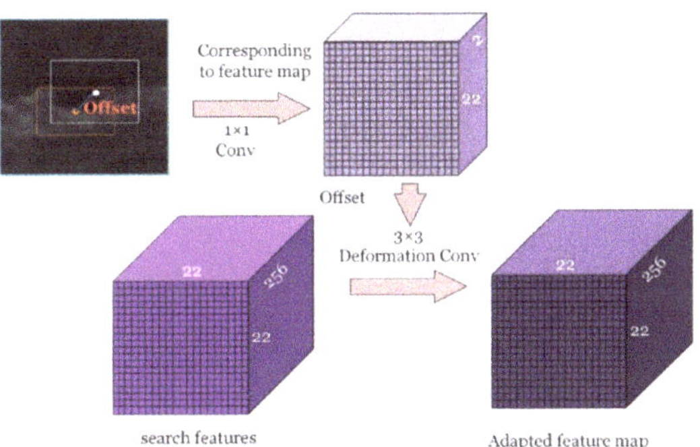

Figure 9. Adapting the feature map guided by anchors.

Step 7: Compare the similarity of the feature maps.

To compare the similarity of the feature map of the search area and the template, the similarity learning subnetwork applies the correlation operation as shown in Figure 10a. The blue cube represents the feature map of the template, and the purple cube represents the feature map of the search area. The feature map of the template changes its channel by the convolution according to the number of anchors k. The correlation can be achieved using the feature map of the template to convolute the feature map of the search area; then, $A^{\mathrm{cls}}_{w \times h \times 2k}$ and $A^{\mathrm{reg}}_{w \times h \times 4k}$ are output, where $2k$ represents the probability of the foreground and background for each anchor, and $4k$ represents the location (x, y) and shape (w, h) of the shadow.

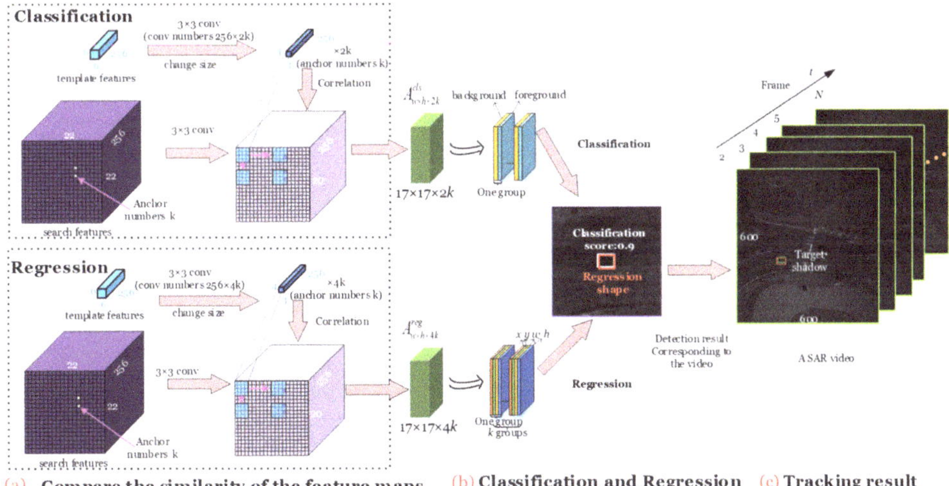

(a) **Compare the similarity of the feature maps** (b) **Classification and Regression** (c) **Tracking result**

Figure 10. The tracking results obtained after comparing the similarity; comparison of the similarity of the feature maps (**a**), classification and regression (**b**), and tracking results (**c**).

Step 8: Classification and regression.

The similarity learning subnetwork is divided into classification and regression branches. In the classification branch, the similarity learning probability map of the foreground and background is obtained, and then the foreground anchor with the highest similarity learning metric is the tracking shadow. The regression branch further regresses the best anchor shape (marked with a white box) to achieve a more accurate shadow shape (marked with a red box) in Figure 10b. Using the trained GASN, the shadow tracking in the Video-SAR image sequence can be achieved only using the shadow's location and shape in the first frame.

Step 9: Tracking results.

As shown in Figure 10c, after searching the whole Video-SAR image sequence, the shadow, i.e., the TOI tracking of Video-SAR, is realized. Because the shadow's location in the first frame is known, only the tracking results of the subsequent frames are shown here, where the green box represents the real location of the shadow, and the red box represents the tracking results.

To make the tracking process easier to read, it is shown in the Algorithm 1 below.

Algorithm 1: GASN tracks arbitrary TOI in Video-SAR

Input: Video-SAR images sequence.
  Begin
1 **do** Pre-process the SAR images.
2     $127 \times 127 \times 3 \leftarrow$ template image, $255 \times 255 \times 3 \leftarrow$ search image
3 **do** Embed features by Siamese subnetwork.
4     $6 \times 6 \times 256 \leftarrow 127 \times 127 \times 3 \otimes$ CNN-1, $22 \times 22 \times 256 \leftarrow 255 \times 255 \times 3 \otimes$ CNN-2
5 **do** Predict anchor location.
6   score $\leftarrow F_1 \otimes 1 \times 1$conv, probabilitymap $\leftarrow$ score $\otimes$ sigmoid
      location $\leftarrow$ probabilitymap > 0.7
7 **do** Predict anchor shape.
8     $\text{IoU} = \frac{\text{area}(P \cap G)}{\text{area}(P \cup G)}$, shape $\leftarrow \max(\text{IoU})$
9 **do** Adapt the feature map guided by anchors.
10   offset $\leftarrow F_1 \otimes 1 \times 1$conv, $F_2 \leftarrow F_1 \otimes 3 \times 3$deformabelconvbased on offset
11 **do** Compare the similarity of the feature maps.
12 $4 \times 4 \times 256 \times 2k \leftarrow 6 \times 6 \times 256, 4 \times 4 \times 256 \times 4k \leftarrow 6 \times 6 \times 256, 20 \times 20 \times 256 \leftarrow 22 \times 22 \times 256$
    $17 \times 17 \times 2k \leftarrow 4 \times 4 \times 256 \times 2k \otimes 20 \times 20 \times 256$
    $17 \times 17 \times 4k \leftarrow 4 \times 4 \times 256 \times 4k \otimes 20 \times 20 \times 256$
13 **do** Classification and regression.
14   Classification $\leftarrow \max(17 \times 17 \times 2k)$, Regression $\leftarrow \max(17 \times 17 \times 4k)$
  End

Output: Tracking results.

---

3. Experiments

All of the experiments were implemented on a personal computer with an Intel Core i7-8700K CPU@3.40 and an NVIDIA GTX1080 graphics card with 8 GB of memory. The software experiment environment was Linux, Ubuntu 16.04, python 3.7, and Pytorch3.0.

3.1. Experimental Data

As existing recognized real Video-SAR data, due to the high resolution, the data of SNL [1] have been used by many scholars for moving target detection and tracking [7–10]. In our experiments, we used both the simulated and real data to verify the effectiveness of GASN for arbitrary TOI tracking in Video-SAR. We produced the simulated Video-SAR data from the echo to approximate real SAR images, and the details of the data are described below.

In the simulated Video-SAR data, two real SAR backgrounds containing roads and six moving targets were simulated, considering the generality. The radar system parameters and the velocity of the moving targets are listed in Tables 1 and 2. Regarding the simulation of the shadow, the scattering coefficient was set to zero because of no reflection. In the experiment, 17 videos were simulated, where 11 videos were utilized for training and 6 for testing. Each video contained 61 frames, and one of the test video sequences is shown in Figure 11. The size of all images was 600 × 600.

The real Video-SAR SNL data contained 50 different moving targets in all 899 frames. When GASN was used for arbitrary TOI tracking, 751 frames with the former 35 targets were set for training, and 148 frames with the latter 15 targets were set for testing. The size of all images was 600 × 600. Compared to the simulated data, there was more noise and clutter in the real Video-SAR data, and the tracking results with clutter are shown in Section 4.2.2.

Table 1. The system parameters of simulated Video-SAR.

Parameter	Value
Center frequency/GHz	35
Platform velocity/m s^{-1}	300
Platform height/m	8000
Pulse repetition frequency/Hz	4000
Total record time/s	10
SNR	40 dB

Table 2. The velocity of the moving targets in the simulated Video-SAR data.

Target	Azimuth Velocity (m s^{-1})	Radial Velocity (m s^{-1})
T1	6	−8
T2	−1	−2
T3	1.5	−3
T4	0.4	−0.8
T5	3	1.5
T6	1.5	−1.5

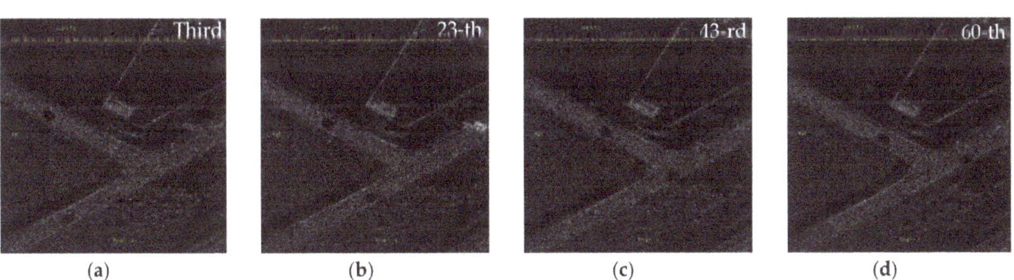

Figure 11. Image sequence of a test video: (**a**) third frame in Video 1; (**b**) 23rd frame in Video 1; (**c**) 43rd frame in Video 1; (**d**) 60th frame in Video 1.

3.2. Implementation Details

To avoid over-fitting, the pre-trained weight of ResNet50 [25] was applied, which was successfully trained from the widely used ImageNet large-scale visual recognition challenge (ILSVRC) data set [26]. Unlike the three-channel RGB for optical images, the SAR images were all gray; therefore, we assigned all three channels to the same gray value to use the pre-trained weights. Due to the limited memory, only conv4 and the upper layers of the pre-trained network weights were fine-tuned for adaptation to the TOI tracking task in Video-SAR. During the training stage, the batch size was four, and the stochastic gradient descent (SGD) [27] was applied, in which the momentum was 0.9, the weight decay was 0.0005, and the learning rate was 0.0001.

Data augmentation techniques were used in our implementation, including translation, scale transformations, blur, and flip. After data augmentation, the amount of data expanded by approximately 10 times, which can better fine-tune the model.

3.3. Loss Function

As shadow occupies a small proportion of the SAR image, we used focal loss [28] as the anchor location loss $loss_{loc}$ to predict the anchor location:

$$loss_{loc} = -(1-p)^{\gamma} \log(p) \tag{3}$$

where p is the probability of the shadow in the location, and $\gamma = 2$ is the hyper-parameter to adjust the drop speed influenced by [29].

Anchor shape loss $loss_{shape}$ uses a smooth $L1$ loss inspired by [12].

$$loss_{shapes} = smoothL_1(1 - \min(\frac{w}{w_g}, \frac{w_g}{w})) + smoothL_1(1 - \min(\frac{h}{h_g}, \frac{h_g}{h})) \tag{4}$$

$$smoothL_1 = \begin{cases} 0.5x & |x| < 1 \\ |x| - 0.5 & otherwise \end{cases} \tag{5}$$

where (w_g, h_g) is the ground truth of the shadow, and (w, h) is the shape of the anchor.

As per Siamese-RPN [17], classification loss $loss_{cls}$ and regression loss $loss_{reg}$ are as follows:

$$loss_{cls} = -\log[p_i{}^* p_i + (1 - p_i^*)(1 - p_i)] \tag{6}$$

$$loss_{reg} = smoothL_1(t_i - t_i^*) \tag{7}$$

where p_i represents the probability of shadow, t represents the ground truth of the center point (x, y) and shape (w, h) of the shadow, and * represents the prediction result.

The total loss function is shown below, where $\lambda_1 = \lambda_2 = 5$ and $\lambda_3 = \lambda_4 = 2$ are the hyper-parameters balancing the four parts.

$$loss = \lambda_1 loss_{loc} + \lambda_2 loss_{shape} + \lambda_3 loss_{cls} + \lambda_4 loss_{reg} \tag{8}$$

By minimizing the loss functions, GASN finally achieves parameter optimization after the iterations.

3.4. Evaluation indicators

To verify the performance of GASN, three general evaluation indicators were used in this paper.

3.4.1. Tracking Accuracy

The expected average overlap (EAO) can represent the tracking accuracy [30], and the greater the EAO, the more accurate the tracking result. EAO is defined as follows:

$$EAO = \frac{\sum_{j=1}^{N_s} mIoU(j)}{N_s}, mIoU = \frac{\sum_{i=1}^{N} IoU(P_i, G)}{N} \tag{9}$$

where IoU is as defined in Equation (1), P is the tracking result, G is the shadow's ground truth, N is the number of images in the Video-SAR sequence, and N_s is the number of videos in the test data. We calculated mIoU, including IoU = 0; therefore, EAO can truly reflect the tracking accuracy.

3.4.2. Tracking Stability

The central location error (CLE) reflects the stability of the tracking method [15]; i.e., the smaller the CLE, the more stable the tracking method, and the CLE is defined as follows:

$$CLE = \sqrt{(x_R - x_G)^2 + (y_R - y_G)^2} \tag{10}$$

where (x_R, y_R) represents the central location of the tracking result, and (x_G, y_G) represents the central location of the shadow's ground truth.

3.4.3. Tracking Speed

The frames per second (FPS) represent the tracking speed, which is defined as follows:

$$FPS = \frac{N}{t} \tag{11}$$

where t represents the total tracking time, and N is the number of images in the Video-SAR sequence.

4. Results

4.1. Results of the Simulated Video-SAR Data

Figure 12 shows the tracking results of the simulated Video-SAR data. In the rest of this paper, the red box represents the tracking results, and the green box represents the ground truths of the shadow. It can be seen that the red and green boxes have a great overlap, which means that GASN can track the target effectively.

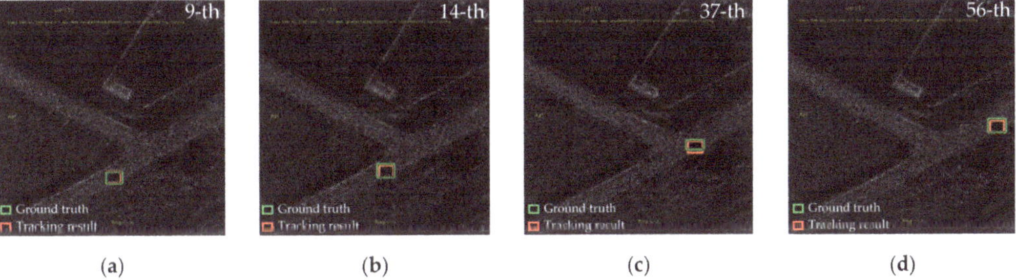

Figure 12. Tracking results of the simulated Video-SAR data: (**a**) 9th frame in Video 2; (**b**) 14th frame in Video 2; (**c**) 37th frame in Video 2; (**d**) 56th frame in Video 2.

We quantitatively analyzed the tracking results of GASN. Because Siamese-FC and Siamese-RPN significantly outperform MOSSE [19] and KCF [20], only the visual comparison results of GASN with Siamese-RPN and Siamese-FC in terms of accuracy, CLE, and FPS indicators are shown.

4.1.1. Comparison with Other Tracking Methods

Figures 13–15 show the results of comparing GASN to Siamese-RPN and Siamese-FC on the six test videos. In the comparative experiments, we retrained Siamese-FC and Siamese-RPN using the same simulated data, and both networks were tuned. Moreover, to ensure the rationality of the experiments, our comparative experiments were all performed under the same conditions, such as the data preprocessing, the hard and soft platforms, and the training mechanism. From the results, we can see that GASN (marked with purple) obtained the highest mIoU (Figure 13) and the lowest CLE (Figure 14) on each video. Moreover, the FPS (Figure 15) of GASN (marked with purple) was almost the same as that of Siamese-RPN (marked with green), which indicates that GASN has almost no speed loss at a higher accuracy. Due to the above phenomenon also applying to real data, we explain the reason in detail in the next section. To reveal the performance of GASN more intuitively, we calculated the average tracking performance of the six testing videos, and the results are shown in Table 3.

In Table 3, for the two traditional methods (MOSSE and KCF), their simple framework leads to two different implications. On the one hand, these methods require low computation (105 FPS for MOSSE and 58 FPS); on the other hand, the simple framework may cause the loss of some information, such as the edges and textures, resulting in the inability to track shadows that are too wide or too long, and, therefore, the accuracy is low (31.21% for MOSSE and 41.80% for KCF). As for the comparison between deep learning methods, the anchors generated by GA-SubNet can better conform to the shape of the shadow in SAR images. Therefore, the accuracy of GASN (60.16%) is better than that of Siamese-RPN (55.61%) or Siamese-FC (43.67%). As for the tracking speed, GASN also slightly improved (32 FPS) compared to Siamese-RPN (31 FPS), because the anchors generated by GA-SubNet are sparse. In addition, GASN achieved the lowest CLE score (6.68) when considering the stability, because GA-SubNet generates anchors based on the probability of the shadow's location. Through the above analysis, we can see that the tracking performance of GASN is better than that of the other methods.

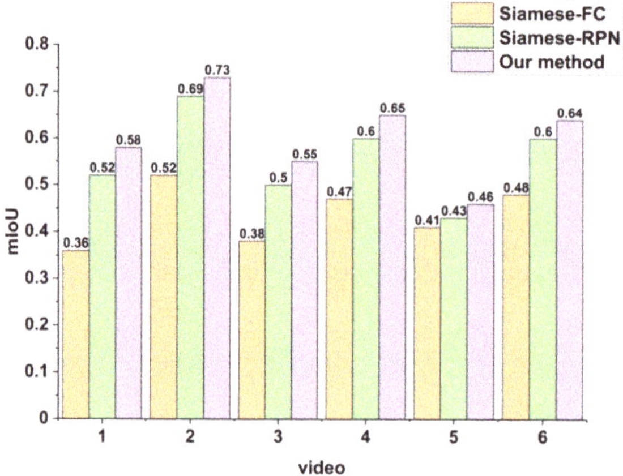

Figure 13. The comparison results of GASN with Siamese-RPN and Siamese-FC on accuracy.

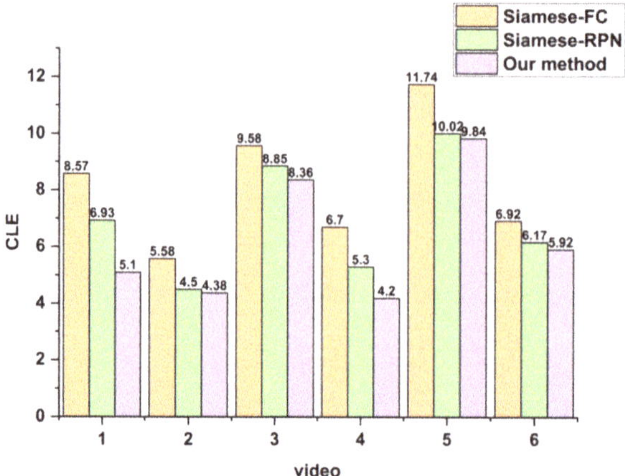

Figure 14. The comparison results of GASN with Siamese-RPN and Siamese-FC on CLE.

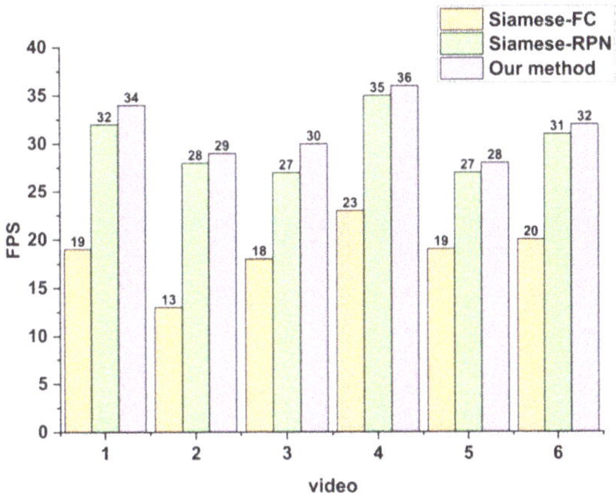

Figure 15. The comparison results of GASN with Siamese-RPN and Siamese-FC on FPS.

Table 3. Average tracking performance of simulated Video-SAR data.

Method	Accuracy	CLE	FPS
MOSSE	31.21%	19.76	105
KCF	41.80%	11.30	58
Siamese-FC	43.67%	8.46	19
Siamese-RPN	55.61%	7.94	31
GASN (ours)	60.16%	6.68	32

4.1.2. Tracking Results with Distractors

To verify that the proposed method only tracks the TOI, we selected two adjacent targets with similar shapes for tracking. Figure 16a,b show the tracking results in the same frame. TOI-2 can be considered a distractor when we want to track TOI-1 in Figure 16a. Similarly, TOI-1 can be considered a distractor when we want to track TOI-2 in Figure 16b. The green box represents the ground truth of the TOI, and the red box represents the tracking results using the proposed method. The overlap between the red and green boxes in both figures is greater than 50%, so the proposed method can accurately track the TOI without errors. The main reasons are as follows: GASN uses the Siamese subnetwork to extract multi-level and more expressive features compared to the traditional methods. In addition, compared to the existing deep learning methods, GASN uses GA-SubNet to provide the general location and shape of the TOI based on the template, which can effectively suppress the distractors. Through the above analysis, we think that the proposed method can accurately track the TOI without errors, although there are distractors in the scene.

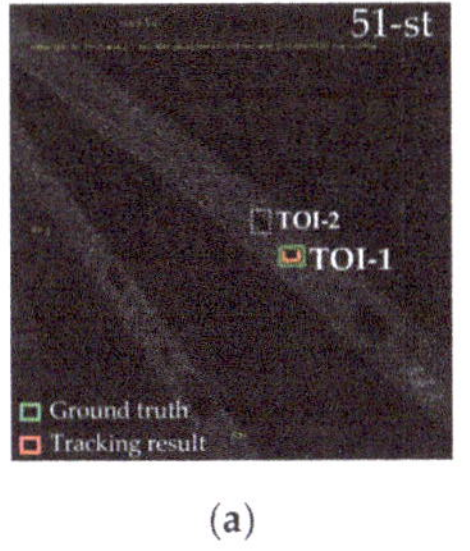

(a)

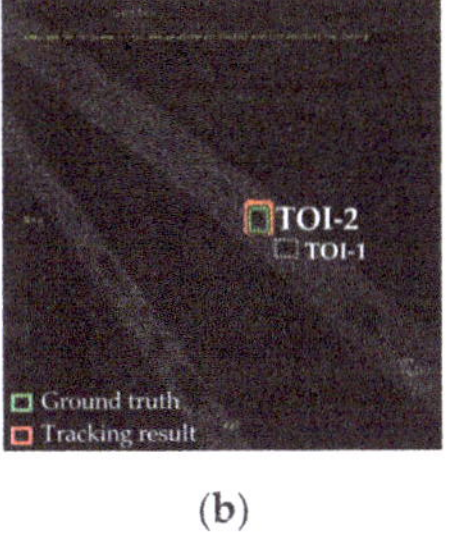

(b)

Figure 16. Tracking results with distractors: (a) TOI-1 in the 51st frame of Video 6; (b) TOI-2 in the 51st frame of Video 6.

4.1.3. Tracking Results of the Target with a Specific Speed

To verify the tracking capability of the proposed method for TOI with a specific speed, we simulated two identical targets, except for the velocity. Figure 17a,b show the tracking results in the same frame. The azimuth velocity of TOI-1 in Figure 17a is 2 m/s and the radial velocity is –2.5m/s, while the azimuth velocity of TOI-2 in Figure 17b is 1.5 m/s and the radial velocity is –1.5m/s. The green box represents the ground truth of the TOI in this tracking process, and the red box represents the tracking result using the proposed method. The overlap between the red and green boxes in both figures is greater than 50%. Therefore, it can be seen that the proposed method can accurately track the TOI with a specific speed.

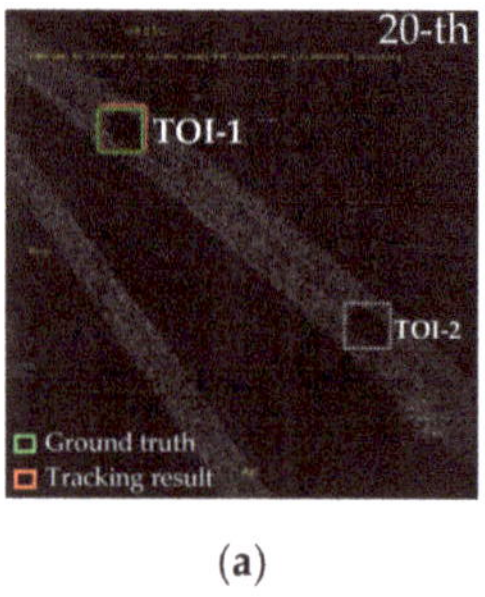

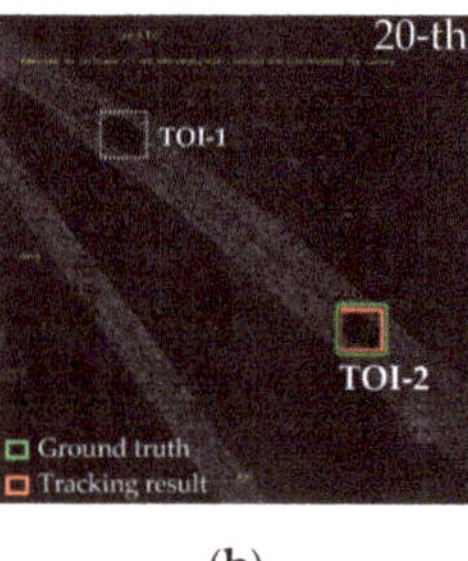

Figure 17. Tracking results of the target with a specific speed: (**a**) TOI-1 in the 20th frame of Video 5; (**b**) TOI-2 in the 20th frame of Video 5.

4.2. Results of Real Video-SAR Data

Figure 18 shows the tracking results using the real Video-SAR data, aiming to verify the effectiveness of GASN using real data. It can be seen that the tracking results (marked with a red box) and the ground truths of the shadow (marked with a green box) have a great overlap (the IoU is greater than 50%), which means that GASN can track the real shadow effectively.

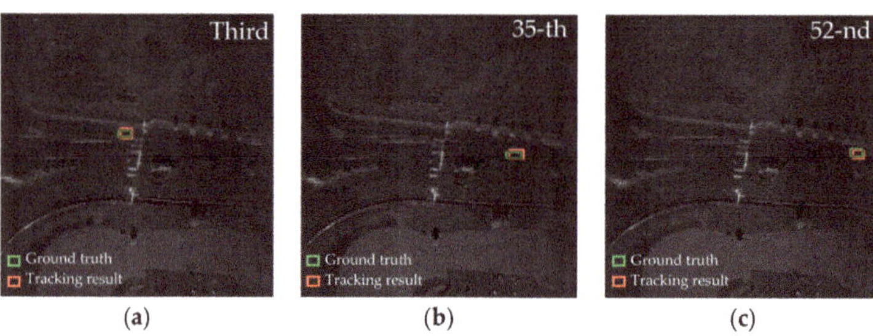

Figure 18. Tracking results of the real Video-SAR data: (**a**) third frame in Video 2; (**b**) 35th frame in Video 2; (**c**) 52nd frame in Video 2.

4.2.1. Comparison with Other Tracking Methods

In the comparative experiments, with the same training mechanism as GASN, we first initialized Siamese-FC and Siamese-RPN using the pre-trained model parameters obtained from the optical image. Then, we adjusted the model parameters using SAR images for tracking in Video-SAR. Moreover, to ensure the rationality of the experiments, our comparative experiments were all performed under the same conditions, such as the data preprocessing and the hard and soft platforms.

Figure 19 shows the accuracy comparison results of the three methods. Siamese-FC (marked with yellow) had the lowest accuracy in each video because it cannot fit the scale transformation of the shadow. For Siamese-RPN (marked with green), the accuracy improved somewhat, because the anchors can handle scale transformation. However, most preset anchors do not perfectly fit the actual shape of the shadow, which results in failure when tracking shadows that are too long or too wide. For GASN (marked with purple), GA-SubNet only locates the anchors containing the center of the shadow to suppress false alarms. GA-SubNet adaptively refines the shape of the anchor to better fit the shadow's shape for further improvement of the tracking accuracy. Therefore, it is obvious that the accuracy of GASN is higher than that of Siamese-RPN and Siamese-FC in Figure 19.

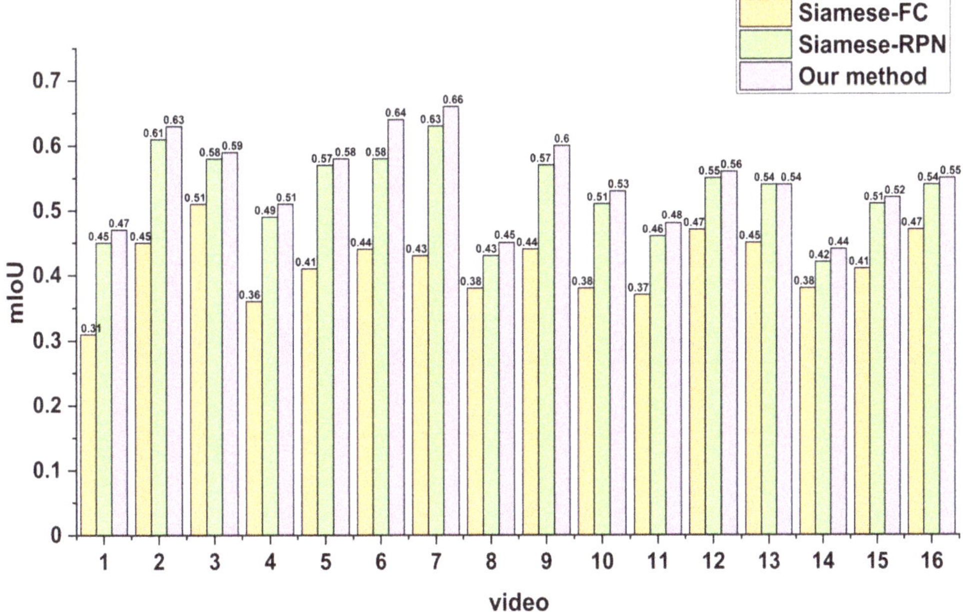

Figure 19. Accuracy comparison of the three methods.

To validate the stability of GASN, we used CLE to compare GASN to Siamese-RPN and Siamese-FC. While GA-SubNet only locates the anchors containing the center of the shadow in advance, GASN can locate the center of the shadow more accurately. As shown in Figure 20, the CLE of GASN (marked with purple) is less than that of Siamese-RPN (marked with green) and Siamese-FC (marked with yellow), which means that TOI tracking using GASN is the most stable.

To validate the speed of GASN, we used FPS to compare GASN to Siamese-RPN and Siamese-FC. Figure 21 shows the comparison results of FPS, from which we can see that GASN (marked with purple) is almost identical to Siamese-RPN (marked with green), while Siamese-FC (marked with yellow) is lower. To the best of our knowledge, Siamese-RPN can satisfy real-time tracking [17]. Compared to Siamese-RPN, on the one hand, GASN needs to calculate the location and shape of the anchors, which reduces the tracking speed. On the other hand, the anchors are sparse, which reduces the computation of subsequent processing. It can be seen from the experimental results that the FPS of GASN is almost the same as that of Siamese-RPN; therefore, our method can achieve real-time tracking.

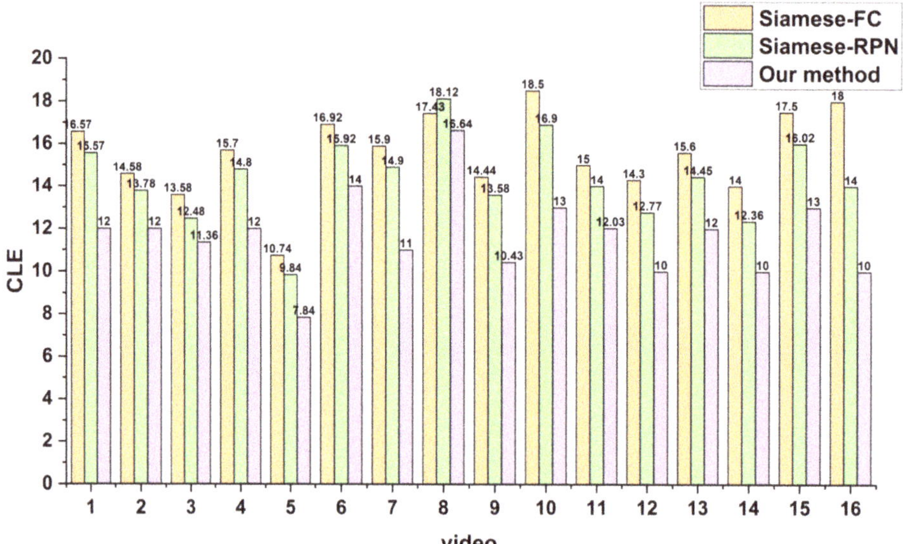

Figure 20. CLE comparison of the three methods.

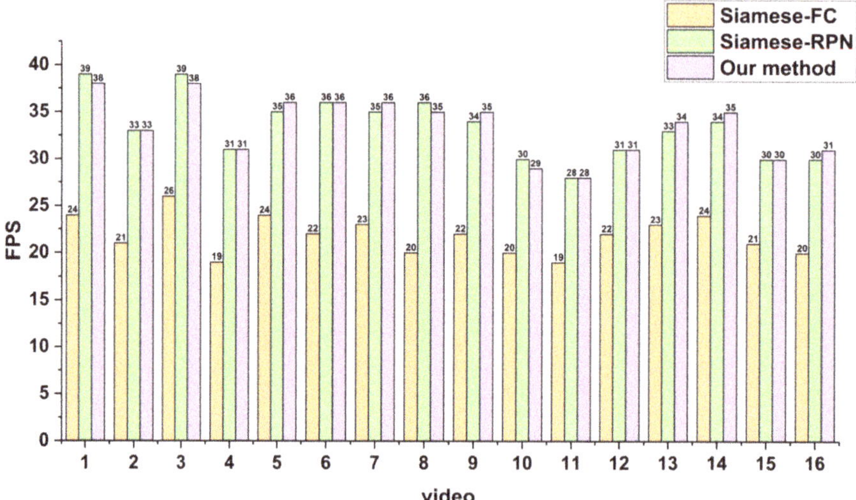

Figure 21. FPS comparison of the three methods.

Table 4 shows the average tracking performance of the real Video-SAR data using the different methods. Due to the simple framework, MOSSE has the lowest performance, with 29.64% accuracy and 37.64 CLE, but the highest speed (125 FPS). Moreover, the deep learning methods improved the accuracy over the traditional correlation filtering methods (MOSSE and KCF), because the networks can extract multi-level and more expressive features. Most importantly, GA-SubNet in GASN only locates the sparse anchors containing the center of the shadow to suppress false alarms. Additionally, GA-SubNet refines the anchor's shape to conform to the shape of the shadow, which further improves the tracking accuracy. Therefore, the accuracy of GASN (54.68%) is better than that of Siamese-RPN (52.75%) and Siamese-FC (41.60%). In addition, because the sparse anchors can reduce the subsequent computation, there is no speed loss in GASN (33 FPS) compared to Siamese-RPN (33 FPS). The above analysis shows that GASN has the highest accuracy (54.68%) without sacrificing speed.

Table 4. Average tracking performance of real Video-SAR data.

Method	Accuracy	CLE	FPS
MOSSE	29.64%	37.64	125
KCF	39.98%	18.79	54
Siamese-FC	41.60%	15.41	21
Siamese-RPN	52.75%	14.69	33
GASN (ours)	**54.68%**	**11.37**	**33**

4.2.2. Tracking Results with Clutter

To verify the suppression ability of the proposed method for clutter, we selected the videos with these two types of interference in the real data for tracking. Because Siamese-RPN has excellent performance in both accuracy and speed in optical tracking, and the proposed method is better than Siamese-RPN, making it applicable to Video-SAR, we compared the proposed method with Siamese-RPN, as shown in Figure 22. Figure 22a,b show the tracking results of the proposed GASN method and Siamese-RPN under background clutter (e.g., road signs), respectively, where the green boxes represent the ground truths of the TOI during this tracking process, and the red boxes represent the tracking results. The comparison clearly shows that the overlap between the tracking results (red) and the labels of the TOI (green) using the proposed GASN method is greater than 50%, while the overlap of Siamese-RPN is less than 30%. Figure 22c,d show the tracking results of the proposed method and Siamese-RPN under environmental clutter (e.g., imaging sidelobe), respectively, and it can be seen that the overlap between the tracking results (red) and the labels of the TOI (green) using the proposed method is higher than the results using Siamese-RPN. Therefore, we believe that the tracking accuracy of the proposed method is higher than that of Siamese-RPN in the presence of clutter.

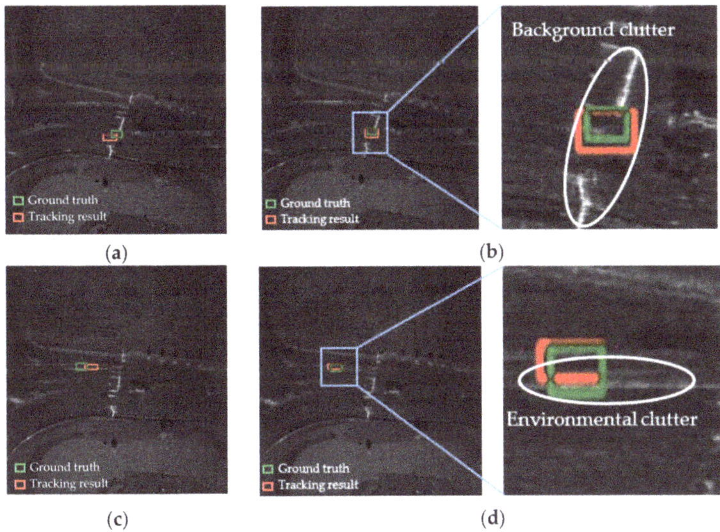

Figure 22. Tracking results with interference: (**a**) Siamese-RPN with background clutter; (**b**) our method with background clutter; (**c**) Siamese-RPN with environmental clutter; (**d**) our method with environmental clutter.

4.2.3. Tracking Results of Different Frame Rates

Figure 23 shows the tracking results of different frame rates. We created Video 16 from Video 15 at a frame rate of 6.4, noting that the frame rate here refers to the rate at which a video is divided into frames. For example, the frame rate of Videos 1–15 was 3.2, which means that an SAR image was captured every 1/3.2 s in the video. The parameters of Video 15 in Figure 23a and of Video 16 in Figure 23b are the same, except for the frame rate. It is obvious that the two boxes in Figure 23b have higher IoUs, i.e., more accurate tracking results. Only the comparison results for frame 5 are shown, showing that the results of almost all frames in Video 16 are more accurate than those of Video 15. The main reason is that the higher the frame rate, the smaller the change in the shadow's location and shape between the adjacent frames. Therefore, it is reasonable to assume that the frame rate is positively correlated with the tracking accuracy.

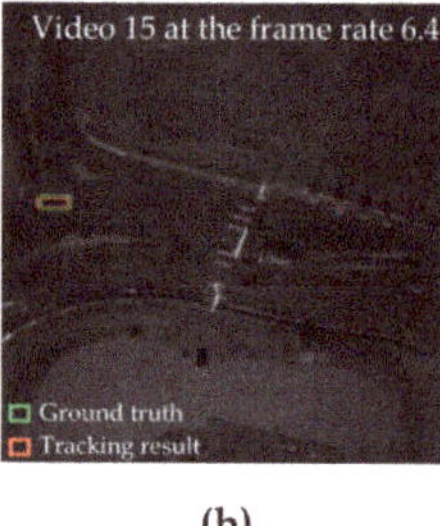

(a)   (b)

Figure 23. True tracking results of different frame rates: (**a**) Video 15 at a frame rate of 3.2; (**b**) Video 15 at a frame rate of 6.4 (Video 16).

4.2.4. Tracking Results of another Real Video-SAR Dataset

We conducted an additional experiment on a new dataset that is derived from [15]. Two videos containing 675 images were used to train the network, and two videos with 389 images were used to test the network. The size of all images was 1000×1000 pixels.

Figure 24 shows the tracking results of another real Video-SAR dataset, and Table 5 shows the average tracking performance. From Table 5, we can see that the accuracy of the proposed method is 1.33% higher than that of Siamese-RPN. Therefore, the proposed method is still more accurate than Siamese-RPN.

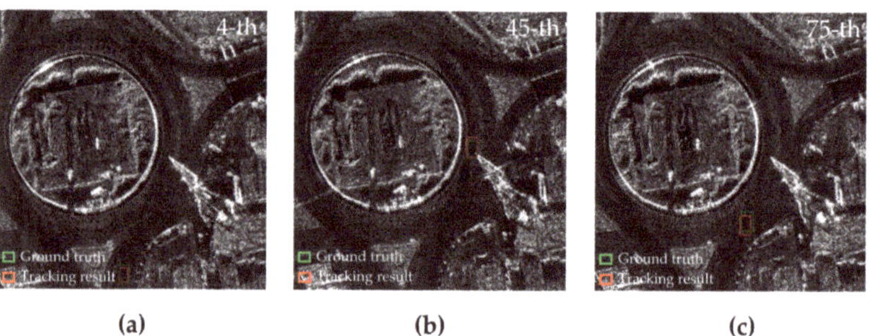

(a)   (b)   (c)

Figure 24. Tracking results of another real Video-SAR data: (**a**) 4th frame; (**b**) 45th frame; (**c**) 75th frame.

Table 5. Average tracking performance of another real Video-SAR data.

Method	Accuracy	CLE	FPS
MOSSE	30.70%	38.73	65
KCF	46.30%	19.03	58
Siamese-FC	51.70%	16.81	21
Siamese-RPN	53.68%	12.04	20
GASN (ours)	55.01%	11.78	19

5. Discussion

5.1. Research on the Transfer

We arranged a set of experiments to verify whether the proposed method entirely relies on the prior information of the TOI, such as the location and shape, rather than the appearance features of the training data. In the first experiment, we used the simulated data for training and the real data for testing, as shown in Figure 25a,b. In the second experiment, we used the real data for training and the simulated data for testing, as shown in Figure 25c,d. We can see that the tracking results (marked with red boxes) and the ground truths of the shadow (marked with green boxes) have a great overlap in the two experiments.

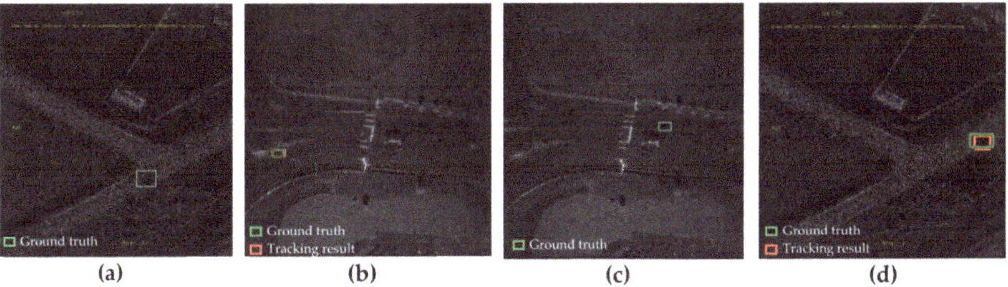

Figure 25. The experimental results of cross-validation: (**a**) simulated Video-SAR data for training; (**b**) real Video-SAR data for testing; (**c**) real Video-SAR data for training; (**d**) simulated Video-SAR data for testing.

To reveal the performance of GASN more intuitively, we evaluated the tracking results using accuracy, and the results are shown in Tables 6 and 7.

Table 6. Cross-validation for testing the simulated Video-SAR data.

Train Data	Test Data	Accuracy
Simulated	Simulated	60.16%
Real	Simulated	59.26%

Table 7. Cross-validation for testing the real Video-SAR data.

Train Data	Test Data	Accuracy
Real	Real	54.68%
Simulated	Real	53.38%

The first set of cross-validation experiments involved training with real data (data B) and testing with simulated data (data A). The results are shown in row 2 of Table 6. For comparison, we also provide the results of both the training and testing using simulated data (see row 1 of Table 6). The experimental results show that their accuracy differs by 0.9%.

The second set of cross-validation experiments involved training with simulation data (data A) and testing with real data (data B). The results are shown in row 2 of Table 7. For comparison, we also provide the results of both the training and testing using real data (row 1 of Table 7). The experimental results show that their accuracy differs by 1.3%.

From the above experiments, we can see that the results of the two cross-validation experiments have little difference in terms of accuracy, which indicates that GASN has good transfer ability.

The proposed GASN in this paper is capable of similarity learning. In other words, GASN is trained with a large number of training samples so that the network has the ability to measure the similarity of two input images (i.e., the template and the search image in the training data). The greater the similarity, the higher the output score of GASN. Therefore, once a template image of TOI is given, the information provided by the template (such as the location and shape) can be used to match the target in the next image based on the similarity measure capabilities of GASN. Then, the target with the highest similarity is determined as the tracking result in the next image. Therefore, GASN can track the TOI using the template information instead of the appearance features of the training data, so the proposed GASN is highly robust.

5.2. Ablation Experiment of GA-SubNet

We explored the effect of GA-SubNet on false alarms. Figure 26 shows the anchors on Siamese-RPN (Figure 26a) and GASN (Figure 26b). It can be seen that after adding GA-SubNet, the anchors are mainly concentrated around the TOI, and the number of anchors is also greatly reduced. Table 8 shows the comparison results of whether to add GA-SubNet or not. Because GA-SubNet discards the useless anchors in the background and improves the imbalance between positive and negative samples, the accuracy is improved by 4.52% after adding GA-SubNet. Therefore, GASN with GA-SubNet can better distinguish the TOI from the background.

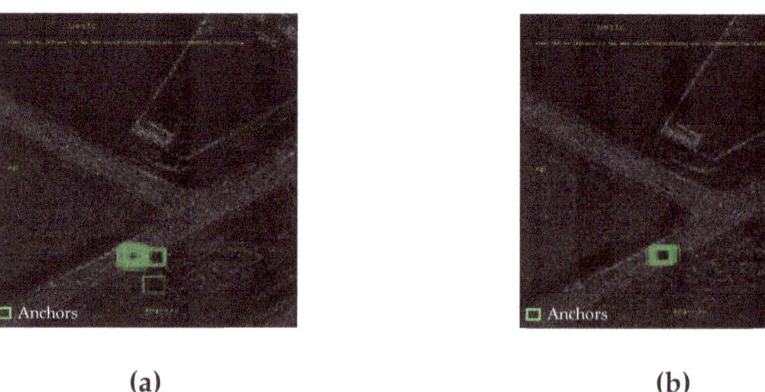

(a) **(b)**

Figure 26. Ablation experiment on GA-SubNet: (**a**) Siamese-RPN; (**b**) GASN.

Table 8. Ablation experiment of GA-SubNet.

Method	GA-SubNet	Accuracy
Siamese-RPN	✗	55.57%
GASN (ours)	✓	60.09%

5.3. Research on Pre-Training

In the deep learning field, in recent years, a common practice is to pre-train a model on some large-scale training data [31–33]. As shown in Figure 6b, the one-channel SAR image needs to be copied three times to use the pre-training parameters of three-channel RGB optical images. This method of copying one-channel SAR images three times has been widely used in SAR image processing tasks [12,15]. For example, to be suitable for

SAR tracking tasks, the pre-training parameters of the optical image are adjusted by the one-channel SNL data copied three times, and the tracking results are good.

To determine whether it is reasonable to apply a model trained on a three-channel RGB image to a one-channel radar image in a completely different domain or not, we arranged a group of experiments. The final tracking results for the simulated data are shown in Table 9. The second row of the results contains the tracking results after pre-training the model using optical images and then fine-tuning the training using SAR images replicated as three channels. The first row contains the tracking results after training using only replicated SAR images without pre-training with optical images. The tracking accuracy is significantly reduced by approximately 4% compared to the second row. This illustrates that it is feasible and reasonable to apply a model trained on three-channel RGB images to one-channel radar images. Therefore, it is wise to use fine-tuning in the absence of sufficient training data.

Table 9. Accuracy indexes of research on pre-training.

Pre-Training	Accuracy
✗	56.73%
✓	60.09%

5.4. Research on the Statistical Analysis

Regarding the statistical analysis of small data, we added an experiment where we trained 10 times and calculated the statistical average (including the mean and variance of the accuracy and the central location error (CLE)). The results are shown in Table 10.

Table 10. The statistical analysis of the tracking result.

Method	Accuracy %	CLE
Siamese-RPN	56.37 ± 0.72	7.49 ± 0.98
GASN (ours)	58.79 ± 0.61	6.56 ± 0.89

From the table, we can see that our method outperforms Siamese-RPN in terms of accuracy (58.79 > 56.37) and the accuracy variance (0.61 < 0.72), which indicates that our method is accurate and that the accuracy is more stable.

Moreover, our method outperforms Siamese-RPN in terms of the central location error (CLE) (7.49 > 6.56) and the CLE variance (0.89 < 0.98), which indicates that the CLE of our method is smaller and that the CLE is more stable.

6. Conclusions

To achieve the tracking of arbitrary TOIs in Video-SAR, this paper proposed a novel GASN. GASN is based on the idea of similarity learning, which uses the feature map of the template as the convolution kernel to slide windows on the feature map of the search image. Then, the output indicates the similarity of the two feature maps. Based on the maximum similarity, GASN can determine the tracking results in the search image. GASN tracks the TOI between the first frame and the next one instead of learning the appearance among all separate frames. Additionally, we established a GA-SubNet, which uses the location information of the template to obtain the location probability in the search image and selects the location with a probability greater than the threshold to exclude false alarms. To improve the tracking accuracy, the anchor that more closely matches the shape of the TOI is obtained by GA-SubNet through adaptive prediction processing. The experimental results showed that the tracking accuracy of the proposed method was 60.16% and 54.68% on the simulated and real Video-SAR data, respectively, which are higher than that of the two deep learning methods Siamese-RPN and Siamese-FC and the two traditional methods MOSSE and KCF.

In the future, we will try to apply scale invariant feature transform (SIFT) [34] and the Lee filter [35] to real Video-SAR for more accurate tracking results and research how to use the accurate tracking trajectory to refocus the moving target.

Author Contributions: Conceptualization, J.B. and X.Z.; methodology, J.B.; software, J.B.; validation, J.B., X.Z. and T.Z.; formal analysis, J.B.; investigation, J.B.; resources, J.S.; data curation, J.S.; writing—original draft preparation, J.B.; writing—review and editing, J.B.; visualization, X.Z.; supervision, T.Z.; project administration, X.Z.; funding acquisition, X.Z., J.S. and S.W. All authors have read and agreed to the published version of the manuscript.

Funding: This work was supported in part by the National Natural Science Foundation of China under grants 61571099, 61501098, and 61671113.

Acknowledgments: The authors thank all reviewers for their comments toward improving our manuscript, as well as the Sandia National Laboratory of the United States for providing SAR images. The authors would also like to thank Durga Kumar for his linguistic assistance during the preparation of this manuscript.

Conflicts of Interest: The authors declare no conflict of interest.

Abbreviations

Symbol	Definition
x	Search image
$\varphi(x)$	Detected feature map
g	Similarity learning function
F_2	New detected feature map for the best anchor shape
$A^{reg}_{w \times h \times 4k}$	Similarity map regression
(w, h)	The weight and height of the shadow
$loss_{loc}$	Anchor location loss
$loss_{cls}$	Classification loss
P	The tracking result
(x_R, y_R)	The center coordinates of the tracking result
t	The total tracking time
z	Template image
$\varphi(z)$	Template feature map
F_1	Original detected feature map
k	The number of anchors
$A^{cls}_{w \times h \times 2k}$	Similarity map for classification
(x, y)	The center point of the shadow in the previous image
$loss_{shape}$	Anchor shape loss
$loss_{reg}$	Regression loss
G	The shadow's ground truth
(x_G, y_G)	The center coordinates of the shadow's ground truth
N	The number of frames of the Video-SAR sequence

References

1. Damini, A.; Balaji, B.; Parry, C.; Mantle, V. A videoSAR mode for the X-band wideband experimental airborne radar. In Proceedings of the Algorithms for Synthetic Aperture Radar Imagery XVII, Orlando, FL, USA, 18 April 2010; p. 76990E.
2. Wells, L.; Sorensen, K.; Doerry, R.B. Developments in SAR and IFSAR systems and technologies at Sandia National Laboratories. In Proceedings of the 2003 IEEE Aerospace Conference Proceedings (Cat. No. 03TH8652.), Big Sky, MT, USA, 8–15 March 2003; pp. 21085–21095.
3. Hawley, R.W.; Garber, W.L. Aperture weighting technique for video synthetic aperture radar. In Proceedings of the Algorithms for Synthetic Aperture Radar Imagery XVIII, Orlando, FL, USA, 4 May 2011; p. 805107.
4. Linnehan, R.; Miller, J.; Bishop, E.; Horndt, V. An autofocus technique for video-SAR. In Proceedings of the Algorithms for Synthetic Aperture Radar Imagery XX, Baltimore, MD, USA, 23 May 2013; p. 874608.
5. Miller, J.; Bishop, E.; Doerry, A. An application of backprojection for Video-SAR image formation exploiting a subaperture circular shift register. In Proceedings of the Algorithms for Synthetic Aperture Radar Imagery XX, Baltimore, MD, USA, 23 May 2013; p. 874609.

6. Wang, H.; Chen, Z.; Zheng, S. Preliminary research of low-RCS moving target detection based on Ka-band Video-SAR. *IEEE Geosci. Remote Sens. Lett.* **2017**, *14*, 811–815. [CrossRef]

7. Henke, D.; Dominguez, E.M.; Small, D.; Schaepman, M.E.; Meier, E. Moving target tracking in single-and multichannel SAR. IEEE Trans. Geosci. *Remote Sens.* **2015**, *53*, 3146–3159. [CrossRef]

8. Yang, X.; Shi, J.; Zhou, Y.; Wang, C.; Wei, S. Ground Moving Target Tracking and Refocusing Using Shadow in Video-SAR. *Remote Sens.* **2020**, *12*, 3083. [CrossRef]

9. Ying, Z.; Daiyin, Z.; Xiang, Y.; Mao, X. Approach to moving targets shadow detection for VideoSAR. *J. Electron. Inf. Technol.* **2017**, *39*, 2197–2202.

10. Zhao, B.; Han, Y.; Wang, H.; Tang, L.; Wang, T. Robust Shadow Tracking for Video-SAR. *IEEE Geosci. Remote Sens. Lett.* **2020**, *18*, 821–825. [CrossRef]

11. Tian, X.; Liu, J.; Mallick, M. Simultaneous Detection and Tracking of Moving-Target Shadows in ViSAR Imagery. *IEEE Trans. Geosci. Remote Sens.* **2020**, *59*, 1182–1199. [CrossRef]

12. Ding, J.; Wen, L.; Zhong, C.; Loffeld, O. Video-SAR Moving Target Indication Using Deep Neural Network. *IEEE Trans. Geosci. Remote Sens.* **2020**, *58*, 7194–7204. [CrossRef]

13. Ren, S.; He, K.; Girshick, R.; Sun, J. Faster r-cnn: Towards real-time object detection with region proposal networks. *arXiv* **2015**, arXiv:150601497. [CrossRef] [PubMed]

14. Gers, F.A.; Schmidhuber, J.; Cummins, F.A. Learning to Forget: Continual Prediction with LSTM. *Neural Comput.* **2000**, *12*, 2451–2471. [CrossRef] [PubMed]

15. Zhou, Y.; Shi, J.; Wang, C.; Hu, H.; Zhou, Z.; Yang, X.; Zhang, X.; Wei, S. SAR Ground Moving Target Refocusing by Combining mRe3 Network and TVβ -LSTM. *IEEE Trans. Geosci. Remote Sens.* **2020**, 1–4. [CrossRef]

16. Wen, L.; Ding, J.; Loffeld, O. Video-SAR Moving Target Detection Using Dual Faster R-CNN. *IEEE J. Sel. Top. Appl. Earth Obs. Remote Sens.* **2021**, *14*, 2984–2994. [CrossRef]

17. Li, B.; Yan, J.; Wu, W.; Zheng, Z.; Hu, X. High performance visual tracking with siamese region proposal network. In Proceedings of the IEEE Conference on Computer Vision and Pattern Recognition, Salt Lake City, UT, USA, 18–23 June 2018; pp. 8971–8980.

18. Bertinetto, L.; Valmadre, J.; Henriques, J.F.; Vedaldi, A.; Torr, P. Fully-convolutional siamese networks for object tracking. In Proceedings of the European conference on computer vision, Amsterdam, The Netherlands, 3 November 2016; pp. 850–865.

19. Bolme, D.S.; Beveridge, J.R.; Draper, B.A.; Lui, Y.M. Visual object tracking using adaptive correlation filters. In Proceedings of the 2010 IEEE Computer Society Conference on Computer Vision and Pattern Recognition, San Francisco, CA, USA, 13–18 June 2010; pp. 2544–2550.

20. Henriques, J.F.; Caseiro, R.; Martins, P.; Batista, J. High-speed tracking with kernelized correlation filters. *IEEE Trans. Pattern Anal. Mach. Intell.* **2014**, *37*, 583–596. [CrossRef] [PubMed]

21. Tao, R.; Gavves, E.; Smeulders, A.W.M. Siamese instance search for tracking. In Proceedings of the IEEE Conference on Computer Vision and Pattern Recognition, Las Vegas, NV, USA, 26 June–1 July 2016; pp. 1420–1429.

22. Held, D.; Thrun, S.; Savarese, S. Learning to track at 100 fps with deep regression networks. In Proceedings of the European conference on computer vision, Amsterdam, The Netherlands, 3 November 2016; pp. 749–765.

23. Cai, Z.; Vasconcelos, N. Cascade r-cnn: Delving into high quality object detection. In Proceedings of the IEEE Conference on Computer Vision and Pattern Recognition, Salt Lake City, UT, USA, 18–23 June 2018; pp. 6154–6162.

24. Dai, J.; Qi, H.; Xiong, Y.; Li, Y.; Zhang, G.; Hu, H.; Wei, Y. Deformable convolutional networks. In Proceedings of the IEEE International Conference on Computer Vision, Venice, Italy, 24–27 October 2017; pp. 764 773.

25. He, K.; Zhang, X.; Ren, S.; Sun, J. Deep residual learning for image recognition. In Proceedings of the IEEE Conference on Computer Vision and Pattern Recognition, Las Vegas, NV, USA, 26 June–1 July 2016; pp. 770–778.

26. Russakovsky, O.; Deng, J.; Su, H.; Krause, J.; Satheesh, S.; Ma, S.; Huang, Z.; Karpathy, A.; Khosla, A.; Bernstein, M. Imagenet large scale visual recognition challenge. *Int. J. Comput. Vis.* **2015**, *115*, 211–252. [CrossRef]

27. Davies, D.L.; Bouldin, D.W. A cluster separation measure. *IEEE Trans. Pattern Anal. Mach. Intell.* **1979**, *2*, 224–227. [CrossRef]

28. Lin, T.-Y.; Goyal, P.; Girshick, R.; He, K.; Dollár, P. Focal loss for dense object detection. In Proceedings of the IEEE international conference on computer vision, Venice, Italy, 24– 27 October 2017; pp. 2980–2988.

29. Wang, J.; Chen, K.; Yang, S.; CL Chen, C.; Lin, D. Region proposal by guided anchoring. In Proceedings of the IEEE/CVF Conference on Computer Vision and Pattern Recognition, Long Beach, CA, USA, 15–20 June 2019; pp. 2965–2974.

30. Kristan, M.; Leonardis, A.; Matas, J.; Felsberg, M.; He, Z. The visual object tracking vot2017 challenge results. In Proceedings of the IEEE International Conference on Computer Vision Workshops, Venice, Italy, 22–29 October 2017; pp. 1949–1972.

31. Van Sloun, R.J.G.; Cohen, R.; Eldar, Y. Deep Learning in Ultrasound Imaging. *Proc. IEEE* **2019**, *108*, 11–29. [CrossRef]

32. Yin, S.; Peng, Q.; Li, H.; Zhang, Z.; You, X.; Fischer, K.; Furth, S.L.; Tasian, G.E.; Fan, Y. Computer-Aided Diagnosis of Congenital Abnormalities of the Kidney and Urinary Tract in Children Using a Multi-Instance Deep Learning Method Based on Ultrasound Imaging Data. In Proceedings of the 2020 IEEE 17th International Symposium on Biomedical Imaging (ISBI), Iowa City, IA, USA, 3–7 April 2020; pp. 1347–1350. [CrossRef]

33. Einsidler, D.; Dhanak, M.; Beaujean, P. A Deep Learning Approach to Target Recognition in Side-Scan Sonar Imagery. In Proceedings of the OCEANS 2018 MTS/IEEE Charleston, Charleston, SC, USA, 22–25 October 2018; pp. 1–4. [CrossRef]

34. Lowe, D.G. Distinctive Image Features from Scale-Invariant Keypoints. *Int. J. Comput. Vis.* **2004**, *60*, 91–110. [CrossRef]
35. Lopes, A.; Touzi, R.; Nezry, E. Adaptive speckle filters and scene heterogeneity. *IEEE Trans. Geosci. Remote Sens.* **1990**, *28*, 992–1000. [CrossRef]

remote sensing

MDPI

Article

A Robust InSAR Phase Unwrapping Method via Phase Gradient Estimation Network

Liming Pu, Xiaoling Zhang *, Zenan Zhou, Liang Li, Liming Zhou, Jun Shi and Shunjun Wei

School of Information and Communication Engineering, University of Electronic Science and Technology of China, Chengdu 611731, China; puliming@std.uestc.edu.cn (L.P.); 201722020918@std.uestc.edu.cn (Z.Z.); liliang@std.uestc.edu.cn (L.L.); zhouliming@std.uestc.edu.cn (L.Z.); shijun@uestc.edu.cn (J.S.); weishunjun@uestc.edu.cn (S.W.)
* Correspondence: xlzhang@uestc.edu.cn

Abstract: Phase unwrapping is a critical step in synthetic aperture radar interferometry (InSAR) data processing chains. In almost all phase unwrapping methods, estimating the phase gradient according to the phase continuity assumption (PGE-PCA) is an essential step. The phase continuity assumption is not always satisfied due to the presence of noise and abrupt terrain changes; therefore, it is difficult to get the correct phase gradient. In this paper, we propose a robust least squares phase unwrapping method that works via a phase gradient estimation network based on the encoder–decoder architecture (PGENet) for InSAR. In this method, from a large number of wrapped phase images with topography features and different levels of noise, the deep convolutional neural network can learn global phase features and the phase gradient between adjacent pixels, so a more accurate and robust phase gradient can be predicted than that obtained by PGE-PCA. To get the phase unwrapping result, we use the traditional least squares solver to minimize the difference between the gradient obtained by PGENet and the gradient of the unwrapped phase. Experiments on simulated and real InSAR data demonstrated that the proposed method outperforms the other five well-established phase unwrapping methods and is robust to noise.

Keywords: interferometric synthetic aperture radar; deep convolutional neural network; phase unwrapping

Citation: Pu, L.; Zhang, X.; Zhou, Z.; Li, L.; Zhou, L.; Shi, J.; Wei, S. A Robust InSAR Phase Unwrapping Method via Phase Gradient Estimation Network. *Remote Sens.* **2021**, *13*, 4564. https://doi.org/10.3390/rs13224564

Academic Editor: João Catalão Fernandes

Received: 1 September 2021
Accepted: 9 November 2021
Published: 13 November 2021

1. Introduction

Synthetic aperture radar interferometry (InSAR) is playing an increasingly important role in the field of surface deformation monitoring and topographic mapping [1–3]. The InSAR system uses two co-registered complex images from different viewing angles to obtain the two-dimensional interferometric phase images. Due to the trigonometric function in the transmitting and receiving models, the obtained interferometric phase is wrapped—that is, its range is in $(-\pi, \pi]$ [4,5]. In order to obtain an accurate elevation measurement of the surveying area, the unwrapped phase must be obtained by adding the correct wrap count to each pixel of the wrapped phase, which is called phase unwrapping. Therefore, in the InSAR data processing pipeline, the measurement accuracy of elevation level is highly correlated with the accuracy of phase unwrapping.

Since phase unwrapping is an ill-posed problem, the phase continuity assumption is usually considered in the process of phase unwrapping: the absolute values of the gradients in the two directions of the unwrapped phase are less than π [6]. Under this assumption, many kinds of phase unwrapping methods have been presented in recent decades, and they can be divided into two categories: path following [5,7,8] and optimization-based methods [9–15]. A path following method selects the integration path for integrating the estimated phase gradient through the residue distribution or the phase quality map, so as to avoid the local error from being propagated globally. Examples are the branch-cut method [5] and the quality-guided method [7]. An optimization-based method minimizes the difference between the estimated gradient and the unwrapping phase gradient through the objective function to obtain the optimal unwrapped phase. Examples are the least squares (LS) method [13] and the statistical-cost, network-flow algorithm for phase unwrapping (SNAPHU) method [15], and phase unwrapping max-flow/min-cut algorithm (PUMA) [10]. Both types of method need to obtain the estimated value of the phase gradient through the phase continuity assumption before unwrapping. Due to the presence of noise and abrupt terrain changes, the phase continuity assumption is not always satisfied—that is, the unwrapped phase may jump above π, which may cause local errors in the unwrapping process. This local error may produce a global error along the integration path, so the estimated phase gradient information will directly affect the final unwrapping accuracy. Therefore, it is a valuable aim to seek a more accurate estimation method of phase gradient information instead of directly relying on the traditional estimation method based on the phase continuity assumption.

In recent years, the deep learning-based phase unwrapping methods have attracted significant interest [16–24]. Most of these methods [16–18] convert the unwrapping problem into a classification problem of the wrap count, and their effectiveness is verified using optical images. In the field of InSAR, the unwrapping problem becomes more difficult because of two characteristics: the complex wrapped phase caused by topography features and the low coherence coefficient. Therefore, combining traditional phase unwrapping methods with deep learning, instead of relying solely on deep learning, is a promising development trend [19–23]. In [19], a modified fully convolutional network was first applied to classify the wrapped phase into normal pixels and layover residues, which can suppress the error propagation of layover residues during the phase unwrapping process. Additionally, a CNN-based unwrapping method was proposed in [20], which feeds the wrapped phase and coherence map into the network at the same time for training to obtain the wrap count gradient. In this method, the wrap count reconstruction is necessary for obtaining the final unwrapping result. A deep learning-based method combined with the minimum cost flow (MCF) unwrapping model was proposed in [21]. In this method, the phase gradient is discretized to match the MCF unwrapping model and treated as a three-classification deep learning problem, but the number of categories may need to change according to the terrain changes, because the three categories cannot cover all situations. In addition, the ambiguity gradient [23] is taken as ground truth for network training, and the MCF model is used as the postprocessing step for final unwrapped phase reconstruction. However, the MCF unwrapping model is usually very complex computationally and requires numerous computational resources [25].

The LS phase unwrapping method is widely used in practical applications and converges quickly [9,26,27]; therefore, we considered combining it and deep learning to improve the unwrapping accuracy while retaining the advantages of the LS method. In the traditional LS method, estimating the phase gradient according to the phase continuity assumption (PGE-PCA) is an essential step. Recent studies [28–32] have indicated that the encoder–decoder architecture based on deep convolutional neural networks (DCNN) can learn the global features from a large number of input images with different levels of noise or other disturbances, which is useful for obtaining the robust phase gradient from noisy wrapped phase images.

In this paper, we propose a robust LS InSAR phase unwrapping method that works via a phase gradient estimation network (PGENet-LS). In this method, we transform the phase gradient estimation into a regression problem and design a phase gradient estimation network based on the encoder–decoder architecture (PGENet) for InSAR. From lots of wrapped phase images with topography features and different levels of noise, PGENet can extract global high-level phase features and recognize the phase gradient between adjacent pixels, so the more accurate and robust phase gradient can be estimated by PGENet than that obtained by PGE-PCA. Finally, the phase unwrapping result is obtained by using the least squares solver to minimize the difference between the gradient obtained by PGENet and the gradient of the unwrapped phase. The phase gradient estimated by PGENet is used to replace the PGE-PCA in the traditional LS unwrapping method. As the accuracy of the phase gradient estimated by PGENet is significantly higher and more robust than that of the phase gradient estimated by PGE-PCA, the proposed method has higher accuracy than the traditional LS phase unwrapping method. A series of experimental results of simulated wrapped phase and real InSAR data demonstrate that the proposed method outperforms the other five well-established phase unwrapping methods and is robust to noise.

This paper is organized as follows. Section 2 introduces the principles of phase unwrapping, problem analysis, PGENet, and the proposed method. In Section 3, the data generation method, loss function, performance evaluation index, and experiment settings are described. In Section 4, a series of experimental results using simulated and real InSAR data are presented. Section 5 and Section 6 present the discussion and conclusions of the paper, respectively.

2. PGENet-LS Phase Unwrapping Method

In this section, we first introduce the principle of phase unwrapping and how to use deep neural networks instead of PGE-PCA to estimate the phase gradient. Then we describe the detailed structure of PGENet, and finally introduce the processing flow of the PGENet-LS phase unwrapping method.

2.1. Principle of Phase Unwrapping

The wrapped phase is distributed in $(-\pi, \pi]$ containing the ambiguities of integral multiples of 2π. For an image pixel point (i, j), the relationship between the wrapped phase $\phi_{i,j}$ and the unwrapped phase $\varphi_{i,j}$ can be expressed as

$$\varphi_{i,j} = \phi_{i,j} + 2\pi k_{i,j} \tag{1}$$

where $k_{i,j}$ is a sequence of integers, which is called the wrap count. When $k_{i,j}$ is known, the unwrapped phase can be recovered from the wrapped phase. However, a unique solution $k_{i,j}$ cannot be obtained because there are two unknowns in (1). Therefore, in the traditional phase unwrapping method, the phase continuity assumption is used to ensure the uniqueness of the phase unwrapping result. Under this assumption, the LS phase unwrapping method can be divided into the following two steps: phase gradient estimation and implementation of the least squares solver. The flowchart is shown in Figure 1.

Step 1: According to the phase continuity assumption, for the two-dimensional phase unwrapping issue, the horizontal gradient and vertical phase gradient can be estimated by

$$\Delta_{i,j}^{x\prime} = \begin{cases} \phi_{i,j+1} - \phi_{i,j}, & -\pi \leq \phi_{i,j+1} - \phi_{i,j} \leq \pi \\ \phi_{i,j+1} - \phi_{i,j} - 2\pi, & \pi < \phi_{i,j+1} - \phi_{i,j} \leq 2\pi \\ \phi_{i,j+1} - \phi_{i,j} + 2\pi, & -2\pi \leq \phi_{i,j+1} - \phi_{i,j} < -\pi \end{cases} \tag{2}$$

$$\Delta_{i,j}^{y\prime} = \begin{cases} \phi_{i+1,j} - \phi_{i,j}, & -\pi \leq \phi_{i,j+1} - \phi_{i,j} \leq \pi \\ \phi_{i+1,j} - \phi_{i,j} - 2\pi, & \pi < \phi_{i,j+1} - \phi_{i,j} \leq 2\pi \\ \phi_{i+1,j} - \phi_{i,j} + 2\pi, & -2\pi \leq \phi_{i,j+1} - \phi_{i,j} < -\pi \end{cases} \tag{3}$$

where $\Delta_{i,j}^{x\prime}$ and $\Delta_{i,j}^{y\prime}$ are the horizontal gradient and vertical gradient, respectively. For brevity, the step is called PGE-PCA.

Step 2: After obtaining the estimated horizontal gradient $\Delta_{i,j}^{x\prime}$ and vertical gradient $\Delta_{i,j}^{y\prime}$, the final unwrapping result $\varphi_{i,j}'$ can be calculated according to the least squares solver of (4).

$$\underset{\varphi_{i,j}'}{\arg\min} \sum_i \sum_j \left| \varphi_{i,j+1}' - \varphi_{i,j}' - \Delta_{i,j}^{x\prime} \right|^2 + \sum_i \sum_j \left| \varphi_{i+1,j}' - \varphi_{i,j}' - \Delta_{i,j}^{y\prime} \right|^2 \tag{4}$$

The meaning of (4) is to minimize the difference between the estimated gradient and the gradient of the unwrapped phase. To obtain the solution of (4), there are mainly two classes of fast algorithms: transformation-based methods [13] and multi-grid methods [12]. We can see that the accuracy of the phase unwrapping result from (4) is directly related to the accuracy of the estimated phase gradient. In other words, if the accuracy of the estimated phase gradient can be improved, we can obtain the more accurate phase unwrapping result.

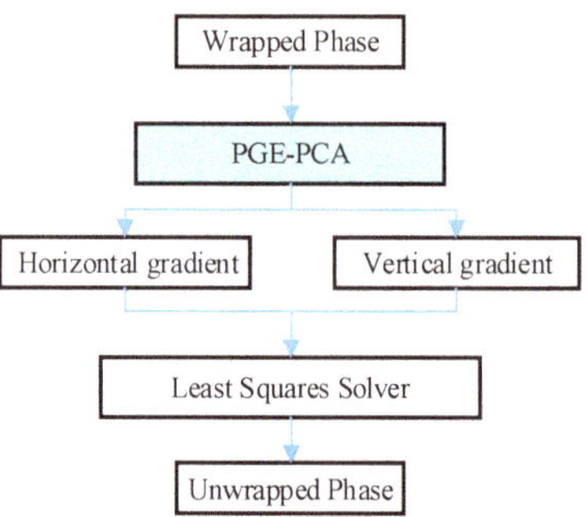

Figure 1. Flowchart of the traditional least squares phase unwrapping method.

2.2. Problem Analysis

In practical applications, the phase continuity assumption is often not satisfied in every pixel due to the presence of noise and abrupt terrain changes. The noise level can be evaluated by the coherence coefficient, which can be expressed as

$$\rho = \left| \frac{E\{S_1 \cdot S_2^*\}}{\sqrt{E\{|S_1|^2\} \cdot E\{|S_2|^2\}}} \right| \tag{5}$$

where S_1 and S_2 are the co-registered master and slave complex images, respectively; $*$ denotes the complex conjugate; and $E\{\cdot\}$ denotes the mathematical expectation.

Figure 2 shows the influences of different coherence coefficients on the wrapped phase and the corresponding gradients in the horizontal and vertical directions. Figure 2a shows the wrapped phase with different coherence coefficients, and Figure 2b,c shows the corresponding phase gradients in two directions obtained by (2) and (3), respectively. We can see that in the case of a coherence coefficient of 1 (no noise), the accurate vertical and horizontal phase gradients can be obtained by (2) and (3), but in the presence of noise, the estimated phase gradients in two directions from (2) and (3) are no longer reliable because the gradient information is destroyed by noise. As the coherence coefficient gets lower and lower, the phase gradients in two directions are destroyed more and more severely. From (2) and (3) and Figure 2, we can see that only considering the relationship between adjacent pixels to calculate the gradient is not enough, which means that more or even global phase information may need to be employed. Therefore, in the presence of noise, we use the global phase information in the gradient estimation process for improving the accuracy of phase gradient estimation. That is to say, the gradient estimation of each pixel does not only depend on the adjacent pixels but on the overall wrapped image.

In recent years, the encoder–decoder architecture based on DCNN has been widely used in image processing in the fields of optics, medicine, and SAR [29–32]. These studies have indicated that the encoder–decoder architecture can learn the global high-level features from input images with different levels of noise or other disturbances. The powerful feature extraction capability is conducive to extract global phase features from the noisy wrapped images to obtain the accurate phase gradient. In addition, according to (2) and (3), it can be seen that the gradient calculation principles in the two directions are the same. Therefore, while the original wrapped phase image and the horizontal gradient image are used as an image pair, the transposed original wrapped phase image and the transposed vertical gradient image of the original wrapped phase image are taken as another equivalent image pair, so that we can get the horizontal and vertical gradient images of the wrapped phase after inputting the original and transposed wrapped phase images to a network, respectively. Based on the above analysis, we designed PGENet based on the encoder–decoder architecture that takes the original or transposed wrapped phase images as inputs and outputs the estimated horizontal or vertical phase gradient images. The deep convolutional neural network can learn global phase features and the phase gradient between adjacent pixels from lots of wrapped phase images with topography features and different coherence coefficients; hence, the accurate and robust horizontal and vertical phase gradient images can be predicted after training.

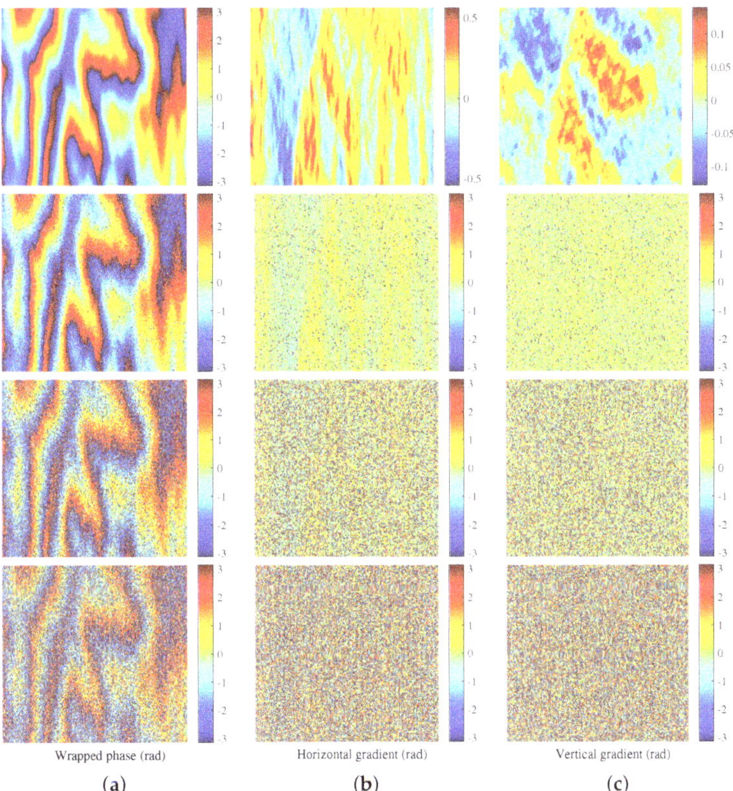

Figure 2. (a) Wrapped phase images, and (b,c) are the corresponding horizontal gradient images and vertical gradient images, respectively. From top to bottom, the coherence coefficients are 1, 0.95, 0.75, and 0.5, respectively.

2.3. PGENet

PGENet is designed based on the encoder–decoder architecture and is mainly composed of two parts: an encoder and a decoder. Its overall structure and the detailed parameters of each layer are shown in Figure 3 and Table 1, respectively. As shown in Figure 3 and Table 1, the encoder part (Encoder) contains eight encoder blocks and converts the input noisy wrapped phase images into more feature maps with smaller sizes, which can enrich the global phase gradient features and reduce memory requirements when adding the number of feature maps. An encoder block is constitutive of two convolution layers (Conv + Relu), and each layer performs two operations, namely, convolution and activation (Relu). After the encoding process, a large number of global phase features are extracted in the feature maps. The decoder part

(Decoder) contains eight decoder blocks and gradually recovers the phase gradient from these extracted global phase feature maps. Each decoder block is constitutive of a convolution layer and a deconvolution layer (Deconv + Relu). Each deconvolution layer performs two operations, namely, deconvolution and activation. During the decoding process, a large number of feature maps are gradually merged into larger images until the output image is the same as the input image. At the same time, due to the fusion of global phase features, the accurate phase gradient is extracted.

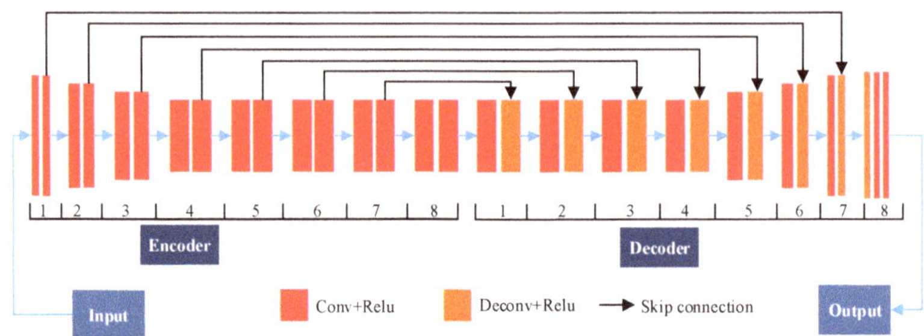

Figure 3. Overall architecture of PGENet.

Table 1. Detailed parameters of each layer in PGENet.

#	Layer Name	Filter Size	# Feature Maps	Padding	Stride	Image Size
Encoder Block 1	Conv + Relu	5×5	64	2	1	$M \times N \times 64$
	Conv + Relu	5×5	64	2	2	$M/2 \times N/2 \times 64$
Encoder Block 2	Conv + Relu	5×5	128	2	1	$M/2 \times N/2 \times 128$
	Conv + Relu	5×5	128	2	2	$M/4 \times N/4 \times 128$
Encoder Block 3	Conv + Relu	5×5	256	2	1	$M/4 \times N/4 \times 256$
	Conv + Relu	5×5	256	2	2	$M/8 \times N/8 \times 256$
Encoder Block 4	Conv + Relu	5×5	512	2	1	$M/8 \times N/8 \times 512$
	Conv + Relu	5×5	512	2	2	$M/16 \times N/16 \times 512$
Encoder Block 5	Conv + Relu	5×5	512	2	1	$M/16 \times N/16 \times 512$
	Conv + Relu	5×5	512	2	2	$M/32 \times N/32 \times 512$
Encoder Block 6	Conv + Relu	5×5	512	2	1	$M/32 \times N/32 \times 512$
	Conv + Relu	5×5	512	2	2	$M/64 \times N/64 \times 512$
Encoder Block 7	Conv + Relu	5×5	512	2	1	$M/64 \times N/64 \times 512$
	Conv + Relu	5×5	512	2	2	$M/128 \times N/128 \times 512$
Encoder Block 8	Conv + Relu	5×5	512	2	1	$M/128 \times N/128 \times 512$
	Conv + Relu	5×5	512	2	2	$M/256 \times N/256 \times 512$
Decoder Block 1	Conv + Relu	5×5	512	2	1	$M/256 \times N/256 \times 512$
	Deconv + Relu	4×4	512	1	2	$M/128 \times N/128 \times 512$
Decoder Block 2	Conv + Relu	5×5	512	2	1	$M/128 \times N/128 \times 512$
	Deconv + Relu	4×4	512	1	2	$M/64 \times N/64 \times 512$
Decoder Block 3	Conv + Relu	5×5	512	2	1	$M/64 \times N/64 \times 512$
	Deconv + Relu	4×4	512	1	2	$M/32 \times N/32 \times 512$
Decoder Block 4	Conv + Relu	5×5	512	2	1	$M/32 \times N/32 \times 512$
	Deconv + Relu	4×4	512	1	2	$M/16 \times N/16 \times 512$
Decoder Block 5	Conv + Relu	5×5	256	2	1	$M/16 \times N/16 \times 256$
	Deconv + Relu	4×4	256	1	2	$M/8 \times N/8 \times 256$
Decoder Block 6	Conv + Relu	5×5	128	2	1	$M/8 \times N/8 \times 128$
	Deconv + Relu	4×4	128	1	2	$M/4 \times N/4 \times 128$
Decoder Block 7	Conv + Relu	5×5	64	2	1	$M/4 \times N/4 \times 64$
	Deconv + Relu	4×4	64	1	2	$M/2 \times N/2 \times 64$
Decoder Block 8	Conv + Relu	5×5	32	2	1	$M/2 \times N/2 \times 64$
	Deconv + Relu	4×4	32	1	2	$M/2 \times N/2 \times 32$
	Conv + Relu	5×5	1	2	1	$M \times N \times 1$

In the process of constructing the Encoder and Decoder, a deep network structure is formed. The deep architecture of PGENet is used to enrich the level of phase gradient features, which can ensure that the network has sufficient phase gradient estimation capabilities when processing noisy wrapped phase images. While increasing the network depth, the feature maps with the same size between the encoder and decoder are added by skip connections [28]. As shown in Figure 3, the skip connections can transfer the extracted global phase features containing more detailed gradient information to the deconvolution layer of the Decoder and accelerate convergence. Therefore, in the decoder process, the low/mid/high-level global phase gradient features containing more detailed information from the Encoder are compensated and fused in the current phase gradient feature maps. Under the effect of the deep structure and skip connections, PGENet can obtain accurate and robust phase gradient estimation results when inputting images with different noise levels.

2.4. PGENet-LS Phase Unwrapping Method

As shown in Figure 4, in the PGENet-LS phase unwrapping method, the horizontal and vertical gradients are estimated by PGENet first, and then the least squares solver is employed to minimize the difference between the gradients obtained by PGENet and the gradients of the unwrapped phase. Therefore, the processing flow can be divided into the following two steps: phase gradient estimation and unwrapping using the least squares solver.

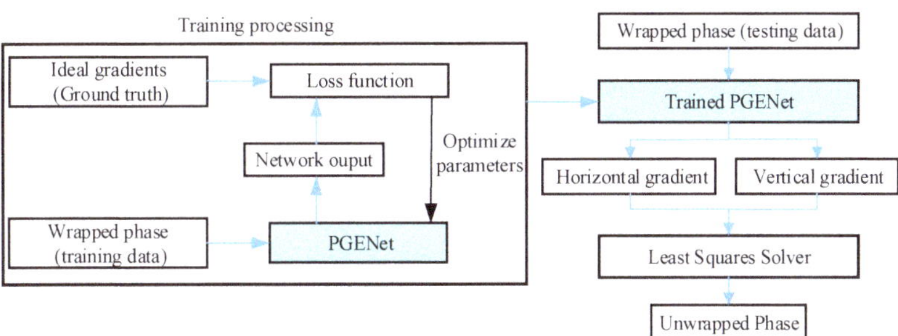

Figure 4. Flowchart of the proposed method.

Step 1. For estimating the phase gradients using PGENet, training and testing are required. PGENet takes the original or transposed wrapped phase images with different coherence coefficients and topography features as input and produces the corresponding horizontal or vertical gradient images. During the training processing, the loss function described in Section 3.2 is selected to update the trainable parameters. During the testing processing, the testing wrapped phase image (simulated or real InSAR data) or its transposed version is fed into the trained PGENet to obtain the estimated horizontal or vertical phase gradient images, respectively.

Step 2: After obtaining the horizontal gradient and vertical gradient estimated by PGENet, all least squares solvers based on phase gradients from (2) and (3) can be used in this step. In this paper, the well-established weighted least squares solver of (6) [26] are selected to get the unwrapping result and can be expressed as

$$\underset{\varphi'_{i,j}}{\arg\min} \sum_i \sum_j \omega^x_{i,j} \left| \varphi'_{i,j+1} - \varphi'_{i,j} - \Delta^{x'}_{i,j} \right|^2 + \sum_i \sum_j \omega^y_{i,j} \left| \varphi'_{i+1,j} - \varphi'_{i,j} - \Delta^{y'}_{i,j} \right|^2 \tag{6}$$

where $w^x_{ij} = \min\left(w^2_{i,j+1}, w^2_{i,j}\right)$, and $w^y_{ij} = \min\left(w^2_{i+1,j}, w^2_{i,j}\right)$. w^x_{ij} are the weights defined as the normalized cross-correlation coefficient.

3. Experiments

In this section, the detailed data generation process, loss function, and performance evaluation index are described.

3.1. Data Generation

To make the trained PGENet have a good generalization capability, a large number of labeled images with topography features are needed for training. Therefore, we used a digital elevation model (DEM) to generate wrapped phase images according to the ambiguity height of the real InSAR system [21,33]. After generating the wrapped phase, the corresponding ideal horizontal and vertical phase gradients were obtained according to (2) and (3). In addition, different levels of noise were added to the ideal (no noise) wrapped phase images, and the noise level is expressed by the coherence coefficient in the field of InSAR [2,33]. A lower coherence coefficient means a higher noise level, and in the absence of noise, the coherence coefficient is 1. The wrapped phase images with different coherence coefficients were used to train PGENet, which ensured that the trained PGENet had good robustness to noise.

The details of the generated training data are as follows. Figure 5a shows the DEM (eastern part of Turkey, 2048 × 2048 pixels) used for training in this study. It was from SRTM 1Sec HGT and can be downloaded from the Sentinel Application Platform (SNAP). The reason for choosing this DEM data was that their topographic features are similar to those of the real wrapped phase in the subsequent experiments. In addition, the ambiguity height of the simulated system (92.13 m) was the same as that of the real wrapped phase image in the subsequent experiments. Similar topographic features and the same ambiguity height made the phase gradient features of the training data and the real data as similar as possible. According to the ambiguity height and the DEM, the corresponding ideal wrapped phase and horizontal and vertical phase gradient images are shown in Figure 5b–d. The examples of the noisy wrapped phase images are shown in Figure 6. Ten noisy wrapped phase images were generated for training. Their coherence coefficients were in the range of [0.5, 0.95], and the interval was 0.05. This range of the coherence coefficients can cover most common InSAR data. In order to reduce the memory requirement, the whole wrapped phase images were cut into image patches with the size of 256 × 256 pixels. To augment the training data, 128 pixels were shifted on the columns or rows in each cropping process to ensure 50% overlap of adjacent image patches. Therefore, the total number of horizontal and vertical gradient image patches for training was 4500.

The details of the generated testing data are as follows. Figure 7a shows the reference DEM (1024 × 1024 pixels) which was used for testing. The 10 noisy wrapped phase images were generated, and the range of the coherence coefficients was the same as that of the training data. Figure 7b shows an example of a noisy wrapped phase with a coherence coefficient of 0.5, and Figure 7c,d shows the ideal horizontal and vertical phase gradients, respectively. The whole wrapped phase images were cut into image patches of 256 × 256 pixels for testing, and 128 pixels were shifted on the columns or rows in each cropping process to ensure 50% overlap of adjacent image patches. Therefore, the total number of horizontal and vertical gradient image patches for testing was 980, accounting for 22% of the sum of training data and testing data.

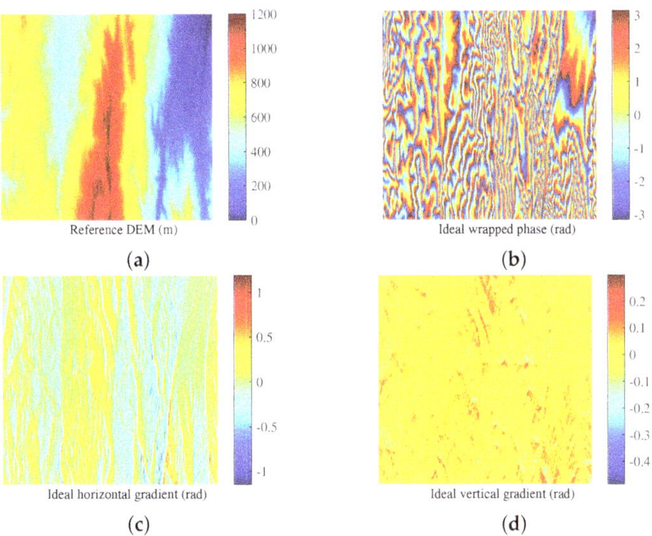

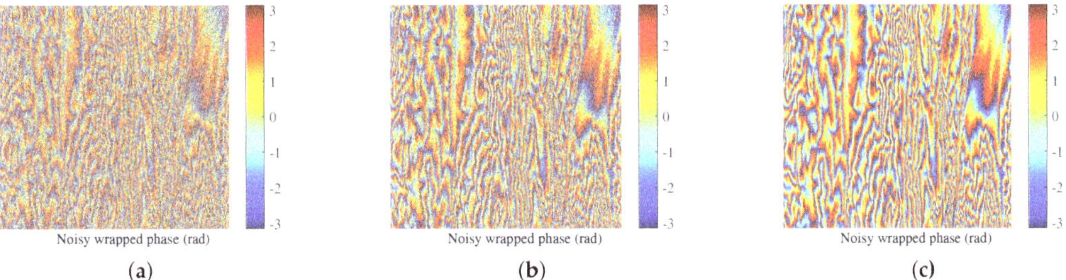

Figure 5. (**a**) Reference DEM used for training. (**b**) Ideal wrapped phase simulated by the reference DEM. (**c**) Ideal horizontal gradient. (**d**) Ideal vertical gradient.

Figure 6. Examples of the noisy wrapped phase with different coherence coefficients. (**a**) 0.5. (**b**) 0.75. (**c**) 0.95.

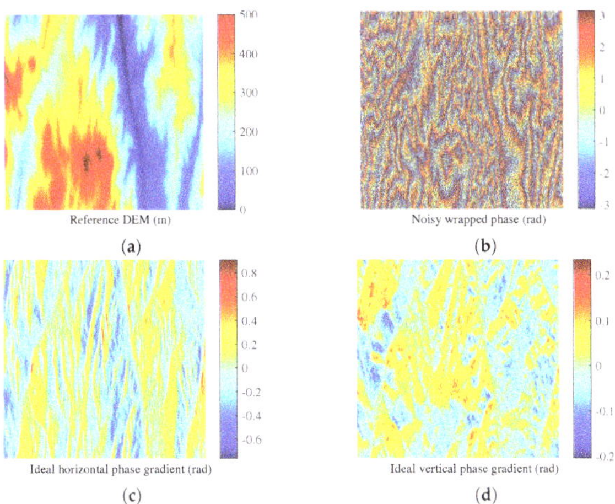

Figure 7. (**a**) Reference DEM used for testing. (**b**) Wrapped phase with a coherence coefficient of 0.5. (**c**) Ideal horizontal phase gradient. (**d**) Ideal vertical phase gradient.

3.2. Loss Function

The widely-used mean-square error (MSE) [33] is taken as the loss function to update the training parameters of PGENet. It is calculated according to the ideal phase gradient (ground truth) and the estimated phase gradient image (network output), and can be expressed as

$$\mathcal{L} = \sum_{i=1}^{N} \frac{\left(\Delta_i - \Delta_i'\right)^2}{N} \tag{7}$$

where N is the number of phase gradient image pixels; Δ_i and Δ_i' are the ideal gradient and the gradient estimated by PGENet, respectively.

3.3. Performance Evaluation Index

To fully evaluate the accuracy and robustness of the proposed method, we employed qualitative and quantitative methods to perform the evaluation. Qualitative evaluation refers to performing the unwrapping accuracy judgment based on the observation of the image by the naked eye. Therefore, the original wrapped images and DEM images obtained by the unwrapped phases of six different unwrapping methods are simultaneously presented in this paper. To more comprehensively and objectively evaluate the unwrapping accuracy and robustness, the unwrapping failure rate [19] and root mean square error (RMSE) were adopted for quantitative evaluation.

The unwrapping failure rate can be defined as

$$\text{UFR} = \frac{1}{MN} \sum_{i=1}^{M} \sum_{j=1}^{N} d_{i,j} \tag{8}$$

where $d_{i,j} = \begin{cases} 1, & \text{if } \left|\varphi_{i,j}' - \varphi_{i,j}\right| \geq \pi \\ 0, & \text{otherwise} \end{cases}$. $\varphi_{i,j}'$ and $\varphi_{i,j}$ are the estimated unwrapped phase and ideal unwrapped phase, respectively. M and N are the width and height of the unwrapped phase image, respectively. A smaller UFR value means better unwrapping accuracy to a certain extent in simulated data processing.

For simulated data processing, in addition to the unwrapping failure rate, the RMSE between the estimated unwrapped phase and ideal unwrapped phase was also employed to evaluate the unwrapping accuracy. For real InSAR data processing, due to the ideal unwrapped phase being unknown and the ultimate goal of wrapped phase processing being to obtain elevation, we employed the RMSE between the reference DEM and the DEM obtained by the unwrapped phases of different unwrapping methods to evaluate the unwrapping accuracy [20,23]. The formulaic expression of RMSE is

$$\text{RMSE} = \sqrt{\sum_{j}^{N} \sum_{i}^{M} \frac{\left(\varphi_{i,j} - \varphi_{i,j}'\right)^2}{MN}} \tag{9}$$

where $\varphi_{i,j}$ is the reference DEM image; $\varphi_{i,j}'$ is the DEM image obtained by the estimated unwrapped phase; M and N are the DEM image width and height, respectively. A small RMSE means that the estimated DEM is close to the reference DEM—that is, the unwrapping accuracy is high.

3.4. General Experiment Settings

We implemented all experiments on a PC with an Intel i7-8700K CPU, a NVIDIA GeForce RTX 2080 GPU, and 64G memory. PGENet was trained for 260 epochs with a batch size of 16 on the TensorFlow platform, and Adam optimizer [34] was selected to accelerate training. The initial learning rate was set to 1×10^{-4} and gradually decayed to 0 exponentially. We chose the early stopping method [35,36] to determine when the network would stop iterating. The trained PGENet was used in the following experiments on simulated and real data.

4. Results

Three experiments were implemented to evaluate the unwrapping accuracy and robustness of the proposed method. In the first experiment, we demonstrated the accuracy of the phase gradient and the robustness of PGENet, and compared PGENet with PGE-PCA. In the second experiment, we evaluated the unwrapping accuracy and robustness of the proposed method on simulated data; and we compared the proposed method with the LS [26], QGPU [7], SNAPHU [14], and PUMA [10] unwrapping methods, and a state-of-the-art deep learning-based method [20]. In the third experiment, the proposed method was compared with the three aforementioned phase unwrapping methods using the real Sentinel-1 InSAR data. For a clear comparison, RMSE was used for performance evaluation in these three experiments.

4.1. Performance Evaluation of PGENet

We first selected a testing sample with a coherence coefficient of 0.5 from the testing samples described in Section 3.1 to visually analyze the performance of phase gradient estimation and then evaluate the estimation accuracy of PGENet from the perspective of the phase gradient error. Meanwhile, PGENet was compared with PGE-PCA.

The reference DEM is shown in Figure 8a, and the corresponding wrapped phase image with a coherence coefficient of 0.5 is shown in Figure 8b. Figure 8c,d shows the ideal horizontal and vertical phase gradients, respectively. The horizontal and vertical phase gradients estimated by PGE-PCA are shown in Figure 9a,b, respectively, and the horizontal and vertical phase gradients estimated by PGENet are shown in Figure 9c,d, respectively. The corresponding gradient errors between the results estimated by PGE-PCA and ideal gradients are shown in Figure 10a,b. The corresponding gradient errors between the results estimated by PGENet and ideal gradients are shown in Figure 10c,d. It can been seen that the horizontal and vertical phase gradients obtained by PGENet are very close to the corresponding ideal horizontal and vertical phase gradients because most pixels of its error image are close to zero. In order to better quantify the gradient errors of the two methods, their error histogram curves were fitted according to their gradient error histograms and are shown in Figure 11. The horizontal axis of Figure 11 is the gradient error, and its vertical axis is the corresponding number of pixels of the gradient error image. The histogram curve can clearly shows the error distribution and is convenient for comparing the errors of different methods, so it was also used for subsequent analysis. From Figure 11, regardless of the horizontal gradient error or the vertical gradient error, it can be seen that the error curve of PGENet is more concentrated near zero and sharper than that of PGE-PCA, so the estimation accuracy of PGENet is significantly better than that of PGE-PCA from the perspective of the phase gradient error.

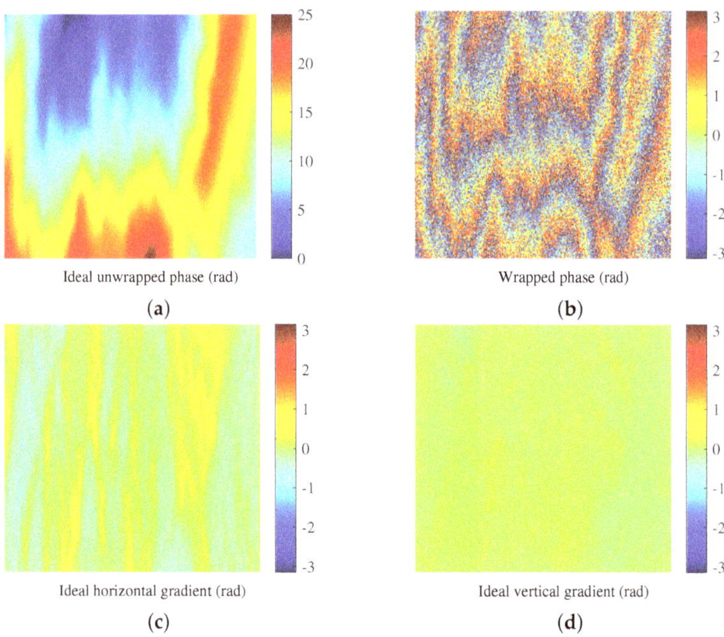

Figure 8. A testing sample. (**a**) Ideal unwrapped phase used for performance evaluation. (**b**) Noisy wrapped phase with a coherence coefficient of 0.5. (**c**) Ideal horizontal phase gradient. (**d**) Ideal Vertical phase gradient.

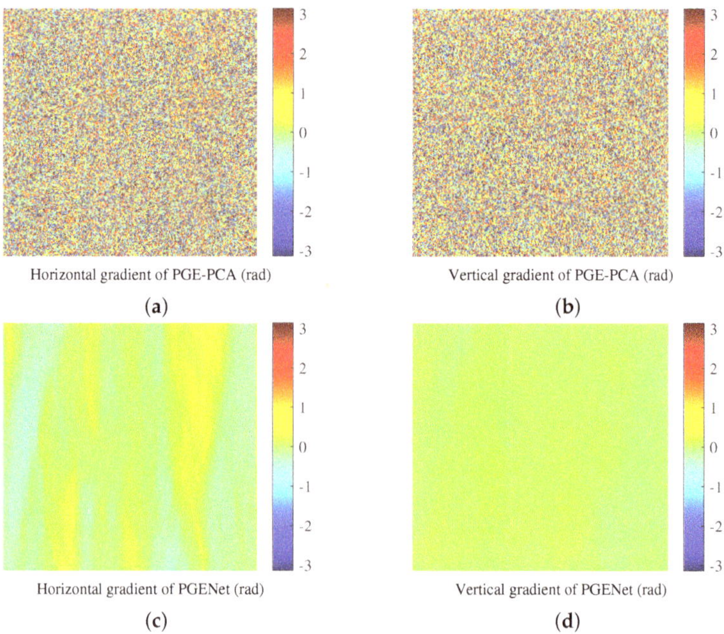

Figure 9. Phase gradient images estimated by two methods. (**a**) Horizontal phase gradient and (**b**) vertical phase gradient estimated by PGE-PCA. (**c**) Horizontal phase gradient and (**d**) vertical phase gradient estimated by PGENet.

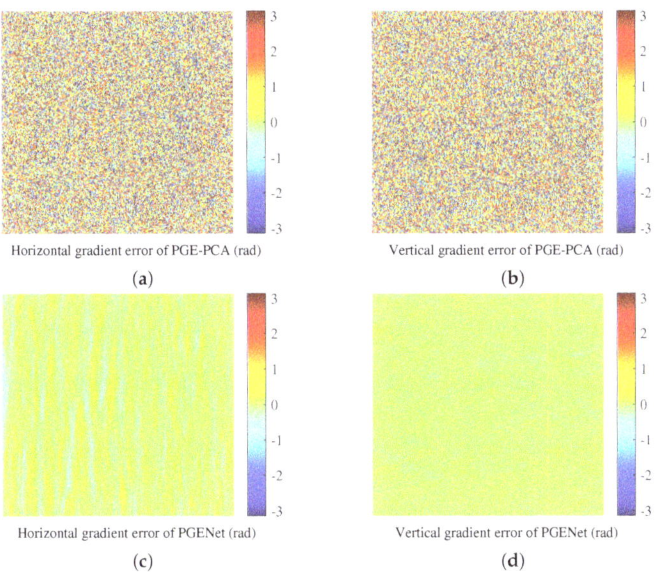

Horizontal gradient error of PGE-PCA (rad)

(**a**)

Vertical gradient error of PGE-PCA (rad)

(**b**)

Horizontal gradient error of PGENet (rad)

(**c**)

Vertical gradient error of PGENet (rad)

(**d**)

Figure 10. Phase gradient estimation errors. (**a**) Horizontal gradient error and (**b**) vertical gradient error of PGE-PCA. (**c**) Horizontal gradient error and (**d**) vertical gradient error of PGENet.

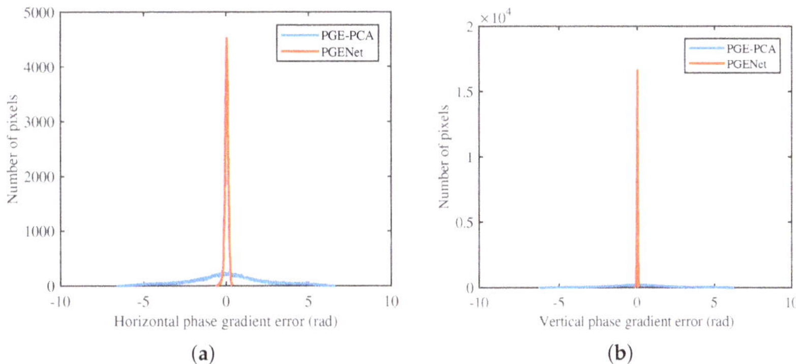

(**a**)

(**b**)

Figure 11. Phase gradient error curves. (**a**) Horizontal phase gradient error. (**b**) Vertical phase gradient error.

4.2. Robustness Testing of PGENet

As described in Section 3.1, the coherence coefficients of all testing samples ranged from 0.5 to 0.95. For the noise robustness testing, we calculated the mean values of the RMSE of PGENet and PGE-PCA for the testing samples with the same coherence coefficients. The results of the horizontal and vertical phase gradients are shown in Figure 12. Regardless of the horizontal gradient estimation result or the vertical gradient estimation result, we can observe that the RMSE of PGENet is smaller, which means that the accuracy of PGENet was significantly higher than that of PGE-PCA for each considered case. Moreover, the RMSE of PGENet did not change significantly with the changes in the coherence coefficient, which means that PGENet is robust to noise. To evaluate the comprehensive performance in response to different coherence coefficient situations, we calculated the mean values of the RMSE of all testing samples and list the results in Table 2. We can see that PGENet is robust to noise and has higher estimation accuracy than PGE-PCA.

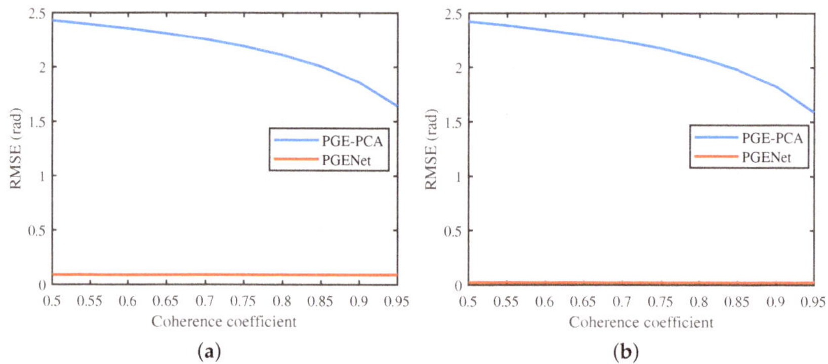

Figure 12. Root mean square errors (RMSE) of two phase gradient estimation methods on simulated data with different coherence coefficients. (**a**) Horizontal phase gradient estimation. (**b**) Vertical phase gradient estimation.

Table 2. A comparison of two phase gradient estimation methods using root mean square error (RMSE) on simulated data with different coherence coefficients.

Methods	Horizontal Phase Gradient	Vertical Phase Gradient
	RMSE (Rad)	RMSE (Rad)
PGE-PCA	2.16	2.14
PGENet	0.09	0.03

4.3. Performance Evaluation of Phase Unwrapping on Simulated Data

We selected a testing sample with a coherence coefficient of 0.5 (Figure 8b) to analyze the unwrapping accuracy of the proposed method from the perspective of the unwrapped phase error. The proposed method is compared with five widely-used phase unwrapping methods, namely, LS, QGPU, PUMA, and SNAPHU methods, and a deep learning-based method.

The unwrapped phases obtained by these six methods are shown in Figure 13. Figure 14 shows the corresponding errors between the estimated unwrapped phases and ideal unwrapped phase (Figure 8a). We can observe that the unwrapped phase obtained by the propose method is close to the ideal unwrapped phase because most pixels of its error image are close to zero. In order to better quantify the unwrapped phase errors of these six methods, their error histogram curves were fitted and are shown in Figure 15. From Figure 15, compared with the other five unwrapping methods, the error curve of the proposed method is more concentrated near zero and sharper, so the unwrapping accuracy of the proposed method was the highest among these six methods.

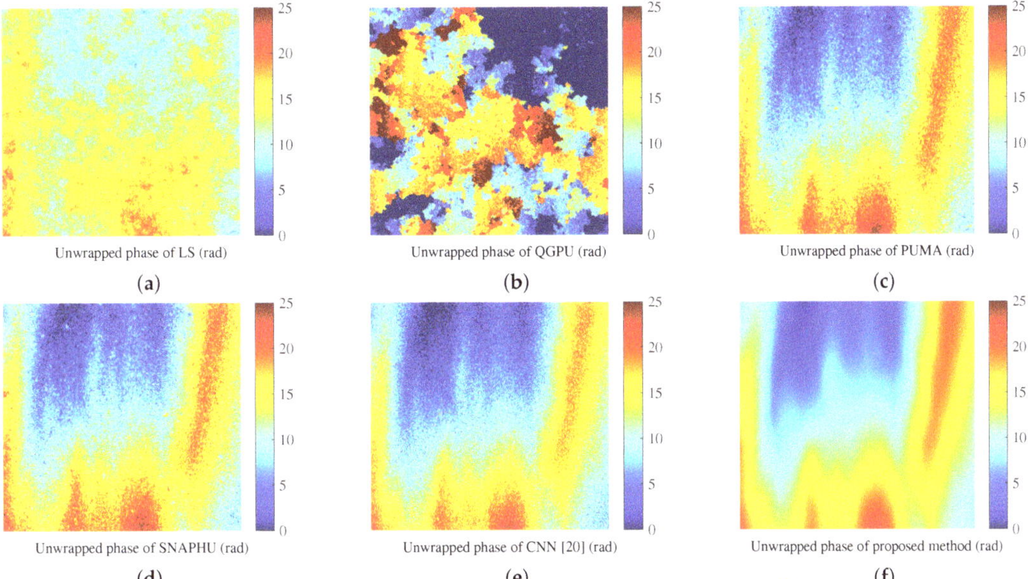

Figure 13. Unwrapped phase images of six phase unwrapping methods on simulated data. (**a**) LS. (**b**) QGPU. (**c**) PUMA. (**d**) SNAPHU. (**e**) CNN [20]. (**f**) Proposed method.

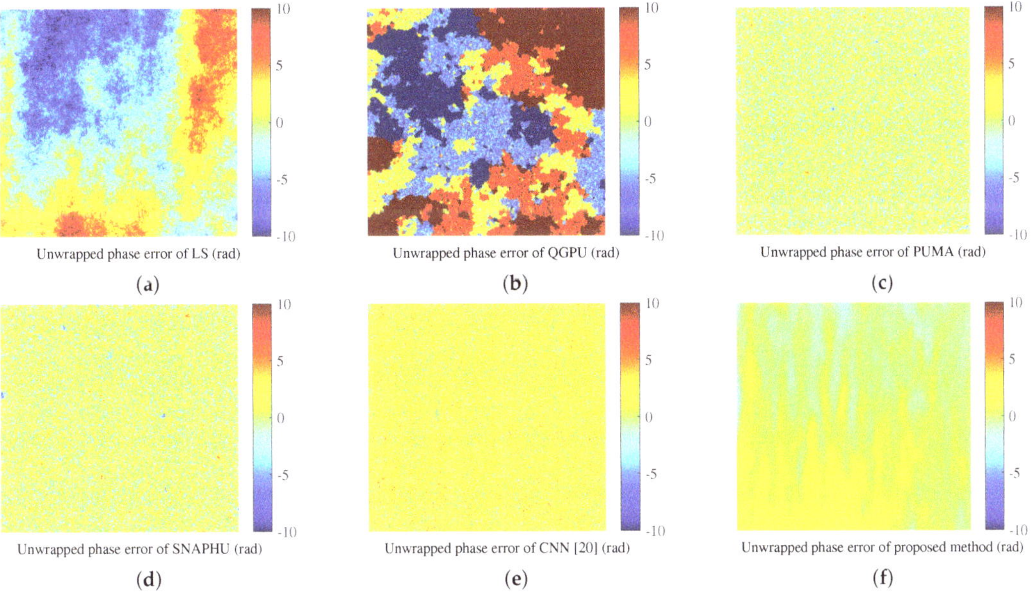

Figure 14. Unwrapped phase errors on simulated data. (**a**) LS. (**b**) QGPU. (**c**) PUMA. (**d**) SNAPHU. (**e**) CNN [20]. (**f**) Proposed method.

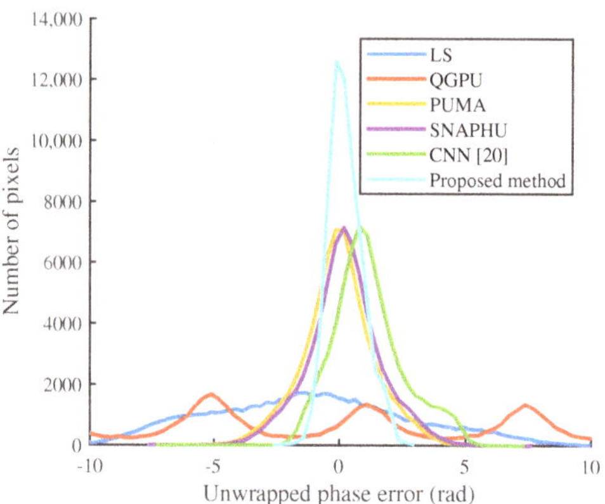

Figure 15. Fitted unwrapped phase error histogram curves of six phase unwrapping methods on simulated data.

4.4. Robustness Testing of Phase Unwrapping on Simulated Data

As described in Section 3.1, the coherence coefficients of all testing samples ranged from 0.5 to 0.95. For the noise robustness testing, we calculated the mean values of the unwrapping failure rate and RMSE of these six unwrapping methods for the testing samples with the same coherence coefficients, and the results are shown in Figure 16. It can been observed that the proposed method had the smallest unwrapping failure rate and RMSE for each considered coherence coefficient case; that is, among these six unwrapping methods, the proposed method has the highest unwrapping accuracy. Moreover, the unwrapping failure rate and RMSE of the proposed method did not change significantly with the changes in the coherence coefficient, which means the proposed method is robust to noise. To evaluate the comprehensive performance in response to different coherence coefficient situations, the mean values of the unwrapping rate and RMSE for all testing samples were calculated, and are listed in Table 3. We can observe that the unwrapping failure rate and RMSE of the proposed method are the smallest among all six unwrapping methods. Compared with the PUMA method, the unwrapping failure rate and RMSE of the proposed method were 89.3% and 49.5% lower, respectively. Compared with the CNN method [20], the unwrapping failure rate and RMSE of the proposed method were 96.3% and 61.2% lower, respectively. The reason for this performance improvement may be that the classification error is further amplified in the post-processing of the CNN method [20] due to the difference between adjacent categories corresponding to an unwrapped phase of 2π. Additionally, in the proposed method, the continuous phase gradient estimation can ensure the unwrapped phase error is within a small range. From Figure 14, we can indeed observe that the unwrapped phase error range of the CNN method [20] is significantly larger than that of the proposed method. In addition, the unwrapping failure rate and RMSE of the CNN method [20] increased as the noise level increased, which may be because the classification error increased as the noise level increased. Furthermore, as the level of noise decreased, the performance gaps among the different unwrapping methods gradually decreased. Based on the above analysis, the proposed method has the highest unwrapping accuracy among these six unwrapping method and is robust to noise.

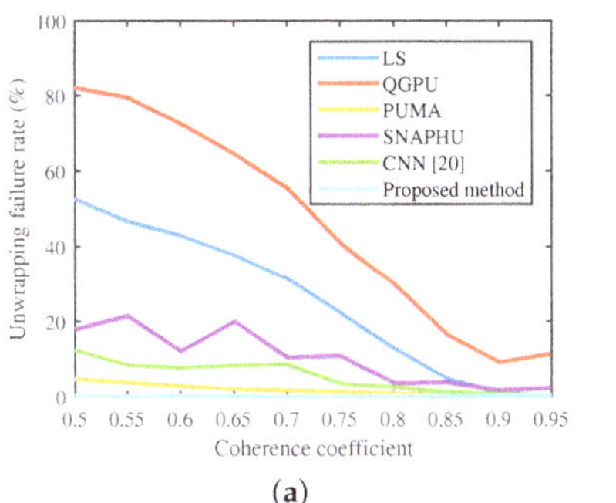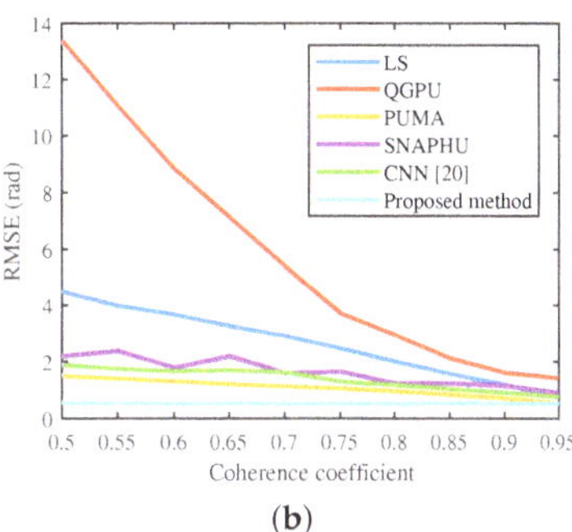

(a)   (b)

Figure 16. Quantitative indexes of six methods for phase unwrapping results on simulated wrapped phase images with different coherence coefficients. (**a**) Unwrapping failure rate. (**b**) Root mean square error (RMSE)

Table 3. Quantitative indexes of six phase unwrapping methods on simulated data.

Methods	Unwrapping Failure Rate (%)	RMSE (Rad)
LS	25.35	2.64
QGPU	46.24	5.77
PUMA	1.87	1.07
SNAPHU	10.49	1.65
CNN [20]	5.45	1.39
Proposed method	0.20	0.54

4.5. Performance Evaluation of Phase Unwrapping on Real InSAR Data

A real wrapped phase image was used to evaluate the phase unwrapping performance of the proposed method. The proposed method's performance is compared with those of other five unwrapping methods. Before unwrapping, the widely-used Goldstein phase filtering algorithm [37] was employed to suppress the noise of the real wrapped phase image. According to the unwrapping results, we performed two operations to obtain the estimated DEM: elevation inversion and terrain correction. This two operations were performed by the standard methods of SNAP software ("Phase to Elevation" and "Range Doppler Terrain Correction").

The wrapped phase image covering the eastern part of Turkey was computed from a pair of SLC images acquired by the SAR satellite Sentinel-1 Interferometric Wide Swath mode on July 2 and 8, 2019. Figure 17a is the wrapped phase image and Figure 17b is the corresponding reference DEM. The DEM is from SRTM 1Sec HGT and can be downloaded with the official SNAP software; and it was processed using the bilinear interpolation method to match the grid size of the estimated DEM (13.93 m).

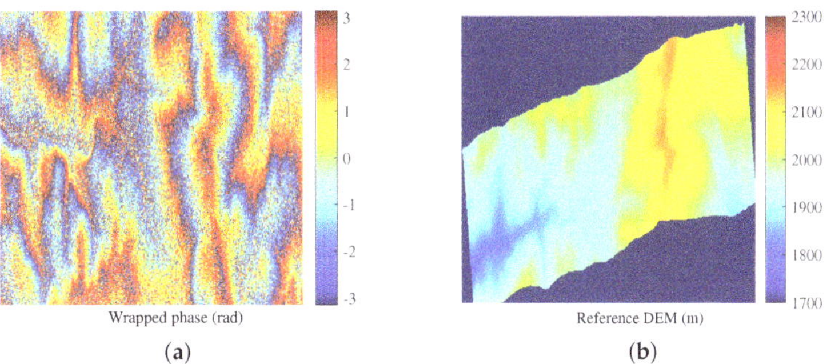

Wrapped phase (rad)   Reference DEM (m)

(a)   (b)

Figure 17. Sentinel-1 InSAR data. (**a**) Wrapped phase image. (**b**) Reference DEM.

In the case of performing phase filtering, the DEM results obtained by the phase unwrapping results of these six unwrapping methods are shown in Figure 18. Figure 19 shows the corresponding errors between the DEM solutions and the reference DEM. We can observe that the DEM obtained by the proposed method is closest to the reference DEM among these six methods because most pixels of its error image are closest to zero. To better quantify the DEM errors of these six unwrapping methods, the histogram error curves were fitted and are shown in Figure 20. From Figure 20, compared with the other five unwrapping methods, the error curve of the proposed method is more concentrated near zero and sharper, so the unwrapping accuracy of the proposed method was the highest among these six methods.

For quantitative evaluation, the RMSE between DEM solutions obtained by these six methods and the reference DEM were calculated and are listed in Table 4. As seen from Table 4, the RMSE of the proposed method is the smallest among these six unwrapping methods. In addition, compared with the PUMA method and CNN method [20], the RMSE of the proposed method is 3.73% or 7.0% lower, respectively. The performance improvement was not as large as with the simulated data, because the noise level was greatly reduced after performing phase filtering. As the level of noise decreases, the performance gaps among different unwrapping methods gradually decreases. Based on the above analysis, the unwrapping accuracy of the proposed method is the highest among these six methods.

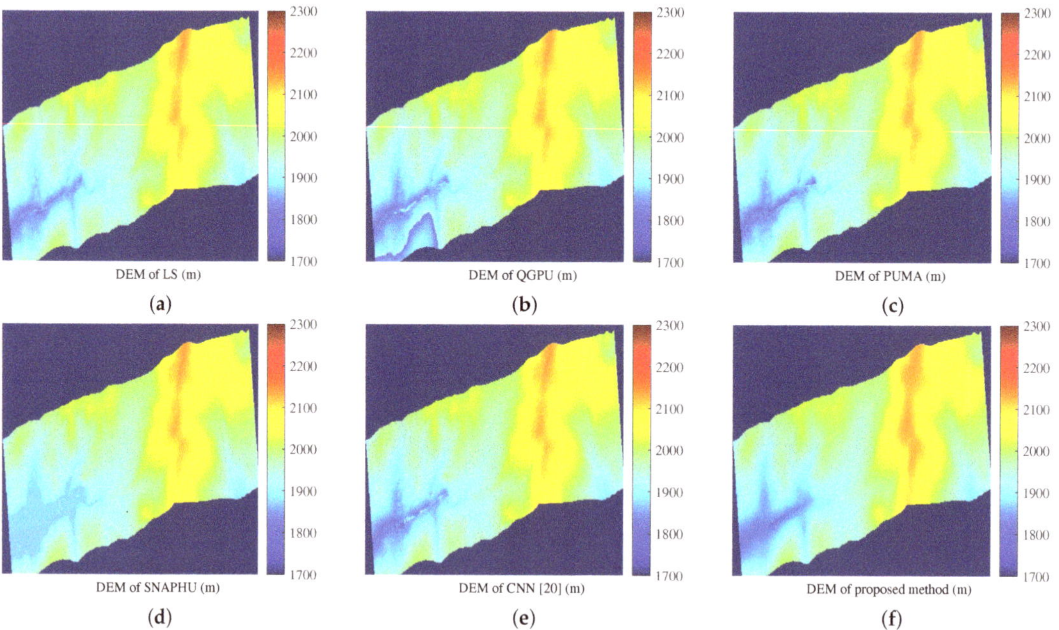

Figure 18. In the case of performing phase filtering, the DEM results of six phase unwrapping methods on real data. (**a**) LS. (**b**) QGPU. (**c**) PUMA. (**d**) SNAPHU. (**e**) CNN [20]. (**f**) Proposed method.

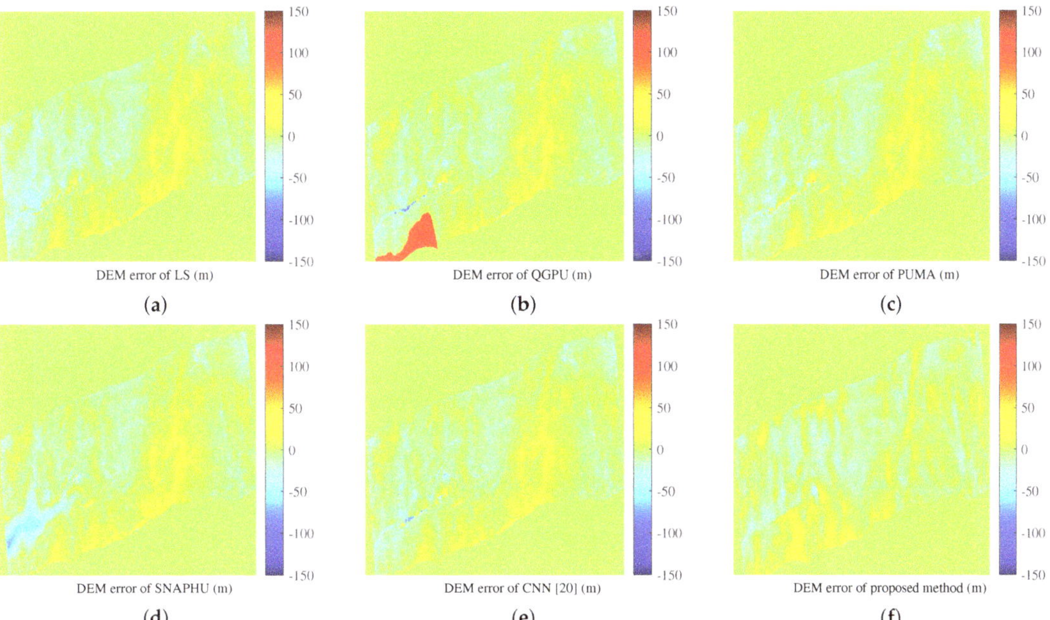

DEM error of LS (m)

(**a**)

DEM error of QGPU (m)

(**b**)

DEM error of PUMA (m)

(**c**)

DEM error of SNAPHU (m)

(**d**)

DEM error of CNN [20] (m)

(**e**)

DEM error of proposed method (m)

(**f**)

Figure 19. In the case of performing phase filtering, the DEM errors on real data. (**a**) DEM errors of the least squares method. (**b**) DEM errors of QGPU. (**c**) DEM errors of PUMA. (**d**) DEM errors of SNAPHU. (**e**) DEM errors of CNN [20]. (**f**) DEM errors of the proposed method.

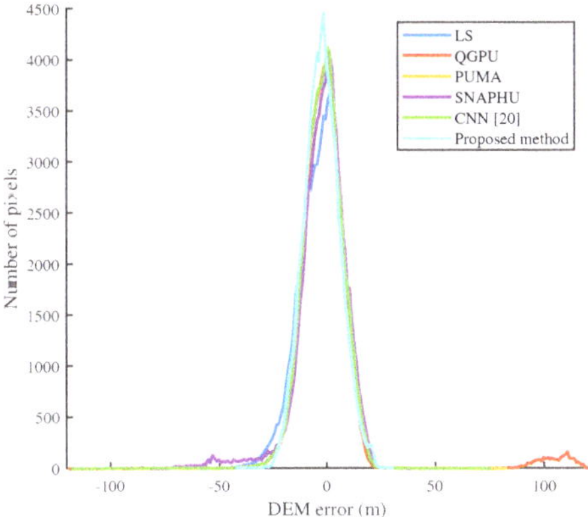

Figure 20. In the case of performing phase filtering, the fitted DEM error histogram curves of six phase unwrapping methods on real InSAR data.

Table 4. In the case of performing phase filtering, the comparison of six phase unwrapping methods using root mean square error (RMSE) on real InSAR data.

Methods	RMSE (m)
LS	8.00
QGPU	15.16
PUMA	7.02
SNAPHU	9.09
CNN [20]	7.27
Proposed method	6.76

4.6. Robustness Testing of Phase Unwrapping on Real InSAR Data

In real data processing, to improve the accuracy of phase unwrapping, phase filtering is often required before unwrapping, but after filtering, there will still be different levels of noise, and its level depends on the noise suppression performance of the filtering algorithm and the quality of the wrapped phase [38]. To evaluate the robustness of the proposed method, this section shows the DEM estimation results of Figure 17a in the case of no phase filtering.

The DEM results obtained by the phase unwrapping results of these six unwrapping methods without phase filtering are shown in Figure 21. Figure 22 shows the corresponding errors between the DEM solutions and the reference DEM. The histogram error curves were fitted and are shown in Figure 23, and the RMSE between DEM solutions and the reference DEM are listed in Table 5. We can see that the unwrapping accuracy of the proposed method was still the highest among these six methods because the RMSE of the proposed method is smallest. In addition, compared with Section 4.5, the unwrapping results of the LS and QGPU methods are failures, and the performances of the PUMA, SNAPHU, and CNN [20] methods decreased significantly when the noise level became larger because their RMSE increased by 85.04%, 39.60%, and 66.16%, respectively. The performance of the proposed method decreased slightly because its RMSE increased by 7.25%. Based on the above analysis, it can be seen that the proposed method has better anti-noise performance than the other five methods in real data processing.

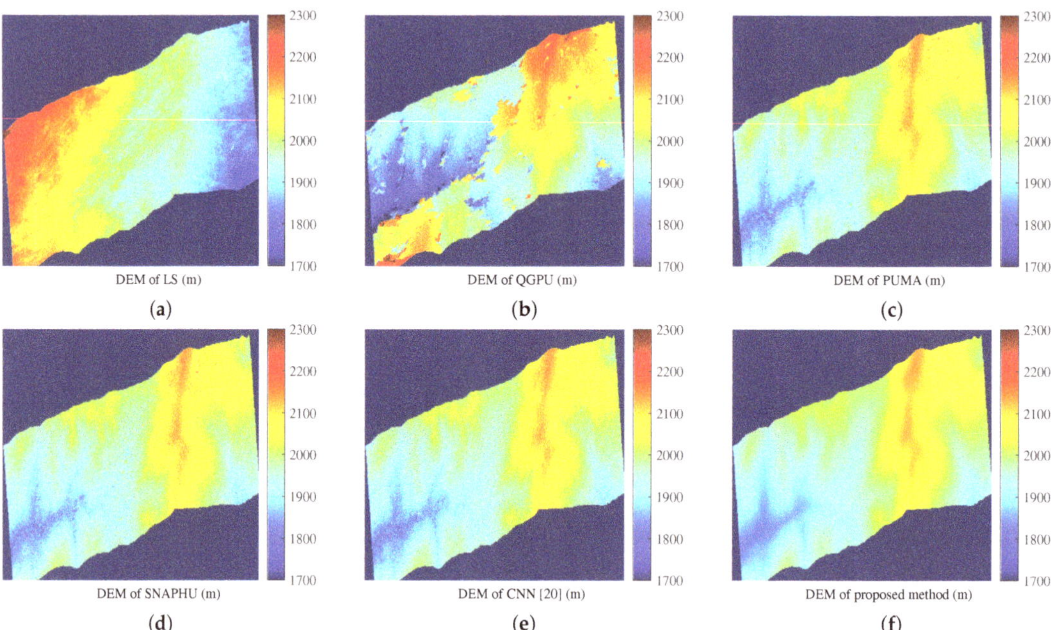

Figure 21. In the case of no phase filtering, the DEM results of six phase unwrapping methods on real data. (**a**) LS. (**b**) QGPU. (**c**) PUMA. (**d**) SNAPHU. (**e**) CNN [20]. (**f**) Proposed method.

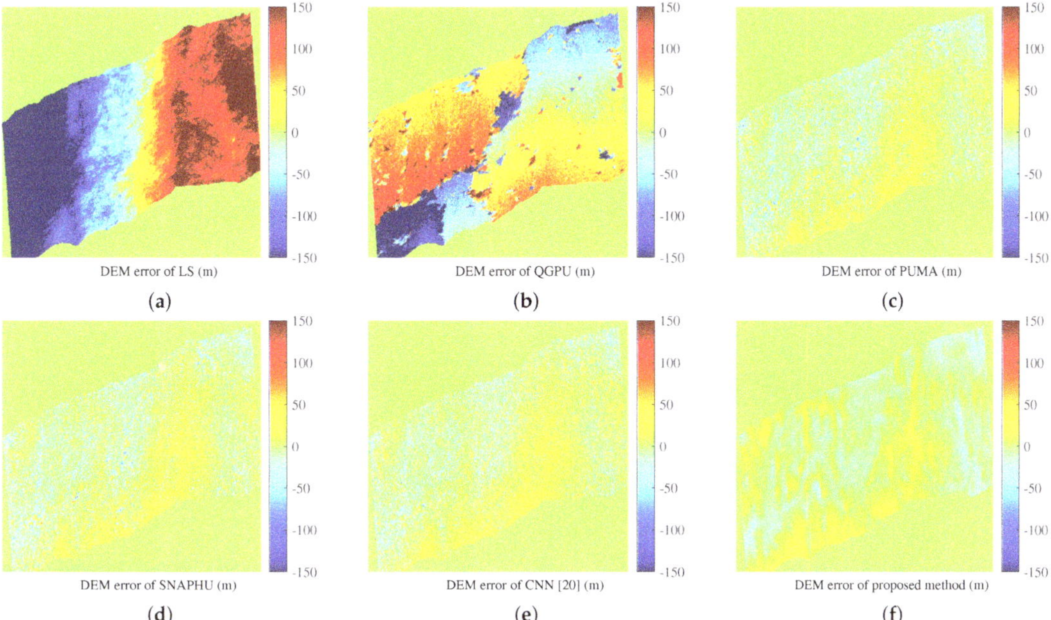

Figure 22. In the case of no phase filtering, the DEM errors on real data. (**a**) DEM errors of the least squares method. (**b**) DEM errors of QGPU. (**c**) DEM errors of PUMA. (**d**) DEM errors of SNAPHU. (**e**) DEM errors of CNN [20]. (**f**) DEM errors of the proposed method.

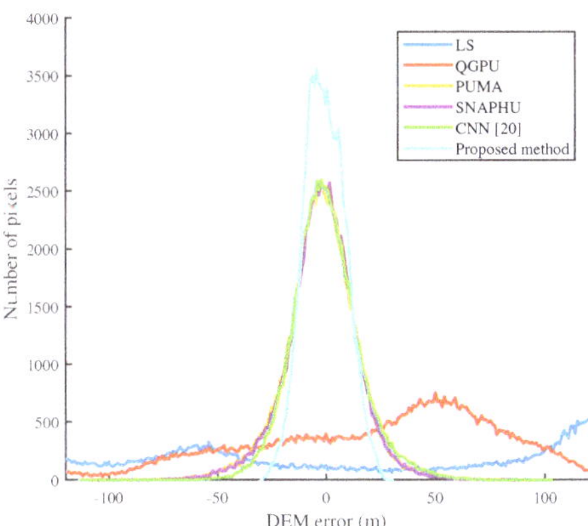

Figure 23. In the case of no phase filtering, the fitted DEM error histogram curves of six phase unwrapping methods on real InSAR data.

Table 5. In the case of no phase filtering, the comparison of six phase unwrapping methods using root mean square error (RMSE) on real InSAR data.

Methods	RMSE (m)
LS	110.89
QGPU	64.08
PUMA	12.99
SNAPHU	12.69
CNN [20]	12.08
Proposed method	7.25

5. Discussion

To analyze the influence of networks with different block numbers on the unwrapping performance, we complemented three experiments with simulated data, and their numbers of blocks were 10, 8, and 5, respectively. Their RMSE were calculated and are listed in Table 6. As the RMSE of the network with eight blocks were smallest, we selected the network with eight blocks. From Tables 3 and 6, we can see that different block numbers led to slight fluctuations in RMSE for our method, but the unwrapping performance was still better than those of the other five unwrapping methods.

Table 6. Root mean square errors (RMSE) of the proposed method in networks with different numbers of blocks.

Number of Block	RMSE (rad)
10	0.6039
8	0.5425
5	0.6150

6. Conclusions

In this paper, a robust InSAR phase unwrapping method combining PGENet and the least squares solver was proposed to improve the accuracy of phase unwrapping. We designed PGENet to estimate the horizontal and vertical gradients first, and then the phase unwrapping result is obtained by using the least squares solver to minimize the difference between the gradient obtained by PGENet and the gradient of the unwrapped phase. The horizontal and vertical gradients estimated by PGENet are used to replace the gradients estimated by PGE-PCA in the traditional LS unwrapping method. PGENet can extract global high-level phase features and recognize the phase gradient between adjacent pixels from lots of wrapped phase images with topography features and different coherence coefficients. Therefore, compared with the phase gradient obtained by PGE-PCA, the more accurate and robust phase gradient can be estimated by PGENet. This is the reason why the proposed method has higher precision and better robustness than the traditional LS unwrapping method. The experimental results on simulated data showed that the proposed method has the highest unwrapping accuracy among six widely-used unwrapping methods and is robust to noise. Furthermore, when processing the real Sentinel-1 InSAR data, the proposed method had the best performance among these six unwrapping methods.

The proposed method successfully combines deep learning and the traditional LS method for InSAR phase unwrapping. The core of this method is the accurate and robust phase gradient estimation based on PGENet, which makes the proposed method have high accuracy and robustness. In future work, to achieve more accurate unwrapping, we will make targeted modifications to PGENet to match more traditional InSAR phase unwrapping methods. In addition, we will use the proposed phase unwrapping method to process more real InSAR data.

Author Contributions: Conceptualization, L.P., X.Z. and J.S.; methodology, L.P. and X.Z.; software, L.P., Z.Z. and S.W.; validation, L.P., Z.Z. and J.S.; formal analysis, L.P., S.W., L.L. and L.Z.; investigation, L.P., Z.Z., L.Z. and L.L.; resources, X.Z. and S.W.; data curation, L.P. and X.Z.; writing—original draft preparation, L.P., X.Z. and J.S.; writing—review and editing, L.P., L.L., L.Z. and S.W.; visualization, L.P., Z.Z., L.L. and L.Z.; supervision, X.Z. and J.S.; project administration, X.Z.; funding acquisition, X.Z. All authors have read and agreed to the published version of the manuscript.

Funding: This research was supported in part by the National Key R&D Program of China under grant 2017YFB0502700, and in part by the National Natural Science Foundation of China under grants 61571099, 61501098, and 61671113.

Institutional Review Board Statement: Not applicable.

Informed Consent Statement: Not applicable.

Data Availability Statement: Not applicable.

Acknowledgments: We thank all editors and reviewers and for their valuable comments and suggestions for improving this manuscript.

Conflicts of Interest: The authors declare no conflict of interest.

References

1. Moreira, A.; Prats-Iraola, P.; Younis, M.; Krieger, G.; Hajnsek, I.; Papathanassiou, K.P. A tutorial on synthetic aperture radar. *IEEE Geosci. Remote Sens. Mag.* **2013**, *1*, 6–43.
2. Yu, H.; Lan, Y.; Yuan, Z.; Xu, J.; Lee, H. Phase Unwrapping in InSAR: A Review. *IEEE Geosci. Remote Sens. Mag.* **2019**, *7*, 40–58.
3. Zhu, X.; Wang, Y.; Montazeri, S.; Ge, N. A Review of Ten-Year Advances of Multi-Baseline SAR Interferometry Using TerraSAR-X Data. *Remote Sens.* **2018**, *10*, 1374.
4. Bamler, R.; Hartl, P. Synthetic aperture radar interferometry. *Inverse Probl.* **1998**, *14*, R1.
5. Goldstein, R.M.; Zebker, H.A.; Werner, C.L. Satellite radar interferometry: Two-dimensional phase unwrapping. *Radio Sci.* **1988**, *23*, 713–720.
6. Itoh, K. Analysis of the phase unwrapping algorithm. *Appl. Opt.* **1982**, *21*, 2470. doi:10.1364/AO.21.002470.
7. Herráez, M.A.; Burton, D.R.; Lalor, M.J.; Gdeisat, M.A. Fast two-dimensional phase-unwrapping algorithm based on sorting by reliability following a noncontinuous path. *Appl. Opt.* **2002**, *41*, 7437–7444.
8. Lin, Q.; Vesecky, J.F.; Zebker, H.A. New approaches in interferometric SAR data processing. *IEEE Trans. Geosci. Remote Sens.* **1992**, *30*, 560–567.
9. Yan, L.; Zhang, H.; Zhang, R.; Xie, X.; Chen, B. A robust phase unwrapping algorithm based on reliability mask and weighted minimum least-squares method. *Opt. Lasers Eng.* **2019**, *112*, 39–45.
10. Bioucas-Dias, J.M.; Valadao, G. Phase unwrapping via graph cuts. *IEEE Trans. Image Process.* **2007**, *16*, 698–709.
11. Zhang, Y.Z.; Wang, W.D.; Li, P. Minimum L 2 norm two dimensional phase unwrapping. *J. Earth Sci. Enivron.* **2005**, *1*, 80–83.
12. Pritt, M.D. Phase unwrapping by means of multigrid techniques for interferometric SAR. *IEEE Trans. Geosci. Remote Sens.* **1996**, *34*, 728–738.
13. Ghiglia, D.C.; Romero, L.A. Robust two-dimensional weighted and unweighted phase unwrapping that uses fast transforms and iterative methods. *JOSA A* **1994**, *11*, 107–117.
14. Chen, C.W.; Zebker, H.A. Phase unwrapping for large SAR interferograms: Statistical segmentation and generalized network models. *IEEE Trans. Geosci. Remote Sens.* **2002**, *40*, 1709–1719.
15. Chen, C.W.; Zebker, H.A. Network approaches to two-dimensional phase unwrapping: intractability and two new algorithms. *JOSA A* **2000**, *17*, 401–414.
16. Liang, J.; Zhang, J.; Shao, J.; Song, B.; Yao, B.; Liang, R. Deep Convolutional Neural Network Phase Unwrapping for Fringe Projection 3D Imaging. *Sensors* **2020**, *20*, 3691.
17. Zhang, T.; Jiang, S.; Zhao, Z.; Dixit, K.; Zhou, X.; Hou, J.; Zhang, Y.; Yan, C. Rapid and robust two-dimensional phase unwrapping via deep learning. *Opt. Express* **2019**, *27*, 23173–23185.
18. Spoorthi, G.; Gorthi, R.K.S.S.; Gorthi, S. PhaseNet 2.0: Phase Unwrapping of Noisy Data Based on Deep Learning Approach. *IEEE Trans. Image Process.* **2020**, *29*, 4862–4872.
19. Li, S.; Xu, H.; Gao, S.; Li, C. A non-fuzzy interferometric phase estimation algorithm based on modified Fully Convolutional Network. *Pattern Recognit. Lett.* **2019**, *128*, 60–69.
20. Sica, F.; Calvanese, F.; Scarpa, G.; Rizzoli, P. A CNN-Based Coherence-Driven Approach for InSAR Phase Unwrapping. *IEEE Geosci. Remote Sens. Lett.* **2020**, early access, doi:10.1109/LGRS.2020.3029565.
21. Zhou, L.; Yu, H.; Lan, Y. Deep Convolutional Neural Network-Based Robust Phase Gradient Estimation for Two-Dimensional Phase Unwrapping Using SAR Interferograms. *IEEE Trans. Geosci. Remote Sens.* **2020**, *58*, 4653–4665. doi:10.1109/tgrs.2020.2965918.
22. Zhou, L.; Yu, H.; Lan, Y.; Xing, m. Artificial Intelligence In Interferometric Synthetic Aperture Radar Phase Unwrapping: A Review. *IEEE Geosci. Remote Sens. Mag.* **2021**, *9*, 10–28. doi:10.1109/mgrs.2021.3065811.
23. Wang, H.; Hu, J.; Fu, H.; Wang, C.; Wang, Z. A Novel Quality-Guided Two-Dimensional InSAR Phase Unwrapping Method via GAUNet. *IEEE J. Sel. Top. Appl. Earth Obs. Remote Sens.* **2021**, *14*, 7840–7856.
24. Ahmad, A.; Lu, Y. Identifying the phase discontinuities in the wrapped phase maps by a classification framework. *Opt. Eng.* **2016**, *55*, 033104.
25. Costantini, M. A novel phase unwrapping method based on network programming. *IEEE Trans. Geosci. Remote Sens.* **1998**, *36*, 813–821.
26. Pritt, M.D. Weighted least squares phase unwrapping by means of multigrid techniques. In Proceedings of the Synthetic Aperture Radar and Passive Microwave Sensing, Paris, France, 25–28 September 1995;International Society for Optics and Photonics: Paris, France, 1995; Volume 2584, pp. 278–288.
27. Guo, Y.; Chen, X.; Zhang, T. Robust phase unwrapping algorithm based on least squares. *Opt. Lasers Eng.* **2014**, *63*, 25–29.
28. Mao, X.; Shen, C.; Yang, Y.B. Image restoration using very deep convolutional encoder-decoder networks with symmetric skip connections. *Adv. Neural Inf. Process. Syst.* **2016**, *29*, 2802–2810.

29. Tao, X.; Gao, H.; Shen, X.; Wang, J.; Jia, J. Scale-recurrent network for deep image deblurring. In Proceedings of the IEEE Conference on Computer Vision and Pattern Recognition, Salt Lake City, UT, USA, 18–23 June 2018; pp. 8174–8182.
30. Ronneberger, O.; Fischer, P.; Brox, T. U-net: Convolutional networks for biomedical image segmentation. In Proceedings of the International Conference on Medical Image Computing and Computer-Assisted Intervention, Munich, Germany, 5–9 October 2015; pp. 234–241.
31. Nah, S.; Hyun Kim, T.; Mu Lee, K. Deep multi-scale convolutional neural network for dynamic scene deblurring. In Proceedings of the IEEE Conference on Computer Vision and Pattern Recognition, Honolulu, HI, USA, 21–26 July 2017; pp. 3883–3891.
32. Zhou, Y.; Shi, J.; Yang, X.; Wang, C.; Kumar, D.; Wei, S.; Zhang, X. Deep multi-scale recurrent network for synthetic aperture radar images despeckling. *Remote Sens.* **2019**, *11*, 2462.
33. Pu, L.; Zhang, X.; Zhou, Z.; Shi, J.; Wei, S.; Zhou, Y. A Phase Filtering Method with Scale Recurrent Networks for InSAR. *Remote Sens.* **2020**, *12*, 3453.
34. Kingma, D.P.; Ba, J. Adam: A method for stochastic optimization. *arXiv* **2014**, arXiv:1412.6980.
35. Yao, Y.; Rosasco, L.; Caponnetto, A. On early stopping in gradient descent learning. *Constr. Approx.* **2007**, *26*, 289–315.
36. Raskutti, G.; Wainwright, M.J.; Yu, B. Early stopping and non-parametric regression: an optimal data-dependent stopping rule. *J. Mach. Learn. Res.* **2014**, *15*, 335–366.
37. Goldstein, R.M.; Werner, C.L. Radar interferogram filtering for geophysical applications. *Geophys. Res. Lett.* **1998**, *25*, 4035–4038.
38. Wang, Y.; Huang, H.; Dong, Z.; Wu, M. Modified patch-based locally optimal Wiener method for interferometric SAR phase filtering. *ISPRS J. Photogramm. Remote Sens.* **2016**, *114*, 10–23.

Article

TCD-Net: A Novel Deep Learning Framework for Fully Polarimetric Change Detection Using Transfer Learning

Rezvan Habibollahi [1], Seyd Teymoor Seydi [1], Mahdi Hasanlou [1,*] and Masoud Mahdianpari [2,3]

[1] School of Surveying and Geospatial Engineering, College of Engineering, University of Tehran, Tehran 14174-66191, Iran; r.habibollahi@ut.ac.ir (R.H.); seydi.teymoor@ut.ac.ir (S.T.S.)

[2] C-CORE, 1 Morrissey Road, St. John's, NL A1B 3X5, Canada; m.mahdianpari@mun.ca

[3] Department of Electrical and Computer Engineering, Memorial University of Newfoundland, St. John's, NL A1C 5S7, Canada

* Correspondence: hasanlou@ut.ac.ir; Tel.: +98-21-6111-4525

Abstract: Due to anthropogenic and natural activities, the land surface continuously changes over time. The accurate and timely detection of changes is greatly important for environmental monitoring, resource management and planning activities. In this study, a novel deep learning-based change detection algorithm is proposed for bi-temporal polarimetric synthetic aperture radar (PolSAR) imagery using a transfer learning (TL) method. In particular, this method has been designed to automatically extract changes by applying three main steps as follows: (1) pre-processing, (2) parallel pseudo-label training sample generation based on a pre-trained model and fuzzy c-means (FCM) clustering algorithm, and (3) classification. Moreover, a new end-to-end three-channel deep neural network, called TCD-Net, has been introduced in this study. TCD-Net can learn more strong and abstract representations for the spatial information of a certain pixel. In addition, by adding an adaptive multi-scale shallow block and an adaptive multi-scale residual block to the TCD-Net architecture, this model with much lower parameters is sensitive to objects of various sizes. Experimental results on two Uninhabited Aerial Vehicle Synthetic Aperture Radar (UAVSAR) bi-temporal datasets demonstrated the effectiveness of the proposed algorithm compared to other well-known methods with an overall accuracy of 96.71% and a kappa coefficient of 0.82.

Keywords: unsupervised change detection; polarimetric synthetic aperture radar (PolSAR); UAVSAR; multi-scale shallow block; multi-scale residual block

Citation: Habibollahi, R.; Seydi, S.T.; Hasanlou, M.; Mahdianpari, M. TCD-Net: A Novel Deep Learning Framework for Fully Polarimetric Change Detection Using Transfer Learning. *Remote Sens.* **2022**, *14*, 438. https://doi.org/10.3390/rs14030438

Academic Editors: Gwanggil Jeon, Tianwen Zhang, Xiaoling Zhang and Tianjiao Zeng

Received: 12 December 2021
Accepted: 17 January 2022
Published: 18 January 2022

1. Introduction

The proliferation of remote sensing (RS) images at different temporal and spatial resolutions have increased its use in a wide range of global environmental and management applications, including change detection [1–5], target detection [6,7], wetland classification [8–10], oil spill detection [11–13], disaster monitoring [14,15] and so on. Detection of change is one of the most important applications of RS, which is essential for better resource management.

Change detection (CD) is the process of identifying changes, caused by manmade or natural factors, in multi-temporal Earth Observation (EO) data [16]. CD algorithms are commonly employed to monitor changes in different applications, including land use and land cover (LULC) [17,18], deforestation [19], urban development [20] and natural disaster [20].

In recent years, Synthetic Aperture Radar (SAR) sensors have become one of the most popular alternatives to other RS techniques because they can provide imaging in all weather conditions, day or night. In addition, SAR sensors are capable of penetrating through clouds, rain, smoke, snow, dust and so on. Therefore, these factors cannot affect the ability of SAR sensors. In addition, SAR sensors use their source of illumination to detect the target. Therefore, the light conditions of the area do not affect their imaging [21].

In general, SAR systems have more advantages than optical sensors in CD applications, because of their ability to acquire periodic images, regardless of weather or daylight [22].

Polarisation is one of the properties of an electromagnetic wave described as a function of time on a plane perpendicular to the direction of propagation based on the geometric location of the electric field vector [22]. PolSAR systems can transmit and receive waves in a variety of linear polarization or circular polarization. This characteristic will provide more scattering information from different aspects of a target. To send waves in linear polarization, two common base polarizations, horizontal linear polarization (H) and vertical linear polarization (V), are used. To send waves in circular polarization, the basic polarizations of right-handed and left-handed circles are used. In PolSAR systems, the transmitted and received waves can be sent and received both in the cross-polarizations (e.g., HV or VH polarization) and in the co-polarizations (e.g., HH or VV polarisation) [23]. Using fully PolSAR data, polarization information can be significantly extracted because it allows phase measurements between different polarization channels [22]. Nevertheless, the microwave imaging mechanism used in PolSAR images makes the background more complicated and the features of the region are mixed up. This is reflected in the structural sensitivity, the geometric distortion of the image, the interference of the imaging systems, and the speckle noise. Therefore, compared to other types of EO data, detecting changes in SAR data is more challenging and thus has been less investigated.

Three main steps for unsupervised CD methods in a SAR image can be summarized as (1) pre-processing, (2) difference image (DI) generation, and (3) analysis of DI to generate the change map [24,25]. In SAR image pre-processing, multi-looking, co-registration of images [26] and speckle filtering [27] are the main techniques. Additionally, DI quality has a significant impact on the final change map. Two common methods for DI production are image difference and image ratio. The main advantage of these methods is simplicity, but they do not consider the edge and neighboring information, and thus have low sensitivity to the speckle noise level [28]. However, the mean operator [29] considers neighboring information and has an excellent inhibitory effect on independent points. To extract more robust features and to improve detection performance, the transformation-based models have been proposed in SAR CD [30]. These approaches transform raw feature vectors to a new feature representation, to reduce the impact of noise, suppress *no-change* areas and highlight the *changes* in a new feature space [31]. For instance, principal component analysis (PCA), multivariate alteration detection (MAD) and iterative reweighted multivariate alteration detection (IR-MAD) were utilized in PolSAR CD [31]. Based on recent studies, transformation-based models have a high ability to extract information [31,32]. However, in these methods, manual feature extraction and identification of information-rich components is an important challenge. Furthermore, these algorithms are pixel-based and do not consider spatial features (e.g., texture).

After DI generation, the analysis of DI is usually done through thresholding or clustering strategies. The key point in the thresholding method is to opt for the threshold value. Several popular methods, such as the Kittler and Illingworth (KI) algorithm [33] and the expectation maximization (EM) algorithm [34], are used in SAR data. In these methods, a model must be established to fit the *no-changed* and *changed* class condition distributions. These methods have weak consequences when the *change* and *no-change* features overlap, or when their statistical distributions are mistakenly modeled and in some cases require frustrating trials and errors. In addition, a generalized KI (GKI) threshold selection algorithm [35], a histogram optimization method [36], and a semi-EM algorithm [37] are used to automatically generate a threshold value in SAR data. Since SAR images are extremely influenced by speckle noise, methods that determine thresholds automatically cannot eliminate it, because noise affects the estimation of parameters of the statistical model. Moreover, choosing a global threshold does not make sense for the entire image and may not cover all sections. Another method for analysis of DI is clustering. This is often based on the k-means [38], multiple kernel k-means [39], and fuzzy c-means (FCM) [40]. Although these algorithms are widely used in SAR CD, there are substantial disadvantages [38–40]. On the one hand, these algorithms are distance-based (Euclidean distance, Mahalanobis distance and so on), which is very sensitive to speckle noise and on the other, these algorithms are presented assuming a balance between *change* and *no-change* classes. In many cases, the *change* pixels are far less than the *no-change* pixels, i.e., the imbalance between the two classes. Traditional clustering methods led to extreme false alarms when challenged with unbalanced data. There are other clustering methods for SAR CD, such as the fuzzy local information C-means algorithm (FLICM) [41] and the reformulated FLICM algorithm (RFLICM) [42], which adds local information to the fuzzy method. Clustering methods have greater flexibility than thresholding methods because there is no need to construct a model. However, they are sensitive to noise because of inadequate attention to spatial information.

Recently, several deep learning (DL) algorithms, such as stacked auto-encoders (SAEs) [43], deep belief networks (DBNs) [44], convolutional neural networks (CNNs) [2], recurrent neural networks (RNNs) [45], pulse coupled neural networks (PCNNs) [46] and generative adversarial networks (GANs) [47] have been proposed for detection changes in EO data. Among these DL methods, the CNN model is commonly employed as a feature extractor for solving visual tasks. One of the most important advantages of CNNs is the automatic extraction of low- to high-level features. Therefore, unlike PCA, MAD and IRMAD algorithms, CNNs do not require manual feature selection and extraction.

1.1. Related Works

The various DL approaches can be divided into several categories. In this study, the DL approaches used to CD are classified into three categories based on the learning technique and the accessibility of labeled or unlabeled training datasets, including (1) supervised, (2) unsupervised, and (3) semi-supervised methods. The first category is supervised methods which train the network by using labeled training datasets. The second category, unsupervised methods that learn from unlabeled datasets. The third category, semi-supervised methods that learn from both labeled and unlabeled datasets.

1.1.1. DL Supervised Methods for EO Data

There are many challenges in training deep supervised neural networks. The most important of which is the need for a large training dataset. The need for large training data, especially in RS applications that sometimes do not have access to the area, remains one of the most substantial challenges. Numerous studies have examined the performance of monitored networks in CD applications. Accordingly, it has been shown that deep neural networks can properly generate change maps if large amounts of labeled training datasets are available.

Mou et al. [48] have proposed a supervised dual-branch end-to-end neural network method for CD. In this network, a CNN and an RNN are joined, therefore developing a Recurrent Convolutional Neural Network (ReCNN) deep architecture. This algorithm was implemented in three main steps: (1) initially, convolutional layers construct feature maps automatically from each image in two separate branches; (2) second, after extracting the feature from both images, a recurrent sub-network is embedded to preserve temporal dependence in the bi-temporal images; and (3) finally, the output of recurrent sub-network is entered as a fully connected layer and a change map is extracted. More specifically, they used three types of the recurrent sub-network, i.e., fully connected RNN, long short-time memory (LSTM) and gated recurrent unit (GRU) to compute the hidden state information for the current input and restore information [48]. Liu, Jiao, Tang, Yang, Ma and Hou [18] have presented a local restricted CNN (LRCNN), which is a new version of CNN, in two main steps: (1) first, they proposed a similarity measure for PolSAR data and produced several Layered Difference Images (LDIs) of PolSAR images. Then, LDIs are improved to Discriminative Enhanced LDIs (DELDIs) for CNN training, and (2) second, the CNN/LRCNN was trained for CD tuning hyperparameters. Finally, based on the optimized trained model [18], a change map was obtained. Jaturapitpornchai et al. [49]

have proposed a supervised method to identify novel building structures in three main steps: (1) first, each 256 × 256-pixel patch at time1 and time2 are concatenated and fed to a U-Net-based network. They used HH polarization ALOS-PALSAR over the same area at different times. (2) Second, a prediction map is derived from the U-Net-based trained model, and (3) at last, by applying a threshold of 0.5, a binary map that indicates the position of newly built constructions is produced [49]. Sun et al. [50] have proposed an end-to-end LU-Net architecture to leverage both spatiality and temporality characteristics simultaneously. This CD method was implemented in two steps: (1) first, they combined the convolution and recurrent structure in a layer and introduced a Conv-LSTM layer, and (2) second, they substituted the standard convolutional layer of U-Net with Conv-LSTM and formed a new architecture, L-UNet [50]. Cao et al. [51] have proposed a CD method for bi-temporal SAR images and introduced a deep denoising network to eliminate the SAR image noise in three main steps: (1) first, a deep denoising model is trained efficiently by using plenty simulated SAR images to estimate the noise constituent. Then, the original SAR image can be cleaned up by removing this noise constituent. (2) Secondly, a denoised DI has been generated from the new image pair after denoising, and (3) finally, using a three-layer CNN, denoised DI has been classified into *changed* and *no-changed* regions [51]. Wang et al. [52] have designed a new deformable residual CNN (DRNet) for SAR images CD. The DRNet was used to adjust the sampling location. Additionally, prior to regular convolution, two stages were added: (1) offset field generation, and (2) deformable feature map generation. Moreover, a new pooling module called residual pooling was designed by replacing the conventional pooling with a set of smaller pooling kernels to discover the multi-scale information of the ground objects.

1.1.2. DL Unsupervised Methods for EO Data

Various supervised DL methods, including CNNs, have demonstrated satisfactory results in computer vision tasks when accompanied by a large labeled dataset [18]. In the case of CD tasks, the training datasets often are insufficient to construct such models. Additionally, constructing a ground truth (GT) map based on real-time *change* information of terrestrial objects takes a lot of time and effort [17]. Consequently, in many cases, it is more effective to learn *change* features from an unsupervised approach.

For instance, Kiana et al. [53] have proposed an unsupervised CD method for SAR images using the Gaussian mixture model (GMM). The CD framework in this study was implemented in two main steps: (1) first, using GMM, three Gaussian distributions were modeled (i.e., positive *change*, negative *change* and *no-change* distribution); (2) then, two thresholds were calculated as injection points of distributions. Before the first threshold, pixels are *negative changes,* between the two thresholds, they are *no-changes,* and after the second threshold, pixels are *positive changes* [53]. Thresholding methods in which the statistical distribution is modeled may be difficult to estimate the statistical parameters when *change* and *no-change* pixels overlap. Moreover, in PolSAR data, this problem can be more pronounced because of the strong effect of speckle noise. Liu et al. [54] proposed an unsupervised symmetric convolutional coupling network (SCCN) for CD based on heterogeneous SAR and optical images. They have defined a coupling function to determine network parameters. This CD method was implemented in two steps: (1) first, each of the two images is fed to one side of the SCCN and transferred to a feature space. In the new feature space, the two input images have more harmonious features and (2) second, a difference map was straight computed through pixel-wise Euclidean distances in feature space [54]. Bergamasco et al. [55] have proposed an unsupervised CD based on convolutional auto-encoder (CAE) feature extraction in two steps: (1) first, to train the CAE, the reconstruction error between the reconstructed output and the input from unlabeled single-time image patches Sentinel 1 was minimized, and (2) second, the trained CAE was used to extract multi-scale features from both the bi-temporal images and extract a change map [55]. Huang et al. [56] have developed a DL unsupervised algorithm that can detect changes in buildings from RS images in two steps: (1) first, a convolutional layer is employed to extract the spatial, texture and spectral features and produce a low-level feature vector for each pixel, and (2) second, a model based on deep belief network and extreme learning machine (DBN-ELM) was applied: a DBN was pre-trained by introducing unlabeled samples and they were then jointly optimized through the use of an ELM classifier [56].

In some cases, first, a pre-train step is performed and the pixels with the greatest likelihood to belong to the *change* and *no-change* classes are extracted. Then, these pixels are utilized to train the model. For instance, Gao et al. [57] proposed a pre-train scheme in two main steps: (1) first, they used a logarithmic ratio operator and a hierarchical FCM classifier to generate pseudo-label training samples, and (2) next, by integrating a CNN model and a dual-tree Complex Wavelet transform, called CWNN, pixels were classified into *change* and *no-change* classes [57]. In addition, Zhang et al. [58] proposed an automated method to detect changes in bi-temporal SAR images based on a pre-train scheme and the PCANet algorithm in two main steps: (1) first, a parameterized pooling algorithm is used to develop a deep difference image (DDI). Following this, Sigmoid nonlinear mapping with two different parameters is applied to DDIs to give two mapped DDIs. Then, the parallel FCM is applied to produce three types of pseudo-label training pixels: *changed*, *no-changed* and *intermediate* pixels. (2) Next, a support vector machine (SVM) was trained using the *changed* and *no-changed* pixels. Finally, the trained model was used to classify *intermediate* pixels and generate a change map [58]. In such methods, the accuracy of the pre-train step is very important. Accordingly, if the pixels are extracted with little precision, the network will not be properly trained. Therefore, training pixels must be extracted with high accuracy.

In a few cases, a fake-GT is generated with unsupervised methods and is used to minimize the DL method's loss function. For instance, Liu et al. [59] developed a CNN-based CD approach. This network was trained based on a two-part loss function. The CD framework in this study was implemented in three main steps: (1) first, a U-Net model was pre-trained using an open-source dataset. Then, the Euclidean distance (ED) is computed between two feature vectors extracted for each pair of pixels in bi-temporal images. (2) Second, based on a fake-GT, the ED is minimized for *no-changed* pixels and maximized for *changed* pixels in the first part of the loss function. The second part of the loss function is designed to transfer the pre-trained model to the target dataset. (3) Finally, after the training is complete, the k-nearest neighbors clustering is applied to extract a change map [59].

1.1.3. DL Semi-Supervised Methods for EO Data

In semi-supervised learning, the little labeled dataset is coupled with large quantities of an unlabeled dataset to form a model. Semi-supervised learning falls between unsupervised learning and supervised learning. One of the most common approaches in semi-supervised is TL fine-tuning [60]. In the case of not enough samples, TL can be used to adapt the features learned in previous tasks, which involves fine-tuning the network pretrained in general images. To achieve this, the final layers of the pre-trained network are usually retrained based on the little data available. Following this approach, Kutlu and Avcı [61] have been proposed a method based on AlexNet fine-tuning. They employed CNN, discrete wavelet transforms (DWT) and LSTM, aiming to obtain the feature vector, translate and strengthen the feature vector and classify the signal, respectively. The framework in this study was implemented in three main steps: (1) first, they used fine-tuning for AlexNet architecture to extract useful features; (2) then, they applied one-dimensional DWT on each feature vector to obtain the approximation coefficients by convolving the signals with the low-pass filter; and (3) finally, the LSTM was used for classification [61]. Venugopal [62] has introduced a semi-supervised CD method based on Resnet-101 fine-tuning in three steps: (1) firstly, two bi-temporal SAR images were converted to grayscale images to compute the similarity between the two images; (2) secondly, a Resnet-101 based multiple dilated deep neural network was fine-tuned to extract the feature sets; and (3) finally, semantic segmentation is applied to detect changes in the two SAR images [62].

Some networks are made up of several sub-networks. In such networks, each of the sub-networks has specific purposes. However, they may cause a large increase in network parameters. To overcome this, some of these sub-networks use the parameters of pre-trained models with zero learning rates. Following this approach, Zhang and Shi [1] proposed an approach based on a deep feature difference CNN (FDCNN) based on two sub-networks named FD-Net and FF-Net, where FD-Net is trained based on sharing parameter from VGG16 and FF-Net is trained based on a few pixel-level samples. The CD framework in this study was implemented in three main steps: (1) first, VGG16 is trained on RS datasets to learn deep features; (2) second, FDCNN is trained based on the proposed change magnitude guided loss function by using a few pixel-level training samples; and (3) third, a binary change map is derived using a threshold value from the change magnitude map inferred using FDCNN [1]. Peng, Bruzzone, Zhang, Guan, Ding and Huang [4] have proposed a new SemiCDNet based on a GAN in two main steps: (1) first, they used both the labeled data and unlabeled data to generate initial predictions (segmentation maps) and entropy maps based on an adopted UNet++ model as a generator. Then, they optimized UNet++ in a supervised manner using a binary cross-entropy loss and (2) second, in the discriminator phase, they introduced two discriminators to apply the distribution compatibility feature of segmentation maps and entropy maps between labeled data and unlabeled data [4]. Although semi-supervised algorithms reduce the need for training data, they can still be challenging in RS applications because they still require high-quality training data.

1.2. Problem Statements and Contribution

As mentioned earlier, the performance of deep learning-based CD methods is highly dependent on the quality and quantity of the training data. Therefore, one of the main challenges of applying DL for CD applications is to provide enough training samples. On the other hand, most of the deep networks that have been developed to detect changes are single-channel or dual-channel. In single-channel architectures, the network only takes one input. Therefore, two images must be converted into one input. This is usually done by differentiating or stacking images. As a result, information is lost. In dual-channel architectures, first, each image has separately entered a channel. Next, the features of each image are extracted. Then, like single-channel architectures, the two feature vectors are converted to a feature vector and entered into a fully connected layer. Because there is usually no information transition and connection between channels, the information is lost.

To overcome these challenges, we proposed a parallel pseudo-label training sample generation method. This method is based on a pre-trained CNN-based model that was trained carefully on two UAVSAR datasets and an FCM algorithm. First, we used a pre-trained model and the TL technique to calculate the probability change map for our datasets. Then, to improve and increase the reliability of the model, it was combined in parallel with the FCM algorithm to select samples that can most likely belong to the *change* and *no-change* classes. Additionally, we introduce a novel end-to-end three-channel deep neural network, called TCD-Net. The three channels of TCD-Net are designed so that the first and third channels independently extract features from each image and identify the objects in each image well, while the second channel identifies distractions and transfers information from the low- to the high levels. Compared with the use of a single- or dual-channel architecture, this three-channel architecture not only provides a feature representation of each image but also identifies changes at various levels. In addition, there are connections between the three channels that prevent data loss. Therefore, the proposed method can learn more strong and abstract representations for the spatial information of a certain pixel. We also utilize an adaptive multi-scale shallow block and an adaptive multi-scale residual block in the TCD-Net architecture to make the network resistant to objects of various sizes with much fewer parameters and to transfer information to the final layer.

In particular, our proposed algorithm consists of three parts: (1) parallel pseudo-label training sample generation, (2) model optimization for TCD-Net, and (3) binary change map generation. Therefore, the main contribution of this study can be summarized as:

1. Developing a new unsupervised DL-based model with three channels for deep feature extraction and evaluating the effectiveness of an intermediate channel by comparing this algorithm with a dual-channel deep network;
2. Introducing an adaptive formula for determining the number of filters in the multi-scale block due to the dependence of the deep features on the kernel size;
3. Proposing high confidence automatic pseudo-label training sample generation framework using a probabilistic parallel scheme based on a pre-trained neural network model and FCM algorithm;
4. Providing highly robust results for PolSAR CD compared with the state-of-the-art (SOTA) unsupervised methods.

The rest of the paper is organized as follows. The methodology is described in Section 2. Section 3 presents the case study. Section 4 presents the experimental results and analyses. Section 5 provides the discussion. Finally, the conclusions and future work is presented in Section 6.

2. Methodology

In this section, we describe the details of the proposed method for CD. According to Figure 1, the general scheme of the proposed method consists of three main steps, including (1) pre-processing, (2) automatic training sample generation, and (3) end-to-end CD learning. We describe these three steps in detail in the following three sub-sections.

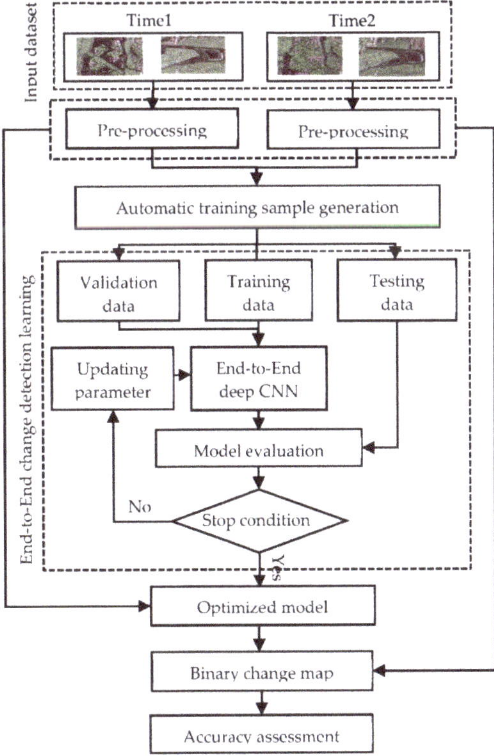

Figure 1. General scheme of the proposed unsupervised binary change detection (CD) method. CNN is convolutional neural network.

2.1. Pre-Processing

Data pre-processing is greatly important in PolSAR CD methods. Multi-looking, co-registration of images [26] and speckle filtering [27] are the main techniques in PolSAR image processing. PolSAR images are always affected by speckle noise, which makes the CD process more challenging. Therefore, there are several methods for speckle filtering; we used the refined Lee filter with a kernel size of 5 [63]. Moreover, geometric correction is used for co-registration of images for comparison and matching. Several GCP points were selected for modeling and a second-order polynomial was used to resample gray values. The final geometric correction accuracy (i.e., RMSE) was approximately 0.4 pixels.

2.2. Automatic Training Sample Generation

The purpose of this section is to produce pseudo-label training samples automatically and without human interference. We use a pre-trained model, which has been trained on large and open-source UAVSAR datasets and extracts a probabilistic change map (PCHM). In fact, we use the TL technique because we apply a pre-trained model instead of training a model. In addition, to improve the reliability and robustness of the results, we use a parallel combination of the results of the pre-trained model and the results of the FCM algorithm. As shown in Figure 2, the proposed method consists of the following main steps:

1. First, we use the CNN-based CD network in [2]. Since this network has previously been trained on UAVSAR data, we call it the pre-trained model. We then calculate the output of the pre-trained model for our datasets. The output of the last *softmax* layer of the pre-trained model gives a PCHM in two classes: *change* and *no-change* classes (i.e., p_c^{CNN} and p_n^{CNN}). Then, by applying a knowledge-based threshold (0.95) on two classes, the pixels that most probably belong to the *change* (w_c^{CNN}) and *no-change* (w_n^{CNN}) classes are separated, i.e.,: $(i,j) \in w_c^{CNN}$ *for* $p_c^{CNN} > 0.95$ *and* $(i,j) \in w_n^{CNN}$ *for* $p_n^{CNN} > 0.95$. By selecting a higher threshold value, fewer training samples are generated, but they are more reliable. The remaining pixels are placed in the *ambiguous* class and are not used in this section.

2. Second, we utilize the log-ratio operator to generate the log-ration image $I_D = \log(I_2/I_1)$. Using the FCM algorithm, we obtain a PCHM in two classes: *change* and *no-change* classes (i.e., p_c^{FCM} and p_n^{FCM}). Similar to the previous approach, by applying a threshold of 0.95, the pixels that most probably belong to the *change* (w_c^{FCM}) and *no-change* (w_n^{FCM}) classes are separated, similarly: $((i,j) \in w_c^{FCM}$ *for* $p_c^{FCM} > 0.95$ *and* $(i,j) \in w_n^{FCM}$ *for* $p_n^{FCM} > 0.95)$.

3. Although we use PCHMs and reliable threshold values, because of the noisy conditions of the SAR images, there may still be pixels that are incorrectly classified. Therefore, to improve accuracy, we aggregate the results of the two methods mentioned in 1 and 2 in parallel, i.e., pixels that both methods labeled as *change* and *no-change* are selected using Equation (1).

$$\begin{cases} (i,j) \in w_c & for\ (i,j) \in w_c^{CNN}\ and\ (i,j) \in w_c^{FCM} \\ (i,j) \in w_n & for\ (i,j) \in w_n^{CNN}\ and\ (i,j) \in w_n^{FCM} \end{cases}, \tag{1}$$

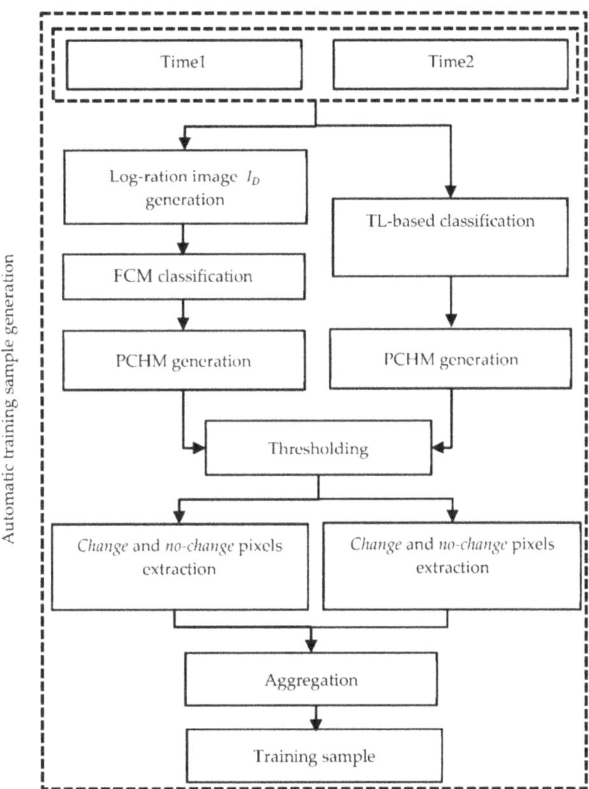

Figure 2. Flowchart of the proposed parallel pseudo-label training sample generation. FCM is fuzzy c-means, PCHM is probabilistic change map, and TL is transfer learning.

2.3. End-to-End Change Detection Learning

2.3.1. Convolutional Layer

The convolutional layer is the core of CNNs. Each layer of convolution in CNNs contains a set of filters and the output of the network is derived from the convolution between the filters and the input layer. Each filter can contain a specific pattern, followed by a specific pattern in the image. In the network training process, these filters are supposed to extract meaningful patterns from each image. Since finding only one pattern does not lead to good results and makes the network limited in terms of performance, the convolutional layer needs to have multiple filters. Therefore, the output of the convolutional layer is a set of different patterns that are called feature maps. The output of a convolutional layer in the *nth* layer is expressed using Equation (2).

$$F^n = g\left(w^n F^{n-1} + b^n\right) \tag{2}$$

where F^{n-1} represents the neuron input from the previous layer, $n-1$; g represents the activation function; b^n represents the bias vector for the current layer; and w^n represents the weighted template for the current layer.

A 2D convolution equation can be used to compute the output of the j th feature map (v) within the i th layer at spatial location (x, y), according to Equation (3).

$$v_{i,j}^{xy} = g\left(b_{i,j} + \sum_m \sum_{r=0}^{R-1} \sum_{s=0}^{S-1} W_{i,j,m}^{r,s} v_{i-1,m}^{(x+r)(y+s)}\right) \tag{3}$$

where g is activation function, b is bias, m is the feature cube connected to the current feature cube in the previous layer and W is the (r, s) th value of the kernel connected to the m th feature cube in the previous layer. Moreover, R and S are the length and width of the convolution kernel size, respectively.

2.3.2. Multi-Scale Block

In RS imagery with meter and sub-meter-level spatial resolution, there are many objects in different sizes. In addition, there are large structures and details in the texture of the objects and ground scenes that need to be extracted. Since small-scale features, like short building edges, typically respond to smaller-sized convolutional filters, but large-scale structures respond better to larger convolutional filters, we use the multi-scale convolutional block. The multi-scale convolutional block extracts helpful dynamic features and improves feature extraction. Using this multi-scale convolutional block, the network can continuously learn a set of features and the related scales at which these features occur with a minimum increase in parameters.

According to Figure 3, in the n th layer of the multi-scale block, three sizes of convolutional filters are set: $1 \times 1, 3 \times 3$ and 5×5. With a 1×1 convolutional kernel, features are extracted from pixels themselves. A 3×3 convolutional kernel extracts features from a small neighborhood. Additionally, a 5×5 convolutional kernel extracts features of a larger range, which is suitable for some continuous large-scale images. In the traditional multi-scale approach, the number of filters is the same for each kernel size, N and the output feature maps have a 3N spectral dimension. Therefore, large kernel size (i.e., 3×3 and 5×5) require more processing time and have high parameters. Therefore, it is better to change the number of filters for each kernel size in the multi-scale block. To achieve this, this research develops an adaptive formula for determining the number of filters (NoFs) in a multi-scale block. To keep a constant total number of filters in each block and to preserve a large increase in the number of parameters, we consider the number of filters that have smaller length and width dimensions more than filters with larger length and width. According to Equation (4), NoFs is the total number of filters of a multi-scale block and is divided into NoF1, NoF2 and NoF3, which are the number of filters for $1 \times 1, 3 \times 3$ and 5×5 kernels, respectively:

$$\begin{cases} NoF_1 = \alpha \times NoF \\ NoF_2 = \beta \times NoF \\ NoF_3 = \gamma \times NoF \\ s.t : \alpha + \beta + \gamma = 1 \end{cases} \tag{4}$$

where α, β, γ are coefficients that determine the number of filters for $1 \times 1, 3 \times 3$ and 5×5 kernels, respectively. To reduce the network parameters, we consider these coefficients in such a way that $\alpha > \beta > \gamma$.

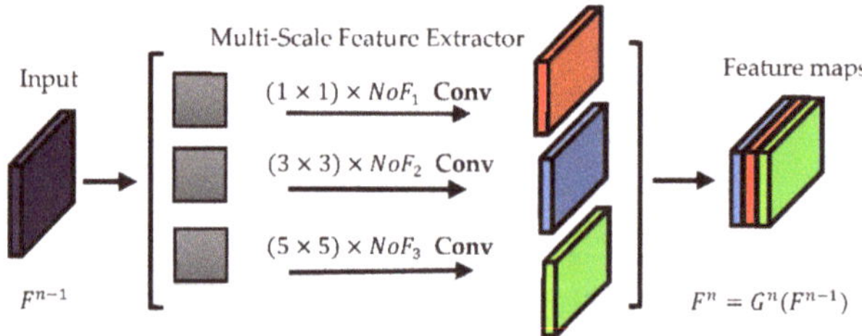

Figure 3. A multi-scale shallow block.

2.3.3. Residual Block

CNNs with deeper layers can generally model more complex patterns and have higher nonlinearity. Visual representation of feature maps shows that a deeper network can lead to the extraction of more robust and abstract features [64]. However, there is a substantial problem in the training process of a deep CNN. As the number of layers' increases, the gradient vanishing problem during back-propagation increases. Therefore, updating the convolutional kernels and bias vectors to achieve optimal allocation of all parameters is very slow. Additionally, it has been seen that as the number of layers gradually increases, the accuracy first increases, then at a point it starts to saturate and finally decreases [64]. For this reason, residual learning has now become one of the most effective solutions available for training deep CNNs. It involves replacing the convolutional filtering process $F^n = G^n(F^{n-1})$ by $F^n = F^{n-1} + G^n(F^{n-1})$, which is called a "skip connection", using the residual $F^{n-1} - F^n$ as a prediction process. This research uses a combination of the multi-scale block and the residual block, called the multi-scale residual block (Figure 4).

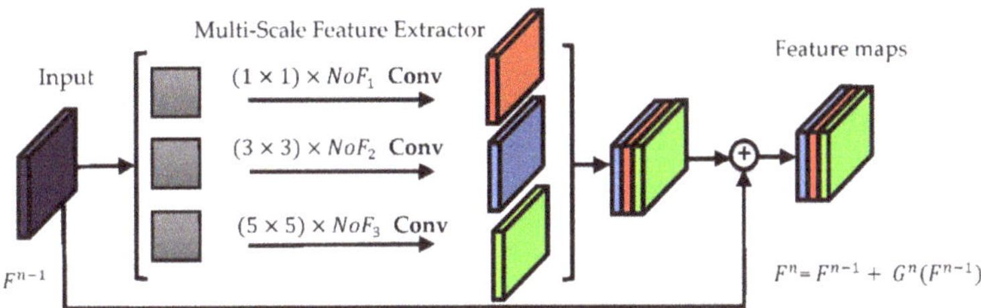

Figure 4. A multi-scale residual block.

2.3.4. TCD-Net for CD

Considering two images I^{t_1} and I^{t_2} taken over the same area at different times t_1 and t_2, the goal is to recognize areas that have changed between the images. Assume that $\hat{CM}$ is the binary change map derived from I^{t_1} and I^{t_2} and $\hat{CM}_{i,j}$ is the change values at location (i, j). Generally, $\hat{CM}_{i,j} \in \{0, 1\}$, $\hat{CM}_{i,j} = 1$ indicates (i, j) is *changed* and otherwise, it indicates (i, j) is *no-changed*. We propose the TCD-Net architecture to generate the binary change map.

- Architecture

As shown in Figure 5, the proposed TCD-Net architecture includes three channels, each of which is a sub-network that extracts feature. The traditional method of DL-based CD requires the conversion of two images to one input for the single-channel networks, leading to missing information. Dual-channel networks extract features from two images and, in the last layer, convert these features into a vector that is then fed to a fully connected layer. Since there is no intermediate channel and there is no information transfer and connection between the channels at different levels, the information is lost. To prevent information loss, we use a three-channel network. Additionally, a multi-channel network converges faster than a single- or dual-channel. In TCD-Net, the first and third channels take bi-temporal images, I^{t_1} and I^{t_2}, separately, and the second channel can learn change information from the features extracted from the first and third channels to obtain DI. In the first and third channels, which are symmetric, there is an adaptive multi-scale shallow block, three adaptive multi-scale residual blocks and two max-pooling layers. The second channel consists of an adaptive multi-scale shallow block, two adaptive multi-scale residual blocks and two max-pooling layers. An adaptive multi-scale shallow/residual block contains one 1×1, one 3×3 and one 5×5 convolutional block, as mentioned before, where the number of their filters for each kernel size is considered adaptive. In multi-scale blocks, after connecting the output of these three convolutional blocks, a 3×3 convolutional block has been installed to adjust the third dimension, allowing the layer input to be added with the output of this section. This no longer requires the number of features extracted from the multi-scale shallow/residual block to be fixed throughout the network. Moreover, each convolutional block includes an activation function (rectified linear unit (ReLU)), batch normalization and many convolutional filters that extract deep features.

We use $f_l^{t_1}$, $f_l^{t_2}$ and $l = \{1, \ldots, L\}$ to represent features in the lth layer of the first and third channels, respectively, corresponding to t_1 and t_2. For instance, the $f_1^{t_1}$ represents the features extracted from the multi-scale shallow block in the first layer of the first channel (corresponding to t_1). Finally, we obtain the features $f_L^{t_1}$ and $f_L^{t_2}$ for I^{t_1} and I^{t_2}, respectively. In the second channel, which we call the intermediate channel, new features are also extracted, which we call intermediate features and represent with f_l^m. At the first layer, the features extracted in the first and third channels are subtracted, $\left(f_1^{t_1} - f_1^{t_2} \right)$ and fed to the second channel. In the second channel, $\left(f_1^{t_1} - f_1^{t_2} \right)$ enters the multi-scale shallow block and f_1^m is extracted. In the next layers, change information is inferred from $\left(f_2^{t_1} - f_2^{t_2} \right) + f_1^m$, $\left(f_3^{t_1} - f_3^{t_2} \right) + f_2^m, \ldots, \left(f_L^{t_1} - f_L^{t_2} \right) + f_{L-1}^m$. That is, at each layer, the features extracted in the first and third channels are subtracted then added to the features extracted in the second channel from the previous layer, thus making our algorithm very powerful in detecting changes. In the last layer, $\left(f_L^{t_1} - f_L^{t_2} \right) + f_{L-1}^m$ is extracted, flattened and fed into a fully connected layer with ReLU activation function. Moreover, we use the *ReLU* after each convolutional layer as a piecewise linear activation function. The ReLU function can be formulated using Equation (5).

$$f(x) = \max(0, x) \tag{5}$$

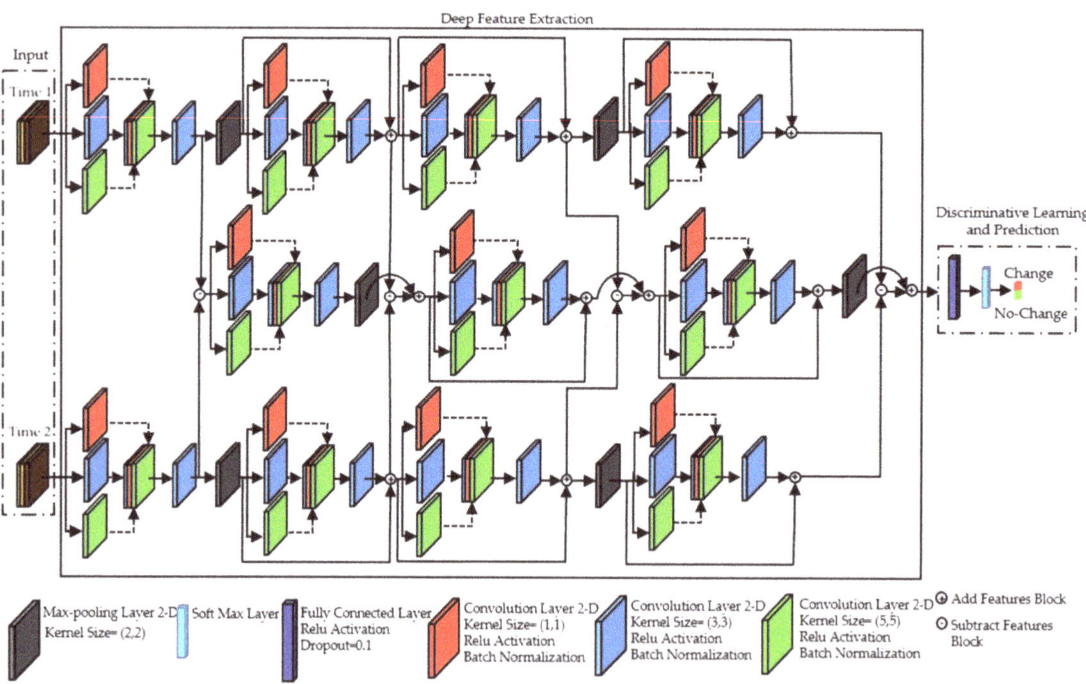

Figure 5. The proposed TCD-Net architecture for CD of remote sensing (RS) datasets.

The latest fully-connected layer is a *softmax* layer. In general, this layer is used to model categorical probability distributions and calculate the probability that each pixel belongs to the *change* and *no-change* classes. Finally, the pixels are divided into two categories of *change* and *no-change*. The *softmax* function is expressed in Equation (6).

$$f(x_i) = \frac{e^{x_i}}{\sum_i e^{x_i}} \tag{6}$$

- Model Optimization

As shown in Figure 1, after the automatic training sample generation phase, the samples generated are divided into three categories: training, testing and validation datasets.

The TCD-Net is trained based on the training dataset. Additionally, the loss value was calculated by the loss function based on the validation dataset. There is no analytical method for optimizing CNN parameters. Thus, optimization is used to adjust the model parameters iteratively. In this research, an Adam optimizer is used to adjust CNN parameters. As a result, the model is trained based on the initial values of the parameters, then the output of the model is compared with the actual value. The error of the training model is fed to the optimizer and is updated the parameters. In an iterative process, the gradient is reduced at this point to minimize the total output error. This process continues until the stop condition is reached, i.e., a certain number of repetitions or a certain error (minimum error). Due to back-propagation, the parameters are updated at each step to decrease the error of comparing the results obtained from the network with the training/validation dataset. Finally, test data is used to evaluate network performance.

In this research, cross-entropy was used to calculate the loss function of the proposed architectures. The performance of the network given the inputs and the labels with optional performance weights and other parameters is calculated by cross-entropy function for inputs (y) and outputs (t) using the following Equation (7):

$$E = \frac{1}{n} \sum_{j=1}^{n} \sum_{i=1}^{k} t_{ij} ln y_{ij} + (1 - t_{ij}) ln(1 - y_{ij}) \tag{7}$$

where n is the number of training samples and k is the number of classes. Additionally, t_{ij} is the ij th entry of the target matrix and y_{ij} is the ij th entry of the training sample matrix.

2.3.5. Accuracy Assessment

Accuracy assessment is an integral part of any RS task and is done in two ways. In the first approach, the results of the proposed method are compared with GT data and in the second approach with sample data. In this study, the final results of the proposed CD method are compared quantitatively as well as qualitatively with the GT data and the results of other SOTA CD methods. The quantitative comparison is based on the metrics described subsequently. Based on the CD results and the GT data, there are four modes: (1) if both the GT data and result are positive, it is considered as True Positive (TP); (2) if the GT data is positive and the result is negative, it is considered as False Negative (FN); (3) if both the GT data and the results are negative, it is considered as True Negative (TN); and (4) if the GT data is negative but the result is positive, it is considered as False Positive (FP). With the help of these four values, the essential criteria such as false-positive rate (FPR) (also called false alarm rate), true-positive rate (TPR) (also called hit rate and recall), false-negative rate (FNR), overall accuracy (OA), precision, detection rate (DR), F1-score, overall error rate (OER), Prevalence (PRE) and kappa coefficient (KC) are calculated by the following relationships shown in Table 1.

Table 1. Formulas for accuracy assessment criteria.

Accuracy Index	Formula
FNR	$\frac{FN}{FN+TP}$
TPR	$\frac{TP}{TP+FN}$
FPR	$\frac{FP}{TN+FP}$
OA	$\frac{TN+TP}{TP+TN+FP+FN}$
Precision	$\frac{TP}{TP+FP}$
F1-Score	$\frac{TP}{TP+\frac{1}{2}(FP+FN)}$
OER	$\frac{FN+FP}{TP+TN+FP+FN}$
PRE	$\frac{(TP+FP)\times(TP+FN)+(FN+TN)\times(FP+TN)}{TP+TN+FP\times FN^2}$
KC	$\frac{OA-PRE}{1-PRE}$

2.3.6. Comparative Methods

To compare the effectiveness of the intermediate layer of TCD-Net, this research compares the TCD-Net algorithm with a dual-channel deep network. This dual-channel network is similar to TCD-Net, except that the intermediate layer is removed. To make the comparison fair, the dual-channel network is trained with the same training samples extracted from the pseudo-label sample generation phase. In addition, the following unsupervised SOTA CD methods are compared and analyzed to confirm the efficiency of TCD-Net. These approaches are PCA_kmeans [65], NR_ELM [66], Gabor_PCANet [67], CWNN [57] and DP_PCANet [58], which are described in brief below:

- PCA_kmeans: Initially, the DI is calculated by using the absolute-value difference between two SAR images. Additionally, the DI is separated into non-overlapping h × h blocks. Then, using PCA, all blocks are projected into the eigenvector space to obtain representation properties. Finally, each pixel is assigned to a cluster based on the minimum Euclidean distance between its feature vector and the cluster's mean feature vector, using the k-means clustering.
- NR_ELM: Initially, a neighborhood-based ratio operator and the hierarchical FCM algorithm are used for generating a DI and identifying pixels of interest in it. Secondly, the ELM classifier is trained using pixel-wise patch features centered on the pixels of interest.
- Gabor_PCANet: Initially, a pre-train step is performed using the Gabor wavelet and the FCM classifier. Secondly, by considering a neighborhood with specific dimensions for each training pixel in the two images and juxtaposing the two image patches, PCA features are extracted from the training patches. Then, the SVM algorithm is used to build a model on PCA features. After completing the training phase, the remaining pixels are divided into two categories: *changed* and *no-changed* pixels.
- CWNN: A convolutional-wavelet neural network (CWNN) method has been applied in bi-temporal SAR images. Firstly, a virtual sample generation scheme is utilized to generate pseudo-label training samples that are likely *changed* or *no-changed*. Secondly, the pseudo-label samples obtained in the previous step are used to train the CWNN network and create a change map.
- DP_PCANet: Firstly, inspired by the convolutional and pooling layers in the CNN, a DDI based on a weighted-pooling kernel has been extracted. Then, using sigmoid nonlinear mapping and parallel FCM, two mapped DDIs are generated. Then, the mapped DDIs are classified into three types of pseudo-label samples, i.e., *changed*, *no-changed* and *ambiguous* samples. Finally, with the SVM model that is trained based on the PCA features, *ambiguous* samples are classified as *changed* or *no-changed*.

These methods are also parameterized using references to the corresponding publications.

3. Case Study

Two co-registered L-band UAVSAR full polarimetric images are utilized to assess the performance of the proposed method. These two images belong to the city of Los Angeles, California, acquired on 23 April 2009 and 11 May 2015, by the JAV Propulsion Laboratory/National Aeronautics and Space Administration UAVSAR. There are 786×300 pixels in the first dataset and 766×300 pixels in the second dataset. Figure 6a–e shows the RGB (Red: |HH–VV|; Green: 2|HV|; Blue: |HH+VV|) Pauli images of the two subsets of the PolSAR scenes. The GT images connected with these subsets, shown in Figure 6c,f, were prepared for the numerical analysis of CD results by using Google Earth images. Actually, the image of GT is a binary image in which the black pixels are *no-change* and the white pixels are *change*. The first and second datasets are called dataset#1 and dataset#2, respectively.

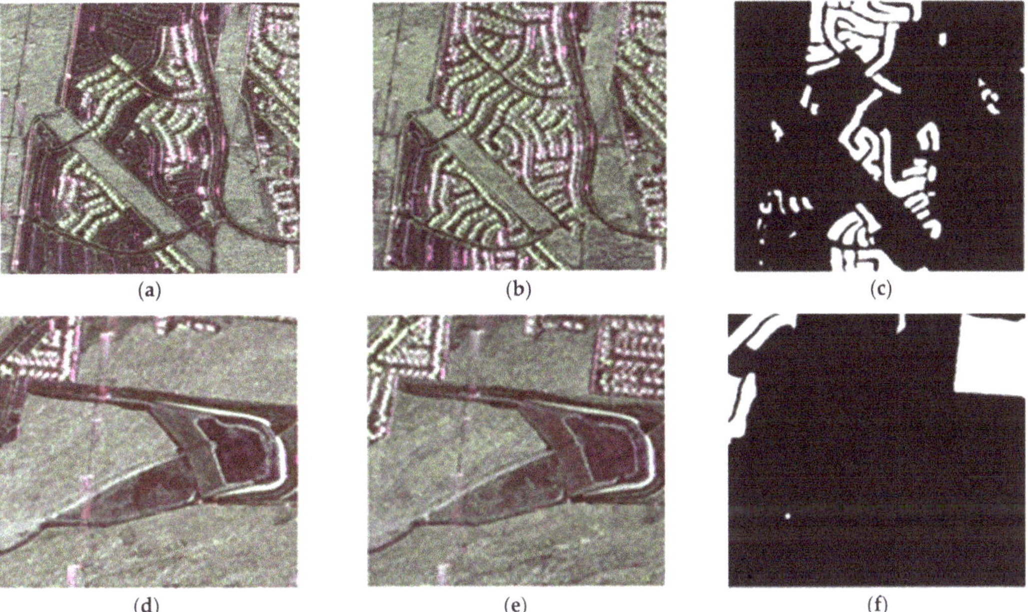

Figure 6. Pauli decomposition of UAVSAR images taken over Los Angeles, California on (**a**,**d**) 23 April 2009; (**b**,**e**) 11 May 2015; (**c**,**f**) ground truths, where white means change area and black means no-change area. Top: dataset#1. Bottom: dataset#2.

4. Experimental Results and Analysis

4.1. Parameter Setting

In NR_ELM and CWNN, parameters are neighborhood size $r = 3 \times 3$ and patch size $w = 7$, respectively. The PCANet parameters are the image patch size $k = 5$, the number of filters $L_1 = L_2 = 8$ and training samples 30% of the total data. In PCA_kmeans, patch size $h = 5$ is used. In Gabor feature extraction, the orientation of Gabor kernel $U = 8$, the scale of Gabor kernel $V = 5$, the maximum frequency $k_{max} = 2\pi$ and the spacing factor between kernels in the frequency domain $f = \sqrt{2}$ are used. To generate DDI and parallel clustering in DP_PCANet, the center bias b in the Sigmoid function $b = 0.1$ and the number of pooled images that are accumulated to generate the DDI, $T = 7$ are used. For TCD-Net, Table 2 lists the details of the configuration settings for each channel. Additionally, Table 3 shows the total number of filters in each multi-scale block. The model parameters are trained based on the mini-batch back-propagation algorithm with a size of 150. The error in 250 epochs is calculated based on the determined objective function and then the parameters are updated. Adam optimizer, with an initial learning rate of 10×10^{-3} with an epsilon value of 10×10^{-10}, is used as the optimization algorithm.

Table 2. TCD-Net configurations of each channel and block.

	Channel 1	Channel 2	Channel 3
Inputs (shape)	$11 \times 11 \times 4$	$11 \times 11 \times 4$	$11 \times 11 \times 4$
Block 1	Multi-Scale Shallow Block: 1×1 Conv1 + BN + RELU(NoF_1^{11}) [1] 3×3 Conv2 + BN + RELU(NoF_2^{11}) 5×5 Conv3 + BN + RELU(NoF_3^{11}) Channel Concat. 3×3 Conv4 + BN + RELU(256)	Multi-Scale Shallow Block: 1×1 Conv1 + BN + RELU(NoF_1^{21}) 3×3 Conv2 + BN + RELU(NoF_2^{21}) 5×5 Conv3 + BN + RELU(NoF_3^{21}) Channel Concat. 3×3 Conv4 + BN + RELU (256)	Multi-Scale Shallow Block: 1×1 Conv1 + BN + RELU(NoF_1^{31}) 3×3 Conv2 + BN + RELU(NoF_2^{31}) 5×5 Conv3 + BN + RELU(NoF_3^{31}) Channel Concat. 3×3 Conv4 + BN + RELU(256)
Output	2×2 Max-Pooling $7 \times 7 \times 256$	2×2 Max-Pooling $7 \times 7 \times 256$	2×2 Max-Pooling $7 \times 7 \times 256$
Block 2	Multi-Scale Residual Block: 1×1 Conv1 + BN + RELU(NoF_1^{12}) 3×3 Conv2 + BN + RELU(NoF_2^{12}) 5×5 Conv3 + BN + RELU(NoF_3^{12}) Channel Concat. 3×3 Conv4 + BN + RELU(256)	Multi-Scale Residual Block: 1×1 Conv1 + BN + RELU(NoF_1^{22}) 3×3 Conv2 + BN + RELU(NoF_2^{22}) 5×5 Conv3 + BN + RELU(NoF_3^{22}) Channel Concat. 3×3 Conv4 + BN + RELU(256)	Multi-Scale Residual Block: 1×1 Conv1 + BN + RELU(NoF_1^{32}) 3×3 Conv2 + BN + RELU(NoF_2^{32}) 5×5 Conv3 + BN + RELU(NoF_3^{32}) Channel Concat. 3×3 Conv4 + BN + RELU(256)
Output	$7 \times 7 \times 256$	$7 \times 7 \times 256$	$7 \times 7 \times 256$
Block 3	Multi-Scale Residual Block: 1×1 Conv1 + BN + RELU(NoF_1^{13}) 3×3 Conv2 + BN + RELU(NoF_2^{13}) 5×5 Conv3 + BN + RELU(NoF_3^{13}) Channel Concat. 3×3 Conv4 + BN + RELU(256) 2×2 Max-Pooling	Multi-Scale Residual Block: 1×1 Conv1 + BN + RELU(NoF_1^{23}) 3×3 Conv2 + BN + RELU(NoF_2^{23}) 5×5 Conv3 + BN + RELU(NoF_3^{23}) Channel Concat. 3×3 Conv4 + BN + RELU(256) 2×2 Max-Pooling	Multi-Scale Residual Block: $1\,1 \times 1$ Conv1 + BN + RELU(NoF_1^{33}) $3\,3 \times 3$ Conv2 + BN + RELU(NoF_2^{33}) $5\,5 \times 5$ Conv3 + BN + RELU(NoF_3^{33}) Channel Concat. 3×3 Conv4 + BN + RELU(256) 2×2 Max-Pooling
Output	$5 \times 5 \times 256$	$5 \times 5 \times 256$	$5 \times 5 \times 256$
Block 4	Multi-Scale Residual Block: 1×1 Conv1 + BN + RELU(NoF_1^{14}) 3×3 Conv2 + BN + RELU(NoF_2^{14}) 5 Conv3 + BN + RELU(NoF_3^{14}) Channel Concat. 3×3 Conv4 + BN + RELU(256)		Multi-Scale Residual Block: 1×1 Conv1 + BN + RELU(NoF_1^{34}) 3×3 Conv2 + BN + RELU(NoF_2^{34}) 5×5 Conv3 + BN + RELU(NoF_3^{34}) Channel Concat. 3×3 Conv4 + BN + RELU(256)
Output	$5 \times 5 \times 256$		$5 \times 5 \times 256$
Classifier	Flatten RELU Fully Connected (350) Softmax Fully Connected (2)		

NoF_k^{ij} is the number of filters for k^{th} convolutional layer of multi-scale block in i^{th} channel of jth block.

Table 3. The total number of filters in each multi-scale block.

Block	Channel		
	1	2	3
1	$NoF_1^{11} + NoF_2^{11} + NoF_3^{11} = 16$	$NoF_1^{21} + NoF_2^{21} + NoF_3^{21} = 64$	$NoF_1^{31} + NoF_2^{31} + NoF_3^{31} = 16$
2	$NoF_1^{12} + NoF_2^{12} + NoF_3^{12} = 32$	$NoF_1^{22} + NoF_2^{22} + NoF_3^{22} = 128$	$NoF_1^{32} + NoF_2^{32} + NoF_3^{32} = 32$
3	$NoF_1^{13} + NoF_2^{13} + NoF_3^{13} = 64$	$NoF_1^{23} + NoF_2^{23} + NoF_3^{23} = 256$	$NoF_1^{33} + NoF_2^{33} + NoF_3^{33} = 64$
4	$NoF_1^{14} + NoF_2^{14} + NoF_3^{14} = 128$		$NoF_1^{34} + NoF_2^{34} + NoF_3^{34} = 128$

4.2. Pseudo-Label Training Sample Generation

As previously mentioned, we first use the pre-trained model introduced in [2]. Based on Table 4, which shows the results of the CD framework proposed in [2] for our case study, it can be seen that this model is not robust for all case studies. On other hand, the performance of this model is dependent on the objects of the study area. For this reason, we generate PCHM using the pre-trained model. Then, by applying a reliable threshold, we extract the pixels that most likely belong to the *change* and *no-change* classes. Quantitative results show that this increases the performance significantly. In addition, we obtain PCHM with the FCM clustering. Finally, we extract the pixels that have been identified in both algorithms as *change* and *no-change* pixels. The quantitative results show that the aggregation of these two algorithms greatly increases accuracy. For dataset#1, the OA and KC in the FCM clustering is 94.10% and 0.66, in the TL-based classification is 93.85% and 0.75 and in the aggregation is 97.58% and 0.84. For dataset#2, the OA and KC in the FCM clustering is 95.84% and 0.52, in the TL-based classification is 97.64% and 0.82, and in the aggregation of these two methods is 99.52% and 0.91. Therefore, the quantitative results show that the aggregation of these two methods improves the pseudo-label generation accuracy. We considered 5% of the total data as the reference data and divided the reference data into 65% for training, 15% for validation, and 20% for testing (Table 5).

Table 4. The accuracy of the CD framework proposed in [2] for dataset#1 and dataset#2.

Method	TN	TP	FP	FN	TPR (%)	FPR (%)	Result FNR (%)	Precision	OA (%)	F1-Score	DR (%)	KC	OER (%)
[2] dataset#1	187,570	27,606	11,581	9043	75.33	5.82	24.67	70.45	91.25	72.86	74.02	0.68	8.76
[2] dataset#2	202,423	10,710	4188	12,479	46.19	2.03	53.81	71.89	92.75	56.24	29.22	0.52	7.25

Table 5. The number of *change* and *no-change* pixels extracted from the parallel pseudo-label generation framework and the number of training, testing and validation pixels used in the training process of TCD-Net.

Dataset	Class	Value Total Number of Pixels	Number of Samples	Percentage (%)	Training	Validation	Testing
dataset#1	*change*	36,649	12,717	34.70	1191	357	366
	no-change	199,151	165,308	83.01	6472	1493	1991
dataset#2	*change*	23,189	11,156	48.11	753	173	231
	no-change	206,611	81,650	39.52	6714	1549	2066

4.3. Comparison of Results for Dataset#1

Figure 7 illustrates the result of CD for dataset#1. As seen, Figure 7a,c,d show the results CD for PCA_kmeans, Gabor_PCANet and DP_PCANet that have many noisy pixels, while other methods provide the low noisy pixels. Furthermore, the NR_ELM, Figure 7b and CWNN, Figure 7e, have miss detection pixels that are evident in the top and bottom of the study areas (red circles). In Figure 7, the red circles show the *no-changed* pixels that have been detected as *changed* pixels by all of the methods except the TCD-Net and dual-channel deep network. The TCD-Net and dual-channel deep network provide significant results compared to other methods in the detection of *no-changed* pixels. However, the TCD-Net (Figure 7g), in detail, discovers subtle *change* pixels better than the dual-channel deep network (Figure 7f). Therefore, the TCD-Net provides a promising result in the detection of both *change* and *no-change* pixels.

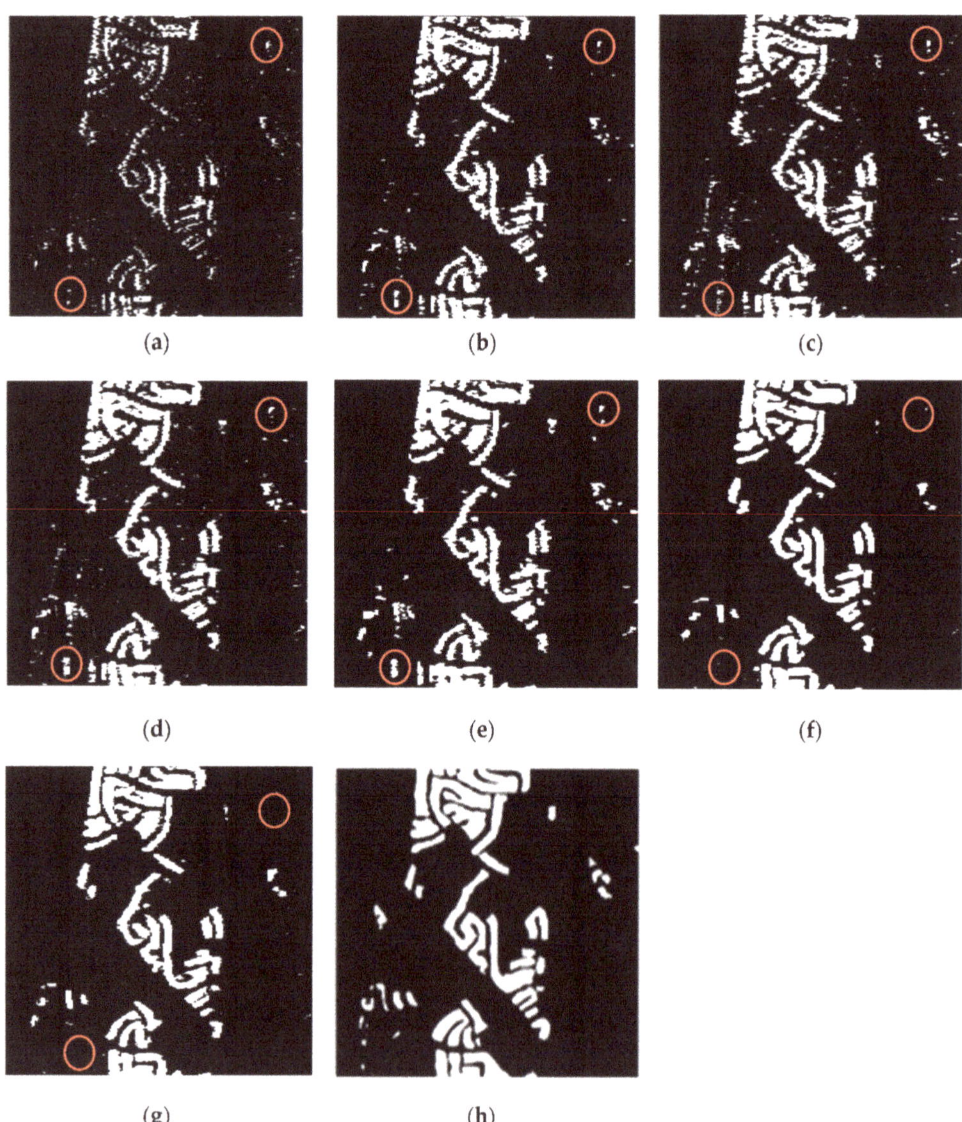

Figure 7. Visualized results of various CD methods on dataset#1; (**a**) PCA_kmeans, (**b**) NR_ELM, (**c**) Gabor_PCANet, (**d**) DP_PCANet, (**e**) CWNN, (**f**) dual-channel deep network, (**g**) TCD-Net, and (**h**) ground truth. The red circles highlight different output performances in *no-change* pixels.

In Table 6, CD quantitative results show that the TCD-Net algorithm performed better than other methods in terms of OA, KC, precision and DR indicators. In particular, the

TCD-Net algorithm has the OA of 95.01% and the KC of 0.80, which is 6.35% and 0.37 more than PCA_kmeans, 2.75% and 0.14 more than NR_ELM, 3.29% and 0.15 more than Gabor_PCANet, 2.43% and 0.11 more than DP_PCANet, 2.24% and 0.10 more than CWNN, and 0.76% and 0.04 more than the dual-channel deep network. Furthermore, the TCD-Net algorithm has a much higher F1-score. Additionally, the OER is much lower in TCD-Net. These results show that TCD-Net is more effective in CD than other algorithms.

Table 6. The accuracy of different change detection methods for dataset#1.

Method	TN	TP	FP	FN	TPR (%)	FPR (%)	FNR (%)	Precision	OA (%)	F1-Score	DR (%)	KC	OER (%)
							Result						
PCA_kmeans	196,839	12,223	2312	24,426	33.35	**1.16**	66.65	84.09	88.66	47.76	33.35	0.43	11.34
NR_ELM	195,411	22,066	3740	14,583	60.21	1.88	39.79	85.51	92.26	70.66	60.21	0.66	7.74
Gabor_PCANet	192,837	23,449	6314	13,200	63.98	3.17	36.02	78.79	91.72	70.62	63.98	0.65	8.28
DP_PCANet	192,994	25,298	6157	11,351	69.03	3.09	30.97	80.43	92.58	74.29	69.03	0.69	7.42
CWNN	193,141	25,603	6010	11,046	69.86	3.02	30.14	80.99	92.77	75.01	69.86	0.70	7.23
Dual-channel Net	195,818	26,424	3333	10,225	72.10	1.67	27.90	88.80	94.25	79.58	72.10	0.76	5.75
TCD-Net	196,646	27,390	2505	9259	74.74	1.26	25.26	91.62	95.01	82.32	74.74	0.80	4.99

4.4. Comparison of Results for Dataset#2

The results of CD for dataset#2 are shown in Figure 8. Similarly, the Gabor_PCANet, NR_ELM and DP_PCANet provide many noisy *changed* pixels while these pixels are *no-change*. Furthermore, most methods have many miss detection pixels in the *no-change* areas, which are more evident at the top and middle of the region of interest. This theme can be seen in the CD results, which are illustrated by red circles in Figure 8. As compared to dataset#1, CD methods perform a little differently. In *change* areas, there are differences among the CD methods, a sample of which is illustrated by green circles in Figure 8. The green circles show that the dual-channel deep network with good performance in *no-change* pixels cannot detect change pixels well. However, the TCD-Net has considerable results compared with other methods in both classes. Additionally, the TCD-Net is more sensitive to the subtly *changed* pixels, while other methods did not detect these in much detail.

In Table 7, we display the values of the mentioned criteria to evaluate the performance of CD methods. The results of dataset#2 also confirm the efficiency of the TCD-Net algorithm. As seen, the TCD-Net has the highest OA, KC, F1-score, precision and DR indicators. The OA and KC are 96.71% and 0.82 of TCD-Net, which is 4.68% and 0.47 higher than PCA_kmeans, 2.59% and 0.16 higher than NR_ELM, 2.31% and 0.15 higher than Gabor_PCANet, 2.11% and 0.14 higher than DP_PCANet, 2.17% and 0.17 higher than CWNN, and 1.26% and 0.1 higher than the dual-channel deep network. The TCD-Net algorithm has a much higher TPR and much lower FNR. In addition, TCD-Net has a much higher KC and DR (approximately 15–60% in DR and 0.1–0.5 in KC). This shows that the proposed methods perform better than the other methods implemented in this paper.

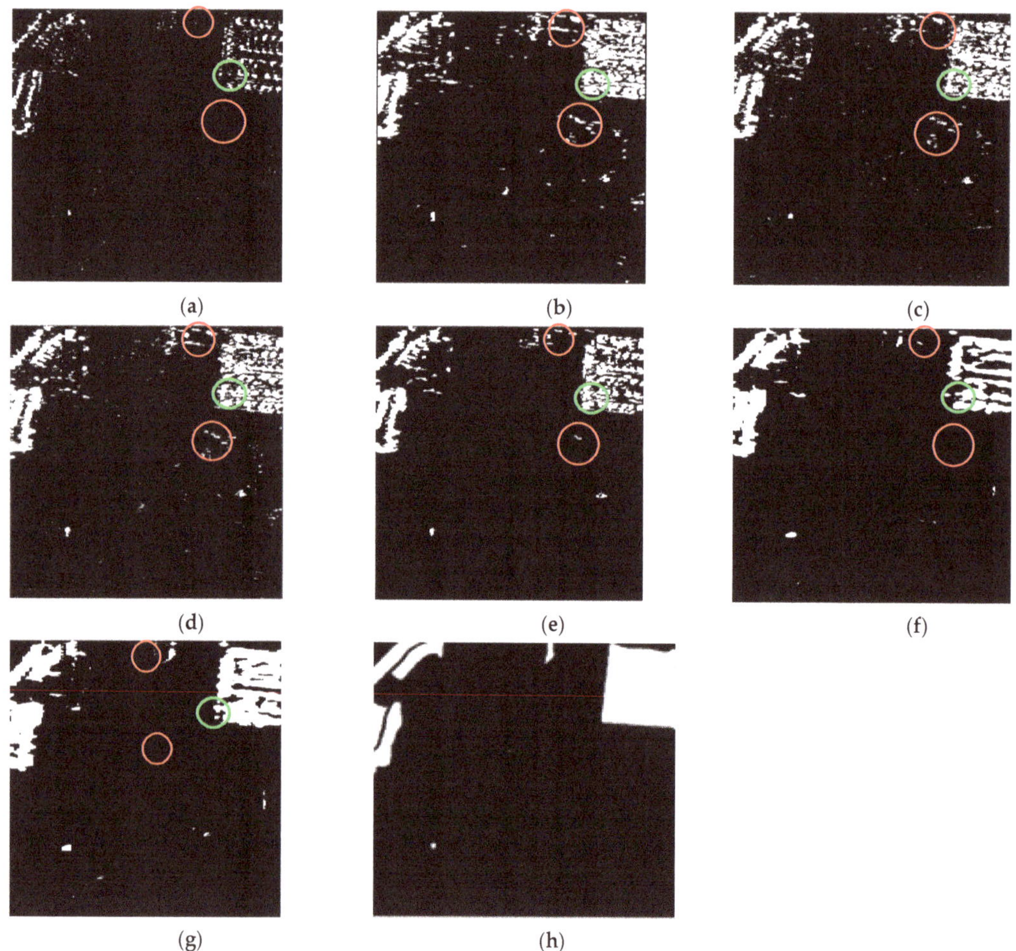

Figure 8. Visualized results of various CD methods on dataset#2; (**a**) PCA_kmeans, (**b**) NR_ELM, (**c**) Gabor_PCANet, (**d**) DP_PCANet, (**e**) CWNN, (**f**) dual-channel deep network, (**g**) TCD-Net and (**h**) ground truth. The red circles highlight different output performances in *no-change* pixels. The green circles highlight different output performances in *change* pixels.

Table 7. The accuracy of different change detection methods for dataset#2.

Method	TN	TP	FP	FN	TPR (%)	FPR (%)	FNR (%)	Precision	OA (%)	F1-Score	DR (%)	KC	OER (%)
PCA_kmeans	205,826	5653	785	17,536	24.38	0.38	75.62	87.81	92.03	57.16	24.38	0.35	7.97
NR_ELM	200,774	15,525	837	7664	66.95	2.83	33.05	72.68	94.12	83.85	66.95	0.66	3.70
Gabor_PCANet	202,050	14,884	4511	8355	64.05	2.18	35.95	76.74	94.40	78.47	64.19	0.67	5.60
DP_PCANet	202,383	15,019	4228	8170	64.77	2.05	35.23	78.03	94.60	80.32	64.77	0.68	5.40
CWNN	204,199	13,060	2412	10,129	56.32	1.17	43.68	84.41	94.54	80.33	56.32	0.65	5.46
Dual-channel Net	203,880	15,460	2731	7729	66.67	1.32	33.33	84.99	95.45	74.72	66.67	0.72	4.55
TCD-Net	203,284	18,949	3327	4240	81.72	1.61	18.28	85.06	96.71	87.86	81.72	0.82	3.29

5. Discussion

In this section, we first compare the TCD-Net in terms of accuracy to other CD methods implemented in this article. Then we compare the TCD-Net with the results of other studies implemented on the UAVSAR datasets. Finally, we mention some of the challenges that the TCD-Net algorithm has resolved.

Most of the CD methods have low efficiency in detecting *change* pixels. In other words, the low value of some indices such as precision, TPR and KC instigated from the low efficiency of the CD algorithms in detecting *change* pixels. However, the TCD-Net simultaneously has high precision, TPR and KC values in two datasets, which indicates high efficiency in detecting *change* pixels. Most algorithms have a reasonable value of OA which indicates that they have been successful in detecting *no-change* pixels. Therefore, the FPR is very low for most CD algorithms. For better evaluation, *change* and *no-change* pixels should be considered together. For this purpose, we consider both the TPR and FPR criteria. The TPR is a criterion defined based on *change* pixels. Therefore, its low value indicates that the algorithm is weak in detecting *change* pixels. Although the PCA_kmeans algorithm has low FPR and detects *no-change* pixels well, its TPR is very low, indicating that it has performed poorly in detecting *change* pixels. The NR_ELM algorithm has a higher TPR than the PCA_kmeans algorithm. It may be because the NR_ELM algorithm uses neighborhood information, but still has higher FPR than TCD-Net. The Gabor_PCANet, DP_PCANet and CWNN algorithms have a much lower TPR for dataset#2 and much higher FPR for dataset#1 than TCD-Net. The dual-channel deep network can discover the *no-changed* pixels well, but there are *changed* pixels, especially in edges in dataset#1 and another area in dataset#2, that the dual-channel deep network cannot detect. However, the TCD-Net detects both the *changed* and *no-changed* pixels well. Comparison of the dual-channel deep network and the TCD-Net shows that the intermediate channel plays a key role in detecting *change* pixels and can improve network performance. These differences are because of the robust and strong architecture of the proposed algorithm (e.g., feature extraction at different levels, separate extraction of features of two images, intermediate connection, sensitivity to different object sizes, and extraction of high-precision training data).

In the following, we quantitatively compare the TCD-Net with the results of other researches applied on the UAVSAR data according to Table 8. Ratha et al. [68] proposed a method based on geodesic distance (GD), which is the distance between an observed Kennaugh matrix and the Kennaugh matrix associated with an elementary target. This algorithm achieved the FPR and KC values equal to 6.9% and 0.73, respectively, for dataset#1, the FPR value of 3.9% and the KC value of 0.75 for dataset#2. By comparing the FPR values between the GD and TCD-Net, it can be found that the TCD-Net performed better than the GD in accurately identifying pixels as *change* or *no-change*. Bouhlel, Akbari and Méric [3] have proposed a determinant ratio test (DRT) statistic for automatic CD in bi-temporal PolSAR images, assuming that the multi-look complex covariance matrix follows the scaled complex Wishart distribution. The DRT algorithm obtained the FPR and DR values of 10.58% and 63.38%, respectively, for dataset#1, and the FPR value of 8.39% and the DR value of 51.49% for dataset#2. The quantitative results demonstrate that the TCD-Net provides an average of 20% higher DR and 8% lower FPR compared to the DRT, which indicates the superiority of TCD-Net. Nascimento et al. [69] have proposed a comparison between the likelihood ratio, Kullback–Leibler (KL) distance, Shannon entropy and Rényi entropy. The results of this research demonstrated that entropy-based algorithms may perform better than algorithms based on the KL distance and probability ratio statistics. Comparison of the TCD-Net algorithm with the best entropy-based algorithms in [69] shows that the TCD-Net algorithm has much higher DR (about 30% in dataset#1 and 20% in dataset#2). In addition, the KC in the TCD-Net algorithm is much higher than the entropy-based algorithm (0.18 in dataset#1 and 0.26 in dataset#2). [68] and [69] are statistical methods and these methods operate in unsupervised manners. Moreover, the TCD-Net acts unsupervised. Nevertheless, the TCD-Net is more effective. One of the important factors in improving the accuracy of the proposed method is the use of deep features while other methods operate on the main polarization channels (i.e., HH, HV, VH and VV). The limited polarization channels and noise conditions cause statistics-based methods to not perform well. In terms of processing time, statistical-based algorithms have less processing time than DL-based algorithms. The training phase of DL-based algorithms is time-consuming, but in general, DL-based algorithms are more accurate than statistical-based algorithms.

Table 8. Comparison of TCD-Net results with other methods developed on UAVSAR images.

Dataset	Indices	Ratha, De, Celik and Bhattacharya [68]	Bouhlel, Akbari and Méric [3]	Nascimento, Frery and Cintra [69]	TCD-Net
			Result		
dataset#1	KC	0.73	N/A	0.62	0.80
	FPR (%)	6.9	10.58	4.14	1.26
	DR (%)	N/A	63.38	42.98	74.74
dataset#2	KC	0.75	N/A	0.56	0.82
	FPR (%)	3.9	8.39	2.17	1.61
	DR (%)	N/A	51.49	62.52	81.72

As mentioned earlier, one of the main challenges of applying DL-based algorithms for CD applications is finding enough training data. Several studies have proposed methods for automatically extracting pseudo-label training samples, which have been employed in this study. In [67], Gabor wavelet features were used to exploit the changed information. In addition, the FCM algorithm was implemented in a coarse-to-fine procedure to obtain enough pseudo-label training samples. In [58], a parallel FCM clustering was developed for SAR images based on combining nonlinear sigmoid mapping, Gabor wavelets and parallel FCM to provide pseudo-label training pixels. These methods are pixel-based and do not take into account spatial information, which may produce isolated pixels as an output. In [66], a pre-classification step was implemented by using a neighborhood-based ratio operator and hierarchical FCM clustering. In addition, some studies have also used trained neural networks and TL techniques. In these methods, pixels are classified based on a global threshold, which can lead to mistakes and less reliability in some cases. In contrast, our pseudo-label sample generation framework is based on probability, it extracts the pixels from the pre-trained model with a probability of more than 95%, and also aggregates with the results of the FCM algorithm for more reliability. In addition, the process of detecting changes in PolSAR images has many challenges. For instance, the process of extracting polarimetric decomposition parameters, which is a common step in conventional PolSAR CD methods, is time-consuming and challenging, especially when dealing with time-series data. In addition, selecting appropriate decompositions with high information content requires optimization algorithms that are also time-consuming. Furthermore, previous studies showed that adding spatial features to scattering information significantly increases the accuracy of CD methods. However, extracting spatial features, such as texture, is challenging because of hardware limitations and long processing time. To overcome these problems, we present the TCD-Net algorithm, which can extract deep features only with four bands and does not require any additional processing (e.g., feature extraction, feature selection and target decomposition). Additionally, DL-based CD methods automatically employ both spatial and spectral features, and because of the simultaneous use of spatial and spectral information, this method is more accurate and robust than other CD methods. In addition, the TCD-Net architecture uses residual and multi-scale blocks. The residual blocks allow information to flow from the initial layers to the final layers, preventing the network depth from increasing too much. Moreover, the multi-scale blocks increase the network sensitivity to objects of different sizes.

6. Conclusions and Future Work

In this study, a novel end-to-end framework based on DL is proposed for detecting changes in the polarimetry UAVSAR datasets. The proposed method can solve the challenges of conventional CD methods (i.e., thresholding, manual feature extraction methods and training data limitation in DL-based CD methods). First, we propose a parallel pseudo-label training sample generation framework, which can generate high-reliability samples for TCD-Net training using a parallel combination of the result of the pre-trained model and FCM algorithm. Numerical analysis shows that the generated samples have provided the OA of 99.52% and KC of 0.91. Second, we construct a TCD-Net architecture with three-channels based on an adaptive multi-scale shallow block and an adaptive multi-scale residual block that are sensitive to objects of different sizes and maintain fundamental information through the transfer of information to higher layers. Therefore, our proposed method has high efficiency in the extraction of deep features. The performance of our proposed method is evaluated using two different UAVSAR datasets. Moreover, the results of our proposed method are compared to other SOTA PolSAR CD methods and a dual-channel deep network to evaluate the effectiveness of an intermediate channel embedded on TCD-Net. The result of CD is evaluated by visual and numerical accuracy assessments indices. Experimental results show that the highest OA of 96.71% and the best KC of 0.82 belong to TCD-Net. In summary, compared to other CD algorithms, the proposed method has several advantages: (1) it is more accurate than other SOTA CD methods; (2) it provides robust results compared with a dual-channel deep network; (3) it is unsupervised and produces appropriate quality and quantity training data; (4) it is strong against noise and complicated and multi-size objects; and (5), its end-to-end framework requires no pre-processing (e.g., manual feature extraction, feature selection and PolSAR target decomposition).

One of the limitations of SAR CD is the complexity and noise conditions of SAR data. The issue can affect the CD and weaken the result of the CD. In this regard, the fusion of multimodal datasets can improve the result of CD and enhance accuracy. The digital elevation model (DEM) is one of the most important datasets that can provide the CD in more detail. In addition, we intend to evaluate the performance of TCD-Net across all single-, dual- and fully-polarized modes in the future.

Author Contributions: Conceptualization, R.H., S.T.S., M.H. and M.M.; methodology, S.T.S.; writing—original draft preparation, S.T.S.; writing—review and editing, S.T.S., M.H. and M.M.; visualization, R.H. and S.T.S.; supervision, M.H. and M.M.; funding acquisition, M.M. All authors have read and agreed to the published version of the manuscript.

Funding: This study received no external funding.

Institutional Review Board Statement: Not applicable.

Informed Consent Statement: Not applicable.

Data Availability Statement: Publicly available datasets were analyzed in this study. These datasets can be found here: [https://rslab.ut.ac.ir] (accessed on 15 January 2022).

Acknowledgments: The authors would like to thank the anonymous reviewers for their valuable comments on our manuscript.

Conflicts of Interest: The authors declare that they have no conflict of interest.

References

1. Zhang, M.; Shi, W. A feature difference convolutional neural network-based change detection method. *IEEE Trans. Geosci. Remote Sens.* **2020**, *58*, 7232–7246. [CrossRef]
2. Seydi, S.T.; Hasanlou, M.; Amani, M. A new end-to-end multi-dimensional CNN framework for land cover/land use change detection in multi-source remote sensing datasets. *Remote Sens.* **2020**, *12*, 2010. [CrossRef]
3. Bouhlel, N.; Akbari, V.; Méric, S. Change Detection in Multilook Polarimetric SAR Imagery With Determinant Ratio Test Statistic. *IEEE Trans. Geosci. Remote Sens.* **2020**, *60*, 5200515. [CrossRef]

4. Peng, D.; Bruzzone, L.; Zhang, Y.; Guan, H.; Ding, H.; Huang, X. SemiCDNet: A semisupervised convolutional neural network for change detection in high resolution remote-sensing images. *IEEE Trans. Geosci. Remote Sens.* **2020**, *59*, 5891–5906. [CrossRef]
5. Sefrin, O.; Riese, F.M.; Keller, S. Deep Learning for Land Cover Change Detection. *Remote Sens.* **2021**, *13*, 78. [CrossRef]
6. Zhang, T.; Zhang, X. ShipDeNet-20: An only 20 convolution layers and <1-MB lightweight SAR ship detector. *IEEE Geosci. Remote Sens. Lett.* **2020**, *18*, 1234–1238.
7. Zhang, T.; Zhang, X.; Ke, X.; Liu, C.; Xu, X.; Zhan, X.; Wang, C.; Ahmad, I.; Zhou, Y.; Pan, D. HOG-ShipCLSNet: A Novel Deep Learning Network with HOG Feature Fusion for SAR Ship Classification. *IEEE Trans. Geosci. Remote Sens.* **2021**, *60*, 5210322. [CrossRef]
8. Hasanlou, M.; Seydi, S.T. Use of multispectral and hyperspectral satellite imagery for monitoring waterbodies and wetlands. In *Southern Iraq's Marshes: Their Environment and Conservation*; Jawad, L.A., Ed.; Springer: Cham, Switzerland, 2021; p. 155.
9. Mohammadimanesh, F.; Salehi, B.; Mahdianpari, M.; Brisco, B.; Gill, E. Full and simulated compact polarimetry sar responses to canadian wetlands: Separability analysis and classification. *Remote Sens.* **2019**, *11*, 516. [CrossRef]
10. Mahdianpari, M.; Jafarzadeh, H.; Granger, J.E.; Mohammadimanesh, F.; Brisco, B.; Salehi, B.; Homayouni, S.; Weng, Q. A large-scale change monitoring of wetlands using time series Landsat imagery on Google Earth Engine: A case study in Newfoundland. *GIScience Remote Sens.* **2020**, *57*, 1102–1124. [CrossRef]
11. Cloude, S.R.; Pottier, E. An entropy based classification scheme for land applications of polarimetric SAR. *IEEE Trans. Geosci. Remote Sens.* **1997**, *35*, 68–78. [CrossRef]
12. Migliaccio, M.; Gambardella, A.; Tranfaglia, M. SAR polarimetry to observe oil spills. *IEEE Trans. Geosci. Remote Sens.* **2007**, *45*, 506–511. [CrossRef]
13. De Maio, A.; Orlando, D.; Pallotta, L.; Clemente, C. A multifamily GLRT for oil spill detection. *IEEE Trans. Geosci. Remote Sens.* **2016**, *55*, 63–79. [CrossRef]
14. Seydi, S.T.; Hasanlou, M.; Chanussot, J. DSMNN-Net: A Deep Siamese Morphological Neural Network Model for Burned Area Mapping Using Multispectral Sentinel-2 and Hyperspectral PRISMA Images. *Remote Sens.* **2021**, *13*, 5138. [CrossRef]
15. Hasanlou, M.; Shah-Hosseini, R.; Seydi, S.T.; Karimzadeh, S.; Matsuoka, M. Earthquake Damage Region Detection by Multitemporal Coherence Map Analysis of Radar and Multispectral Imagery. *Remote Sens.* **2021**, *13*, 1195. [CrossRef]
16. Bai, Y.; Tang, P.; Hu, C. kCCA transformation-based radiometric normalization of multi-temporal satellite images. *Remote Sens.* **2018**, *10*, 432. [CrossRef]
17. Cao, C.; Dragićević, S.; Li, S. Land-use change detection with convolutional neural network methods. *Environments* **2019**, *6*, 25. [CrossRef]
18. Liu, F.; Jiao, L.; Tang, X.; Yang, S.; Ma, W.; Hou, B. Local restricted convolutional neural network for change detection in polarimetric SAR images. *IEEE Trans. Neural Netw. Learn. Syst.* **2018**, *30*, 818–833. [CrossRef] [PubMed]
19. De Bem, P.P.; de Carvalho Junior, O.A.; Fontes Guimarães, R.; Trancoso Gomes, R.A. Change detection of deforestation in the Brazilian Amazon using landsat data and convolutional neural networks. *Remote Sens.* **2020**, *12*, 901. [CrossRef]
20. Asokan, A.; Anitha, J. Change detection techniques for remote sensing applications: A survey. *Earth Sci. Inform.* **2019**, *12*, 143–160. [CrossRef]
21. Lee, J.-S.; Pottier, E. *Polarimetric Radar Imaging: From Basics to Applications*; CRC Press: Boca Raton, FL, USA, 2017.
22. Verma, R. Polarimetric Decomposition Based on General Characterisation of Scattering from Urban Areas and Multiple Component Scattering Model. Master's Thesis, University of Twente, Enschede, The Netherlands, 2012.
23. Lee, J.-S.; Pottier, E. *Polarimetric Radar Imaging: From Basics to Applications*, 2nd ed.; CRC Press: Boca Raton, FL, USA, 2009.
24. Bruzzone, L.; Prieto, D.F. An adaptive semiparametric and context-based approach to unsupervised change detection in multitemporal remote-sensing images. *IEEE Trans. Image Processing* **2002**, *11*, 452–466. [CrossRef]
25. Gong, M.; Su, L.; Jia, M.; Chen, W. Fuzzy clustering with a modified MRF energy function for change detection in synthetic aperture radar images. *IEEE Trans. Fuzzy Syst.* **2013**, *22*, 98–109. [CrossRef]
26. Inglada, J.; Giros, A. On the possibility of automatic multisensor image registration. *IEEE Trans. Geosci. Remote Sens.* **2004**, *42*, 2104–2120. [CrossRef]
27. Dekker, R. Speckle filtering in satellite SAR change detection imagery. *Int. J. Remote Sens.* **1998**, *19*, 1133–1146. [CrossRef]
28. Li, Y.; Peng, C.; Chen, Y.; Jiao, L.; Zhou, L.; Shang, R. A deep learning method for change detection in synthetic aperture radar images. *IEEE Trans. Geosci. Remote Sens.* **2019**, *57*, 5751–5763. [CrossRef]
29. Inglada, J.; Mercier, G. A new statistical similarity measure for change detection in multitemporal SAR images and its extension to multiscale change analysis. *IEEE Trans. Geosci. Remote Sens.* **2007**, *45*, 1432–1445. [CrossRef]
30. Deng, J.; Wang, K.; Deng, Y.; Qi, G. PCA-based land-use change detection and analysis using multitemporal and multisensor satellite data. *Int. J. Remote Sens.* **2008**, *29*, 4823–4838. [CrossRef]
31. Seydi, S.T.; Shahhoseini, R. Transformation Based Algorithms for Change Detection in Full Polarimetric remote SENSING Images. *Int. Arch. Photogramm. Remote Sens. Spat. Inf. Sci.* **2019**, *42*, 963–967. [CrossRef]
32. Hasanlou, M.; Seydi, S.T. Hyperspectral change detection: An experimental comparative study. *Int. J. Remote Sens.* **2018**, *39*, 7029–7083. [CrossRef]
33. Kittler, J.; Illingworth, J. Minimum error thresholding. *Pattern Recognit.* **1986**, *19*, 41–47. [CrossRef]
34. Dempster, A.P.; Laird, N.M.; Rubin, D.B. Maximum likelihood from incomplete data via the EM algorithm. *J. R. Stat. Soc. Ser. B (Methodol.)* **1977**, *39*, 1–22.

35. Moser, G.; Serpico, S.B. Generalized minimum-error thresholding for unsupervised change detection from SAR amplitude imagery. *IEEE Trans. Geosci. Remote Sens.* **2006**, *44*, 2972–2982. [CrossRef]

36. Hu, H.; Ban, Y. Unsupervised change detection in multitemporal SAR images over large urban areas. *IEEE J. Sel. Top. Appl. Earth Obs. Remote Sens.* **2014**, *7*, 3248–3261. [CrossRef]

37. Su, L.; Gong, M.; Sun, B.; Jiao, L. Unsupervised change detection in SAR images based on locally fitting model and semi-EM algorithm. *Int. J. Remote Sens.* **2014**, *35*, 621–650. [CrossRef]

38. Zheng, Y.; Zhang, X.; Hou, B.; Liu, G. Using combined difference image and k-means clustering for SAR image change detection. *IEEE Geosci. Remote Sens. Lett.* **2013**, *11*, 691–695. [CrossRef]

39. Jia, L.; Li, M.; Zhang, P.; Wu, Y.; Zhu, H. SAR image change detection based on multiple kernel K-means clustering with local-neighborhood information. *IEEE Geosci. Remote Sens. Lett.* **2016**, *13*, 856–860. [CrossRef]

40. Li, H.-C.; Celik, T.; Longbotham, N.; Emery, W.J. Gabor feature based unsupervised change detection of multitemporal SAR images based on two-level clustering. *IEEE Geosci. Remote Sens. Lett.* **2015**, *12*, 2458–2462.

41. Krinidis, S.; Chatzis, V. A robust fuzzy local information C-means clustering algorithm. *IEEE Trans. Image Processing* **2010**, *19*, 1328–1337. [CrossRef] [PubMed]

42. Gong, M.; Zhou, Z.; Ma, J. Change detection in synthetic aperture radar images based on image fusion and fuzzy clustering. *IEEE Trans. Image Processing* **2011**, *21*, 2141–2151. [CrossRef] [PubMed]

43. Liu, G.; Li, L.; Jiao, L.; Dong, Y.; Li, X. Stacked Fisher autoencoder for SAR change detection. *Pattern Recognit.* **2019**, *96*, 106971. [CrossRef]

44. Samadi, F.; Akbarizadeh, G.; Kaabi, H. Change detection in SAR images using deep belief network: A new training approach based on morphological images. *IET Image Processing* **2019**, *13*, 2255–2264. [CrossRef]

45. Saha, S.; Bovolo, F.; Bruzzone, L. Change detection in image time-series using unsupervised lstm. *IEEE Geosci. Remote Sens. Lett.* **2020**, *19*, 8005205. [CrossRef]

46. Petrou, M.; Sturm, P. Pulse Coupled Neural Networks for Automatic Urban Change Detection at Very High Spatial Resolution. In *Iberoamerican Congress on Pattern Recognition*; Springer: Berlin/Heidelberg, Germany, 2009.

47. Hou, B.; Liu, Q.; Wang, H.; Wang, Y. From W-Net to CDGAN: Bitemporal change detection via deep learning techniques. *IEEE Trans. Geosci. Remote Sens.* **2019**, *58*, 1790–1802. [CrossRef]

48. Mou, L.; Bruzzone, L.; Zhu, X.X. Learning spectral-spatial-temporal features via a recurrent convolutional neural network for change detection in multispectral imagery. *IEEE Trans. Geosci. Remote Sens.* **2018**, *57*, 924–935. [CrossRef]

49. Jaturapitpornchai, R.; Matsuoka, M.; Kanemoto, N.; Kuzuoka, S.; Ito, R.; Nakamura, R. Newly built construction detection in SAR images using deep learning. *Remote Sens.* **2019**, *11*, 1444. [CrossRef]

50. Sun, S.; Mu, L.; Wang, L.; Liu, P. L-UNet: An LSTM Network for Remote Sensing Image Change Detection. *IEEE Geosci. Remote Sens. Lett.* **2020**, *19*, 8004505. [CrossRef]

51. Cao, X.; Ji, Y.; Wang, L.; Ji, B.; Jiao, L.; Han, J. SAR image change detection based on deep denoising and CNN. *IET Image Processing* **2019**, *13*, 1509–1515. [CrossRef]

52. Wang, J.; Gao, F.; Dong, J. Change detection from SAR images based on deformable residual convolutional neural networks. In Proceedings of the 2nd ACM International Conference on Multimedia in Asia, Online, 7 March 2021; pp. 1–7.

53. Kiana, E.; Homayouni, S.; Sharifi, M.; Farid-Rohani, M. Unsupervised Change Detection in SAR images using Gaussian Mixture Models. *Int. Arch. Photogramm. Remote Sens. Spat. Inf. Sci.* **2015**, *40*, 407. [CrossRef]

54. Liu, J.; Gong, M.; Qin, K.; Zhang, P. A deep convolutional coupling network for change detection based on heterogeneous optical and radar images. *IEEE Trans. Neural Netw. Learn. Syst.* **2016**, *29*, 545–559. [CrossRef] [PubMed]

55. Bergamasco, L.; Saha, S.; Bovolo, F.; Bruzzone, L. Unsupervised change-detection based on convolutional-autoencoder feature extraction. In Proceedings of the Image and Signal Processing for Remote Sensing XXV, Strasbourg, France, 9–11 September 2019; p. 1115510.

56. Huang, F.; Yu, Y.; Feng, T. Automatic building change image quality assessment in high resolution remote sensing based on deep learning. *J. Vis. Commun. Image Represent.* **2019**, *63*, 102585. [CrossRef]

57. Gao, F.; Wang, X.; Gao, Y.; Dong, J.; Wang, S. Sea ice change detection in SAR images based on convolutional-wavelet neural networks. *IEEE Geosci. Remote Sens. Lett.* **2019**, *16*, 1240–1244. [CrossRef]

58. Zhang, X.; Su, H.; Zhang, C.; Atkinson, P.M.; Tan, X.; Zeng, X.; Jian, X. A Robust Imbalanced SAR Image Change Detection Approach Based on Deep Difference Image and PCANet. *arXiv* **2020**, arXiv:2003.01768.

59. Liu, J.; Chen, K.; Xu, G.; Sun, X.; Yan, M.; Diao, W.; Han, H. Convolutional neural network-based transfer learning for optical aerial images change detection. *IEEE Geosci. Remote Sens. Lett.* **2019**, *17*, 127–131. [CrossRef]

60. Khelifi, L.; Mignotte, M. Deep learning for change detection in remote sensing images: Comprehensive review and meta-analysis. *IEEE Access* **2020**, *8*, 126385–126400. [CrossRef]

61. Kutlu, H.; Avcı, E. A novel method for classifying liver and brain tumors using convolutional neural networks, discrete wavelet transform and long short-term memory networks. *Sensors* **2019**, *19*, 1992. [CrossRef] [PubMed]

62. Venugopal, N. Sample selection based change detection with dilated network learning in remote sensing images. *Sens. Imaging* **2019**, *20*, 31. [CrossRef]

63. Yommy, A.S.; Liu, R.; Wu, S. SAR image despeckling using refined Lee filter. In Proceedings of the 2015 7th International Conference on Intelligent Human-Machine Systems and Cybernetics, Hangzhou, China, 26–27 August 2015; pp. 260–265.

64. Yuan, Q.; Wei, Y.; Meng, X.; Shen, H.; Zhang, L. A multiscale and multidepth convolutional neural network for remote sensing imagery pan-sharpening. *IEEE J. Sel. Top. Appl. Earth Obs. Remote Sens.* **2018**, *11*, 978–989. [CrossRef]
65. Celik, T. Unsupervised change detection in satellite images using principal component analysis and k-means clustering. *IEEE Geosci. Remote Sens. Lett.* **2009**, *6*, 772–776. [CrossRef]
66. Gao, F.; Dong, J.; Li, B.; Xu, Q.; Xie, C. Change detection from synthetic aperture radar images based on neighborhood-based ratio and extreme learning machine. *J. Appl. Remote Sens.* **2016**, *10*, 046019. [CrossRef]
67. Gao, F.; Dong, J.; Li, B.; Xu, Q. Automatic change detection in synthetic aperture radar images based on PCANet. *IEEE Geosci. Remote Sens. Lett.* **2016**, *13*, 1792–1796. [CrossRef]
68. Ratha, D.; De, S.; Celik, T.; Bhattacharya, A. Change detection in polarimetric SAR images using a geodesic distance between scattering mechanisms. *IEEE Geosci. Remote Sens. Lett.* **2017**, *14*, 1066–1070. [CrossRef]
69. Nascimento, A.D.; Frery, A.C.; Cintra, R.J. Detecting changes in fully polarimetric SAR imagery with statistical information theory. *IEEE Trans. Geosci. Remote Sens.* **2018**, *57*, 1380–1392. [CrossRef]

remote sensing

MDPI

Article

A Transformer-Based Coarse-to-Fine Wide-Swath SAR Image Registration Method under Weak Texture Conditions

Yibo Fan, Feng Wang * and Haipeng Wang

Key Laboratory for Information Science of Electromagnetic Waves (Ministry of Education),
School of Information Science and Technology, Fudan University, Shanghai 200433, China;
ybfan19@fudan.edu.cn (Y.F.); hpwang@fudan.edu.cn (H.W.)
* Correspondence: fengwang@fudan.edu.cn

Abstract: As an all-weather and all-day remote sensing image data source, SAR (Synthetic Aperture Radar) images have been widely applied, and their registration accuracy has a direct impact on the downstream task effectiveness. The existing registration algorithms mainly focus on small sub-images, and there is a lack of available accurate matching methods for large-size images. This paper proposes a high-precision, rapid, large-size SAR image dense-matching method. The method mainly includes four steps: down-sampling image pre-registration, sub-image acquisition, dense matching, and the transformation solution. First, the ORB (Oriented FAST and Rotated BRIEF) operator and the GMS (Grid-based Motion Statistics) method are combined to perform rough matching in the semantically rich down-sampled image. In addition, according to the feature point pairs, a group of clustering centers and corresponding images are obtained. Subsequently, a deep learning method based on Transformers is used to register images under weak texture conditions. Finally, the global transformation relationship can be obtained through RANSAC (Random Sample Consensus). Compared with the SOTA algorithm, our method's correct matching point numbers are increased by more than 2.47 times, and the root mean squared error (RMSE) is reduced by more than 4.16%. The experimental results demonstrate that our proposed method is efficient and accurate, which provides a new idea for SAR image registration.

Keywords: synthetic aperture radar; image registration; transformer

Citation: Fan, Y.; Wang, F.; Wang, H. A Transformer-Based Coarse-to-Fine Wide-Swath SAR Image Registration Method under Weak Texture Conditions. *Remote Sens.* **2022**, *14*, 1175. https://doi.org/10.3390/rs14051175

Academic Editors: Tianwen Zhang, Tianjiao Zeng and Xiaoling Zhang

Received: 19 January 2022
Accepted: 23 February 2022
Published: 27 February 2022

Publisher's Note: MDPI stays neutral with regard to jurisdictional claims in published maps and institutional affiliations.

1. Introduction

Synthetic aperture radar (SAR) has the advantages of working in all weather, at all times, and having strong penetrability. SAR image processing is developing rapidly in civilian and military applications. There are many practical scenarios for the joint processing and analysis of multiple remote sensing images, such as data fusion [1], change detection [2], and pattern recognition [3]. The accuracy of the image matching affects the performance of the above downstream tasks. However, SAR image acquisition conditions are diverse, such as different polarizations, incident angles, imaging methods, time phases, and so on. At the same time, defocusing problems caused by motion errors degrade the image quality. Besides this, the time and spatial complexity of traditional methods are unacceptable for large images. Thus, for the mass of scenes where multiple SAR images are processed simultaneously, SAR image registration is a real necessity. The nonlinear distortion and inherent speckle noise of SAR images leave wide-swath SAR image registration as a knot to be solved.

The geographical alignment of two SAR images, under different imaging conditions, is based on the mapping model, which is usually solved by the relative relationship of the corresponding parts from images. The two images are reference images and sensed images to be registered. Generally speaking, conventional geometric transformation models include affine, projection, rigid body, and nonlinear transformation models. In this paper, we focus on the most pervasive affine transformation model.

The registration techniques in the Computer Vision (CV) field have continued to spring up for decades. The existing normal registration methods can be divided mainly into traditional algorithms and learning-based algorithms. The traditional methods mainly include feature-based and region-based methods. The region-based method finds the best transformation parameters based on the maximum similarity coefficient, and includes mutual information methods [4], Fourier methods [5], and cross-correlation methods [6,7]. Stone et al. [5] presented a Fourier-based algorithm to solve translations and uniform changes of illumination in aerial photos. Recently, in the field of SAR image registration, Luca et al. [7] used cross-correlation parabolic interpolation to refine the matching results. This series of methods only use plain gray information and risk mismatch under speckle noise and radiation variation.

Another major class of registration techniques in the CV field is the feature-based method. It searches for geometric mappings such as points, lines, contours, and regional features based on the stable feature correspondences across two images. The most prevalent method is SIFT (Scale Invariant Feature Transform) [8]. SIFT has been widely used in the field of image registration due to the following invariances: rotation, scale, grayscale, and so on. PCA-SIFT (Principal Component Analysis-SIFT) [9] applies dimensionality reduction to SIFT descriptors to improve the matching efficiency. Slightly different from the classical CV field, a series of unique image registration methods appear in the SAR image processing field. Given the characteristics of SAR speckle noise, SAR-SIFT [10] adopts a new method of gradient calculation and feature descriptor generation to improve the SAR image registration performance. KAZE-SAR [11] uses the nonlinear diffusion filtering method KAZE [12] to build the scale space. Xiang [13] proposed a method to match large SAR images with optical images. To be specific, the method combines dilated convolutional features with epipolar-oriented phase correlation to reduce horizontal errors, and then fine-tunes the matching part. Feature-based methods are more flexible and effective; as such, they are more practical under complex spatial change. Coherent speckle noise has consequences on the conventional method's precision, and the traditional matching approach fails to achieve the expected results under complex and varied scenarios.

Deep learning [14] (DL) has exploded in CV fields over the past decade. With strong abilities of feature extraction and characterization, deep learning is in wide usage across remote sensing scenarios, including classification [15], detection [16], image registration [17], and change detection [18]. More and more methods [19,20] use learning-based methods in the registration of the CV field. He et al. [19] proposed a Siamese CNN (Convolutional Neural Networks) to evaluate the similarity of patch pairs. Zheng et al. [20] proposed SymReg-GAN, which achieves good results in medical image registration by using a generator that predicts the geometric transformation between images, and a discriminator that distinguishes the transformed images from the real images. Specific to the SAR image (remote sensing image) registration field, Li et al. [21] proposed a RotNET to predict the rotation relationship between two images. Mao et al. [22] proposed a multi-scale fused deep forest-based SAR image registration method. Luo et al. [23] used pre-trained deep residual neural features extracted from CNN for registration. The CMM-Net (cross modality matching net) [24] used CNN to extract high-dimensional feature maps and build descriptors. DL often requires large training datasets. Unlike optical natural images, it is difficult to accurately label SAR images due to the influence of noise. In addition, most DL-based SAR image registration studies generally deal with small image blocks with a fixed size, but in practical applications, wide-swath SAR images cannot be directly matched.

As was outlined earlier in this article, Figure 1 lists some of the registration methods for SAR (remote sensing) image domains. Although many SAR image registration methods exist, there are still some limitations:

- Feature points mainly exist in the strong corner and edge areas, and there are not enough matching point pairs in weak texture areas.
- Due to the special gradient calculation and feature space construction method, the traditional method runs slowly and consumes a lot of memory.

- The existing SAR image registration methods mainly rely on the CNN structure, and lack a complete relative relationship between their features due to the receptive field's limitations.

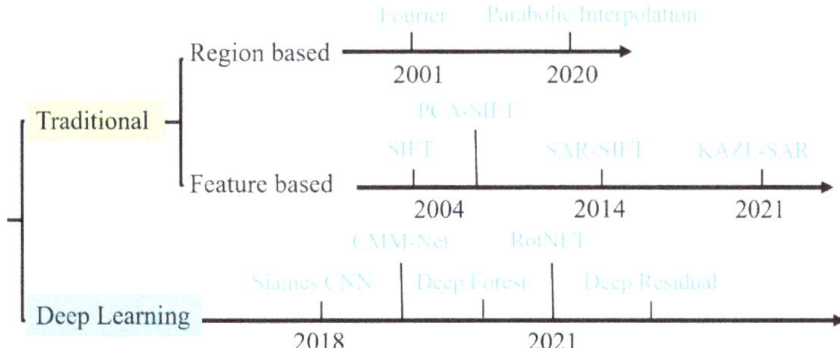

Figure 1. Remote sensing image registration milestones in the last two decades.

Based on the above analysis, this paper proposes a wide-swath SAR image fine-level registration framework that combines traditional methods and deep learning. The experimental results show that, compared with the state of the art, the proposed method can obtain better matching results. Under the comparison and analysis of the matching performance in different data sources, the method in this paper is more effective and robust for SAR image registration.

The general innovations of this paper are as follows:

1. A CNN and Transformer hybrid approach is proposed in order to accurately register SAR images through a coarse-to-fine form.
2. A stable partition framework from the full image to sub-images is constructed; in this method, the regions of interest are selected in pairs.

The remainder of this paper is organized as follows. In Section 2. Methods, the proposed framework of SAR image registration and the learning-based sub-image matching method are discussed in detail. In Section 3. Experimental Results and Analyses, specified experiments, as well as quantitative and qualitative results, are given. In Section 4. Discussion, the conclusion is provided.

2. Methods

In this study, we propose a phased SAR image registration framework that combines traditional and deep learning methods. The framework is illustrated in Figure 2; the proposed method mainly consists of four steps. First, the ORB [25] and GMS [26] are used to obtain the coarse registration result via the downsampled original image. Second, K-means++ [27] select cluster centers of registration points from the previous step, and a series of corresponding original-resolution image slices are obtained. Third, we register the above image pairs through deep learning. The fourth step is to integrate the point pair subsets and obtain the final global transformation result after RANSAC [28].

As a starting point for our work, we first introduce the existing deep learning mainstream.

2.1. Deep Learning-Related Background

As AlexNet [29] won first place in 2012 ImageNet, deep learning had begun to play a leading role in CV, NLP (natural language processing), and other fields. The current mainstream of deep learning includes two categories: CNN and Transformer. CNN does well in the extraction of local information from two-dimensional data, such as images. Because the deep neural network can extract key features from massive data, deep CNN is performed outstandingly in image classification [30], detection [16,31], and segmentation [32].

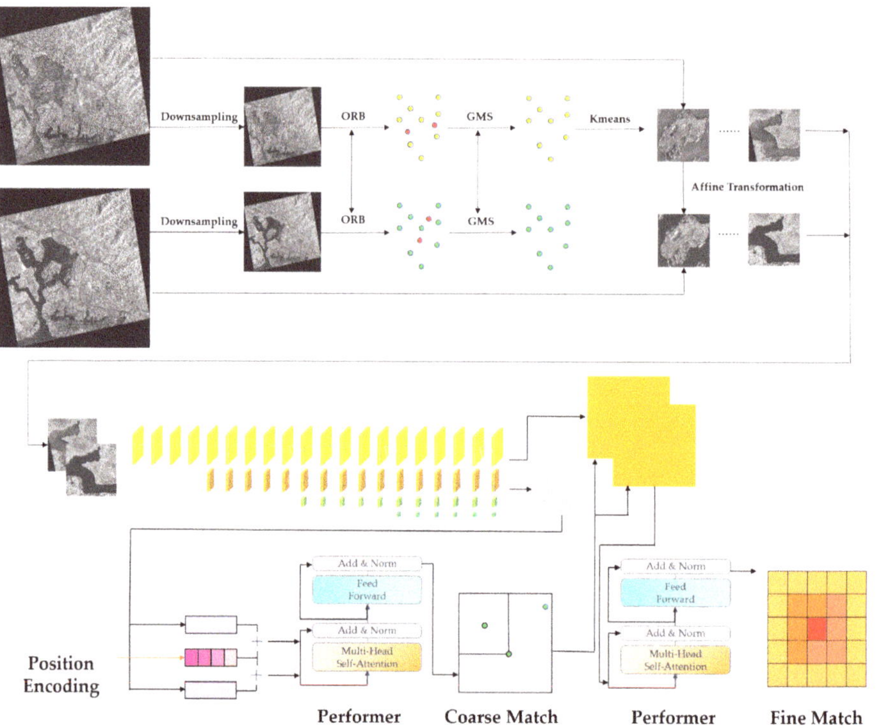

Figure 2. The pipeline of the proposed method.

Corresponding to text and other one-dimensional sequence data, currently, the most widely used processing method is Transformer [33], which solves the long-distance relying problem using a unique self-attention mechanism. It is sweeping NLP, CV, and related fields.

Deep learning has been widely used in SAR image processing over the past few years. For example, Hou et al. [16] proposed ship classification using CNN in an SAR ship dataset. Guo et al. [31] applied an Attention Pyramid Network for aircraft detection. Transformer is also used in recognition [34], detection [35] and segmentation [36]. LoFTR (Local Feature TRansformer) [37] has been proposed as a coarse-to-fine image matching method based on Transformers. However, to our knowledge, Transformer has not been applied to SAR image registration. Inspired by [37], in this article we use Transformer and CNN to improve the performance of SAR image registration.

The method proposed in this paper is mainly inspired by LoFTR. The initial consideration is that in the SAR image registration scene, due to the weak texture information, traditional CV registration methods based on gradient, statistical information, and other classical methods cannot obtain enough matching point pairs. LoFTR adopts a two-stage matching mechanism and features coding with Transformer, such that each position in the feature map contains the global information of the whole image. It works well in natural scenes, and also has a good matching effect even in flat areas with weak texture information. However, considering that SAR images have weaker texture information than optical images, it is difficult to obtain sufficient feature information.

In order to obtain more matching feature point pairs and give consideration to model complexity and algorithm accuracy, this paper adopts several modification schemes for SAR image scenes. (1) Feature Pyramid Network is used as a feature extraction network in LoFTR; in this paper, an advanced convolutional neural network, is adopted as a feature extraction part in order to obtain more comprehensively high- and low-resolution features with feature fusion. (2) This paper analyzes the factors that affect the number of matching point pairs, and finds that the size of the low-resolution feature map has an obvious direct impact on the number of feature point pairs. The higher the resolution, the higher the number of correct matching points that are finally extracted. Therefore, (1/2,1/5) resolution is adopted to replace the original (1/2,1/8) or (1/4,1/16); such a change leads to the number of matching point pairs increasing significantly. (3) In order to further reduce the algorithm complexity and improve the algorithm speed, this paper combines the advanced linear time complexity method to encode features, such that the location features at the specific index of the feature map can be weighted by the full image information, which can further improve the efficiency while ensuring the algorithm accuracy. The detailed expansion and analysis of the above parts are in the following sections.

2.2. Rough Matching of the Down-Sampled Image

The primary reasons that the traditional matching methods SIFT and SURF (Speeded-Up Robust Features) [38] cannot be applied directly to SAR images are the serious coherent speckle noise and the weak discontinuous texture. It is often impossible to obtain sufficient matching points on original-resolution SAR images by the traditional method. At the same time, the semantic information of the original-size image is relatively scarce. Therefore, we do not simply use traditional methods to process the original image. Considering that the down-sampled image is similar to a high-level feature map in deep CNN with rich semantic information, we use the down-sampled image (the rate is 10 almost) to perform rough pre-matching, as shown in Figure 3. The most representative method is SIFT. However, it runs slowly, especially for large images. The ORB algorithm is two orders of magnitude more rapid [25] than SIFT. ORB is a stable and widely used feature point detection description method.

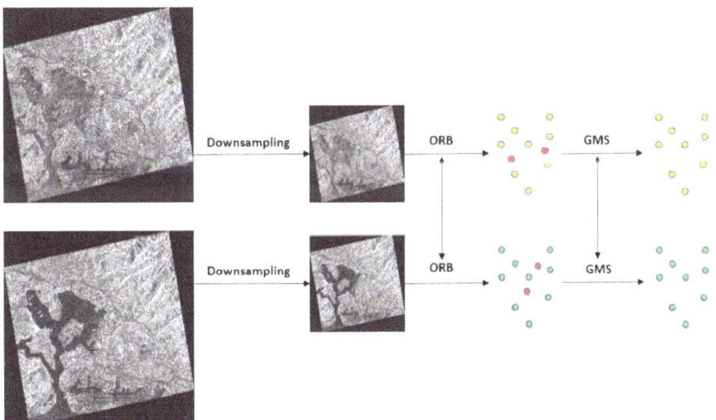

Figure 3. The pipeline of rough matching.

ORB combines and improves the FAST (Features from Accelerated Segment Test) [39] keypoint detector and the BRIEF (Binary Robust Independent Elementary Features) [40] descriptor. FAST's idea is that if the pixel's gray is distinguished from the surrounding neighborhood (i.e., it exceeds the threshold value), it may be a feature point. To be specific, FAST uses a neighborhood of 16 pixels to select the initial candidate points. Non-maximum suppression is used to eliminate the adjacent points. The gaussian blurring of different scales is performed on the image in order to achieve scale invariance.

The intensity weighted sum of a patch is defined as the centroid, and the orientation is obtained via the angle between the current point and the centroid. Orientation invariance can be enhanced by calculating moments. BRIEF is a binary coded descriptor that uses binary and bit XOR operations to speed up the establishment of feature descriptors and reduce the time for feature matching. Steered BRIEF and rBRIEF are applied for rotation invariance and distinguishability, respectively. Overall, FAST accelerates the feature point detection, and BRIEF reduces the spatial redundancy.

GMS is applied after ORB to obtain more matching point pairs; here is a brief description. If the images I_a and I_b, respectively, have N and M feature points, the set of feature points is written as {M, N}, the feature matching pair in the corresponding two images is $X_{a \to b} = \{x_1, x_2, \cdots x_n\}$, $x_i = \{m, n\}$, and a and b are the neighborhoods of the feature points from two images I_a and I_b. For a correct matching point pair, there are more matching points as support for its correctness. For the matching pair x_i, $S_i = |x_i| - 1$ is used to represent the support of its neighboring feature points, where x_i is the number of matching pairs in the neighborhood of x_i. Because the matching of each feature point is independent, it can be considered that S_i approximately obeys the binomial distribution, and can be defined as

$$S_i \sim \begin{cases} B(n, p_t) & x_i \text{ matches correctly} \\ B(n, p_f) & x_i \text{ matches wrongly} \end{cases} \tag{1}$$

n is the average number of feature points in each small neighborhood. Let f_a be one of the supporting features belonging to region a. p_t is the probability that region b includes the nearest neighbor of f_a, and similarly, p_f can be defined, and p_t and p_f can be obtained by the following formulae:

$$\begin{aligned} p_t &= p\left(f_a^t\right) + p\left(f_a^f\right)p\left(f_a^b \mid f_a^f\right) = t + (1-t)\beta m/M \\ p_f &= p\left(f_a^f\right)p\left(f_a^b \mid f_a^f\right) = (1-t)\beta m/M \end{aligned} \tag{2}$$

f_a^t, f_a^f, and f_a^b correspond to events: f_a is correctly matched, f_a is incorrectly matched, and f_a's matching point appears in region b. m represents the number of all of the feature points in region b in image I_b, and M represents the number of all of the feature points in image I_b. In order to further improve the discriminative ability, the GMS algorithm uses the multi-neighborhood model to replace the single-neighborhood model:

$$S_i = \sum_{k=1}^{K} |X_{a_k b_k}| - 1 \tag{3}$$

K is the number of small neighborhoods near the matching point, $X_{a_k b_k}$ is the number of matching pairs in the two matching neighborhoods, and S_i can be extended to

$$S_i \sim \begin{cases} B(Kn, p_t) & x_i \text{ matches correctly} \\ B(Kn, p_f) & x_i \text{ matches wrongly} \end{cases} \tag{4}$$

According to statistics, an evaluation score P is defined to measure the ability of the function S_i to discriminate between right and wrong matches, as follows:

$$P = \frac{m_t - m_f}{s_t - s_f} = \sqrt{Kn} \frac{p_t - p_f}{\sqrt{p_t(1-p_t)} + \sqrt{p_f(1-p_f)}} \tag{5}$$

Among them, s_t and s_f are the standard deviations of S_i in positive and false matches, respectively, and m_t and m_f are the mean values, respectively. It can be seen from Formula (5) that the greater the feature points' number, the higher the matching accuracy. If we set $S_{ij} = \sum_{K=1}^{K=9} |X_{i_k j_k}|$ for grid pair {i,j} and $\tau \approx 6\sqrt{n}$ for the threshold, then {i,j} is regarded as a correctly matched grid pair when $S_{ij} > \tau$.

In order to reduce the computational complexity, GMS replaces the circular neighborhood with a non-overlapping square grid to speed up S_{ij}'s calculation. Experiments have shown that when the number of feature points is 10,000, the image is divided into a

20×20 grid. The GMS algorithm scales the grid size for image size invariance, introduces a motion kernel function to process the image, and converts the rotation changes into the rearrangement of the corresponding neighborhood grid order to ensure rotation invariance.

2.3. Sub-Image Acquisition from the Cluster Centers

The existing image matching methods mostly apply to small-size images, which have lower time and storage requirements. Although some excellent methods can reach sub-pixels in local areas, they cannot be extended to a large scale due to their unique gradient calculation method and scale-space storage. Take the representative algorithm SAR-SIFT, for example; its time and memory consumption vary with the size, as shown in Figure 4, and when the image size reaches 5000–10,000 pixels or more, the memory reaches a certain peak. This computational consumption is unacceptable for ordinary desktop computers. The test experiment here was performed on high-performance workstations. Even so, the memory consumption caused by the further expansion of the image size is unbearable.

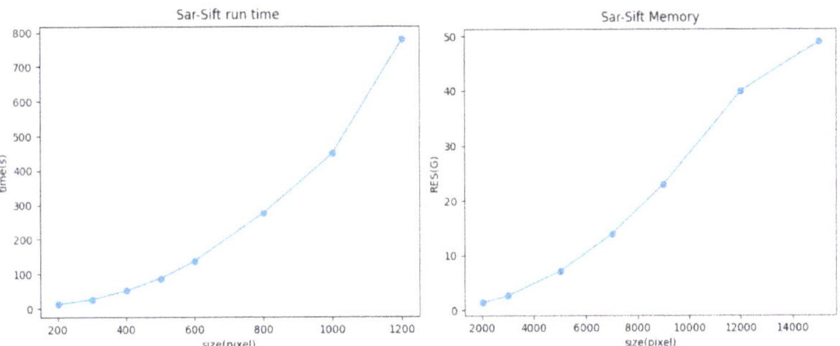

Figure 4. Trends in time and space consumption along with the image size, with SAR-SIFT as the method.

Storage limitation is also one of the key considerations. In addition to this, the time complexity of the algorithm also needs to be taken seriously. This is because, in specific practical applications, most scenarios are expected to be processed in quasi-real time. It can be seen that, for small images, SAR-SIFT can be processed within seconds, and for medium-sized images, it takes roughly minutes. For larger images, although better registration results may be obtained, the program running time of several hours or even longer cannot be accepted. Parallel optimization processing was tried here, but it did not speed the process up significantly.

According to the above analysis, due to the special gradient calculation method and the storage requirements of the scale space, wide-swath SAR image processing will risk the boom of the time and space complexity. As a comparison, we also tried the method of combining ORB with GMS for large image processing, but the final solution turned out to be wrong. The above has shown the time and spatial complexity from a qualitative point of view. The following uses SIFT as an example to analyze the reasons for the high time complexity from a formula perspective. The SIFT algorithm mainly covers several stages, as shown in Figure 5.

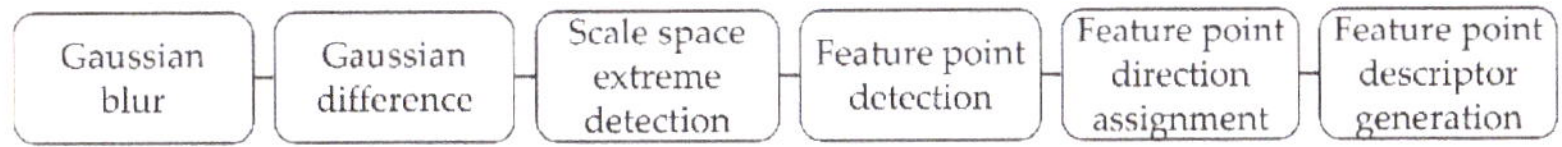

Figure 5. The pipeline of the SIFT algorithm.

The overall time complexity is composed of the sum of the complexity for each stage. Assume that the size of the currently processed image is NxN.

1. Regarding Gaussian blur, there are a total of $\hat{s}$ groups of images, and each group consists of s scales; for the original-resolution image NxN, Gaussian filter group $G(x, y, \sigma)$ is

$$G(x, y, \sigma) = \frac{1}{2\pi\sigma^2} e^{-\frac{x^2+y^2}{2\sigma^2}} \tag{6}$$

The corresponding time complexity is $O\left(N^2 w^2 s\right)$. For each pixel, a weighted sum of the surrounding Gaussian filtering $(w \times w)$ is required, with complexity:

$$L(x, y, \sigma) = \sum_{u=-\frac{w-1}{2}}^{\frac{w-1}{2}} \sum_{v=-\frac{w-1}{2}}^{\frac{w-1}{2}} G(u, v) I(x + u, y + v) \tag{7}$$

The complexity of all of the groups is

$$O\left(\sum_{j=0}^{\hat{s}-1} \frac{N^2}{2^j} w^2 s\right) = O\left(N^2 w^2 s\right) \tag{8}$$

2. To calculate the Gaussian difference, subtract each pixel of adjacent scales once in one direction.

$$D_i^j = L_{i+1}^j - L_i^j \tag{9}$$

$$O\left(\sum_{j=0}^{\hat{s}-1} \frac{sN^2}{2^j}\right) = O\left(sN^2\right) \tag{10}$$

3. To calculate the extremum detection in scale space, each point is compared with 26 adjacent points in the scale space. If the whole points are larger or smaller than the point, it is regarded as an extreme point; the complexity is

$$O\left(\sum_{j=0}^{\hat{s}-1} \frac{(s + 2)N^2}{2^j}\right) = O\left(sN^2\right) \tag{11}$$

4. For keypoint detection, the principal curvature needs to be calculated. The computational complexity of each point is $O(1)$, so the total time complexity of all of the groups is $O\left(\alpha N^2 s\right)$ considering αN^2 extrema and $\alpha\beta N^2$ keypoints.

5. For the keypoint orientation distribution, keypoint amplitude, and direction

$$m_i^j(x, y) = \sqrt{\left(L_i^j(x + 1, y) - L_i^j(x - 1, y)\right)^2 + \left(L_i^j(x, y + 1) - L_i^j(x, y - 1)\right)^2} \tag{12}$$

$$\theta_i^j(x, y) = \tan^{-1} \frac{\left(L_i^j(x, y + 1) - L_i^j(x, y - 1)\right)}{\left(L_i^j(x + 1, y) - L_i^j(x - 1, y)\right)} \tag{13}$$

Non-keypoint points with magnitudes close to the peak are added as newly added keypoints. The total number of output points is

$$\alpha\beta N^2 + \gamma\left(N^2 - \alpha\beta N^2\right) = \alpha\beta N^2(1 - \gamma) + \gamma N^2 \cong N^2(\alpha\beta + \gamma) \tag{14}$$

The computational complexity of each point is $O(1)$, and the total complexity is $O\left(N^2 s\right)$.

6. For the feature point descriptor generation, the complexity of each point is $O(x^2)$, and the total complexity is $O\left(x^2 N^2(\alpha\beta + \gamma)\right)$.

Based on the analysis of the above results, we believe that the reasons for the failure of the above algorithm in actual wide-swath SAR image registration are as follows: (1) It is inefficient to calculate the scale space of the entire image. For most areas, it is not easy to find feature points that can establish a mapping relationship, which leads to potential ineffective calculations. Not only are these time-consuming, the corresponding scale space and feature points descriptor also consume a lot of storage resources. (2) For the feature point sets obtained from the two images, one-to-one matching needs to be carried out by the brute force calculation of the Euclidean distance, etc. Most points are not possible candidate points, and the calculated Euclidean distance needs to be stored, such that again there is an invalid calculation during the match.

From the perspective of the algorithm's operation process, we will discuss the reason why algorithms such as SAR-SIFT are good at sub-image registration but fail in wide-swath images. The most obvious factors are time and space consumption. The reason can be found from a unified aspect: calculation and storage are not directional. There is redundancy in the calculation of the scale space. Some areas can be found beforehand in order to reduce the calculation range of the scale space, and the calculation amount of subsequent mismatch can also be reduced. At the same time, feature point matching does not have certain directivity because, for feature points in a small area, points from most of the area in another image are not potential matching ones. Therefore, redundant calculation and storage can be omitted.

In this paper, the idea of improving the practicality of wide-swath SAR image registration is to reduce the calculation range of the original image and the range of candidate points according to certain criteria. Based on the coarse registration results of the candidate regions, we determine the approximate spatial correspondences, and then perform more refined feature calculations and matching in the corresponding image slice regions.

In this work, the corresponding slice areas with a higher probability of feature points are selected. K-means++ is used to obtain the clustering centers of coarse matching points in the first step. The clustering center is marked as the geometric center in order to obtain the image slices. By using the geometric transformation relationship, a set of image pairs corresponding approximately to the same geographic locations are obtained. Adopting this approach has the following advantages:

- There are often more candidate regions of feature points near the cluster center.
- There is usually enough spatial distance between the clustering centers
- The clustering center usually does not fall on the edge of the image.

K-means++ is an unsupervised learning method which is usually used in scenarios such as data mining. K-means++ needs to cluster N observation samples into K categories. Here, K = 4. the cluster centers are used as the slice geometric centers, and the slice size is set to 640×640. According to the above process, a series of rough matching image groups are obtained within an error of about ten pixels.

2.4. Dense Matching of the Sub-Image Slices

After the above processing is performed on the original-resolution SAR image, a set of SAR image slices are obtained. As is known, compared with optical image registration, an SAR image meets many difficulties: it has a low resolution and signal-to-noise ratio, overlay effects, perspective shrinkage, and a weak texture. Therefore, the original-resolution SAR image's alignment is more difficult than the optical alignment.

This article uses Transformer. Based on the features extracted by CNN, Transformers are used to obtain the feature descriptors of the two images. The global receptive field provided by Transformer enables the method in this article to fuse the local features and contextual location information, which can produce dense matching in low-texture areas (usually, in low-texture areas, it is difficult for feature detectors to generate repeatable feature points).

The overall process consists of several steps, as shown in the lower half of Figure 2:

1. The feature extraction network HRNet (High-Resolution Net) [41]: Before this step, we combine ORB and GMS to obtain the image rough matching results, use the K-means++ method to obtain the cluster centers of the rough matching feature points, and obtain several pairs of rough matching image pairs. The input of the HRNet is every rough matching image pair, and the output of the network is the high and low-resolution feature map after HRNet's feature extraction and fusion.
2. The low-resolution module: The input is a low-resolution feature map obtained from HRNet, which is expanded into a one-dimensional form and added with positional encoding. The one-dimensional feature vector after position encoding is processed by the Performer [42] to obtain the feature vector weighted by the global information of the image.
3. The matching module: The one-dimensional feature vector obtained from the two images in the previous step is operated to obtain a similarity matrix. The confidence matrix is obtained after softmax processing on the similarity matrix. The pairs that are greater than a threshold in the confidence matrix and satisfy the mutual proximity criterion are selected as the rough matching prediction.
4. Refine module: For each coarse match obtained by the matching module, a window of size wxw is cut from the corresponding position of the high-resolution feature map. The features contained in the window are weighted by the Performer, and the accurate matching coordinates are finally obtained through cross-correlation and softmax. For each pair of rough matching images, the outputs of the above step are matched point pairs with precise coordinates, and after the addition of the initial offset of rough matching, all of the point pairs are fused into a whole matched point set. After the implementation of the RANSAC filtering algorithm, the final overall matching point pair is generated, and then the spatial transformation solution is completed.

2.4.1. HRNet

Traditional methods such as VGGNet [43] and ResNets (Residual Networks) [44] include a series of convolution and pooling, which loses a lot of spatial detail information. The HRNet structure maintains high-resolution feature maps, and combines high- and low-resolution subnet structures in parallel to obtain multi-scale information.

HRNet is used as a network model for multi-resolution feature extraction in this method. At the beginning of this paper, we tried a variety of convolutional neural network models, including ResNets, EfficientNet [45] and FPN [46]; we found that HRNet has the best effect. The HRNet's structure is shown in Figure 6; the network is composed of multiple branches, including the fusion layer with different resolution branches' information interactions, and the transition layer, which is used to generate the 1/2 resolution downsampling branch. By observing the network input and output of HRNet at different stages, it can be seen that multi-resolution feature maps with multi-level information will be output after the full integration of the branches with different resolutions.

Figure 6. HRNet architecture diagram.

The Transformer part of this paper requires high and low resolution to complete the coarse matching of the feature points and the more accurate positioning of specific areas. HRNet, as a good backbone, outputs feature maps with a variety of resolutions to choose from and full interaction between the feature maps, such that it contains high-level semantic information with a low resolution, and low-level detail information with a high resolution.

In addition, the subsequent part of this paper makes further attempts to combine different resolutions. It can be seen in subsequent chapters that improving the resolution of the feature maps in the rough matching stage can significantly increase the number of matching point pairs. Default HRNet outputs 1/2, 1/4, 1/8 resolution feature maps. Under the constraints of the experimental environment, we chose 1/5 and 1/2 as the low and high resolutions. The reason will be discussed in Section 3. Experimental Results and Analyses. The 1/5 resolution can be obtained from other resolutions by interpolation. The 1/2 resolution feature map cascades the 1×1 convolutional layer and the output works as the fine-level feature map. The 1/4 and 1/8 resolution feature maps are all interpolated to 1/5 resolution. After stitching, the coarse-level feature map is obtained through the 1×1 convolutional layer.

2.4.2. Performer

The Transformer has outstanding performance in many fields of CV, such as classification and detection. With the help of a multi-head self-attention mechanism, the Transformer can capture richer characteristic information. Generally speaking, Transformer complexity is squared with sequence length. In order to improve the speed of training and inference, a linear time complexity Transformer was also proposed recently, i.e., Performer [42]. It can achieve faster self-attention through the positive Orthogonal Random features approach.

$$\text{Attention}\ (Q, K, V) = \text{softmax}\left(\frac{QK^{T}}{\sqrt{d_k}}\right) V \tag{15}$$

Self-attention (as shown in Figure 7) performs an attention-weighted summation of the current data and global data, and realizes a special information aggregation by calculating the importance of the current location feature relative to other location features. The feedforward part contains the linear layer and the GELU (Gaussian Error Linear Unit) activation function. Each layer adopts Layer Normalization in order to ensure the consistency of the feature distribution, and to accelerate the convergence speed of the model training.

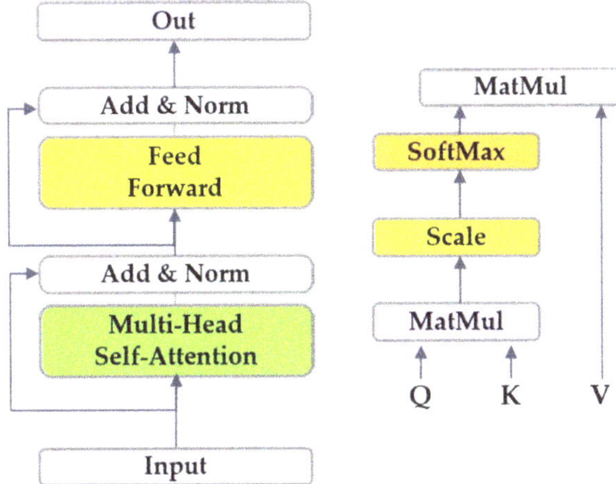

Figure 7. Performer (Transformer) encoder architecture and self-attention schematic diagram.

As is shown in reference [42], Performer can achieve space complexity O(Lr + Ld + rd) and time complexity O(Lrd), but the original Transformer's regular attention is O(L^2 + Ld) and $O(L^2d)$, respectively. The sinusoidal position encoding formula used in this work is as follows:

$$\begin{cases} \mathbf{p}_{k,2i} = \sin\left(k/10000^{2i/d}\right) \\ \mathbf{p}_{k,2i+1} = \cos\left(k/10000^{2i/d}\right) \end{cases} \tag{16}$$

Based on the features extracted by CNN, Performers are used to obtain the feature descriptors of the two images. The global receptive field provided by Performer enables our method to fuse local features and contextual location information, which can produce dense matching in low-texture areas (usually, in low-texture areas, it is difficult for feature detectors to generate repeatable feature points).

2.4.3. Training Dataset

Due to the influence of noise, it is difficult to accurately annotate the control points of an SAR image, and the corresponding matching dataset of an SAR image is not common. MegaDepth [47] contains 196 groups of different outdoor scenes; it applies SOTA (State-of-the-Art) methods to obtain depth maps, camera parameters, and other information. The dataset contains different perspectives and periodic scenes. Considering the dataset size and GPU memory, 1500 images were selected as a validation set, and the long side of the image was scaled to 640 during the training and 1200 during the verification.

2.4.4. Loss Function

$$L = L_c + L_f = -\frac{1}{\left|M_c^{gt}\right|} \sum_{(\tilde{i},\tilde{j}) \in M_C^{gt}} \log P_c\left(\tilde{i},\tilde{j}\right) + \frac{1}{|M_f|} \sum_{(\tilde{i},\tilde{j}') \in M_f} \frac{1}{\sigma^2\left(\hat{i}\right)} \|\hat{j}' - \tilde{j}'_{gt}\|_2 \tag{17}$$

As in [37], this article uses a similar loss function configuration. Here is a brief explanation. Pc is the confidence matrix returned by dual softmax. The true label of the confidence matrix is calculated by the camera parameters and depth maps. The nearest neighbors of the two sets of low-resolution grids are used as the true value of the coarse matching Mc, and the low resolution uses negative log-likelihood as the loss function.

The high resolution adopts the L2 norm. For a point, the uncertainty is measured by calculating the overall variance in the corresponding heatmap. The real position of the current point is calculated from the reference point, camera position, and depth map. The total loss is composed of low- and high-resolution items.

2.5. Merge and Solve

After obtaining the corresponding matching point sets of each image slice pair, the final solution requires mapping point sets of the entire image. Considering that the registration mapping geometric relationship solved by each set of slices is not necessarily the same, this work merges all of the point sets. The RANSAC method is used here to obtain the final result, i.e., a set of corresponding subsets describing the two large images. The corresponding point numbers must be less than the sum of the independent one. Without bells and whistles, the affine matrix of the entire image is solved.

3. Experimental Results and Analyses

In this section, we design several experiments to validate the performance of our methods from three perspectives: (1) the comparative performance tests with SOTA methods for different data sources, (2) the checkerboard visualization of the matching, (3) scale, rotation and noise robustness tests, and (4) the impact of the network's high- and low-resolution settings on the results. First, a brief introduction to the experimental datasets is given.

3.1. Experimental Data and Settings

In this work, datasets from five sources were used to verify the algorithm's effectiveness, which contains GF-3, TerraSAR-X, Sentinel-1, ALOS, and SeaSat. These data include a variety of resolutions, polarization modes, orbital directions, and different terrains. Table 1 and Figure 8 contain detailed information. DEC and ASC mean "descending" and "ascending", respectively.

Table 1. Experimental datasets.

Pair	Sensor	Size	Resolution	Polar	Orbit Direction	Data	Location
1	GF-3	15,470 × 11,093	1 m	VV	DEC	20180420	USA
		15,276 × 11,498	1 m	VV	DEC	20180425	New Jersey
2	GF-3	17,110 × 11,635	1 m	VV	DEC	20180804	China
		15,986 × 11,718	1 m	HH	DEC	20180814	Shaanxi
3	GF-3	28,334 × 11,868	1 m	HH	DEC	20190201	USA
		30,752 × 12,384	1 m	HH	DEC	20190208	Alaska
4	GF-3	14,736 × 11,391	1 m	HH	ASC	20180825	USA
		13,840 × 11,349	1 m	HH	ASC	20180820	Hawaii
5	GF-3	13,102 × 10,888	1 m	HH	ASC	20181119	Philippines
		14,554 × 12,287	1 m	HH	DEC	20180715	Bagan
6	GF-3	20,792 × 11,602	1 m	VV	ASC	20180609	Russia
		20,660 × 11,382	1 m	VV	ASC	20180705	Saratov
7	TerraSAR-X	8208 × 5572	1 m	HH	DEC	20130314	China
		8208 × 5562	1 m	HV	DEC	20130303	Shanghai
8	TerraSAR-X	23,741 × 28,022	1 m	HH	ASC	20160912	China
		23,998 × 29,505	1 m	HH	ASC	20161004	Liaoning
9	Sentinel-1	25,540 × 16,703	20 × 22 m	VH	DEC	20211211	USA
		25,540 × 16,704	20 × 22 m	VH	DEC	20211129	St. Francis
10	Sentinel-1	25,649 × 16,722	20 × 22 m	VH	ASC	20211129	China
		25,649 × 16,722	20 × 22 m	VH	ASC	20211211	Guangdong
11	Sentinel-1	25,336 × 16,707	20 × 22 m	VH	ASC	20211210	China
		25,335 × 16,707	20 × 22 m	VH	ASC	20211128	Liaoning
12	ALOS	5600 × 4700	20 × 10 m	HH	ASC	20100717	USA
		5600 × 4700	20 × 10 m	HH	ASC	20100601	Montana
13	ALOS	6454 × 5729	20 × 10 m	HH	ASC	20080416	China
		6502 × 5715	20 × 10 m	HH	ASC	20080115	Jiangsu
14	ALOS	6291 × 5508	20 × 10 m	HH	ASC	20081121	China
		6464 × 5712	20 × 10 m	HH	ASC	20110221	Shandong
15	SeaSat	11,611 × 11,094	12.5 m	HH	DEC	19780922	Norway
		11,399 × 10,952	12.5 m	HH	DEC	19781010	
16	SeaSat	11,493 × 11,371	12.5 m	HH	DEC	19780811	Russia
		11,717 × 11,135	12.5 m	HH	DEC	19780722	
17	SeaSat	11,191 × 10,653	12.5 m	HH	ASC	19780902	UK
		11,155 × 10,753	12.5 m	HH	ASC	19780926	

In order to verify the effectiveness of the proposed matching method, several evaluation criteria were used to evaluate the accuracy of the SAR image registration, as shown below:

1. The root mean square error, RMSE, is calculated by the following formula:

$$\text{RMSE} = \sqrt{\frac{1}{N}\sum_{i=1}^{N}\left(x_i^{2\prime}-x_i^1\right)^2 + \left(y_i^{2\prime}-y_i^1\right)^2} \tag{18}$$

2. NCM stands for the number of matching feature point pairs filtered by the RANSAC algorithm, mainly representing the number of feature point pairs participating in the calculation of the spatial transformation model. It is a filtered point subset of the

matching point pairs output by algorithms such as SAR-SIFT. For the solution of the affine matrix, the larger the value, the better the image registration effect.

3.2. Performance Comparison

In this section, we compare the proposed method with several methods: SAR-SIFT, HardNet [48], TFeat [49], SOSNet [50], LoFTR, KAZE-SAR, and CMM-Net. HardNet, TFeat, and SOSNet use GFTT [51] as the feature point detector, and the patch size of the above methods is 32 × 32. GFTT will pick the top N strongest corners as feature points. The comparison methods are introduced briefly as follows:

1. SAR-SIFT uses SAR-Harris space instead of DOG to find the key points. Unlike the square descriptor of SIFT, SAR-SIFT uses the circular descriptor to describe neighborhood information.
2. HardNet proposes the loss that maximizes the nearest negative and positive examples' interval in a single batch. It uses the loss in metric learning, and outputs feature descriptors with 128 dimensionalities, like SIFT.
3. SOSNet adds second-order similarity regularization for local descriptor learning. Intuitively, first-order similarity aims to give descriptors of matching pairs a smaller Euclidean distance than descriptors of non-matching pairs. The second-order similarity can describe more structural information; as a regular term, it helps to improve the matching effect.
4. TFeat uses triplets to learn local CNN feature representations. Compared with paired sample training, triplets containing both positive and negative samples can generate better descriptors and improve the training speed.
5. LoFTR proposes coarse matching and refining dense matches by a self-attention mechanism. It combines high- and low-resolution feature maps extracted by CNN to determine rough matching and precise matching positions, respectively.
6. KAZE-SAR uses a nonlinear diffusion filter to build the scale space.
7. CMM-Net uses VGGNet to extract high-dimensional feature maps and build descriptors. It uses triplet margin ranking loss to balance the universality and uniqueness of the feature points.

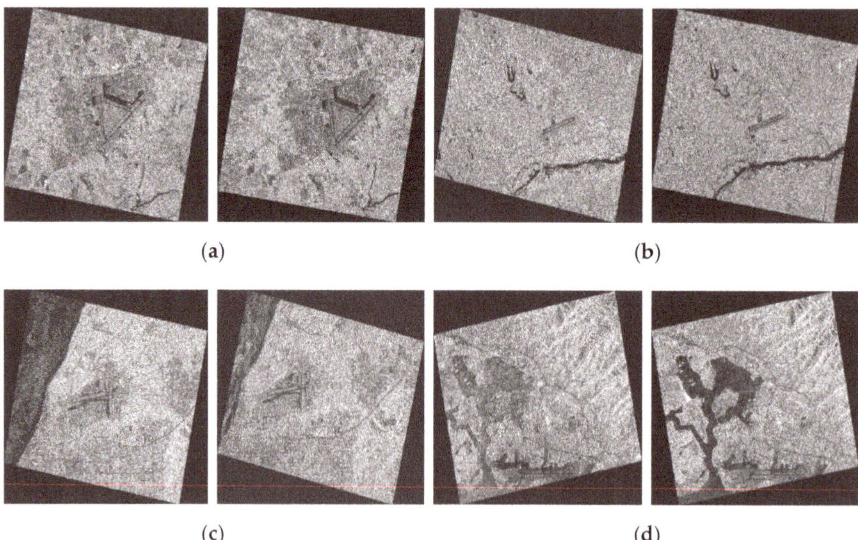

Figure 8. *Cont.*

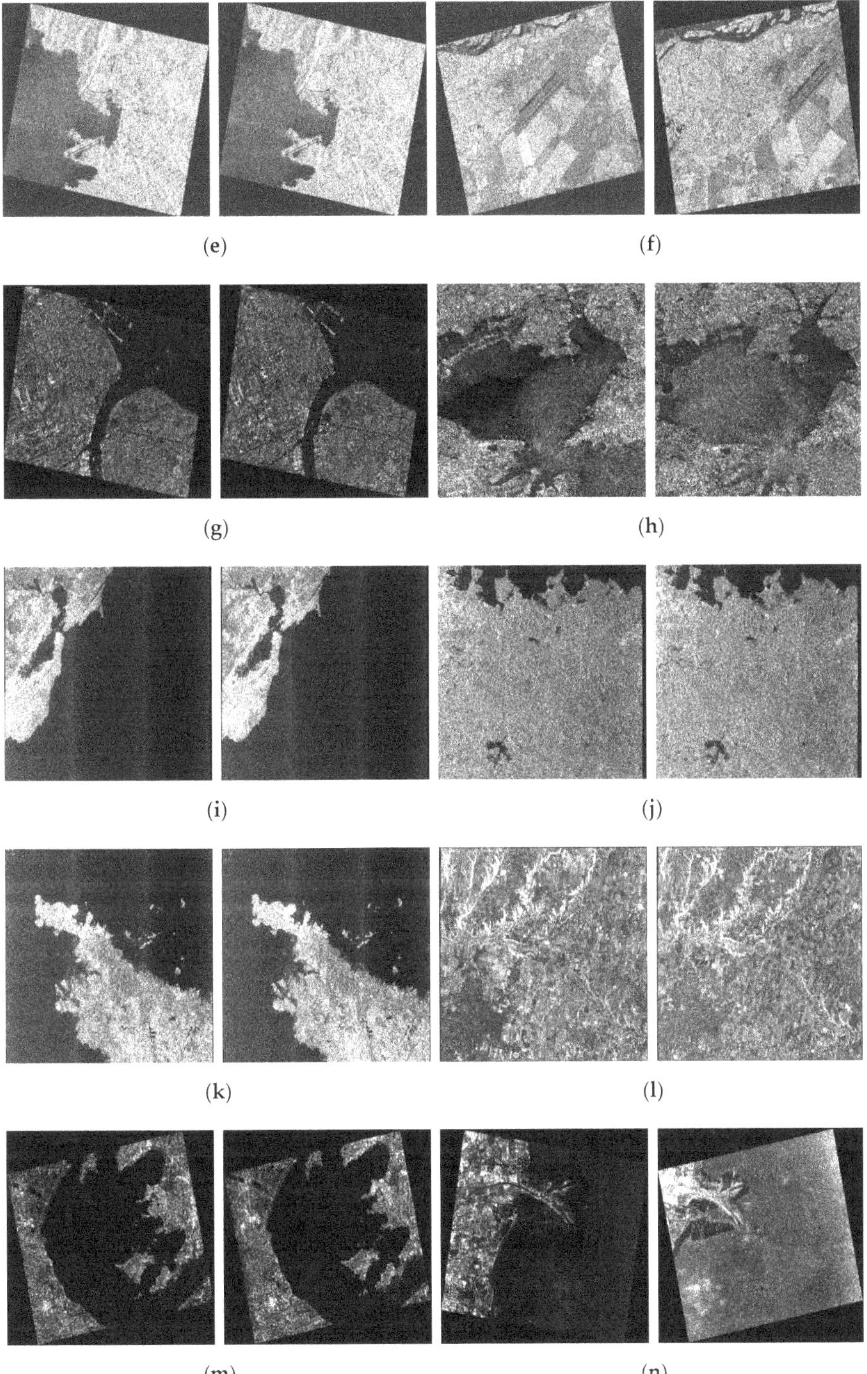

Figure 8. *Cont.*

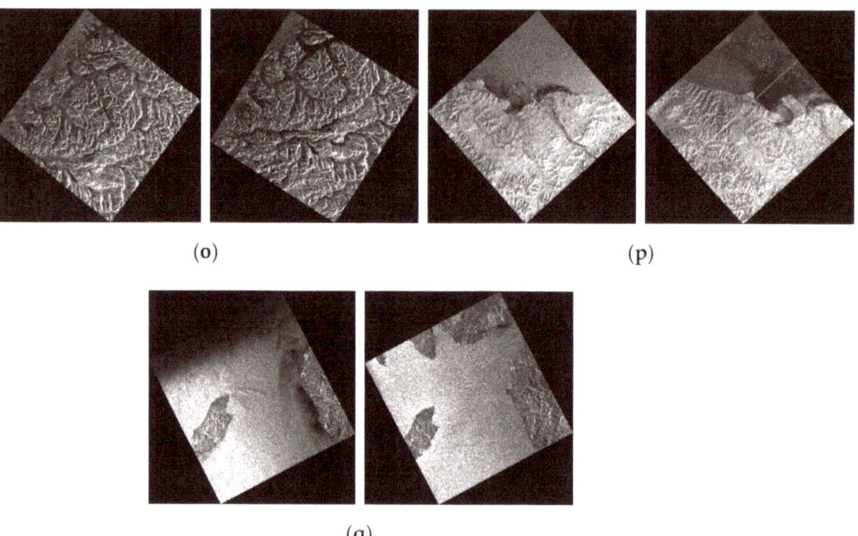

Figure 8. The SAR image datasets (**a**–**q**) of datasets 1–17.

As for our method, we set K = 4 and window size = 640 for the sub-image. SAR-SIFT and KAZE-SAR are traditional methods, and Hardnet, Tfeat, SOSNet, LoFTR, and CMM-Net are deep learning methods. The algorithm in this paper was trained and tested on a server with a GPU of NVIDIA TITAN_X (12GB), a CPU of Intel(R) Core(TM) I7-5930K @3.50GHZ, and a memory size of 128 GB. The comparison experiment was carried out using the same hardware. As will be discussed later (Section 3.4), under existing hardware conditions, (1/2,1/5) resolution was adopted in this paper in order to achieve the best effect, and was used as the final network model to calculate the speed and accuracy of the algorithm. Like diverse methods, in addition to feature point detection and feature point description, other processing steps are consistent with our method, including the rough matching of sub-sampling images and the acquisition of subimages. The other settings were based on the original settings of the algorithm in order to ensure the fairness of the comparison.

Table 2 shows the performance results of several methods on the above dataset, in which the best performance corresponding to each indicator is shown in bold. '-' in the table means that the matching result of the corresponding algorithm is incorrect. It can be seen that the performance of our method on RMSE is better than the comparison methods in more than half of the datasets. For all of the SAR image registration datasets, the performance of our method reaches the sub-pixel level. Considering NCM, our method obtains the best performance for all of the total datasets, as well as a better spatial distribution of points, i.e., dense matching. Figure 9 shows that our method's NCMs are higher than those of other methods, while the RMSEs are lower in most cases.

3.3. Visualization Results

In order to display the matching accuracy more intuitively, we added the checkerboard mosaic images. In Figure 10, the continuity of the lines can reflect the matching accuracy. As the pictures show, the areas and lines overlap well, indicating the high accuracy of the proposed method.

Table 2. RMSE and NCM of diverse methods on the datasets.

Pair	HardNet [48]		SOSNet [50]		TFeat [49]		SAR-SIFT [10]		LoFTR [37]		KAZE-SAR [11]		CMM-Net [24]		Ours	
	RMSE	NCM	RMSE	NCM	RMSE	NCM	RMSE	NCM	RMSE	NCM	RMSE	NCM	RMSE	NCM	RMSE	NCM
1	0.629	111	0.603	107	0.561	78	0.628	30	0.658	1063	0.685	169	0.645	22	0.572	8506
2	0.589	38	0.670	51	0.658	65	0.624	50	0.702	298	0.653	74	0.622	19	**0.569**	1763
3	0.588	25	0.660	27	0.455	38	0.472	14	0.663	178	0.664	52	**0.415**	13	0.609	2410
4	0.655	83	0.648	109	0.652	60	0.607	22	0.678	156	0.665	109	**0.384**	10	0.528	6923
5	0.674	10	0.632	8	-	-	0.547	7	0.620	133	0.592	7	**0.501**	9	0.661	223
6	-	-	-	-	-	-	**0.453**	7	0.664	204	0.552	6	0.492	11	0.571	758
7	0.594	1343	0.610	1441	0.604	1738	0.708	1045	0.588	11,816	0.613	4253	0.570	47	**0.546**	12,190
8	0.620	85	0.668	74	0.603	91	0.682	50	0.659	891	0.659	209	0.605	21	**0.484**	3121
9	0.681	255	0.636	256	0.660	216	0.601	50	0.655	3152	0.653	778	0.596	30	**0.503**	20,319
10	0.637	446	0.640	472	0.631	398	0.715	141	0.650	4073	0.642	1142	0.691	30	**0.485**	18,270
11	0.643	297	0.630	405	0.623	315	0.691	82	0.664	3626	0.641	850	0.656	31	**0.518**	23,865
12	0.654	1083	0.663	932	0.657	1076	0.670	105	0.607	12,226	0.615	2836	0.666	66	**0.537**	24,515
13	0.632	920	0.653	946	0.634	854	0.618	483	0.634	4577	0.624	2173	0.583	35	**0.560**	21,782
14	0.641	22	-	-	0.658	8	0.561	15	0.654	220	0.551	15	0.686	10	**0.538**	949
15	0.643	635	0.682	520	0.664	661	0.669	128	0.642	446	**0.629**	1069	0.628	27	0.696	4099
16	**0.458**	152	0.596	118	0.510	182	0.645	47	0.663	883	0.638	180	0.599	23	0.555	4401
17	0.628	415	0.669	395	0.629	625	0.632	82	0.663	4588	0.670	755	0.583	63	**0.577**	14,446

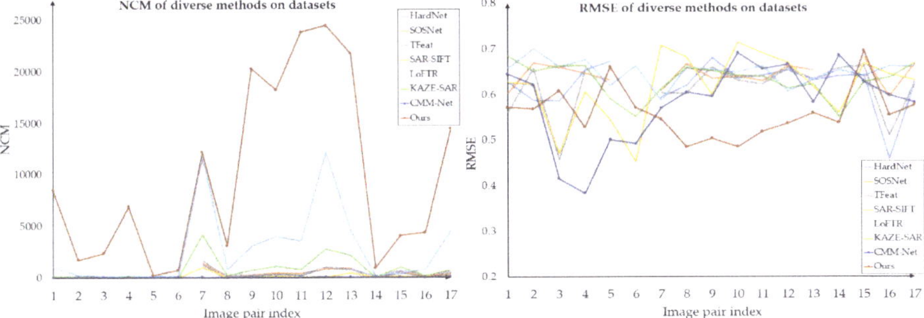

Figure 9. NCM and RMSE of diverse methods on the datasets.

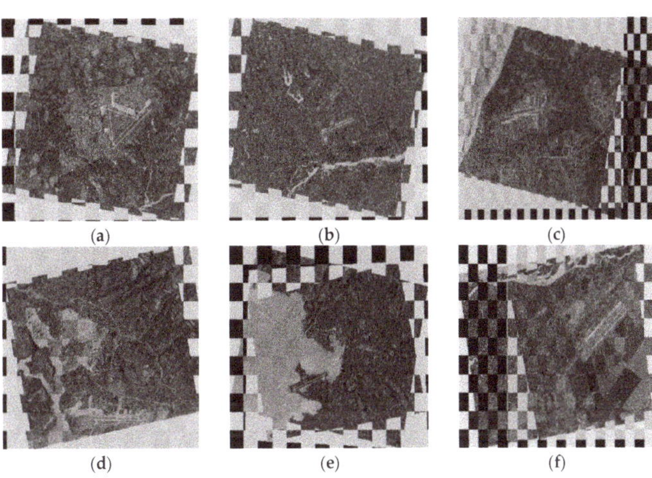

Figure 10. *Cont.*

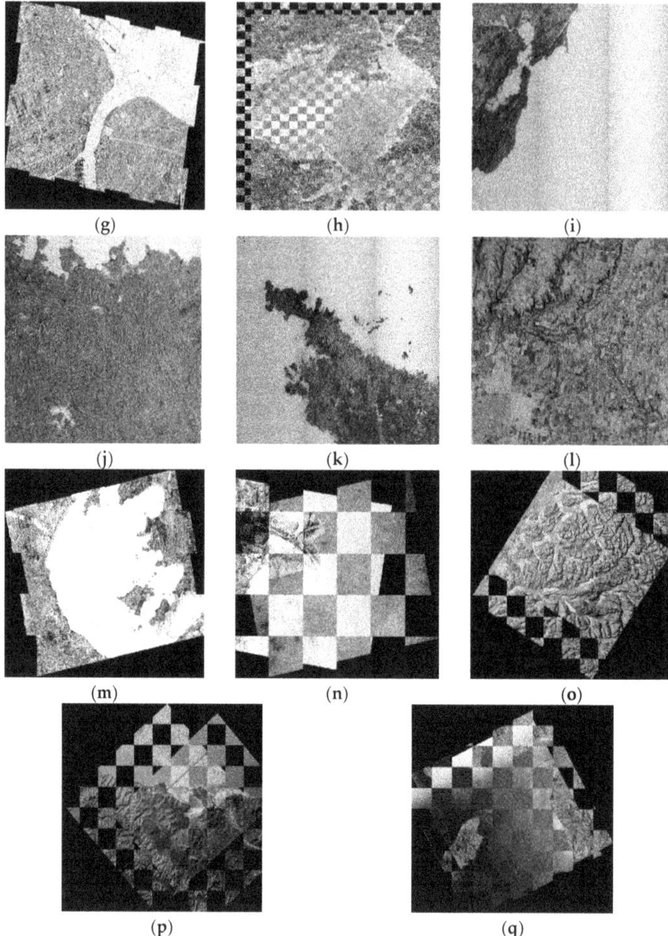

Figure 10. The checkerboard mosaic images (**a**–**q**) of datasets 1–17.

Considering that our method adopts the strategy of fusing local and global features, it can fully extract matching point pairs in the selected local area, which leads to better solutions that are closer to the affine transformation relationship of real images. In the data pairs 14 and 17, the two images have relatively strong changes in their radiation intensities, and the method in this paper can still achieve good matching results, which proves that the method has certain robustness to changes in radiation intensity. Dataset 5 contains two images of different orbit directions. It can be seen from the road and other areas in the figure that the matching is precise. Datasets 2 and 7 contain multi-temporal images of different polarizations. Due to the scattering mechanism, the same objects in different polarizations may be different in the images. Our method demonstrates the stability in multi-polarization.

3.4. Analysis of the Performance under Different Resolution Settings

Considering that our proposed method has a coarse-to-fine step, we analyzed the registration performance of different resolution settings to find the best ratio. In this experiment, we tested several different high and low resolutions. Table 3 shows the corresponding performances, respectively, and the method of the best performance for each data is indicated in bold. All of the resolution parameter combinations include (1/4, 1/16), (1/2,1/8), (1,1/8), and (1/2,1/5).

Table 3. RMSE and NCM of the different resolution settings on the datasets.

Pair	Ours_16_4		Ours_8_2		Ours_8_1		Ours_5_2	
	RMSE	NCM	RMSE	NCM	RMSE	NCM	RMSE	NCM
1	0.683	323	0.638	3789	0.665	3885	**0.572**	**8506**
2	0.680	68	0.669	533	0.680	763	**0.569**	**1763**
3	0.703	95	0.627	1012	0.673	895	**0.609**	**2410**
4	0.676	261	0.637	1288	0.671	1706	**0.528**	**6923**
5	0.638	29	**0.624**	149	0.636	157	0.661	**223**
6	**0.504**	26	0.618	328	0.653	487	0.571	**758**
7	0.685	1389	0.624	8925	0.593	12,152	**0.546**	**12,190**
8	0.646	222	0.634	2165	0.643	1955	**0.484**	**3121**
9	0.680	658	0.641	6253	0.633	7568	**0.503**	**20,319**
10	0.658	819	0.644	2104	0.624	3348	**0.485**	**18,270**
11	0.659	986	0.646	3025	0.642	4731	**0.518**	**23,865**
12	0.642	1336	0.635	9616	0.612	11,253	**0.537**	**24,515**
13	0.636	1187	0.631	4382	0.639	4906	**0.560**	**21,782**
14	0.638	61	0.620	558	0.668	603	**0.538**	**949**
15	0.690	558	0.709	1203	**0.667**	2478	0.696	**4099**
16	0.711	195	0.643	1679	0.648	2063	**0.555**	**4401**
17	0.654	850	0.621	3339	0.668	3719	**0.577**	**14,446**

It can be seen from Table 3 that, for the datasets, the resolution of (1/2,1/5) has the best performance. It is shown in Table 3 that the size of the low resolution directly affects the overall matched point's number. It can be found intuitively that the total potential points of resolution 1/16 are a quarter of 1/8's point number $(1/2 \times 1/2 = 1/4)$. As such, configuration (1/2,1/5) can obtain more matched point pairs. Take configurations (1/2,1/8) and (1,1/8) for an example; for high-resolution feature maps, the number of matched points increases to a certain extent with the increase of the resolution. However, here, the GPU memory requested by configuration (1,1/5) exceeds the upper limit of the machine used in this work, so we chose a compromise configuration, (1/2,1/5).

4. Discussion

The experimental results corroborate the accuracy and robustness of our method. There are three main reasons for this: First, the features extracted based on Transorfmer are richer, including the local gray information of the image itself, and global information such as the context. Second, the down-sampled image has stronger semantic information, and is suitable for traditional registration methods. The subsequent registration is provided with a better initial result of coarse matching. Third, according to the K-means++ clustering method, the relationship between the original image and the sub-images to be registered is constructed, and representative sub-images are obtained in order to reduce time and space consumption.

From the performance analysis and model hyperparameter comparison experiments, it can be seen that our proposed method achieved stable and accurate matching results under different ground object scenes and various sensors' data conditions. Now, we will further examine the rotation, scaling and noise robustness of the proposed method. Furthermore, another vital criterion—the execution time—needs to be compared. Finally, we will show the matching accuracy's impact on downstream tasks.

4.1. Rotation and Scale Test

In practical applications, there are often resolution inconsistencies and rotations between the sensed image and the reference image. In order to test the rotation and scale robustness of the proposed subimage registration method, we experimented on the data with a simulated variation. The RADARSAT SAR sensor collected the data of Ottawa, in May and August 1997, respectively. The size of the two original images is 350 × 290, as shown in Figure 11.

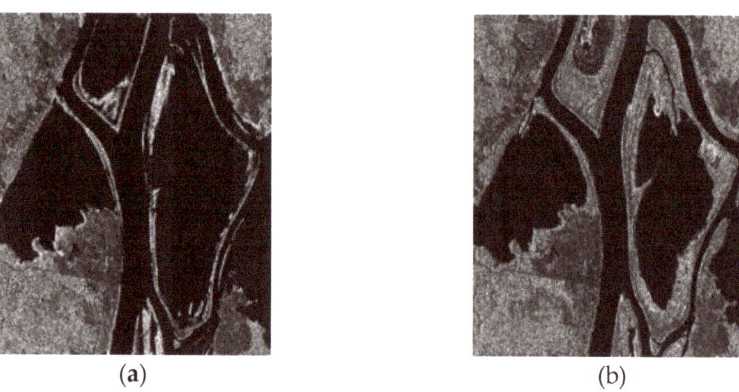

(a) **(b)**

Figure 11. The Ottawa data: (**a**) May 1997 and (**b**) August 1997.

A matching test between the two images at 5° intervals from –15° to 15° was carried out to verify the rotation robustness. In addition, two scaling ratios of 0.8 and 1.2 were tested to simulate the stability of the image registration algorithm at different resolutions. In all of the above cases, more than 100 matching points could be extracted between the two SAR images with an RMSE of around 0.7 (the subpixel level). As Figures 12 and 13 show, the proposed method has rotation and scale robustness.

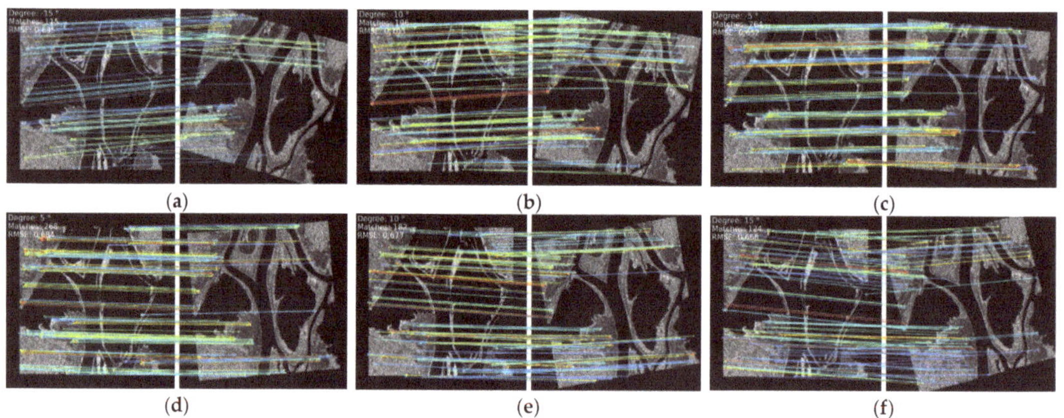

(a) **(b)** **(c)**

(d) **(e)** **(f)**

Figure 12. The registration performance of SAR images at varying rotations: (**a**) –15°, (**b**) –10°, (**c**) –5°, (**d**) 5°, (**e**) 10°, and (**f**) 15°.

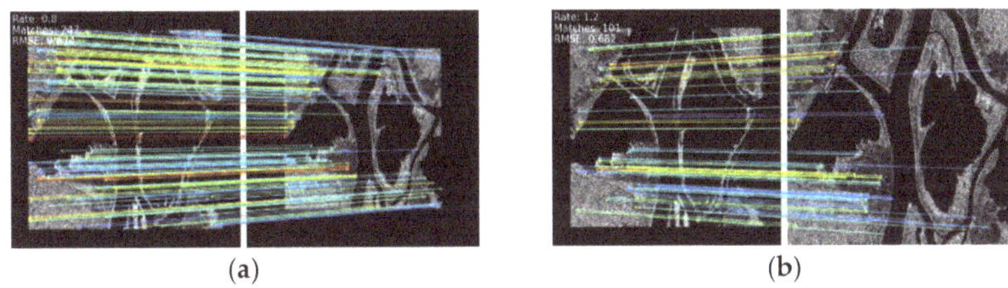

(a)   (b)

Figure 13. The registration performance of SAR images at varying sizes: (**a**) 0.8, and (**b**)1.2.

In addition, we can see that, under different rotation and scaling conditions, a large number of matching points can be obtained not only in the strong edge area but also in the weak texture area. This is significantly helpful for SAR image registration including low-texture areas. The simulation experiments reflect the validity of image registration with various changes in real scenes.

4.2. Robustness Test of the Algorithm to Noise

Previously, we discussed the robustness of the algorithm to scale and rotation. Considering that there is often a high degree of noise in SAR images, taking motion error as an example, SAR images in practical applications will possess some unfocused positions. Whether or not it has stable performance in noisy scenes, this paper refers to the method and results of [52]; here, two sets of images—before and after autofocus—are tested in order to verify the stability of the algorithm.

Here, we used our sub-image method to register the defocusing data; the results can be seen in Figure 14, and although the image data has a high degree of noise due to error motion defocusing, our method can still obtain a good matching result between two images, and matching points can also be maintained at a high level with error at the subpixel level. It is thus proven that the proposed method is robust to noise.

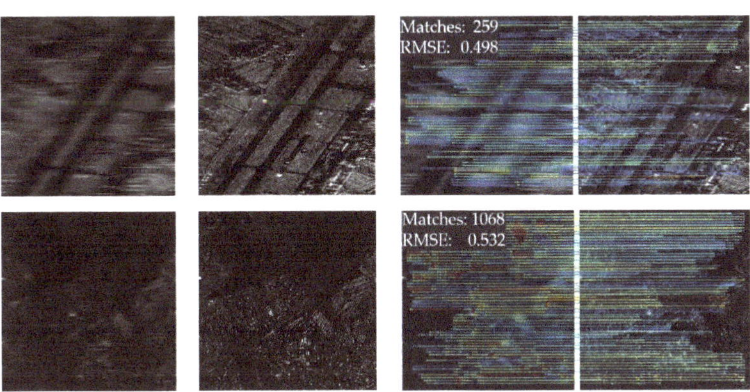

Figure 14. Algorithm robustness test in a noise scenario.

4.3. Program Execution Time Comparison

In actual tasks, the registration process needs to achieve real-time or quasi-real-time analysis; as such, in addition to the accuracy of the algorithm, timeliness is also a focus of measurement. We also compared several representative methods on a selection of characteristic dataset pairs, i.e., 2, 8, 12, 17. The algorithm in this paper selected the resolution configuration of (1/2, 1/5) as a comparison. As Table 4 shows, our method is significantly faster than the traditional method, SAR-SIFT, and slightly slower than other deep learning methods.

Table 4. Execution time (s) comparison of the different methods.

Pair	HardNet [48]	SOSNet [50]	TFeat [49]	SAR-SIFT [10]	LoFTR [37]	Ours
2	42.868	24.543	21.962	513.152	**19.496**	21.321
8	43.142	43.129	40.710	1999.966	**38.633**	40.553
12	26.514	26.980	**24.730**	948.620	28.974	45.687
17	**22.611**	23.085	20.772	829.606	16.344	25.033

Methods like TFeat use a shallow convolutional neural network so that the feature extraction phase is faster. Given that our approach is multi-stage, the model is relatively complex. Except for the model itself, our method obtains the most matching points and consumes exponentially more time in both the feature point matching and filtering stages. Although the running time is slightly longer, the SAR image registration performance is significantly improved. In further work, we will consider improving the efficiency by adjusting the distillation learning of the feature extraction module in order to obtain a lightweight network with similar performance.

4.4. Change Detection Application

In some applications—such as SAR image change detection—the simultaneous analysis of SAR images with different acquisition conditions is inevitable. We carried out a simple analysis, and the registration result was applied to the task of SAR image change detection. In this project, we use the previously mentioned Ottawa dataset.

In this experiment, we rotate one of the images to achieve a relative image offset. The Ottawa data of two SAR images were matched first, and the change detection results were obtained after they were processed by two registration methods: SAR-SIFT and ours. The PCA-Kmeans [53] method was used as the basic change detection method. Kappa was used as the change detection performance metric; the formula is as follows:

$$\kappa = \frac{2 \times (TP \times TN - FN \times FP)}{(TP + FP) \times (FP + TN) + (TP + FN) \times (FN + TN)} \tag{19}$$

The Kappa coefficient can be used to measure classification accuracy. The higher the value is, the more accurate the classification result is. Compared with SAR-SIFT, our proposed method improved the kappa indicator from 0.307 to 0.677, which shows that accurate registration can lead to better change detection results. Intuitively, from Figure 15, we can see that our method results (**b**) are more similar to the ground truth (**c**) than SAR-SIFT (**a**). The deviation of the image registration will cause different objects to be mistaken for the same area during change detection, which will be mistaken for obvious changes.

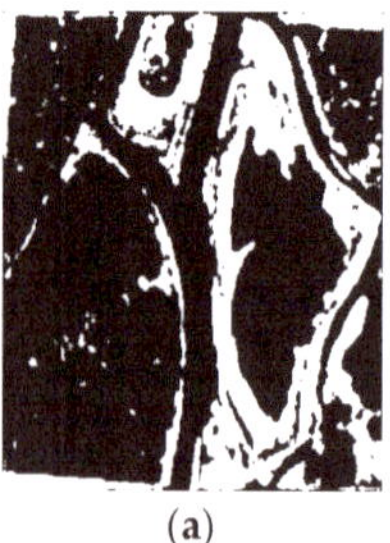  

(**a**)　　　　　(**b**)　　　　　(**c**)

Figure 15. The change map: (**a**) SAR-SIFT, (**b**) the proposed method, and (**c**) the ground truth.

5. Conclusions

This paper proposes a novel wide-swath SAR image registration method which uses a combination of traditional methods and deep learning to achieve accurate registration. Specifically, we combined the clustering methods and traditional registration methods to complete the stable extraction of representative sub-image slices containing high-probability regions of feature points. Inspired by Performer's self-attention mechanism, a coarse-to-fine sub-image dense-matching method was adopted for the SAR image matching under different terrain conditions, including weak texture areas.

The experimental results demonstrate that our method achieved good performance for different datasets which include multi-temporal, multi-polarization, multi-orbit direction, rotation, scaling, noise changes. At the same time, the combination of CNN and Performer verified the effectiveness of the strong representation in SAR image registration. Under the framework of sub-images matching to original images matching, stable dense matching can be obtained in high-probability regions. This framework overcomes the time-consuming problem of the traditional method of matching. Compared with existing methods, more matching point pairs can be obtained by adjusting the model parameter settings in our method. Rotation, scaling and noise experiments were also carried out to verify the robustness of the algorithm. The results showed that a large number of matching point pairs can be obtained even in regions with weak textures, which shows that our method can combine local and global features to characterize feature points more effectively.

In addition, the experimental results suggest that the running time is significantly less than those of traditional methods but slightly longer than those of similar deep learning methods; as such, the way in which to further simplify the network model will be the focus of the next step. Meanwhile, the matching between heterogeneous images is also a topic that can be discussed further.

Author Contributions: Conceptualization, Y.F., H.W. and F.W.; validation, Y.F., H.W. and F.W.; formal analysis, Y.F., H.W. and F.W.; investigation, Y.F., H.W. and F.W.; resources, Y.F., H.W. and F.W.; data curation, Y.F., H.W. and F.W.; writing—original draft preparation, Y.F., H.W. and F.W.; writing—review and editing, Y.F., H.W. and F.W.; visualization, Y.F., H.W. and F.W.; funding acquisition, Y.F., H.W. and F.W. All authors have read and agreed to the published version of the manuscript.

Funding: This work was supported in part by the National Natural Science Foundation of China (Grant No. 61901122), the Natural Science Foundation of Shanghai (Grant No. 20ZR1406300, 22ZR1406700), and the China High-resolution Earth Observation System (CHEOS)-Aerial Observation System Project (30-H30C01-9004-19/21).

Institutional Review Board Statement: Not applicable.

Informed Consent Statement: Not applicable.

Data Availability Statement: Data sharing not applicable.

Conflicts of Interest: The authors declare no conflict of interest.

References

1. Kulkarni, S.C.; Rege, P.P. Pixel level fusion techniques for SAR and optical images: A review. *Inf. Fusion* **2020**, *59*, 13–29. [CrossRef]
2. Song, S.L.; Jin, K.; Zuo, B.; Yang, J. A novel change detection method combined with registration for SAR images. *Remote Sens. Lett.* **2019**, *10*, 669–678. [CrossRef]
3. Tapete, D.; Cigna, F. Detection of Archaeological Looting from Space: Methods, Achievements and Challenges. *Remote Sens.* **2019**, *11*, 2389. [CrossRef]
4. Suri, S.; Schwind, P.; Reinartz, P.; Uhl, J. Combining mutual information and scale invariant feature transform for fast and robust multisensor SAR image registration. In Proceedings of the American Society of Photogrammetry and Remote Sensing (ASPRS) Annual Conference, Baltimore, MD, USA, 9–13 March 2009.
5. Stone, H.S.; Orchard, M.T.; Chang, E.; Martucci, S.A.; Member, S. A Fast Direct Fourier-Based Algorithm for Sub-pixel Registration of Images. *IEEE Trans. Geosci. Remote Sens.* **2001**, *39*, 2235–2243. [CrossRef]
6. Xiang, Y.; Wang, F.; You, H. An Automatic and Novel SAR Image Registration Algorithm: A Case Study of the Chinese GF-3 Satellite. *Sensors* **2018**, *18*, 672. [CrossRef]

7. Pallotta, L.; Giunta, G.; Clemente, C. Subpixel SAR image registration through parabolic interpolation of the 2-D cross correlation. *IEEE Trans. Geosci. Remote Sens.* **2020**, *58*, 4132–4144. [CrossRef]

8. Lowe, D.G. Distinctive image features from scale-invariant keypoints. *Int. J. Comput. Vis.* **2004**, *60*, 91–110. [CrossRef]

9. Yan, K.; Sukthankar, R. PCA-SIFT: A more distinctive representation for local image descriptors. In Proceedings of the 2004 IEEE Computer Society Conference on Computer Vision and Pattern Recognition, CVPR 2004, Washington, DC, USA, 27 June–2 July 2004; Volume 2, pp. 506–513.

10. Dellinger, F.; Delon, J.; Gousseau, Y.; Michel, J.; Tupin, F. SAR-SIFT: A SIFT-like algorithm for SAR images. *IEEE Trans. Geosci. Remote Sens.* **2014**, *53*, 453–466. [CrossRef]

11. Pourfard, M.; Hosseinian, T.; Saeidi, R.; Motamedi, S.A.; Abdollahifard, M.J.; Mansoori, R.; Safabakhsh, R. KAZE-SAR: SAR Image Registration Using KAZE Detector and Modified SURF Descriptor for Tackling Speckle Noise. *IEEE Trans. Geosci. Remote Sens.* **2021**, *60*, 5207612. [CrossRef]

12. Alcantarilla, P.F.; Bartoli, A.; Davison, A.J. KAZE features. In Proceedings of the European Conference on Computer Vision, Florence, Italy, 7–13 October 2012; Springer: Berlin, Germany, 2012; pp. 214–227.

13. Xiang, Y.; Jiao, N.; Wang, F.; You, H. A Robust Two-Stage Registration Algorithm for Large Optical and SAR Images. *IEEE Trans. Geosci. Remote Sens.* **2021**, in press. [CrossRef]

14. LeCun, Y.; Bengio, Y.; Hinton, G. Deep learning. *Nature* **2015**, *521*, 436–444. [CrossRef]

15. Chen, S.; Wang, H.; Xu, F.; Jin, Y.Q. Target Classification using the Deep Convolutional Networks for SAR Images. *IEEE Trans. Geosci. Remote Sens.* **2016**, *54*, 4806–4817. [CrossRef]

16. Hou, X.; Ao, W.; Song, Q.; Lai, J.; Wang, H.; Xu, F. FUSAR-Ship: Building a high-resolution SAR-AIS matchup dataset of Gaofen-3 for ship detection and recognition. *Sci. China Inf. Sci.* **2020**, *63*, 140303. [CrossRef]

17. Wang, S.; Quan, D.; Liang, X.; Ning, M.; Guo, Y.; Jiao, L. A deep learning framework for remote sensing image registration. *ISPRS J. Photogramm. Remote Sens.* **2018**, *145*, 148–164. [CrossRef]

18. Geng, J.; Ma, X.; Zhou, X.; Wang, H. Saliency-Guided Deep Neural Networks for SAR Image Change Detection. *IEEE Trans. Geosci. Remote Sens.* **2019**, *57*, 7365–7377. [CrossRef]

19. He, H.; Chen, M.; Chen, T.; Li, D. Matching of remote sensing images with complex background variations via Siamese convolutional neural network. *Remote Sens.* **2018**, *10*, 355. [CrossRef]

20. Zheng, Y.; Sui, X.; Jiang, Y.; Che, T.; Zhang, S.; Yang, J.; Li, H. SymReg-GAN: Symmetric Image Registration with Generative Adversarial Networks. *IEEE Trans. Pattern Anal. Mach. Intell.* **2021**, in press. [CrossRef]

21. Li, Z.; Zhang, H.; Huang, Y. A Rotation-Invariant Optical and SAR Image Registration Algorithm Based on Deep and Gaussian Features. *Remote Sens.* **2021**, *13*, 2628. [CrossRef]

22. Mao, S.; Yang, J.; Gou, S.; Jiao, L.; Xiong, T.; Xiong, L. Multi-Scale Fused SAR Image Registration Based on Deep Forest. *Remote Sens.* **2021**, *13*, 2227. [CrossRef]

23. Luo, X.; Lai, G.; Wang, X.; Jin, Y.; He, X.; Xu, W.; Hou, W. UAV Remote Sensing Image Automatic Registration Based on Deep Residual Features. *Remote Sens.* **2021**, *13*, 3605. [CrossRef]

24. Lan, C.; Lu, W.; Yu, J.; Xu, Q. Deep learning algorithm for feature matching of cross modality remote sensing images. *Acta Geodaetica et Cartographica Sinica.* **2021**, *50*, 189.

25. Rublee, E.; Rabaud, V.; Konolige, K.; Bradski, G.R. ORB: An efficient alternative to SIFT or SURF. In Proceedings of the IEEE International Conference on Computer Vision (ICCV), Barcelona, Spain, 6–13 November 2011; pp. 2564–2571.

26. Bian, J.; Lin, W.Y.; Matsushita, Y.; Yeung, S.K.; Nguyen, T.D.; Cheng, M.M. Gms: Grid-based motion statistics for fast, ultra-robust feature correspondence. In Proceedings of the IEEE Conference on Computer Vision and Pattern Recognition (CVPR), Honolulu, HI, USA, 25–30 June 2017; pp. 4181–4190.

27. Arthur, D.; Vassilvitskii, S. K-means++: The Advantages of Careful Seeding. In Proceedings of the Eighteenth Annual ACM-SIAM Symposium on Discrete Algorithms, New Orleans, LA, USA, 7–9 January 2007; Society for Industrial and Applied Mathematics: Philadelphia, PA, USA; pp. 1027–1035.

28. Fischler, M.A.; Bolles, R.C. Random sample consensus—A paradigm for model-fitting with applications to image-analysis and automated cartography. *Commun. ACM* **1981**, *24*, 381–395. [CrossRef]

29. Krizhevsky, A.; Sutskever, I.; Hinton, G.E. ImageNet Classification with Deep Convolutional Neural Networks. In *Advances in Neural Information Processing Systems 25*; Pereira, F., Burges, C.J.C., Bottou, L., Weinberger, K.Q., Eds.; Curran Associates, Inc.: Red Hook, NY, USA, 2012; pp. 1097–1105.

30. Zhang, Z.; Wang, H.; Xu, F.; Jin, Y.Q. Complex-valued convolutional neural network and its application in polarimetric SAR image classification. *IEEE Trans. Geosci. Remote Sens.* **2017**, *55*, 7177–7188. [CrossRef]

31. Guo, Q.; Wang, H.; Xu, F. Scattering Enhanced attention pyramid network for aircraft detection in SAR images. *IEEE Trans. Geosci. Remote Sens.* **2020**, *99*, 1–18. [CrossRef]

32. Duan, Y.; Liu, F.; Jiao, L.; Zhao, P.; Zhang, L. SAR Image segmentation based on convolutional-wavelet neural network and markov random field. *Pattern Recognit.* **2016**, *64*, 255–267. [CrossRef]

33. Vaswani, A.; Shazeer, N.; Parmar, N.; Uszkoreit, J.; Jones, L.; Gomez, A.N.; Kaiser, Ł.; Polosukhin, I. Attention is all you need. In *Advances in Neural Information Processing Systems*; MIT Press: Cambridge, MA, USA, 2017; pp. 5998–6008.

34. Wang, Z.; Zhao, J.; Zhang, R.; Li, Z.; Lin, Q.; Wang, X. UATNet: U-Shape Attention-Based Transformer Net for Meteorological Satellite Cloud Recognition. *Remote Sens.* **2022**, *14*, 104. [CrossRef]

35. Zhao, C.; Wang, J.; Su, N.; Yan, Y.; Xing, X. Low Contrast Infrared Target Detection Method Based on Residual Thermal Backbone Network and Weighting Loss Function. *Remote Sens.* **2022**, *14*, 177. [CrossRef]
36. Xu, X.; Feng, Z.; Cao, C.; Li, M.; Wu, J.; Wu, Z.; Shang, Y.; Ye, S. An Improved Swin Transformer-Based Model for Remote Sensing Object Detection and Instance Segmentation. *Remote Sens.* **2021**, *13*, 4779. [CrossRef]
37. Sun, J.; Shen, Z.; Wang, Y.; Bao, H.; Zhou, X. LoFTR: Detector-free local feature matching with transformers. In Proceedings of the IEEE/CVF Conference on Computer Vision and Pattern Recognition, Montreal, QC, Canada, 11–17 October 2021; pp. 8922–8931.
38. Bay, H.; Tuytelaars, T.; Van Gool, L. SURF: Speeded Up Robust Features. In Proceedings of the 9th European Conference on Computer Vision, Graz, Austria, 7–13 May 2006; pp. 404–417.
39. Rosten, E.; Drummond, T. Machine learning for high-speed corner detection. In Proceedings of the 9th European Conference on Computer Vision, Graz, Austria, 7–13 May 2006; pp. 430–443.
40. Calonder, M.; Lepetit, V.; Strecha, C.; Fua, P. BRIEF: Binary Robust Independent Elementary Features. In Proceedings of the 11th European Conference on Computer Vision, Heraklion, Greece, 5–11 September 2010; pp. 778–792.
41. Sun, K.; Xiao, B.; Liu, D.; Wang, J. Deep high-resolution representation learning for human pose estimation. In Proceedings of the IEEE Conference on Computer Vision and Pattern Recognition, Long Beach, CA, USA, 16–20 June 2019; pp. 5693–5703.
42. Choromanski, K.; Likhosherstov, V.; Dohan, D.; Song, X.; Gane, A.; Sarlos, T.; Weller, A. Rethinking attention with performers. In Proceedings of the International Conference on Learning Representations, Virtual, 26 April–1 May 2020.
43. Simonyan, K.; Zisserman, A. Very deep convolutional networks for large-scale image recognition. In Proceedings of the International Conference on Learning Representations, Banff, AB, Canada, 14–16 April 2014.
44. He, K.; Zhang, X.; Ren, S.; Sun, J. Deep Residual Learning for Image Recognition. In Proceedings of the 2016 IEEE Conference on Computer Vision and Pattern Recognition (CVPR), Las Vegas, NV, USA, 27–30 June 2016; pp. 770–778.
45. Tan, M.; Le, Q. Efficientnet: Rethinking model scaling for convolutional neural networks. In Proceedings of the International Conference on Machine Learning, Long Beach, CA, USA, 9–15 June 2019; pp. 6105–6114.
46. Lin, T.Y.; Dollar, P.; Girshick, R.; He, K.; Hariharan, B.; Belongie, S. Feature Pyramid Networks for Object Detection. In Proceedings of the 2017 IEEE Conference on Computer Vision and Pattern Recognition (CVPR), Honolulu, HI, USA, 21 July–26 July 2017; IEEE Computer Society: Washington, DC, USA, 2017.
47. Li, Z.; Snavely, N. MegaDepth: Learning Single-View Depth Prediction from Internet Photos. In Proceedings of the 2018 IEEE/CVF Conference on Computer Vision and Pattern Recognition (CVPR), Salt Lake City, UT, USA, 18–23 June 2018; pp. 2041–2050.
48. Mishchuk, A.; Mishkin, D.; Radenovic, F.; Matas, J. Working hard to know your neighbor's margins: Local descriptor learning loss. In *Advances in Neural Information Processing Systems*; MIT Press: Cambridge, MA, USA, 2017; pp. 4826–4837.
49. Balntas, V.; Riba, E.; Ponsa, D.; Mikolajczyk, K. Learning Local Feature Descriptors with Triplets and Shallow Convolutional Neural Networks. In Proceedings of the British Machine Vision Association (BMVC) 2016, York, UK, 19–22 September 2016; Volume 1, p. 3.
50. Tian, Y.; Yu, X.; Fan, B.; Wu, F.; Heijnen, H.; Balntas, V. SOSNet: Second Order Similarity Regularization for Local Descriptor Learning. In Proceedings of the Computer Vision and Pattern Recognition, Long Beach, CA, USA, 16–20 June 2019; pp. 11008–11017.
51. Shi, J.; Tomasi, C. Good features to track. In Proceedings of the 1994 Proceedings of IEEE Conference on Computer Vision and Pattern Recognition, Seattle, WA, USA, 21–23 June 1994; pp. 593–600.
52. Pu, W. SAE-Net: A Deep Neural Network for SAR Autofocus. *IEEE Trans. Geosci. Remote Sens.* **2022**, in press. [CrossRef]
53. Celik, T. Unsupervised change detection in satellite images using principal component analysis and k-means clustering. *IEEE Geosci. Remote Sens. Lett.* **2009**, *6*, 772–776. [CrossRef]

Article

CRTransSar: A Visual Transformer Based on Contextual Joint Representation Learning for SAR Ship Detection

Runfan Xia [1,2,3], Jie Chen [1,2,3,*], Zhixiang Huang [1,2], Huiyao Wan [1,2,3], Bocai Wu [3], Long Sun [3,4,5], Baidong Yao [3], Haibing Xiang [3] and Mengdao Xing [4,5]

[1] Information Materials and Intelligent Sensing Laboratory of Anhui Province, Anhui University, Hefei 230601, China; p20301160@stu.ahu.edu.cn (R.X.); zxhuang@ahu.edu.cn (Z.H.); p19201033@stu.ahu.edu.cn (H.W.)

[2] Key Laboratory of Intelligent Computing & Signal Processing, Ministry of Education, Anhui University, Hefei 230601, China

[3] 38th Research Institute of China Electronics Technology Group Corporation, Hefei 230601, China; 18110995593@189.cn (B.W.); sl99goal@163.com (L.S.); yao1984@mail.ustc.edu.cn (B.Y.); xianghb@irsa.ac.cn (H.X.)

[4] National Lab of Radar Signal Processing, Xidian University, Xi'an 710126, China; xmd@xidian.edu.cn

[5] Collaborative Innovation Center of Information Sensing and Understanding, Xidian University, Xi'an 710126, China

* Correspondence: jiechen@ustc.edu

Citation: Xia, R.; Chen, J.; Huang, Z.; Wan, H.; Wu, B.; Sun, L.; Yao, B.; Xiang, H.; Xing, M. CRTransSar: A Visual Transformer Based on Contextual Joint Representation Learning for SAR Ship Detection. *Remote Sens.* **2022**, *14*, 1488. https://doi.org/10.3390/rs14061488

Academic Editors: Tianwen Zhang, Tianjiao Zeng and Xiaoling Zhang

Received: 3 February 2022
Accepted: 14 March 2022
Published: 19 March 2022

Publisher's Note: MDPI stays neutral with regard to jurisdictional claims in published maps and institutional affiliations.

Abstract: Synthetic-aperture radar (SAR) image target detection is widely used in military, civilian and other fields. However, existing detection methods have low accuracy due to the limitations presented by the strong scattering of SAR image targets, unclear edge contour information, multiple scales, strong sparseness, background interference, and other characteristics. In response, for SAR target detection tasks, this paper combines the global contextual information perception of transformers and the local feature representation capabilities of convolutional neural networks (CNNs) to innovatively propose a visual transformer framework based on contextual joint-representation learning, referred to as CRTransSar. First, this paper introduces the latest Swin Transformer as the basic architecture. Next, it introduces the CNN's local information capture and presents the design of a backbone, called CRbackbone, based on contextual joint representation learning, to extract richer contextual feature information while strengthening SAR target feature attributes. Furthermore, the design of a new cross-resolution attention-enhancement neck, called CAENeck, is presented to enhance the characterizability of multiscale SAR targets. The mAP of our method on the SSDD dataset attains 97.0% accuracy, reaching state-of-the-art levels. In addition, based on the HISEA-1 commercial SAR satellite, which has been launched into orbit and in whose development our research group participated, we released a larger-scale SAR multiclass target detection dataset, called SMCDD, which verifies the effectiveness of our method.

Keywords: transformer; deep learning; SAR target detection; multiscale learning; ship detection

1. Introduction

Synthetic-aperture radar (SAR) is an active microwave sensor that produces all-weather earth observations without being restricted by light and weather conditions. Compared with optical remote sensing images, SAR has significant application value. In recent years, SAR target detection and recognition have been widely used in military and civilian fields, such as military reconnaissance, situational awareness, agriculture, forestry management and urban planning. In particular, future war zones will extend from the traditional areas of land, sea and air to space. As a reconnaissance method with unique advantages, synthetic-aperture radar satellites may be used to seize the right to control information on future war zones and even play a decisive role in the outcome of these wars. SAR image target detection and recognition is the key technology with which to realize these military and civilian applications. Its core idea is to efficiently filter out regions and targets of interest through detection algorithms, and accurately identify their category attributes.

By contrast, from optical images, the imaging mechanism of SAR images is very different. SAR targets have characteristics such as strong scattering, unclear edge contour information, multiscale, strong sparseness, weak, small, sidelobe interference, and complex background. The SAR target detection and recognition tasks present huge challenges. In recent years, many research teams have also conducted extensive research on the above-mentioned difficulties. For SAR target imaging problems, phase modulation from a moving target's higher-order movements severely degrades the focusing quality of SAR images, because the conventional SAR ground moving target imaging (GMTIm) algorithm assumes a constant target velocity in high-resolution GMTIm with single-channel SAR. To solve this problem, a novel SAR-GMTIm algorithm [1] in the compressive sensing (CS) framework is proposed to obtain high-resolution SAR images with highly focused responses and accurate relocation. To improve moving target detectors, one study proposed a new moving target indicator (MTI) scheme [2] by combining displaced-phase-center antenna (DPCA) and along-track interferometry (ATI) sequentially to reduce false alarms compared to MTI via either DPCA or ATI. As shown by the simulation results, the proposed method can not only reduce the false alarm rate significantly, but can also maintain a high detection rate. Another study proposed a synthetic-aperture radar (SAR) change-detection approach [3] based on a structural similarity index measure (SSIM) and multiple-window processing (MWP) The work proposed by focusing on SAR imaging [2] can be found in [1]. The main focus of these studies is on the detection of moving SAR targets [3] and changes in SAR images, while that of our study is SAR target detection.

The use of a detector with constant false-alarm rate (CFAR) [4] is common in radar target detection. Constant false-alarm rate detection is an important part of automatic radar target detection. It can be used as the first step in extracting targets from SAR images and it is the basis for further target identification. However, traditional methods rely too much on expert experience to design manual features, which have great feature limitations. The traditional methods are also difficult to adapt to SAR target detection in complex scenes and cannot be used for large-scale practical applications. Based on traditional feature-extraction target-detection methods, the histogram-of-oriented-gradient (HOG) feature is a feature descriptor used for object detection in computer vision and image processing. HOG calculates histograms based not on color values but on gradients. It constructs features by calculating and counting the histograms of gradient directions in local areas of the image. HOG features combined with support-vector-machine (SVM) classifiers have been widely used in SAR image recognition. In recent years, with the development of computer vision, convolutional neural networks have been applied to SAR image detection, and a large number of deep neural networks have been developed, including AlexNet [5], VGGNet [6], ResNet [7], and GoogLeNet [8]. Additionally, methods such as Faster R-CNN [9], SSD [10], and YOLO V3 [11] are also widely used in SAR image recognition. Moreover, we mainly rely on the advantages of CNN because it is highly skilled in extracting local feature information from images with more refined local attention capabilities. However, because of the large downsampling coefficient used in CNN to extract features, the network misses small targets. In addition, a large number of studies has shown that the actual receptive field in CNN is much smaller than the theoretical receptive field, which is not conducive to making full use of context information. CNN's feature capturability is unable to extract global representations. Although we can enhance CNN's global capturability by continuously stacking deeper convolutional layers, this results in a number of layers that are too deep, too many parameters for the model to learn, difficulty in effectively converging, and the possibility that the accuracy may not be greatly improved. Additionally, the model is too large, the amount of calculation increases sharply, and it becomes difficult to guarantee timeliness.

In recent years, the use of a classification and detection framework with a transformer [12] as the main body has received widespread attention. Since Google proposed bidirectional encoder representation from transformers (BERT) [13], the BERT model has also been developed, and the structure that plays an important role in BERT includes a transformer. Generalized autoregressive pretraining for language understanding (XLNET) [14] and other models have since emerged. BERT's core has not changed and still includes a transformer. The first vision transformer (ViT) for image classification was proposed in [15] and obtained the best results in optical natural scene recognition. Network models, such as detection transformer (DETR) [16] and Swin Transformer [17], with a transformer utilized for the main body, have appeared in succession.

Swin Transformer is currently mainly used in image classification, optical object detection, and the instance segmentation of natural scenes in the field of computer vision. In the field of remote sensing, the Swin Transformer is mainly used in image segmentation [18] and semantic segmentation [19]. We investigated the papers in this area in detail, and did not find any research work in the field of SAR target detection. We can transfer the entire framework to target segmentation and transfer work, which is also a focus of our future work, at a later date.

The successful application of a transformer in the field of image recognition is mainly due to three advantages. The first advantage includes the ability to break through the RNN model's limitation, enabling it to be calculated in parallel. The second advantage is that compared with CNN, the number of operations required to calculate the association between two positions does not increase with distance. The third advantage is that self-attention enables it to produce more interpretable models. We can check the attention distribution from the model. Each attention head can learn to perform different tasks. Compared with the CNN architecture, the transformer has better global feature capturabilities. Therefore, due to the key technical difficulties in the above-mentioned SAR target detection task, this paper combines the global context information perception of a transformer and the local information feature extractability of CNN that is oriented to the SAR target detection task, and innovatively proposes a context-based joint visual transformer framework for representation learning, referred to as CRTransSar. This is the first framework attempt in the field of SAR target detection. The experimental results from the SSDD and self-built SAR target dataset show that our method achieves higher precision. This paper focuses on the optimization design of the backbone and neck parts of the target detection framework. Therefore, we take the cascaded mask r-cnn framework as the basic framework of our method, and our method can be used as a functional module that is flexibly embedded in any other target detection frame. The main contributions of this paper include the following:

1. First, to address the lack of global long-range modeling and perception capabilities of existing CNN-based SAR target detection methods, we designed an end-to-end SAR target detector with a visual Transformer as the backbone.
2. Secondly, we incorporated strategies such as multi-dimensional hybrid convolution and self-attention, and constructed a new visual transformer backbone based on contextual joint representation learning, called CRbackbone, to improve the contextual salient feature description of multi-scale SAR targets.
3. In addition, to better adapt to multi-scale changes and complex background disturbances, we constructed a new cross-resolution attention enhancement neck, called CAENeck, which can guide the multi-resolution learning of dynamic attention modules with little increase in computational complexity.
4. Furthermore, we constructed a large-scale multi-class SAR target detection benchmark dataset. The source data were mainly from HISEA-1, China's first commercial remote sensing SAR satellite, developed by our research group.

2. Related Work

2.1. SAR Target Detection Algorithm

In traditional ship SAR target detection, the use of a detector with a constant false-alarm rate (CFAR) [4] is a common method for radar target detection. It can be used as the first step in extracting targets from SAR images and is the basis for further target identification. Chen et al. [20] proposed a histogram-based CFAR (H-CFAR) method, which directly uses the gray histogram of the SAR image and combines it with CFAR to successfully achieve ship target detection. Li et al. [4] proposed an improved super-pixel level CFAR detection method, which uses weighted information entropy (WIE) to describe the statistical characteristics of super-pixels and better distinguishes between targets and cluttered super-pixels. With the development of computer vision, convolutional neural networks have been applied to the detection of SAR images, and a large number of deep neural networks have emerged, such as AlexNet, VGGNet, ResNet, and GoogLeNet, which also enable Faster R-CNN, SSD, and YOLO V3. These neural networks are widely used in SAR image recognition. Roughly divided into two-stage detection methods, such as Mask R-CNN [21] and Faster R-CNN, and single-stage detection methods, such as YOLO V3 and SSD, transformers are added to combine with CNN, and deep-level extraction features are more suitable for SAR targets.

The two-stage detection method joins the fully connected segmentation subnet after the basic feature network and the original classification and regression task are divided into three tasks: classification, regression, and segmentation. These tasks are applied to SAR target detection to improve ship recognition accuracy. Its working principle is divided into four stages. First, a set of basic volumes, relu activation functions and pooling layers are used to extract features. Next, the feature maps are passed into the subsequent RPN and the fully connected layer. The RPN network is used to generate region proposals. Next, roi pooling of this layer collects the feature maps input by the convolutional layer and the proposals generated by the RPN network before passing them into the following fully connected layer to determine the target category. Finally, the proposal and features are used. The maps calculate the category of the proposal, and the bounding box regression is again used to obtain the detection frame's final precise position.

Single-stage detection methods, such as SSD [10], mainly detect specific targets directly from many dense anchor points and use features of different scales to predict the object. The main idea is to uniformly conduct dense sampling at different positions of the picture. Different sampling approaches can be used, including scale and aspect ratio. Next, CNN is used to extract features and directly perform classification and regression. The introduction of the YOLO series improves the detection speed.

Existing deep learning-based SAR ship detection algorithms have huge model sizes and very deep network scales. A series of algorithms are proposed by Xiaoling Zhang's team, ShipDeNet-20 [22] is a novel SAR ship detector, built from scratch; it is lighter than most algorithms and can be applied effectively to hardware transplantation. The detection accuracy of SAR ships is reduced due to the huge imbalance in the number of samples in different scenarios. Thus, to solve this problem, the authors of [23] proposed a balance scene learning mechanism (BSLM) for offshore and inshore ship detection in SAR images. In addition, the authors of [24] proposed a novel approach for high-speed ship detection in SAR images based on a grid convolutional neural network (G-CNN). This method improves the detection speed by meshing the input image, inspired by the basic principle of YOLO, and using depthwise separable convolution. However, existing most studies improve detection accuracy at the expense of detection speed. Thus, to solve this problem, HyperLi-Net was proposed [25] for high-accuracy and high-speed SAR ship detection. In addition, a novel high-speed SAR ship detection approach mainly using a depthwise separable convolution neural network (DS-CNN) was proposed [26]. In this approach, we integrated multi-scale detection mechanism, concatenation mechanism and anchor box mechanism to establish a brand-new light-weight network architecture for high-speed SAR ship detection. There are still some challenges hindering accuracy improvements for

SAR ship detection, such as complex background interferences, multi-scale ship feature differences, and indistinctive small ship features. Therefore, to address these problems, a novel quad feature pyramid network (Quad-FPN) [27] is proposed for SAR ship detection.

2.2. Transformer

Since the emergence of the attention mechanism and its high-quality performance in natural language processing, researchers have tried to introduce this attention mechanism into computer vision. Currently, however, research is mainly focused on optical natural image scenes in the field of image detection. Applying a transformer in the field of vision has recently become increasingly popular. Vision transformers [15] enable it to simultaneously learn low-level features and high-level semantic information by combining convolutional and regular transformers [12]. Experiments have proven that after replacing the last convolution module of Resnet [6] with a visual transformer, the number of parameters is reduced, and the accuracy is improved. DETR [16] uses a complete transformer to build an end-to-end target detection model. The largest highlight is the decoder. The original decoder is used to predict and generate sentence sequences, but in the target detection task, the input of the decoder is 0. The object vector outputs the object category and coordinates after FFN. The DETR model is simple and straightforward, except that the model abandons the manual method of designing anchors. Small objects have less pixel information and are easily lost in the downsampling process. For example, ships at sea are small in size. To address the detection problem of such obvious object size differences, the classic method is the image pyramid used for multiscale change enhancement, but this involves a considerable amount of calculation. However, multiscale detection is becoming increasingly important in target detection, especially for small targets.

In general, there are two main model architectures in the related work that use a transformer in computer vision (CV). One is a transformer-only structure [11], and the other is a hybrid structure that combines the backbone network of CNN with a transformer. In [15], a vision transformer was proposed for image classification for the first time. This research shows that dependence on CNN is not necessary. When directly applied to a sequence of image blocks, the transformer can also perform image classification tasks well. The research is based on a large amount of data for model pretraining and migration to multiple image recognition benchmark datasets. The results show that the vision transformer (ViT) model is comparable to the current optimal convolutional network. As a result, the computing resources required for its training are greatly reduced. The research specifically divides the image into multiple image patches and uses the linear embedding sequence of these image patches as the input for the transformer. Subsequently, the token processing method in the natural language processing (NLP) field is used to process the image block and train the image classification model in a supervised manner. When training with a medium-scale dataset (such as ImageNet), the model produces unsatisfactory results. This seemingly frustrating result is predictable. The transformer lacks some of the inherent inductive biases of CNN, such as translation, degeneration, and locality. Thus, after training with insufficient data, the transformer cannot generalize well. However, if the model is trained on a large dataset (14–300 m image), the situation is quite different. The study found that large-scale training outperforms inductive bias. When pretraining on a large enough data scale and migrating to tasks with fewer data points, the transformer can achieve excellent results.

2.3. Related Datasets in the Field of SAR Target Detection

On 1 December 2017, at the BIGSARDATA conference held in Beijing, China, a dataset SSDD [28] for ship target detection in SAR images was disclosed. SSDD is the first public dataset in this field. As of 25 August 2021, from 161 papers on deep learning-based SAR ship detection, 75 used SSDD as the training and test data, accounting for 46.6%, which shows the popularity and significance of SSDD in the SAR remote sensing community. The datasets used in other papers are the other five public datasets proposed in recent years, namely the SAR-Ship dataset released by Wang et al. in 2019, the AIR SARShip-1.0 released by Sun et al. [29] HRSID released in 2020, and LS SSDD-v1.0, released by Zhang et al. [30] in 2020. The original paper of SSDD used a random ratio of 7:1:2 to divide the dataset into training set, validation set, and test set. However, this random partitioning mechanism leads to great uncertainty over the samples in the test set, resulting in different results when using the same detection algorithm for multiple training and testing. This is because the number of samples in SSDD is too small, only 1160, and random division may destroy the distribution consistency between the training and test sets. Similar to HRSID [31] and LS-SSDD-v1.0, here, images containing land are considered as near-shore samples, while other images are considered as far-sea samples. The numbers of near-shore and far-ocean samples were highly unbalanced (19.8% and 80.2%, respectively), a phenomenon consistent with the fact that the oceans cover much more of the Earth's surface than land. In the SSDD dataset, there are a total of 1160 images and 2456 ships, with an average of 2.12 ships per image, and the dataset will continue to expand in the future. Compared with the PASCAL VOC [32] dataset, which features 20 categories of objects, SSDD has fewer pictures, but the category is only ships, so it is enough to train the detection model.

The HRSID dataset was released by Su Hao from the University of Electronic Science and Technology of China in January 2020. HRSID is a dataset for ship detection, semantic segmentation, and instance segmentation tasks in high-resolution Sar images. The dataset contains a total of 5604 high-resolution SAR images and 16,951 ship instances. The ISSID dataset borrows from the construction process of the Microsoft common objects in context (COCO) [33] dataset, including SAR images at different resolutions, polarization, sea state, sea area, and coastal ports. This dataset is the benchmark against which the researchers evaluate their methods. For HRSID, the resolutions of the SAR images are: 0.5 m, 1 m, and 3 m, respectively.

3. The Proposed Method

This paper combines the respective advantages of the transformer [12] and CNN architectures, and is oriented to the SAR target detection task. Thus, we innovatively propose a visual transformer SAR target detection framework based on contextual joint-representation learning, called CRTransSar. The overall framework is shown in Figure 1. This is the first framework attempt in the field of SAR target detection.

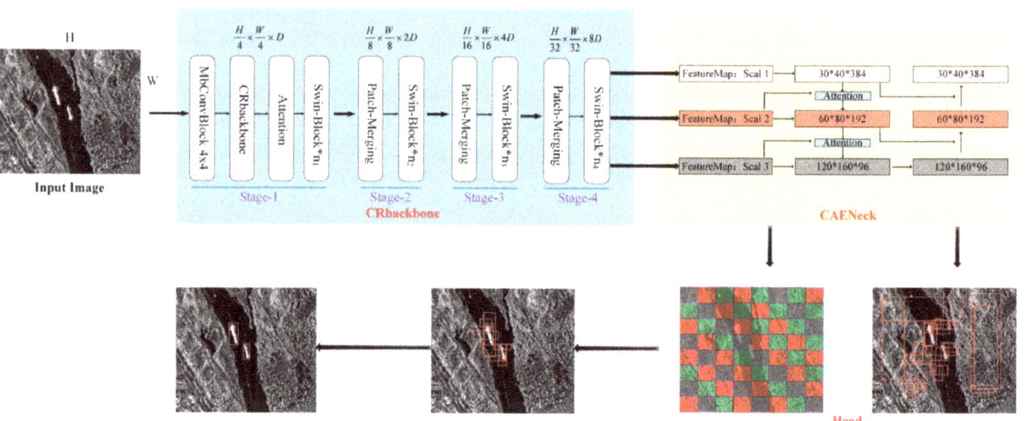

Figure 1. Overall architecture of the CRTransSar network.

3.1. The Overall Framework of Our CRTransSar

First, based on the cascade mask r-cnn two-stage model as the basic architecture, this paper innovatively introduces the latest Swin Transformer architecture as the backbone, introduces the local feature extraction module of CNN, and redesigns a target detection framework. The design of the framework can fully extract and integrate the global and local joint representations.

Furthermore, this paper combines the respective advantages of a Swin Transformer and CNN to design a brand-new backbone, referred to as CRbackbone. Thus, the model can make full use of contextual information, perform joint-representation learning, extract richer contextual feature information, and improve the multi-characterization and description of multiscale SAR targets.

Finally, we designed a new cross-resolution attention enhancement Neck, CAENeck. A feature pyramid network [34] is used to convey strong semantic features from top to bottom, enhancing the two-way multiscale connection operation through top-down and bottom-up attention, while also aggregating the parameters from different backbone layers to different detection layers, which can guide the multi-resolution learning of dynamic attention modules with little increase in computational complexity.

As shown in Figure 1, CRTransSar is mainly composed of four parts: CRbackbone, CAENeck, RPN-Head, and Roi-Head. First, we used our designed CRbackbone to extract features from the input image and performed a multiscale fusion of the obtained feature maps. The bottom feature map is responsible for predicting small targets and the high-level feature map is responsible for predicting large targets. The RPN module receives the multiscale feature map and starts to generate anchor boxes, generating nine anchors corresponding to each point on the feature map, which can cover all possible objects on the original image. Using a 1×1 convolution to make prediction scores and prediction offsets for each anchor frame, all the anchor frames and labels were matched. Next, we calculated the value of IOU to determine whether the anchor frame belonged to the background or the foreground. Here, we establish a standard to distinguish the samples. The positive sample and the negative sample, after the above steps, obtain a set of suitable proposals. The received feature map and the above proposal are passed into ROI pooling for unified processing, and then finally passed to the fully connected RCNN network for classification and regression.

3.2. Backbone Based on Contextual Joint Representation Learning: CRbackbone

Aiming at the strong scattering, sparseness, multiscale, and other characteristics of SAR targets, this paper combines the respective advantages of transformer and CNN architectures to design a target detection backbone based on contextual joint representation learning, called CRbackbone. It performs joint representation learning, extracts richer contextual feature salient information, and improves the feature description of multiscale SAR targets.

First, we used the Swin Transformer, which currently performs best in NLP and optical classification tasks, as the basic backbone. Next, we incorporated CNN's multiscale local information acquisition and redesigned the architecture of a Swin Transformer. Influenced by the latest EfficientNet [35] and inspired by the architecture of CoTNet [36], we introduced multidimensional hybrid convolution in the patchembed part to expand the receptive field, depth, and resolution, which enhanced the feature perception domain. Furthermore, the self-attention module was introduced to strengthen the comparison between different windows on the feature map, and for contextual information exchange.

3.2.1. Swin Transformer Module

For SAR images, small target ships in large scenes easily lose information in the process of downsampling. Therefore, we use a Swin Transformer [17]. The framework is shown in Figure 2. The transformer has general modeling capabilities and it is complementary to convolution. It also has powerful modeling capabilities, better connections between vision and language, a large throughput, and large-scale parallel processing capabilities. When a picture is input into our network, first the transformer [11] is used to process the image because we need to use all of the means that can be processed to divide the picture into tokens similar to NLP with the high-resolution characteristics of the image. The language difference leads to a layered transformer whose representation is calculated by moving the window. By limiting self-attention calculations to non-overlapping partial windows while allowing cross-window connections, the shifted window scheme leads to higher efficiency. This layered architecture has the flexibility of modeling at various scales and has linear computational complexity relative to the image size. This is an improvement to the vision transformer.

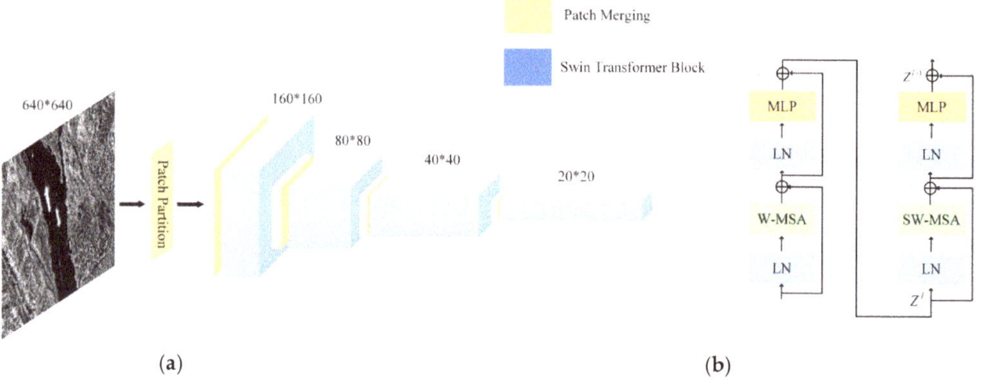

Figure 2. Overall architecture of Swin Transformer. (**a**) Swin Transformer structure diagram. (**b**) Swin Transformer blocks.

The vision transformer always focuses on the patch that is segmented at the beginning and does not perform any operations on the patch in the subsequent process. Thus, it does not affect the receptive field. A Swin Transformer is processed when a window is enlarged; subsequently, the calculation of self-attention is calculated in units of windows. This is equivalent to introducing locally aggregated information, which is very similar to the convolution process of CNN. The step size is the same as the size of the convolution kernel; thus, the windows do not overlap. The difference is that CNN performs the calculation of convolution in each window, and each window finally obtains a value, which represents the characteristics of this window. The Swin Transformer performs the self-attention calculation in each window and obtains an updated window. Next, through the patch merging operation, the window is merged, and the merged window continues to perform self-calculation. The Swin Transformer places the patches of the surrounding four windows together in the process of continuous downsampling, and the number of patches decreases. In the end, the entire image has only one window and seven patches. Therefore, we believe that downsampling means reducing the number of patches, but the size of the patches increases, which increases the receptive field.

As illustrated in Figure 3, the first module uses a regular window partitioning strategy, which starts from the top-left pixel, and the 8 × 8 feature map is evenly partitioned into 2 × 2 windows o 4 × 4 (M = 4) in size. Next, the next module adopts a windowing configuration that is shifted from that of the preceding layer by displacing the windows by (M/2, M/2) pixels from the regularly partitioned windows. With the shifted window-partitioning approach, consecutive Swin Transformer blocks are computed as:

$$\hat{z}^l = \text{W-MSA}\left(\text{LN}\left(z^{l-1}\right)\right) + z^{l-1} \quad z^l = \text{MLP}(\text{LN}(\hat{z}^l)) + \hat{z}^l \quad \hat{z}^{l+1} = \text{SW-MSA}(\text{LN}(z^l)) + z^l \quad z^{l+1} = \text{MLP}(\text{LN}(\hat{z}^{l+1})) + \hat{z}^{l+1} \tag{1}$$

where $\hat{z}^l$ and z^l denote the output features of the SW-MSA module and the MLP module for block, respectively; W-MSA and SW-MSA denote window-based multi-head self-attention using regular and shifted window partitioning configurations, respectively.

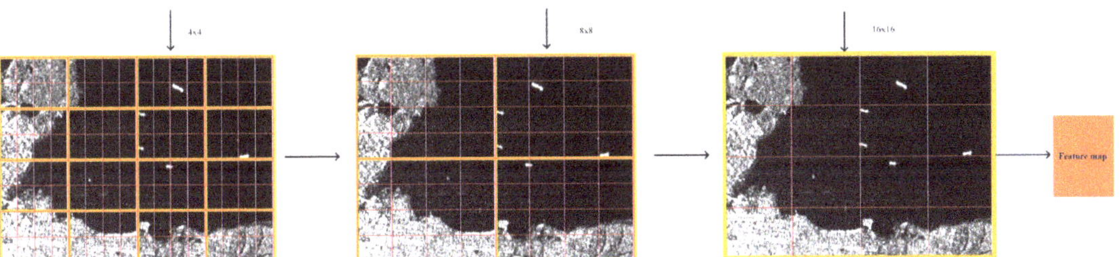

Figure 3. Swin Transformer sliding window. In the figure, 4, 8, 16 represent the number of patches.

A Swin Transformer performs self-attention in each window. Compared with the global attention calculation performed by a transformer, we assume that the complexity of a known MSA is the square of the image size. According to the complexity of an MSA, we can conclude that the complexity is $(3 \times 3)^2 = 81$. The Swin Transformer calculates self-attention in each local window (the red part). According to the complexity of MSA, we can see that the complexity of each red window is 1×1 squared, which is 1 to the fourth power. When there are nine windows, the complexity of these windows is summed, and the final complexity is nine, which is a greatly reduced figure. The calculation formulas for the complexity of an MSA and W-MSA are expressed by Formulas (2) and (3).

$$\Omega(MSA) = 4hwC^2 + 2(hw)^2C \tag{2}$$

$$\Omega(W\text{-}MSA) = 4hwC^2 + 2M^2hwC \tag{3}$$

Although computing self-attention inside the window may greatly reduce the complexity of the model, different windows cannot interact with each other, resulting in a lack of expressiveness. To better enhance the performance of the model, shifted-windows attention is introduced. Shifted windows alternately move between successive Swin Transformer blocks.

3.2.2. Self-Attention Module

Due to its spatial locality and other characteristics in computer vision tasks, CNN can only model local information and lacks the ability to model and perceive long distances. The use of a Swin Transformer introduces a shifted-window partition to improve this defect. The problem of information exchange between different windows is not limited to the exchange of local information. Furthermore, based on multihead attention, this paper takes into account the CotNet [36] contextual attention mechanism and proposes to integrate the attention module block into the Swin Transformer. The independent Q and K matrices are connected to each other. After the feature extraction network moves to patchembed, the feature map of the input network is 640*640*3. However, the length and width of the input data are not all 640*640. Next, we determine whether it is an integer multiple of 4 according to the length and width of the feature map to determine whether to pad the length and width of the feature map, followed by two convolutional layers. The feature channel changes from the previous 3 channels to 96 channels, and the feature dimension also changes to 1/4 of the previous dimension. Finally, the size of the attention module is 160*160*96, and the size of the convolution kernel is 3×3, as shown in Figure 4. The feature dimension and feature channel of the module remain unchanged, which strengthens the information exchange between the different windows on the feature map.

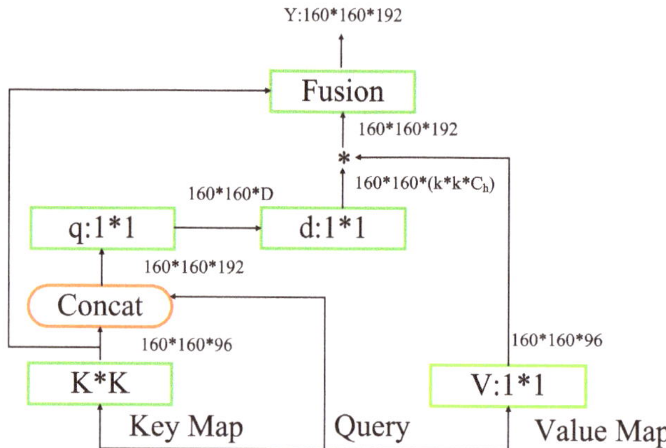

Figure 4. Self-attention module block.

The first step is to define three variables: Q = X, K = X, and V = XW$_v$. V is subjected to 1×1 convolution processing, then K is the grouped convolution operation of K $\times$ K and is recorded as the Q matrix and concat operation. The result after the concat performs two 1×1 convolutions and the calculation are shown in formula 4.

The self-attention module first encodes the contextual information of the input keys through 3×3 convolution to obtain a static contextual expression K^1 about the input; the encoded keys are further concatenated with the input query and the dynamic multi-head attention matrix is learned through two consecutive 1×1 convolutions. The resulting attention matrix is multiplied by the input values to obtain a dynamic contextual representation K^2 about the input. The fusion result of static context and dynamic context expression is used as output O. The architecture of the self-attention module is shown in Figure 5.

$$A = [K^1, Q]W_\theta W\delta \tag{4}$$

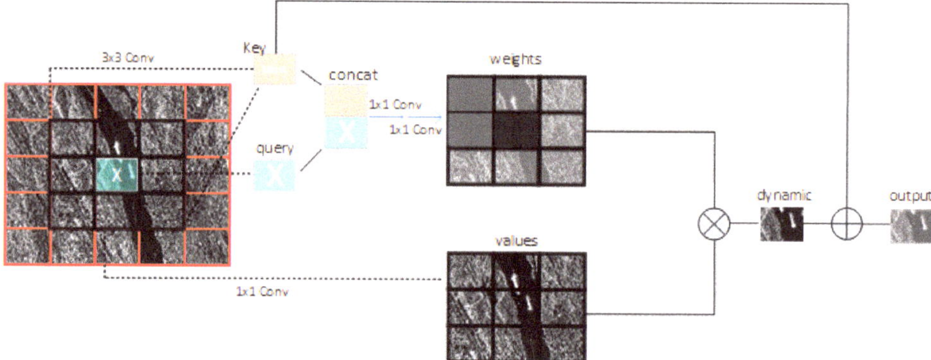

Figure 5. Architecture of self-attention module.

Here, A does not just model the relationship between Q and K. Thus, through the guidance of context modeling, the communication between each part is strengthened, and the self-attention mechanism is enhanced.

$$K^2 = V \otimes A \tag{5}$$

$$O = K^2 \oplus K^1 \tag{6}$$

Unlike the traditional self-attention mechanism, the self-attention module block structure of this paper combines contextual information and self-attention. Unlike the latest global self-attention mechanism, HOG-ShipCLSNet [37] and PFGFE-Net [38] are distinguish the characteristics of different scales and different polarization modes, so as to ensure sufficient global responses to comprehensively describe SAR ships. The specific process is to first calculate Φ and θ through two 1×1 convolutional layers, and it is used to characterize feature A through a 1×1 convolutional layer. The 1×1 convolutional layer is used to characterize feature $g(\cdot)$. We then obtain the similarity f by matrix multiplication $\theta^T\Phi$, and, finally, f through a softmax function/layer with a sigmoid activation is multiplied by g to obtain the self-attention output. Furthermore, to make the output y_i match the dimension of the input x to facilitate the follow-up element-wise adding operation, we add an extra 1×1 convolutional layer to achieve the dimension shaping. This is because in the embedded space, the number of convolution channels is c/2, which is not equal to the number of input channels c. This process is similar to the function of the residual or skip connections in ResNet, which can be described by 1×1. The weight matrix of the convolution layer is multiplied by y_i and then added to the input.

3.2.3. Multidimensional Hybrid Convolution Module

To increase the receptive field according to the characteristics of the SAR target, this section describes the proposed method in detail. The feature extraction network proposed in this paper is based on a Swin Transformer in order to improve the backbone. The CNN convolution is integrated into the patchembed module with the attention mechanism, and it is reconstructed. The entire feature extraction network structure diagram is shown in Figure 6. Affected by the efficient network [35], a multidimensional hybrid convolution module is introduced in the patchembed module. The reason why we introduced this network is that according to the mechanism characteristics of CNN, the more convolutional layers are stacked, the larger the receptive field of the feature maps. We used this approach to expand the receptive field and the depth of the network, and to increase the resolution to improve the performance of the network.

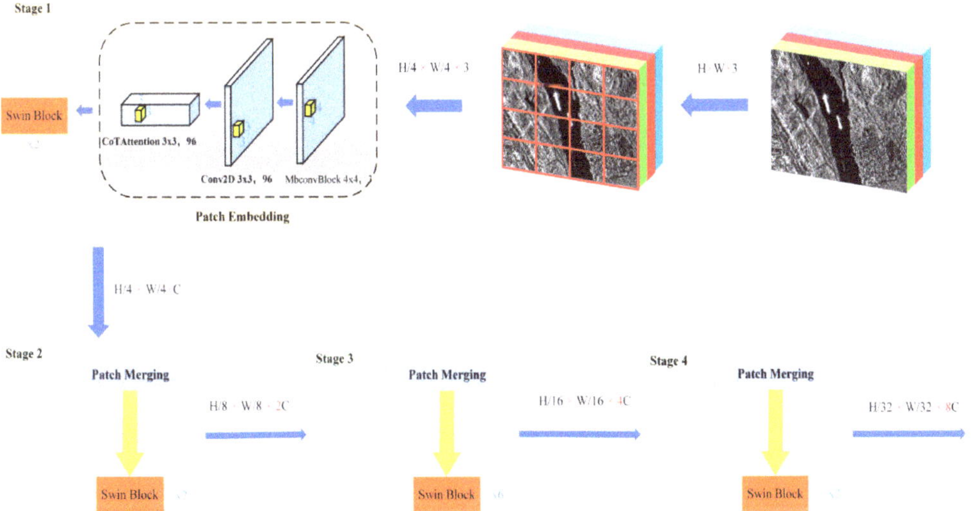

Figure 6. Overall architecture of CRbackbone.

When computing resources increase, if we thoroughly search for various combinations of the three variables of width, depth, and image resolution, the search space is infinite and the search efficiency is very low. The key to obtaining higher accuracy and efficiency is to balance the scaling ratios (d, r, w) of the three dimensions of network width, network depth, and image resolution using the combined zoom method:

$$\text{depth}: d = \alpha^{\phi} \quad \text{width}: w = \beta^{\phi} \quad \text{resolution}: r = \gamma^{\phi} \text{s. t.} \alpha \cdot \beta^2 \cdot \gamma^2 \approx 2 \alpha \geq 1, \beta \geq 1, \gamma \geq 1 \tag{7}$$

α, β, γ are constants (not infinite because the three correspond to the amount of computation), which can be obtained by grid search. The mixing coefficient ϕ can be adjusted manually. If the network depth is doubled, the corresponding calculation amount will double, and the network width or image resolution will double the corresponding calculation amount, that is, the calculation amount of the convolution operation (FLOPS) is proportional to d, ω γ^2. There are two square terms in the condition. Under this constraint, after specifying the mixing coefficient ϕ, the network calculation amount is about 2^{Φ} times what it was before.

Now, we can integrate the above three methods and integrate the hybrid parameter expansion method. Although there is no lack of research in this direction about models, such as MobileNet [39], ShuffleNet [40], M-NasNet [41], etc., the model is compressed by reducing the amount of parameters and calculations. The model is also applied to mobile devices and edge devices, but the amount of parameters and calculations are considerably reduced at the same time. However, the accuracy of the model is greatly improved. The patchembed module mainly increases the channel dimensions of each patch, which are divided into non-overlapping patch sets by patch partitioning processing the input picture H × W × 3, which reduces the size of the feature map and sends it to the Swin Transformer block for processing. When each feature map is sent to patchembed's dimension 2 × 3 × H × W and then finally sent to the next module, the dimension is 2 × 3 × 96. When four downsamplings are achieved through the convolutional layer and the number of channels becomes 96, a layer of a multidimensional hybrid convolution module is stacked before the 3 × 3 convolution. The size of the convolution kernel is 4, keeping the number of channels fed into the convolution unchanged, which also increases the depth of the receptive field and the network. This improves the efficiency of the model.

3.3. Cross-Resolution Attention Enhancement Neck: CAENeck

This paper, inspired by the structure of SGE [42] and PAN [43], addresses the small targets in large scenes, including the strong scattering characteristics of SAR imaging and the characteristics of low discrimination between targets and backgrounds. This paper designs a new cross-resolution attention enhancement neck, called CAENeck.

The specific step is to divide the feature map into G groups according to the channel, and then to calculate the attention of each group. After global average pooling is performed on each group, g is obtained, and then g is a matrix multiplied with the original grouped feature map. Next, we proceed to the norm. Additionally, sigmoid was used to perform the operation, and the result obtained was the matrix multiplied by the original grouping feature map. The specific steps are shown in Figure 7.

The attention mechanism is added to connect the context information, and attention is incorporated at the top-to-bottom connection. This is to better integrate the shallow and deep feature map information and to better extract the features of small targets, along with the goals of the target. The positioning is shown in Figure 1. We upsample during the transfer process of the feature map from top to bottom. The size of the feature map increases. The deepest layer is strengthened by the attention module and concatenated with the feature map of the middle layer, which then passes through the attention module. A concat connection is formed with the most shallow feature map. The specific steps are as follows. The neck receives the feature maps of three scales; 30*40*384, 60*80*192, 120*160*96, and 30*40*384 are the deepest features, which are upsampled and pay attention to the force enhancement operation, before being connected with 60*80*192. Finally, upsampling and attention enhancement are carried out to connect with the shallowest feature map.

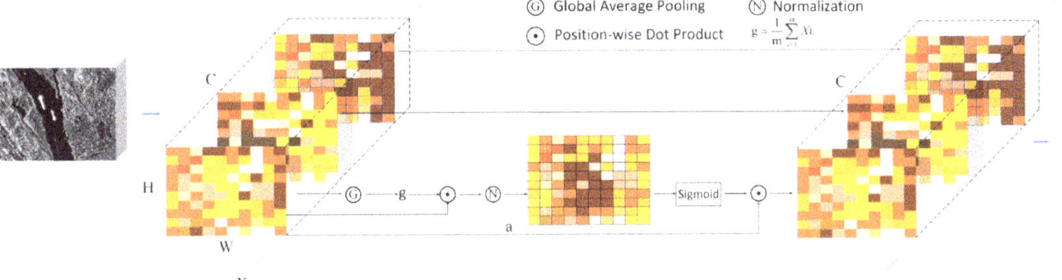

Figure 7. Neck attention enhancement module.

This series of operations is carried out from top to bottom. Next, bottom-up multiscale feature fusion is performed. Figure 1 shows the neck part. The SAR target is a very small target in a large scene, especially the marine ship target of the SSDD dataset. At sea, the ship itself has very little pixel information, and it easily loses the information of small objects in the process of downsampling. Although the high-level feature map is rich in semantic information for prediction, it is not conducive to the positioning of the target. The low-level feature map has little semantic information but is beneficial to the location of the target. The FPN [34] structure is a fusion of high-level and low-level from top to bottom. It is achieved through upsampling. The attention module is added during the upsampling process to integrate contextual information mining and self-attention mechanisms into a unified body. The ability to extract the information of the target location is enhanced. Furthermore, the bottom-up module has a pyramid structure from the bottom to the high level, which realizes the fusion of the bottom level and the high level after downsampling, while enhancing the extraction of semantic feature information. The small feature map is responsible for detecting large ships, and the large feature map is responsible for detecting small ships. Therefore, attention enhancement is very suitable for multiscale ship detection in SAR images.

3.4. Loss

The loss function is used to estimate the gap between the model output y and the true value y to guide the optimization of the model. This paper uses different loss functions in the head part. The specific formulas for the loss of the category in the RPN-head use cross-entropy loss, and the regression loss utilization function is as follows:

$$L(\{P_i\},\{t_i\}) = \frac{1}{N_{class}}\sum_i L_{cls}(p_i, p_i^*) + \lambda\frac{1}{N_{reg}}\sum_i p_i^* L_{reg}(t_i, t_i^*) \tag{8}$$

where $\sum_i L_{cls}(p_i, p_i^*)$ represents the filtered anchor's classification loss, P_i is the true value of each anchor's category, and $\sum_i p^* L_{reg}(t_i, t_i^*)$ is the predicted category of each anchor. Representing the loss of the regression, the function formula used for the regression loss is as follows:

$$L_{reg}(t_i, t_i^*) = \sum_{i \in x,y,w,h} smooth_{L1}(t_i - t_i^*) \tag{9}$$

$$smooth_{L1}(x) = \begin{cases} 0.5x^2 & if\,|x| < 1 \\ |x| - 0.5 & otherwise \end{cases} \tag{10}$$

4. Experiments and Results

This section evaluates our proposed detection method through experimental results. First, we use the SSDD dataset and the SMCDD dataset as experimental data. The SMCDD dataset provides some of the parameter settings of the experiment. The next part describes the influence of the attention enhancement backbone, the reconstruction of the patchembed module, and the multiscale attention enhancement neck. Finally, we compare our method with other methods to verify the effectiveness of our method. Our hardware platform is a personal computer with an Intel i5 CPU based on the mmdet [44] framework, an NVIDIA RTX2060 GPU, 8 GB of video memory, and an Ubuntu 18.04 operating system.

4.1. Dataset
4.1.1. SSDD Dataset

We used a remote sensing SAR image dataset. The SAR ship detection established in 2017 used a SSDD ship dataset, which sets the baseline of the SAR ship detection algorithm and is used by many other scholars. The SSDD dataset contains data in a variety of scenarios, including different polarization modes and scenarios. We used the same labeling method as the most popular PASCAL VOC dataset. A total of 1160 images and 2456 ships were included, with an average of 2.12 ships per image. Although the number of images was small, we used it as a benchmark for ship target detection performance. The ratio of the training image, verification image, and test image was 7:2:1. The SSDD dataset is shown in Table 1.

Table 1. SSDD dataset description.

Category	Indicator
Scenes	RadarSat-2, TerraSAR-X, Sentinel-1
Polarization	HH, VV, HV, VH
Resolution	1–15 m
Number of pictures	1160
Number of ships	2456

4.1.2. SMCDD Dataset

Our research group will soon release the SAR dataset, which contains data from the satellite HISEA-1 called SMCDD, as shown in Figure 8.

The HISEA-1 satellite is China's first commercial SAR synthetic-aperture radar satellite, jointly developed by the 38th Research Institute of China Electronics Technology Group Corporation, China Changsha Tianyi Space Science and Technology Research Institute Co., Ltd. (Changsha, China), as well as other units. Since its entry into orbit, the HISEA-1 has performed more than 1880 imaging tasks, obtaining 2026 striped images, 757 spotlight images, and 284 scanned images. The HISEA-1 has the ability to provide stable data services. The slice data of the SMCDD dataset we constructed are all from the SAR large scene image captured by HISEA-1.

Our SMCDD dataset contains four types of data: ship data, airplane data, bridge data, and oil tank data, as shown in Figure 9. The images we used were all large. There were four polarization modes, and, as a result, we cut them into 256, 512, 1024, and 2048 sizes. We used slices of 1024 and 2048 sizes and finally passed our screening and cleaning, leaving 1851 bridges, 39,858 ships, 12,319 oil tanks, and 6368 aircraft, as shown in the figure. Although the current version of the dataset is unbalanced, we will continue to expand the dataset in the future. We also verified the effectiveness of the method proposed in this paper through our dataset. The data information is shown in Table 2.

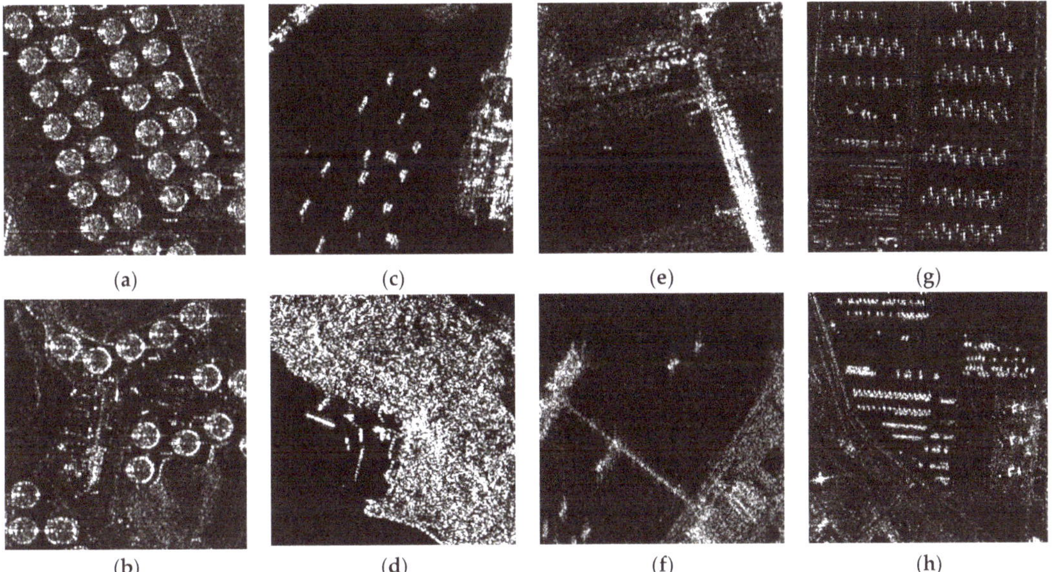

Figure 8. Some examples of the dataset SMCDD to be released by the research group. (**a**,**b**) are oil tanks; (**c**,**d**) are ship; (**e**,**f**) are bridges; (**g**,**h**) are aircraft.

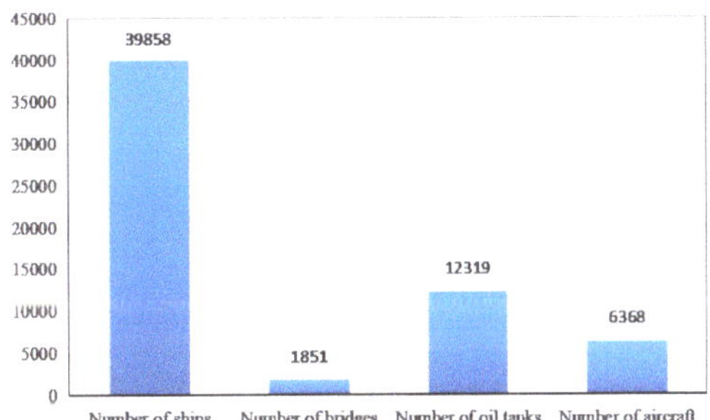

Figure 9. Description of the research group dataset SMCDD.

Table 2. Description of the research group dataset SMCDD.

Category	Indicator
Scenes	HISEA-1
Polarization	HH, VV, HV, VH

In contrast to the existing open-source SAR target detection dataset, our SMCDD dataset has the following advantages:

(1) The existing SAR target detection dataset is only for ship detection, and our dataset categories are richer, covering aircraft, ships, bridges, and oil tanks.
(2) The existing SAR target detection data collection is small, which makes it difficult to support the effective training of large-scale models. Our data collection is larger, which can support the training and verification of large-scale models.
(3) Since our dataset covers different types of SAR target data, our dataset can be used as a verification library for research directions, such as multiclass detection and

recognition, long-tailed distribution (class imbalance), small sample detection and recognition, etc. This will greatly promote the overall development of the SAR target detection field.

(4) We cut our large-scene images into various sizes, such as 2048*2048, 1024*1024, 512*512, and 256*256. We filtered and cleaned the data, leaving 1851 bridges, 39,858 ships, 12,319 oil tanks, and 6368 aircraft. Although the current version of the dataset is unbalanced, we will continue to expand the dataset in the future. For larger targets, such as bridges, we need to choose more large slice samples, generally 1024*1024 or 2048*2048, so that the network can better train the data.

(5) We have large-scene SAR images of HISEA-1, which can provide a large amount of training data to improve the SAR target detection performance of the network in large scenes.

4.2. Setup and Implementation Details

This study used Python 3.7.10, PyTorch 1.6.0, CUDA 10.1, CUDNN 7.6.3, and MMCV1.3.1, and the results of our network pretraining model were Swin-T on ImageNet. A total of 500 epochs was set up for training in the entire network. Due to the limitations of computer hardware and the size of the network itself, the batch size was set to 2. Each training sent two images to the network for processing, and the AdamW optimizer was selected as the model. The initial learning rate was set to 0.0001, the weight attenuation was 0.0001, and strategies such as LoadImageFromFile, LoadAnnotations, RandomFlip, and AutoAugment were used to optimize the training pipeline, as well as to enhance the online data, which enhanced the robustness of the algorithm. We adjusted the image size and finally selected the most suitable size for the network proposed in this paper to be 640*640.

4.3. Evaluation Metric

To quantitatively evaluate the performance of the proposed cascade mask rcnn with the improved Swin Transformer as the backbone and CAENeck as the neck detection algorithm, the accuracy, recall rate, average accuracy (mAP) and F-measure (F1) were used as evaluation indicators. Accuracy refers to the rate of correct detection of ships in all detection results, and recall refers to the rate of correct detection of ships in all ground facts. The definition of precision and recall is as follows:

$$P = \frac{TP}{TP + FP} \tag{11}$$

$$R = \frac{TP}{TP + FN} \tag{12}$$

In the formula, TP, FP, and FN represent the positive samples predicted by the model as positive, the negative samples predicted by the model as positive, and the positive samples predicted by the model as negative, respectively. In addition, if the IoU between the predicted bounding box and the real bounding box is higher than the threshold of 0.5, the bounding box is recognized as a correctly detected ship. The precision recall (PR) curve shows the precision recall rate under different confidence thresholds. MAP is a comprehensive metric that calculates the average precision under the recall range [0, 1]. The definition of mAP is as follows:

$$mAP = \int_0^1 P(R)\,dR \tag{13}$$

In the formula, R is the recall value, which represents the precision corresponding to the recall. $F1$ evaluates the comprehensive performance of the detection network proposed in this paper by considering the accuracy and recall rate. $F1$ is defined as:

$$F1 = \frac{2 \times P \times R}{P + R} \tag{14}$$

4.4. Analysis of Experimental Results

4.4.1. Ablation Experiments

A. The Influence of CRbackbone on the Experimental Evaluation Index

During the experiment, we first added the improved network backbone to the network, and the neck part was consistent with the baseline. We compared the benchmark Swin Transformer as the backbone, and PAN as the neck, and evaluated the test indicators. The comparison results are shown in Table 3. We observed that adding the optimized backbone to the percentage of mAP (0.5) led to an improvement of 0.4%; the improvement of mAP (0.75) was 6.5%, and the recall rate was also improved, by 1.3%. Therefore, the improved backbone had a propelling effect on the optimization of the network.

Table 3. The influence of CRbackbone on the experimental evaluation index.

Method	P	R	F1	mAP_{50}	mAP_{75}
CRbackbone	0.920	0.982	0.950	0.965	0.766
Baseline	0.912	0.969	0.940	0.961	0.701

B. The Influence of the CAENeck Module on the Experimental Evaluation Index

During the experiment, first added the improved network neck into the network, and the backbone part was consistent with the baseline. The lightweight attention-enhancement neck module was discarded to study its influence on the experiment, and an evaluation of the experimental indicators was carried out. The comparison results are shown in Table 4. We observed a 0.1% improvement in the percentage of mAP (0.5), a 2.5% improvement in mAP (0.75), and a 0.2% improvement in the recall rate. In the neck part, the detection performance also improved.

Table 4. The influence of CAENeck module on experimental evaluation index.

Method	P	R	F1	mAP_{50}	mAP_{75}
CAENeck	0.918	0.971	0.944	0.962	0.726
Baseline	0.912	0.969	0.940	0.961	0.701

4.4.2. Experimental Comparison with Current Methods

To compare traditional methods and advanced methods, we adopted the same parameter settings to test and verify them. We propose to use the improved Swin Transformer as the backbone's cascade mask RCNN target detection network to verify and compare the SSDD dataset and the dataset to be released by our research group. The experimental results are shown in the following table.

The target detection model proposed in this paper achieved a substantial improvement over the SSDD dataset. The accuracy of mAP (0.5) reached 97.0%, the accuracy of mA (0.75) reached 76.2%, and the F1 was 95.3. It can be seen that through the improvement of the Swin Transformer, the integration of the CotNet attention mechanism, and the lightweight EfficientNet module in patchembed promoted the optimization of the backbone. The cross-resolution attention enhancement neck strengthened the characteristics of the different scales. The fusion of the maps and these several methods are of great help for detecting ships. We compared the two-stage, single-stage, and anchor-free methods. The experiments showed that the detection accuracy of the method proposed in this paper is generally higher than that of the two-stage methods, such as Faster RCNN (88.5%) and Cascade R-CNN [45] (89.3%). We also compared our results with those of single-stage yolov3 (95.1%), SSD (84.9%), and RetinaNet (90.5%) [46]. The experimental results were also higher than those of the single-stage detection algorithm, which shows that the transformer uses the attention mechanism. Powerful functions, cascaded local information, and enhanced multiscale fusion is more conducive to the detection of inshore vessels without the perception of noise or the identification of ships of different sizes. We also compared the most advanced

FCOS [47] and CenterNet [48] without anchor frame detection and Cascade R-CNN [45] and Libra R-CNN [49] with anchor frame detection, as shown in Tables 5 and 6. This paper also draws the PR curve to compare the difference between the different networks, in Figure 10.

Table 5. Comparison with the latest anchor-free target detection method.

Method	P	R	F1	mAP$_{50}$	mAP$_{75}$
FCOS [47]	0.872	0.925	0.898	0.951	–
CenterNet [48]	0.803	0.933	0.863	0.945	–
CRTransSar (Ours)	0.925	0.983	0.953	0.970	0.762

Table 6. Comparison with the latest anchor-based target detection method.

Method	P	R	F1	mAP$_{50}$	mAP$_{75}$
Faster R-CNN [9]	0.810	0.942	0.871	0.885	0.488
Cascade Mask R-CNN	0.840	0.940	0.887	0.902	–
Cascade R-CNN [45]	0.819	0.930	0.871	0.893	–
YOLOV3 [11]	0.873	0.960	0.914	0.951	0.659
SSD [10]	0.770	0.93	0.883	0.849	–
RetinaNet [26]	0.870	0.945	0.906	0.905	0.526
Libra R-CNN [49]	0.808	0.924	0.862	0.887	–
CRTransSar (Ours)	0.925	0.983	0.953	0.970	0.762

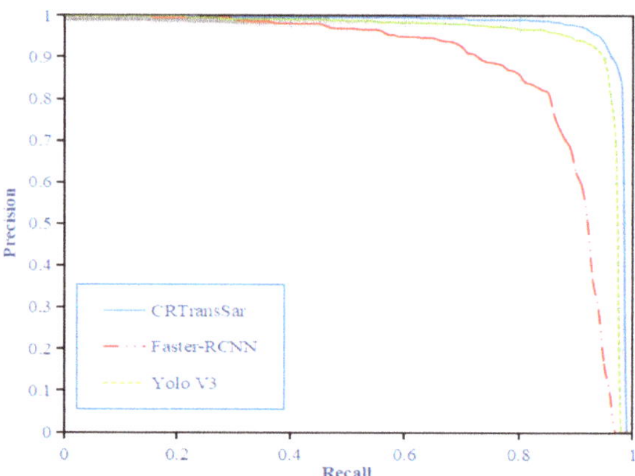

Figure 10. Comparison with the PR curve of the classic method.

The basic principle of CenterNet is that each target object is modeled as a center point to represent it. No candidate frame is required, nor is postprocessing, such as non-maximum suppression. CenterNet uses a fully convolutional network to generate a high-resolution feature map, classifies and judges each pixel of the feature map, and determines whether it is the center point of the target category or the background. This feature map gives each target the position of the center point of the object, the processing confidence in the center point of the target is 1, and the confidence of the background point is 0. Now, since there is no anchor box, there is no need to calculate the IoU between the anchor box and the bounding box to obtain positive samples to directly train the regressor. Instead, each point (located within the bounding box and having the correct class label) that is determined to be a positive sample is part of the regression of the bounding box size parameter.

This paper quotes the latest SAR target detection methods, FBR-Net [50], Center-Net++ [51], NNAM [52], DCMSNM [53], and DAPN [54]. Since the relevant papers do not have specific data divisions, this paper has no way to fully reproduce the results from other relevant papers. Therefore, this paper can only quote them. The results are compared horizontally, as shown in Table 7.

Table 7. Comparison with the latest SAR target detection method.

Method	P	R	F1	mAP
FBR-Net [50]	0.928	0.940	0.934	0.941
CenterNet++ [51]	0.833	0.952	0.889	0.951
NNAM [52]	0.843	0.851	0.849	0.798
DCMSNM [53]	0.836	0.834	0.835	0.896
DAPN [54]	0.711	0.909	0.798	0.898
CRTransSar (Ours)	0.925	0.983	0.953	0.970

To demonstrate the robustness of our proposed algorithm, we conducted comparative experiments with low SNR on salt and pepper noise, random noise, and Gaussian noise. In the salt and pepper noise experiments, our method led to mAP of 94.8. The map of our method was 5.5% higher than Yolo v3 and 9.7% higher than Faster R-CNN. In the random noise experiments, our method led to mAP of 96.7. The map of our method was 3% higher than Yolo v3 and 11.1% higher than Faster R-CNN. In the Gaussian noise experiments, our method led to mAP of 95.8. The map of our method was 1% higher than Yolo v3 and 10.7% higher than Faster R-CNN. The experimental results, presented in Table 8, show that we produced a reliable performance for SAR target detection tasks in low SNR.

Table 8. Comparison with low SNR of other advanced methods.

Method	Salt and Pepper Noise				Gaussian Noise				Random Noise			
	P	R	mAP	F1	P	R	mAP	F1	P	R	mAP	F1
Faster R-CNN	80.0	92.7	85.1	85.9	79.2	92.4	85.1	85.3	79.8	92.5	85.6	85.7
YOLO	84.3	93.4	89.3	88.6	89.4	97.2	94.8	93.1	88.7	97.2	93.7	92.7
CRTransSar (Ours)	90.2	96.7	94.8	93.3	90.0	97.8	95.8	93.7	91.3	98.2	96.7	94.6

In order for our proposed method to effectively solve the SAR target detection task, we also made corresponding experimental comparisons for the computational cost of the Swin Transformer. The FPS and parameter statistics of several representative target detection algorithms are shown in Table 9. Compared with the single-stage target detection algorithm, our parameter was 34M higher than YOLO V3, and the FPS was 28.5M lower, but the mAP was 1.9% higher than YOLO v3. Compared with two-stage target detection, the parameter amount was 52M higher than Faster R-CNN, and the FPS was 11.5M lower. Compared with Cascade R-CNN, the parameter amount was 8M higher and the FPS was 4.5M lower. Compared with Cascade Mask R-CNN, the number of parameters was 19M higher and the FPS was 7.5M lower. However, our mAP was 8.5% higher than that of Faster R-CNN, 7.7% higher than that of Cascade R-CNN, and 6.8% higher than that of Cascade Mask R-CNN. Because the overall architecture of the Swin Transformer is still relatively large, the large volume of Transformer is a general problem in this field, and we plan to make further improvements in model lightweighting and compression in the future.

Table 9. Compared with computational cost of other advanced methods.

Method	mAP	R	FPS (img/s)	Parameter (M)
yolo v3	0.951	0.960	36	62
Faster R-CNN [9]	0.885	0.942	3	44
RetinaNet	0.905	0.945	3	77
Cascade R-CNN	0.893	0.930	12	88
Cascade Mask R-CNN	0.902	0.940	14	77
CRTransSar (Ours)	0.970	0.983	7.5	96

4.4.3. Comparison between Experimental Results of the SMCDD Data Set

We used state-of-the-art object detection methods to evaluate our self-built SMCDD dataset. We chose CRTransSar, RetinaNet, and YOLOV3 as our benchmark algorithms, as shown in Table 10. There was a large number of dense targets in the oil tanks and aircraft in the SMCDD data set, which posed great challenges to the detection. It can be seen from the data that CRTransSar's mAP reached 16.3, which was better than RetinaNet and yolov3, and it was also higher than these two models in Recall.

Table 10. Comparison results on the SMCDD data set.

Method	mAP	R
RetinaNet [46]	0.161	0.203
YOLOV3 [11]	0.150	0.128
CRTransSar(Ours)	0.163	0.250

4.5. Visualization Result Verification and Analysis

To verify the effectiveness of the method in this paper, we visualized the SSDD dataset and the dataset to be released by our own research group, and obtained satisfactory results. We randomly selected some near-shore and far-shore ships for inspection. It can be seen from the figure that the use of multiscale fusion feature maps can more effectively improve the results of SAR images in different scenes, meaning that the method proposed in this paper can extract features with rich semantic information, even from complex backgrounds near shore. This method can also eliminate the interference and accurately identify the place where the naked eye has difficulty distinguishing between the noise and the ship. It can also eliminate some marine object noise, such as ships in the distant sea, and can be accurately distinguished. We also accurately verified the ships photographed by HISEA-1.

(1) This section visually verifies the performance of the network from two datasets, which are divided into inshore and offshore sets. Figure 11 shows the visual verification of SSDD inshore ships. When there is a relatively small amount of dense ships, the network's detection performance is better, and it is not disturbed by shore noise. Figure 12 is the dataset to be released by our research group, which contains inshore ships photographed by the HISEA-1 satellite and high-resolution satellites.

(2) Figures 13 and 14 are the SSDD dataset of the far sea and the results of the identification of the offshore ships of the dataset to be released by our research group, respectively. Offshore, because the surrounding environment receives less noise interference, the recognition accuracy is higher than it is inshore. Therefore, almost all target ships can be accurately identified in the offshore scene.

(3) To demonstrate the object detection performance of our proposed method for large scenes, we selected our self-built SMCDD dataset as the inference dataset. Our original data were obtained from the 38th Research Institute of China Electronics Technology Group Corporation. Because the data belonged to a secret military institution, we signed a confidentiality agreement with them. The original image of the large scene was obtained by HISEA-1. However, to further demonstrate the effectiveness of our method, we used the sliced data of some large scenes with a size of 2048*2048. As shown in Figure 15, from the

visualization results, it can be seen that Figure 15a,f are missing detections in three places; Figure 15b,d feature one false detection.

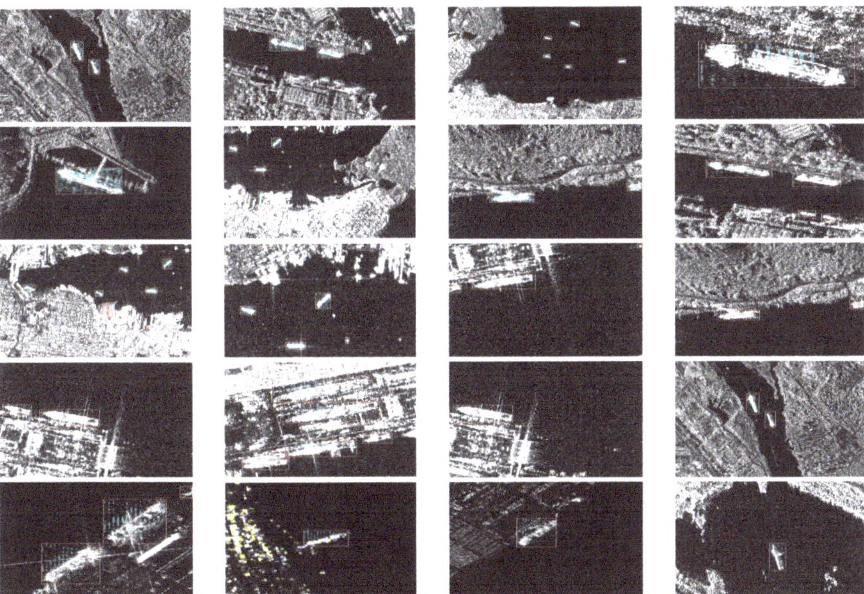

Figure 11. SSDD inshore inspection results. The red rectangular box is the correct visualization result of the CRTransSar method inshore on the SSDD dataset.

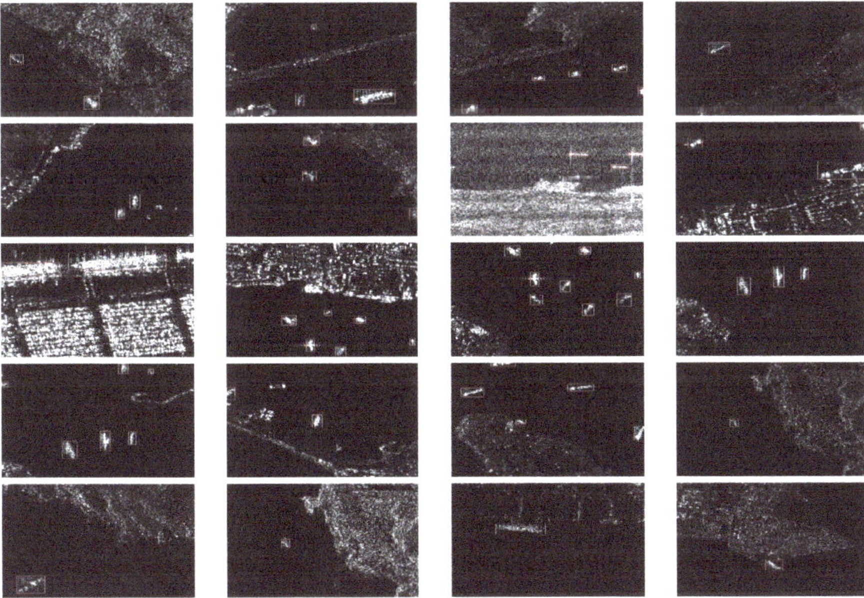

Figure 12. The results of verification of inshore taken by HISEA-1 Satellite. The red rectangular box is the correct visualization result of the CRTransSar method in the inshore scene of a port captured by HISEA-1 Satellite.

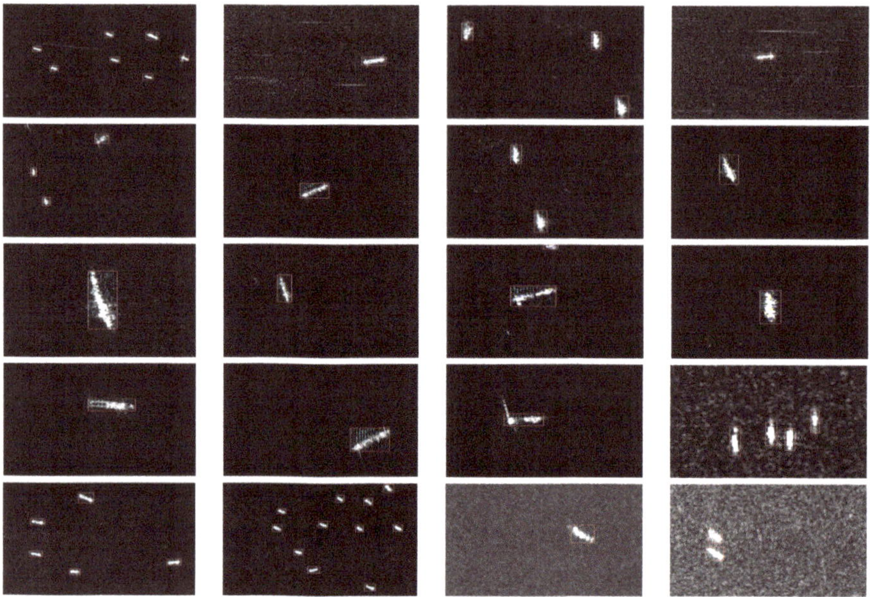

Figure 13. SSDD offshore ship identification result. The red rectangular box is the correct visualization result of the CRTransSar method offshore on the SSDD dataset.

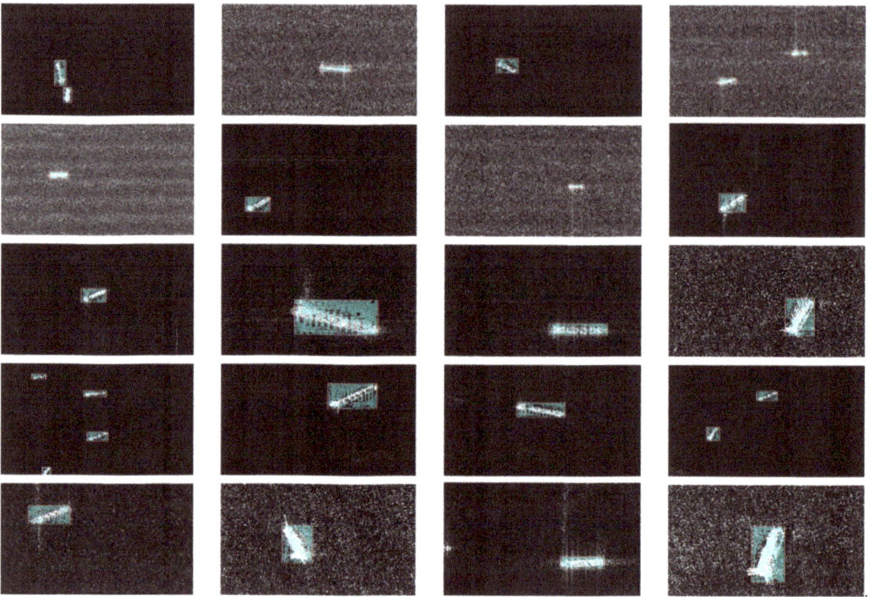

Figure 14. The results of verification of offshore ships taken by HISEA-1 satellites. The red rectangular box is the correct visualization result of the CRTransSar method in the offshore scene of a port captured by HISEA-1 Satellite.

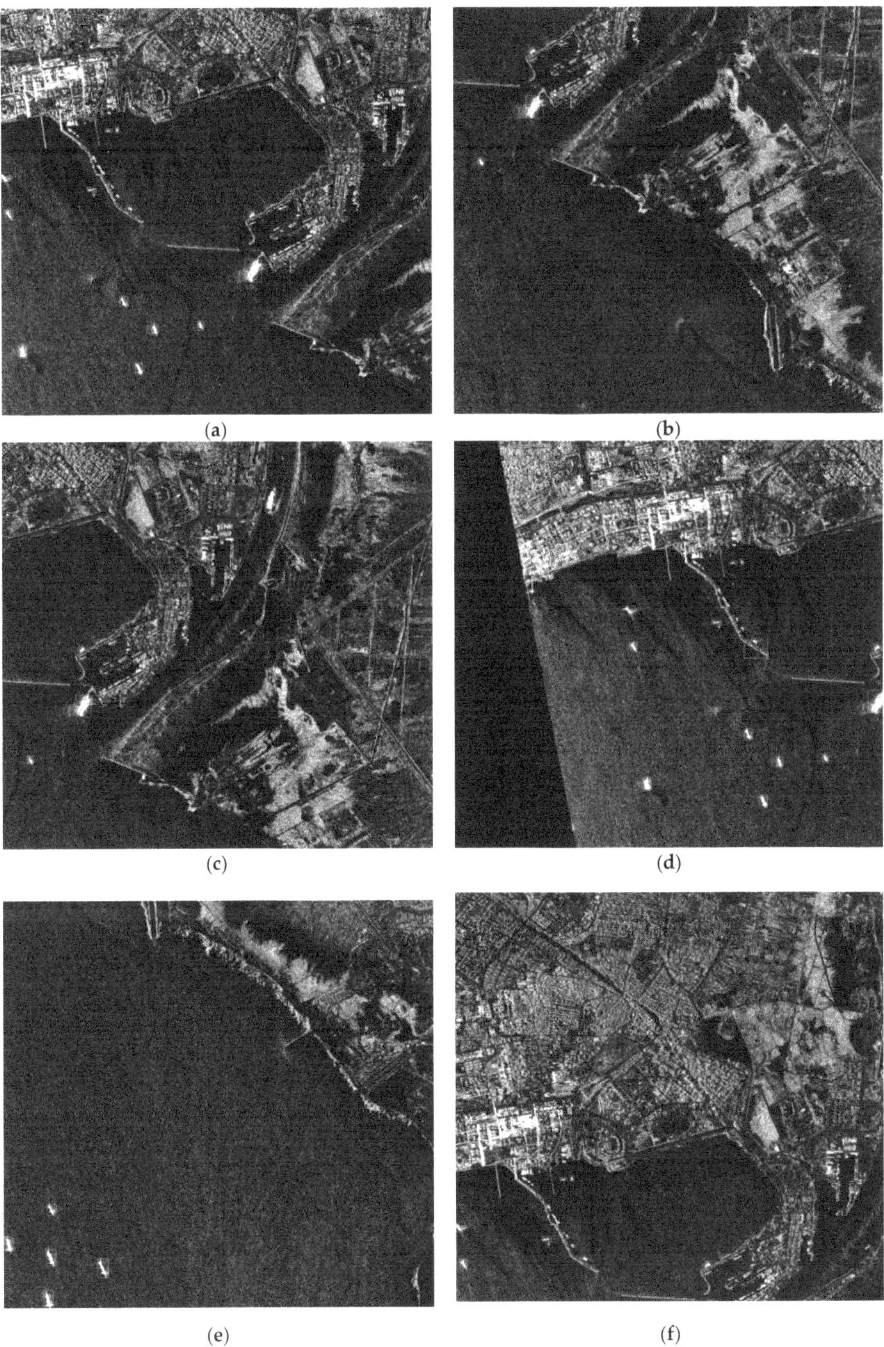

Figure 15. Visualization results of large scenes. (**a**–**f**) are slices of a large scene taken by HISEA-1 Satellite in a port. The red mark in the figure is the visualization result of the ships detected by CRTransSar in this scene.

5. Discussion

The four graphs in Figure 16 show some errors in the visualized results. It can be seen that (a) there are obvious detection frames to identify a ship, picture (b) has obvious undetected ships detected, and picture (c) has obvious detection frames to identify multiple ships, while one ship is detected by multiple ships. In (d), there are multiple ships that have not been recognized. We can solve the problem of the difficult identification of neighboring ships by segmentation, and introduce nonlocal mean models to highlight edge information.

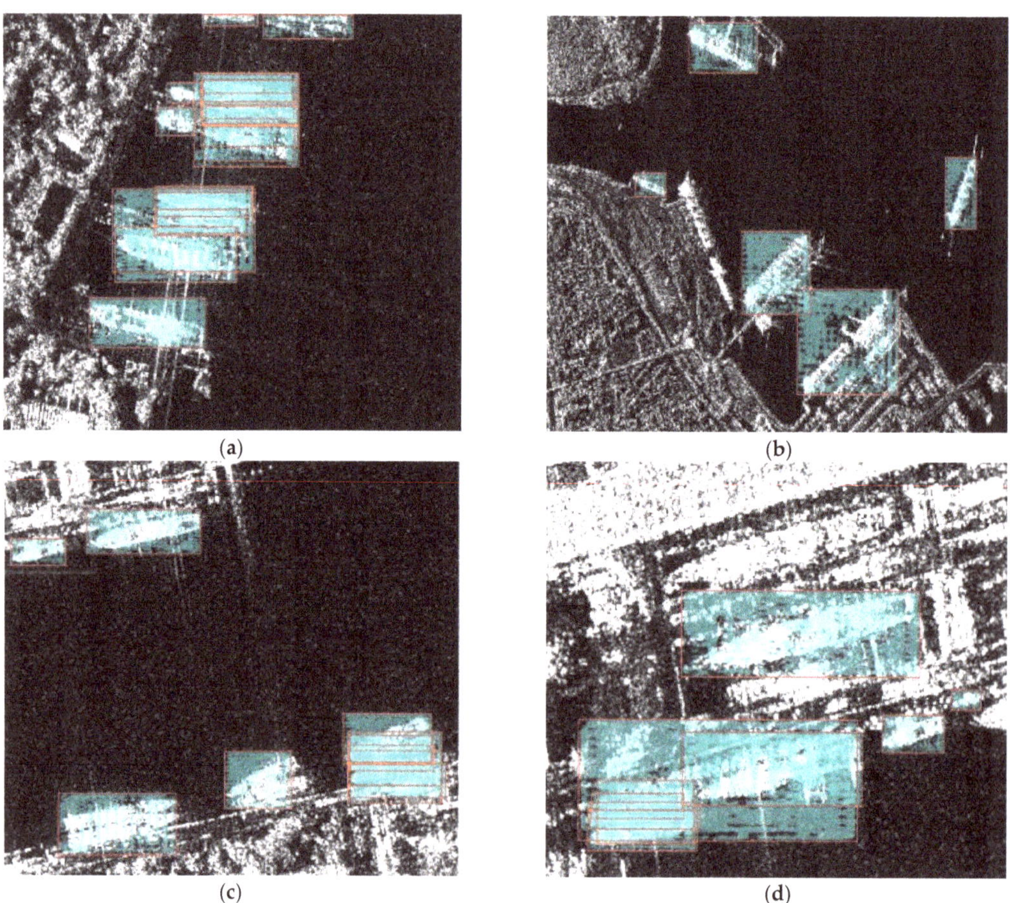

Figure 16. Visualization results of misdetected samples. (**a–d**) are the inshore visualization results of the CRTransSar method under the SSDD dataset. (**a,c,d**) are the visualization results of false-alarms. (**b,d**) are the visualization results of missed-detection.

6. Conclusions

SAR target detection has important application value in military and civilian fields. Aiming to overcome the difficulties of SAR targets, such as strong scattering, sparseness, multiscale, unclear contour information, and complex interference, we propose a visual transformer SAR target detection framework based on contextual joint representation learning, called CRTransSar. In this paper, CNN and the transformer are innovatively combined to improve the feature representation and the detectability of SAR targets in a balanced manner. This study was based on the use of a Swin Transformer and integrates CNN architecture ideas. We also redesigned a new backbone, named CRbackbone, which makes full use of contextual information, conducts joint-representation learning, and extracts richer context-feature salient information. Furthermore, we constructed a new cross-resolution attention enhancement neck, called CAENeck, which is used to enhance the ability to characterize SAR targets at different scales.

We conducted related experiments on the SSDD dataset and SMCDD dataset, as well as verification experiments on the SSDD dataset and the SMCDD dataset to be released by our research group. We performed visual verification of the classification of near-shore vessels and high-water vessels in the verification experiment. The high-quality results prove the robustness and practicability of our method. In the comparison experiment on the two-stage and no-anchor frames, higher precision was achieved. The method proposed in this paper achieves 97.0% mAP (0.5) and 76.2% mAP (0.75). In future work, we will first standardize the SMCDD dataset of our research group and release it for download and use. In addition, we will introduce segmentation to detect densely adjacent ships and explore more efficient distillation methods that do not require time-consuming training. Combined with pruning methods, model compression will be more diversified and easier to transplant, as will the lightweight development of the network.

Author Contributions: Conceptualization, R.X.; methodology, R.X.; software, R.X.; validation, R.X.; formal analysis, R.X.; investigation, R.X.; resources, R.X.; data curation, R.X.; writing—original draft preparation, R.X.; writing—review and editing, R.X., J.C., Z.H. and H.W.; visualization, R.X.; supervision, J.C., Z.H., B.W., L.S., B.Y., H.X. and M.X.; project administration, J.C.; funding acquisition, J.C. All authors have read and agreed to the published version of the manuscript.

Funding: This work was supported in part by the National Natural Science Foundation of China under Grant 62001003, in part by the Natural Science Foundation of Anhui Province under Grant 2008085QF284, and in part by the China Postdoctoral Science Foundation under Grant 2020M671851.

Institutional Review Board Statement: Not applicable.

Informed Consent Statement: Not applicable.

Data Availability Statement: The data used in this study are open data sets. The dataset can be downloaded at https://pan.baidu.com/s/1paex4cEYdTMjAf5R2Cy9ng (2 February 2022).

Acknowledgments: We would like to thank the anonymous reviewers for their constructive and valuable suggestions on the earlier drafts of this manuscript.

Conflicts of Interest: The authors declare no conflict of interest.

References

1. Kang, M.-S.; Kim, K.-T. Ground moving target imaging based on compressive sensing framework with single-channel SAR. *IEEE Sens. J.* **2019**, *20*, 1238–1250. [CrossRef]
2. Lee, M.-J.; Kang, M.-S.; Ryu, B.-H.; Lee, S.-J.; Lim, B.-G.; Oh, T.-B.; Kim, K.-T. Improved moving target detector using sequential combination of DPCA and ATI. *J. Eng.* **2019**, *2019*, 7834–7837. [CrossRef]
3. Kang, M.; Baek, J. SAR Image Change Detection via Multiple-Window Processing with Structural Similarity. *Sensors* **2021**, *21*, 6645. [CrossRef] [PubMed]
4. Robey, F.C.; Fuhrmann, D.R.; Kelly, E.J.; Nitzberg, R. A CFAR adaptive matched filter detector. *IEEE Trans. Aerosp. Electron. Syst.* **1992**, *28*, 208–216. [CrossRef]
5. Krizhevsky, A.; Sutskever, I.; Hinton, G.E. Imagenet classification with deep convolutional neural networks. *Adv. Neural Inf. Process. Syst.* **2012**, *25*, 1097–1105. [CrossRef]
6. Simonyan, K.; Zisserman, A. Very deep convolutional networks for large-scale image recognition. *arXiv* **2014**, arXiv:1409.1556.
7. He, K.; Zhang, X.; Ren, S.; Sun, J. Deep residual learning for image recognition. In Proceedings of the IEEE Conference on Computer Vision and Pattern Recognition, Las Vegas, NV, USA, 27–30 June 2016; pp. 770–778.
8. Szegedy, C.; Liu, W.; Jia, Y.; Sermanet, P.; Reed, S.; Anguelov, D.; Erhan, D.; Vanhoucke, V.; Rabinovich, A. Going deeper with convolutions. In Proceedings of the IEEE Conference on Computer Vision and Pattern Recognition, Boston, MA, USA, 7–12 June 2015; pp. 1–9.
9. Ren, S.; He, K.; Girshick, R.; Sun, J. Faster r-cnn: Towards real-time object detection with region proposal networks. *Adv. Neural Inf. Process. Syst.* **2015**, *28*, 91–99. [CrossRef] [PubMed]

10. Liu, W.; Anguelov, D.; Erhan, D.; Szegedy, C.; Reed, S.; Fu, C.-Y.; Berg, A.C. Ssd: Single shot multibox detector. In Proceedings of the European Conference on Computer Vision, Amsterdam, The Netherlands, 11–14 October 2016; Springer: Berlin/Heidelberg, Germany, 2016; pp. 21–37.

11. Redmon, J.; Farhadi, A. Yolov3: An incremental improvement. *arXiv* **2018**, arXiv:1804.02767.

12. Vaswani, A.; Shazeer, N.; Parmar, N.; Uszkoreit, J.; Jones, L.; Gomez, A.N.; Kaiser, Ł.; Polosukhin, I. Attention is all you need. *Adv. Neural Inf. Process. Syst.* **2017**, *30*, 5998–6008.

13. Devlin, J.; Chang, M.-W.; Lee, K.; Toutanova, K. Bert: Pre-training of deep bidirectional transformers for language understanding. *arXiv* **2018**, arXiv:1810.04805.

14. Yang, Z.; Dai, Z.; Yang, Y.; Carbonell, J.; Salakhutdinov, R.R.; Le, Q.V. Xlnet: Generalized autoregressive pretraining for language understanding. *Adv. Neural Inf. Process. Syst.* **2019**, *32*.

15. Dosovitskiy, A.; Beyer, L.; Kolesnikov, A.; Weissenborn, D.; Zhai, X.; Unterthiner, T.; Dehghani, M.; Minderer, M.; Heigold, G.; Gelly, S. An image is worth 16x16 words: Transformers for image recognition at scale. *arXiv* **2020**, arXiv:2010.11929.

16. Carion, N.; Massa, F.; Synnaeve, G.; Usunier, N.; Kirillov, A.; Zagoruyko, S. End-to-end object detection with transformers. In Proceedings of the European Conference on Computer Vision, Glasgow, UK, 23–28 August 2020; Springer: Berlin/Heidelberg, Germany, 2020; pp. 213–229.

17. Liu, Z.; Lin, Y.; Cao, Y.; Hu, H.; Wei, Y.; Zhang, Z.; Lin, S.; Guo, B. Swin transformer: Hierarchical vision transformer using shifted windows. *arXiv* **2021**, arXiv:2103.14030.

18. Xu, X.K.; Feng, Z.J.; Cao, C.Q.; Li, M.Y.; Wu, J.; Wu, Z.Y.; Shang, Y.J.; Ye, S.B. An Improved Swin Transformer-Based Model for Remote Sensing Object Detection and Instance Segmentation. *Remote Sens.* **2021**, *13*, 4779. [CrossRef]

19. Panboonyuen, T.; Jitkajornwanich, K.; Lawawirojwong, S.; Srestasathiern, P.; Vateekul, P. Transformer-Based Decoder Designs for Semantic Segmentation on Remotely Sensed Images. *Remote Sens.* **2021**, *13*, 5100. [CrossRef]

20. Chen, L.M.; Meng, J.M.; Yang, X.Z.; Lang, H.T.; Zhang, T. H-cfar based rapid ship targets detection in sar imagery. *J. Hefei Univ. Technol. (Nat. Sci.)* **2012**, *12*, 1633–1636+1643.

21. He, K.; Gkioxari, G.; Dollár, P.; Girshick, R. Mask r-cnn. In Proceedings of the IEEE International Conference on Computer Vision, Venice, Italy, 22–29 October 2017; pp. 2961–2969.

22. Zhang, T.; Zhang, X. ShipDeNet-20: An only 20 convolution layers and 1-MB lightweight SAR ship detector. *IEEE Geosci. Remote Sens. Lett.* **2020**, *18*, 1234–1238. [CrossRef]

23. Zhang, T.; Zhang, X.; Shi, J.; Wei, S.; Wang, J.; Li, J.; Su, H.; Zhou, Y. Balance scene learning mechanism for offshore and inshore ship detection in SAR images. *IEEE Geosci. Remote Sens. Lett.* **2020**, *19*, 1–5. [CrossRef]

24. Zhang, T.; Zhang, X. High-speed ship detection in SAR images based on a grid convolutional neural network. *Remote Sens.* **2019**, *11*, 1206. [CrossRef]

25. Zhang, T.; Zhang, X.; Shi, J.; Wei, S. HyperLi-Net: A hyper-light deep learning network for high-accurate and high-speed ship detection from synthetic aperture radar imagery. *ISPRS J. Photogramm. Remote Sens.* **2020**, *167*, 123–153. [CrossRef]

26. Zhang, T.; Zhang, X.; Shi, J.; Wei, S. Depthwise separable convolution neural network for high-speed SAR ship detection. *Remote Sens.* **2019**, *11*, 2483. [CrossRef]

27. Zhang, T.; Zhang, X.; Ke, X. Quad-FPN: A novel quad feature pyramid network for SAR ship detection. *Remote Sens.* **2021**, *13*, 2771. [CrossRef]

28. Zhang, T.; Zhang, X.; Li, J.; Xu, X.; Wang, B.; Zhan, X.; Xu, Y.; Ke, X.; Zeng, T.; Su, H. Sar ship detection dataset (ssdd): Official release and comprehensive data analysis. *Remote Sens.* **2021**, *13*, 3690. [CrossRef]

29. Xian, S.; Zhirui, W.; Yuanrui, S.; Wenhui, D.; Yue, Z.; Kun, F. AIR-SARShip-1.0: High-resolution SAR ship detection dataset. *J. Radars* **2019**, *8*, 852–862.

30. Zhang, T.; Zhang, X.; Ke, X.; Zhan, X.; Shi, J.; Wei, S.; Pan, D.; Li, J.; Su, H.; Zhou, Y. LS-SSDD-v1. 0: A deep learning dataset dedicated to small ship detection from large-scale Sentinel-1 SAR images. *Remote Sens.* **2020**, *12*, 2997. [CrossRef]

31. Wei, S.; Zeng, X.; Qu, Q.; Wang, M.; Su, H.; Shi, J. HRSID: A high-resolution SAR images dataset for ship detection and instance segmentation. *IEEE Access.* **2020**, *8*, 120234–120254. [CrossRef]

32. Everingham, M.; Winn, J. The pascal visual object classes challenge 2012 (voc2012) development kit. *Pattern Anal. Stat. Model. Comput. Learn. Tech. Rep.* **2011**, *8*, 5.

33. Lin, T.-Y.; Maire, M.; Belongie, S.; Hays, J.; Perona, P.; Ramanan, D.; Dollár, P.; Zitnick, C.L. Microsoft coco: Common objects in context. In Proceedings of the European Conference on Computer Vision, Zurich, Switzerland, 6–12 September 2014; Springer: Berlin/Heidelberg, Germany, 2014; pp. 740–755.

34. Lin, T.-Y.; Dollár, P.; Girshick, R.; He, K.; Hariharan, B.; Belongie, S. Feature pyramid networks for object detection. In Proceedings of the IEEE Conference on Computer Vision and Pattern Recognition, Honolulu, HI, USA, 21–26 July 2017; pp. 2117–2125.

35. Tan, M.; Le, Q. Efficientnet: Rethinking model scaling for convolutional neural networks. In Proceedings of the International Conference on Machine Learning, PMLR, Long Beach, CA, USA, 9–15 June 2019; pp. 6105–6114.

36. Li, Y.; Yao, T.; Pan, Y.; Mei, T. Contextual transformer networks for visual recognition. *arXiv* **2021**, arXiv:2107.12292.

37. Zhang, T.; Zhang, X.; Ke, X.; Liu, C.; Xu, X.; Zhan, X.; Wang, C.; Ahmad, I.; Zhou, Y.; Pan, D. HOG-ShipCLSNet: A novel deep learning network with hog feature fusion for SAR ship classification. *IEEE Trans. Geosci. Remote Sens.* **2021**, *60*, 1–22. [CrossRef]

38. Zhang, T.; Zhang, X. A polarization fusion network with geometric feature embedding for SAR ship classification. *Pattern Recognit.* **2022**, *123*, 108365. [CrossRef]

39. Howard, A.G.; Zhu, M.; Chen, B.; Kalenichenko, D.; Wang, W.; Weyand, T.; Andreetto, M.; Adam, H. Mobilenets: Efficient convolutional neural networks for mobile vision applications. *arXiv* **2017**, arXiv:1704.04861.
40. Zhang, X.; Zhou, X.; Lin, M.; Sun, J. Shufflenet: An extremely efficient convolutional neural network for mobile devices. In Proceedings of the IEEE Conference on Computer Vision and Pattern Recognition, Salt Lake City, UT, USA, 23–28 June 2018; pp. 6848–6856.
41. Tan, M.; Chen, B.; Pang, R.; Vasudevan, V.; Sandler, M.; Howard, A.; Le, Q.V. Mnasnet: Platform-aware neural architecture search for mobile. In Proceedings of the IEEE/CVF Conference on Computer Vision and Pattern Recognition, Long Beach, CA, USA, 16–17 June 2019; pp. 2820–2828.
42. Li, X.; Hu, X.; Yang, J. Spatial group-wise enhance: Improving semantic feature learning in convolutional networks. *arXiv* **2019**, arXiv:1905.09646.
43. Bochkovskiy, A.; Wang, C.-Y.; Liao, H.-Y.M. Yolov4: Optimal speed and accuracy of object detection. *arXiv* **2020**, arXiv:2004.10934.
44. Chen, K.; Wang, J.; Pang, J.; Cao, Y.; Xiong, Y.; Li, X.; Sun, S.; Feng, W.; Liu, Z.; Xu, J. MMDetection: Open mmlab detection toolbox and benchmark. *arXiv* **2019**, arXiv:1906.07155.
45. Cai, Z.; Vasconcelos, N. Cascade R-CNN: Delving into High Quality Object Detection. In Proceedings of the IEEE Conference on Computer Vision and Pattern Recognition (CVPR), Salt Lake City, UT, USA, 18–23 June 2018.
46. Lin, T.-Y.; Goyal, P.; Girshick, R.; He, K.; Dollár, P. Focal loss for dense object detection. In Proceedings of the IEEE International Conference on Computer Vision, Venice, Italy, 22–29 October 2017; pp. 2980–2988.
47. Tian, Z.; Shen, C.; Chen, H.; He, T. Fcos: Fully convolutional one-stage object detection. In Proceedings of the IEEE/CVF International Conference on Computer Vision, Seoul, Korea, 28–29 October 2019; pp. 9627–9636.
48. Duan, K.; Bai, S.; Xie, L.; Qi, H.; Huang, Q.; Tian, Q. Centernet: Keypoint triplets for object detection. In Proceedings of the IEEE/CVF International Conference on Computer Vision, Seoul, Korea, 28–29 October 2019; pp. 6569–6578.
49. Pang, J.; Chen, K.; Shi, J.; Feng, H.; Ouyang, W.; Lin, D. Libra r-cnn: Towards balanced learning for object detection. In Proceedings of the IEEE/CVF Conference on Computer Vision and Pattern Recognition, Long Beach, CA, USA, 16–17 June 2019; pp. 821–830.
50. Fu, J.; Sun, X.; Wang, Z.; Fu, K. An anchor-free method based on feature balancing and refinement network for multiscale ship detection in SAR images. *IEEE Trans. Geosci. Remote Sens.* **2020**, *59*, 1331–1344. [CrossRef]
51. Guo, H.; Yang, X.; Wang, N.; Gao, X. A CenterNet++ model for ship detection in SAR images. *Pattern Recognit.* **2021**, *112*, 107787. [CrossRef]
52. Chen, C.; He, C.; Hu, C.; Pei, H.; Jiao, L. A deep neural network based on an attention mechanism for SAR ship detection in multiscale and complex scenarios. *IEEE Access* **2019**, *7*, 104848–104863. [CrossRef]
53. Jiao, J.; Zhang, Y.; Sun, H.; Yang, X.; Gao, X.; Hong, W.; Fu, K.; Sun, X. A densely connected end-to-end neural network for multiscale and multiscene SAR ship detection. *IEEE Access* **2018**, *6*, 20881–20892. [CrossRef]
54. Cui, Z.; Li, Q.; Cao, Z.; Liu, N. Dense attention pyramid networks for multi-scale ship detection in SAR images. *IEEE Trans. Geosci. Remote Sens.* **2019**, *57*, 8983–8997. [CrossRef]

Article

A Lightweight Position-Enhanced Anchor-Free Algorithm for SAR Ship Detection

Yun Feng [1,2,3], Jie Chen [1,2,3,*,†], Zhixiang Huang [1,2,3,†], Huiyao Wan [1,2,3], Runfan Xia [1,2,3], Bocai Wu [3], Long Sun [3,4,5] and Mengdao Xing [4,5,†]

1   Information Materials and Intelligent Sensing Laboratory of Anhui Province, Anhui University, Hefei 230601, China; p19301125@stu.ahu.edu.cn (Y.F.); zxhuang@ahu.edu.cn (Z.H.); p19201033@stu.ahu.edu.cn (H.W.); p20301160@stu.ahu.edu.cn (R.X.)
2   Key Laboratory of Intelligent Computing and Signal Processing of Ministry of Education, School of Electronics and Information Engineering, Anhui University, Hefei 230601, China
3   38th Research Institute of China Electronics Technology Group Corporation, Hefei 230601, China; 18110995593@189.cn (B.W.); sl99goal@163.com (L.S.)
4   National Lab of Radar Signal Processing, Xidian University, Xi'an 710071, China; xmd@xidian.edu.cn
5   Collaborative Innovation Center of Information Sensing and Understanding, Xidian University, Xi'an 710071, China
*   Correspondence: jiechen@ustc.edu
†   Jie Chen is Member of IEEE; Zhixiang Huang is Senior Member of IEEE; Mengdao Xing is Fellow of IEEE.

Abstract: As an active microwave device, synthetic aperture radar (SAR) uses the backscatter of objects for imaging. SAR image ship targets are characterized by unclear contour information, a complex background and strong scattering. Existing deep learning detection algorithms derived from anchor-based methods mostly rely on expert experience to set a series of hyperparameters, and it is difficult to characterize the unique characteristics of SAR image ship targets, which greatly limits detection accuracy and speed. Therefore, this paper proposes a new lightweight position-enhanced anchor-free SAR ship detection algorithm called LPEDet. First, to resolve unclear SAR target contours and multiscale performance problems, we used YOLOX as the benchmark framework and redesigned the lightweight multiscale backbone, called NLCNet, which balances detection speed and accuracy. Second, for the strong scattering characteristics of the SAR target, we designed a new position-enhanced attention strategy, which suppresses background clutter by adding position information to the channel attention that highlights the target information to more accurately identify and locate the target. The experimental results for two large-scale SAR target detection datasets, SSDD and HRSID, show that our method achieves a higher detection accuracy and a faster detection speed than state-of-the-art SAR target detection methods.

Keywords: deep learning; SAR ship detection; position-enhanced attention; lightweight backbone

Citation: Feng, Y.; Chen, J.; Huang, Z.; Wan, H.; Xia, R.; Wu, B.; Sun, L.; Xing, M. A Lightweight Position-Enhanced Anchor-Free Algorithm for SAR Ship Detection. *Remote Sens.* **2022**, *14*, 1908. https://doi.org/10.3390/rs14081908

Academic Editors: Tianwen Zhang, Tianjiao Zeng and Xiaoling Zhang

Received: 6 March 2022
Accepted: 13 April 2022
Published: 15 April 2022

1. Introduction

SAR is one of the main ways of imaging Earth's surface for civilian and military research purposes at any time of day and is not affected by the weather or other imaging characteristics. With rapid updates of tools, information and technology, a large number of SAR images have been obtained. Due to the particularities of SAR imaging, the artificial interpretation of SAR images is a time-consuming and labor-intensive process and so a considerable amount of data have not been fully utilized. SAR image target detection aims to automatically locate and identify specific targets from images and has wide application prospects in real life. For example, in a military context, location detection of specific military targets is conducive to tactical deployment and coastal defense early warning capabilities. In a civil context, the detection of smuggling and illegal fishing vessels is helpful for the monitoring and management of maritime transport.

Since optical images are widely used in daily life, many researchers have developed numerous target detection algorithms based on optical images, but there are relatively few studies on SAR images. Due to the long imaging wavelength and complex imaging mechanism of SAR images, their targets are discontinuous; that is, they are composed of multiple discrete and irregular bright spots of scattering centers. Therefore, SAR images are difficult to interpret intuitively. At the same time, SAR images have the characteristics of an uneven target distribution and great sparsity. These characteristics make SAR image target detection very different from common optical image target detection. When target detection models used for optical images are directly used for SAR image detection without considering the particularity of SAR images, the advantages of the algorithm are not fully manifested. The development of SAR image target detection technology can be introduced via the following two aspects: traditional SAR target detection and SAR ship detection using deep learning.

Traditional SAR image target detection algorithms mainly include the constant false alarm rate (CFAR) [1] detection algorithm based on the background clutter statistical distribution and artificial image texture feature detection algorithms. The method based on the CFAR uses the background units around the target and selects the constant false alarm probability to determine the detection threshold. There are two main reasons for its poor detection rate: one is that the same statistical model is used for all the clutter in the sliding window, which easily leads to a mismatch of the statistical model in the maladaptive regions. Second, the algorithm does not make full use of the feature information in the image, but only uses the statistical distribution characteristics of the image gray values. Huang et al. [2] proposed a CFAR algorithm based on target semantic features, which has a lower false alarm rate when detecting targets in high-resolution SAR images. The detection algorithm based on artificial extraction of image texture features has good performance for some kinds of target detection; however, in the case of large differences in target features, the performance drops significantly. Stein et al. [3] proposed a target detection method based on the rotation-invariant wavelet transform. Compared with the CFAR detection algorithm, the texture feature-based algorithm utilizes more image information and has higher detection accuracy. However, texture features need to be extracted by manual design, and the design process is complicated and time-consuming, so it is difficult to ensure the timeliness of detection.

SAR ship detection methods based on deep learning have become a research priority and a large number of methods based on convolutional neural networks have emerged. Zhang et al. [4] proposed a learning mechanism for marine balanced scenes when the number of SAR image samples was extremely unbalanced and which extracted features from images by establishing a generative adversarial network, using the k-means algorithm for clustering and expansion of the number of samples to train the mode. The model has achieved good results. The lightweight SAR ship detector "ShipDeNet-20" [5] greatly reduces the size of the model and combines the feature fusion, feature enhancement and scale sharing feature pyramid modules to further improve the accuracy, which is conducive to hardware transplantation. HyperLi-Net [6] achieves high accuracy and high speed in SAR image ship detection. The high accuracy is achieved by the multi-receptive field, dilated convolution, channel and spatial attention, feature fusion and feature pyramid modules. High speed is achieved by fusion of region-free models, small kernels, narrow channels, separable convolutions and batch normalization. Its model is also more lightweight, which is more conducive to hardware porting. Tz et al. [7] solved the four imbalance problems in the SAR ship detection process and proposed corresponding solutions, which were combined into the model to obtain a new balanced learning network. Zhang et al. [8] mainly used depth-wise separable convolution to constitute a new SAR ship detection method. By integrating the multi-scale detection, connection and anchor box mechanisms, this method makes the model more lightweight and the detection speed is also improved to a certain extent. Zhang et al. [9] gridded the input image and used depthwise separable convolution operations. The backbone convolutional neural network and the detection convolutional neural network are combined to form a new grid convolutional neural network, which has achieved good results in SAR ship detection. RetinaNet [10] is essentially composed of resnet + FPN + two FCN sub-networks. The design idea is that the backbone selects effective feature extraction networks such as vgg and resnet. FPN is intended to strengthen the use of multi-scale features formed in resnet, to obtain a feature map with stronger expressiveness and include multi-scale target area information, and finally use two FCN sub-networks with the same structure but no shared parameters on the feature map set of FPN so as to complete the target box category classification and bbox position regression tasks. The SSD [11] model completely eliminates proposal generation and subsequent pixel or feature resampling stages and encapsulates all computations in a single network. This makes SSD easy to train and directly integrated into systems that require detection components. The core of the SSD approach is to use small convolutional filters to predict class scores and position offsets for a fixed set of default bounding boxes on the feature map. The network model of YOLOv3 [12] is mainly composed of 75 convolutional layers. Since the fully connected layer is not used, the network can correspond to input images of any size. In addition, the pooling layer does not appear in YOLOv3. Instead, the stride of the convolutional base layer is set to 2 to achieve the effect of down sampling and the scale-invariant features are transferred to the next layer. In addition, YOLOv3 also uses structures similar to ResNet and FPN networks, which are also beneficial for improving detection accuracy. YOLOv3 is mainly aimed at small targets and the accuracy has been significantly improved. YOLOX [13] is the first model to apply the anchor-free mode in the YOLO series. The specific operation is to explicitly define the 3×3 region of the truth frame projected to the center of the feature graph as the positive sample region and predict the four values of the target position (the offset distance of the upper left corner and the height and width of the frame). The AFSar [14] network model redesigns the backbone network, replaces the original Darknet-53 with MobileNetV2 and improves it. At the same time, the detection head and neck are newly designed, making it a lightweight network model. The RFB-net [15] algorithm introduces a receptive field block (RFB) into the SSD [11] network and strengthens the feature extraction ability, influenced by the way the human visual system works.

In summary, the following problems still need to be resolved:

(1) The existing algorithms of SAR image detection are mainly based on the design of anchors. However, setting the hyperparameters of an anchor heavily relies on human experience and a generated anchor greatly reduces model training speed. In addition, a detection algorithm with anchors mostly focuses on the capture of target edge information, while the unclear contour information of SAR images, especially with respect to small- and medium-sized SAR targets, greatly limits its detection efficiency.

(2) In order to further improve accuracy, most of the existing work blindly adds model structure and skills, resulting in a large number of model parameters, slow inference speed and low efficiency in practical applications, which is not conducive to the deployment of a model using mobile devices and greatly reduces the practicality of the model.

(3) The existing work does not consider the scattering of SAR images and the unclear target profile, which results in an algorithm being unable to better suppress the background clutter to emphasize the salient information of the target, which greatly reduces model performance.

To this end, we propose a new lightweight position-enhanced anchor-free SAR ship detection algorithm called LPEDet which improves the accuracy and speed of SAR ship detection from a more balanced perspective. The main contributions are as follows:

(1) To solve the problems that occur because anchor-based detection algorithms are highly dependent on design frameworks based on expert experience and the difficulties that occur in solving problems such as unclear contour information and complex backgrounds of SAR image ship targets, we introduced an anchor-free target detection algorithm. We introduced the latest YOLOX as the base network and, inspired by

the latest lightweight backbone, LCNet [16], replaced the backbone Darknet-53 with LCNet and then optimized the design according to the SAR target characteristics.

(2) To balance speed and model complexity, we constructed a new lightweight backbone called NLCNet through the ingenious design of depthwise separable convolutional modules and the novel structural construction of multiple modules. Experiments show that our proposed lightweight backbone greatly improved inference speed and detection accuracy.

(3) In order to improve the SAR target localization ability against complex backgrounds, inspired by coordinate attention [17], we designed a position-enhanced attention strategy. The strategy is to add target position awareness information to guide attention to better highlight the target area, effectively suppress the problem of insufficient feature extraction caused by SAR target strong scattering and better detect targets against complex backgrounds, thereby improving detection accuracy.

2. Related Work

The development process for SAR image target detection technology ranges from traditional SAR target detection to SAR target detection using deep learning. In the target detection task based on deep learning, the main task of target detection is to take the image as the input and output the characteristic image of the corresponding input image through the backbone network. Therefore, the performance of target detection is closely related to the feature extraction of the backbone network. Many studies have designed different feature extraction backbone networks for different application scenarios and detection tasks.

(1) Traditional SAR target detection algorithm.
     The traditional SAR target detection algorithm is as follows. Ai et al. [18] proposed a joint CFAR detection algorithm based on gray correlation by utilizing the strong correlation characteristics of adjacent pixels inside the target SAR images. The CFAR algorithm only considers the gray contrast and ignores target structure information, which causes poor robustness and anti-interference ability and poor detection performance under complex background clutter. Kaplan et al. [19] used the extended fractal (EF) feature to detect vehicle targets in SAR images. This feature is sensitive not only to the contrast of the target background but also to the target size. Compared with the CFAR algorithm, the false alarm rate of detection is reduced. Charalampidis [20] proposed the wavelet fractal (WF) feature, which can effectively segment and classify different textures in images.

(2) Common SAR image backbone networks based on deep learning.
     It can be seen from VGG [21] that a deeper network can be formed by stacking modules with the same dimension. For a given receptive field, it is shown that compared with using a large convolution kernel for convolution, the effect of using a stacked small convolution kernel is preferable. GoogLeNet [22] adopts a modular structure (inception structure) to enrich network receptive fields with convolutional kernels of different sizes. ShuffleNetV1 [23] and ShuffleNetV2 [24] adopt two core operations: pointwise group convolution and channel shuffling, and they exchange information through channel shuffling. GhostNet [25] divides the original convolution layer into two parts. First, a traditional convolution operation is applied to the input to generate feature maps, then these feature maps are transformed using a linear operation, merging all the features together to get the final result. In DarkNet-53, the poolless layer, the fully connected layer and the reduction of the feature graph are achieved by increasing the step size of the convolution kernel. Using the idea of feature pyramid networks (FPNs), the outputs of three scale feature layers are 13×13, 26×26 and 52×52. Among them, 13×13 is suitable for detecting large targets and 52×52 is suitable for detecting small targets. Although the above backbone network greatly improves detection accuracy, it also introduces a large number of parameters into the model and the detection speed is relatively slow. MobileNetV1 [26]

constructed a network by utilizing depth-separable convolution, which consists of two steps: depthwise convolution and pointwise convolution. MobileNetV2 [27] introduced a residual structure on the basis of MobileNetV1, which first raised the dimension and then reduced the dimension. Although the model is lightweight, it is suitable only for large models and it provides no significant improvement in accuracy in small networks. The characteristic of a remote sensing image target is density and it is difficult to distinguish between target contours and the background environment. A new algorithm [28] is proposed for the above difficulties which can also be used for video target recognition. It mainly uses the visual saliency mechanism to extract the target of the region of interest and experiments show the effectiveness of its results. In addition to SAR image target detection, the research on images captured by UAVs should continue to advance because the use of UAV images for target detection has broad application prospects in real life. The target detection of UAV images is the subject of [29], which combines the deep learning target detection method with existing template matching and proposes a parallel integrated deep learning algorithm for multi-target detection.

(3)   SAR image detection algorithm based on deep learning.

Jiao et al. [30] considered that the multi-scale nature of SAR image ship targets and the background complexity of offshore ship targets were not conducive to monitoring and the authors innovatively proposed a model based on the faster-RCNN framework. Improvements have been made and a new training strategy has also been proposed so that the training process focuses less on simple targets and is more suitable for the detection of ship targets with complex backgrounds in SAR images, improving detection performance and thereby solving the problem of multiple scales and multiple scenes. Chen et al. [31] mainly focused on indistinguishable ships on land and densely arranged ships at sea and combined a model with an attention mechanism. The purpose was to better solve the above two problems frequently encountered in ship target detection. The application of an attention mechanism can better enable the location of the ship targets we need to detect. At the same time, the loss function is also improved, that is, generalized cross loss is introduced, and soft non-maximum suppression is also used in the model. Therefore, the problem of densely arranged ship targets can be better solved and detection performance can be improved. Cui et al. [32] considered the multi-scale problem of ship targets in SAR images and used a densely connected pyramid structure in their model. At the same time, a convolution block attention module was used to refine the feature map, highlight the salient features, suppress the fuzzy features and effectively improve the accuracy of the SAR image ship target. Although the above algorithms generally have high detection accuracy, model size is large, inference speed is slow and they do not take the characteristics of SAR images into account, which greatly limits the performance of these algorithms. Wan et al. [14] proposed an anchor-free SAR ship detection algorithm, the backbone network of which is the more lightweight MobileNetV2S network, and further improved the neck and head, so that the overall model effect is optimal. However, their improved strategy did not fully consider the characteristics of SAR targets against complex backgrounds, which is an issue worthy of further exploration.

Therefore, we propose a new SAR image detection method that comprehensively considers the tradeoff between algorithm accuracy and speed.

3. Methods

This paper proposes a position-enhanced anchor-free SAR ship detection algorithm called LPEDet which generally includes the benchmark anchor-free detection benchmark network YOLOX, the lightweight feature enhancement backbone NLCnet and a position-enhanced attention strategy. The overall framework is shown in Figure 1. The model proposed in this paper will be explained in detail from three aspects.

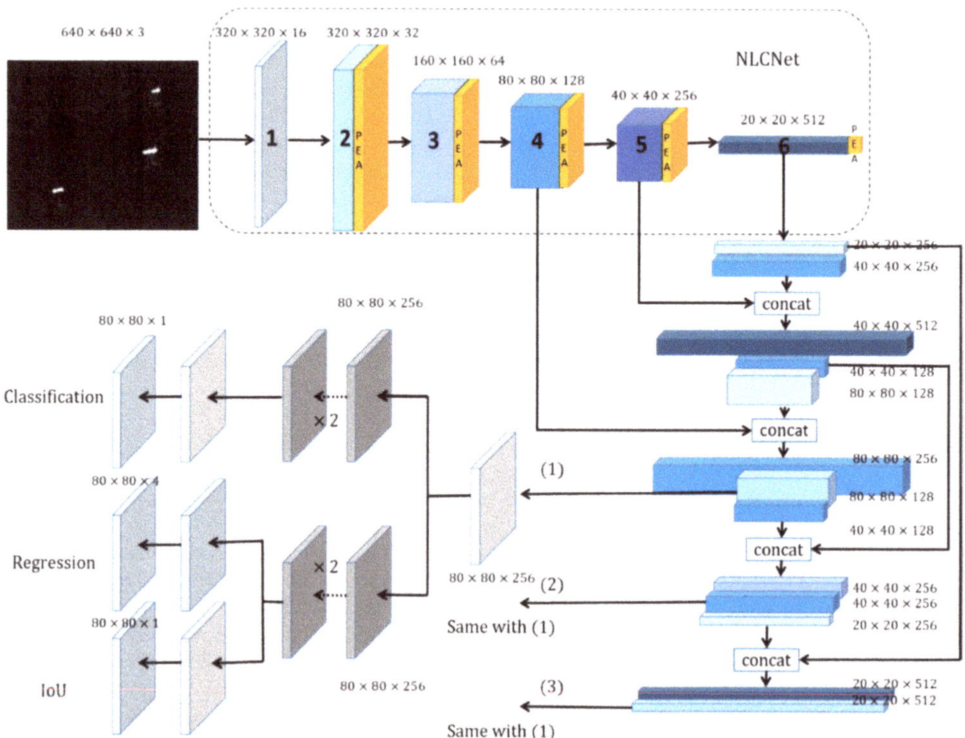

Figure 1. Overall framework of the model. PEA = position-enhanced attention. (The numbers (1–6) represent the output feature maps of blocks 1–6, respectively, and PEA is added to the adjacent blocks. The subsequent operations of (2), (3) are the same with (1)).

3.1. Benchmark Network

Considering the unclear edge information of SAR targets and avoiding the short-comings of traditional anchor-based methods, inspired by the latest anchor-free detection framework YOLOX [13], we use YOLOX as the benchmark network. YOLOX is the first to apply the anchor-free mode in the YOLO series. The specific operation is to explicitly define the 3×3 region of the truth frame projected to the center of the feature graph as the positive sample region and predict the four values of the target position (the offset distance of the upper left corner and the height and width of the frame). To better allocate fuzzy samples, YOLOX uses the simOTA algorithm for positive and negative sample matching. The general process of the simOTA algorithm is as follows: First, we calculated the matching degree of each pair. Then, we selected the top k prediction boxes with the smallest cost in a fixed central area. Finally, the grids associated with these positive samples were marked as positive.

Since YOLOX represents various improvements to the YOLO series, including a decoupling head, a new tag allocation strategy and an anchor-free mechanism, it is a high-performance detector subject to a trade-off between accuracy and speed. In the face of the SAR ship detection problem, these characteristics of YOLOX precisely match SAR image target sparsity, small sample characteristics and target scattering, so we chose YOLOX as the baseline of our network. Although YOLOX has achieved the performance of SOTA in optical image detection, its model size is too large and its model complexity is too high such that it cannot be applied in SAR image detection. Therefore, we redesigned the backbone network of YOLOX.

3.2. Lightweight Feature Enhancement Backbone: NLCNet

Most of the existing YOLO series backbones use DarkNet-53 and CSPNet architectures. Such backbones are usually excellent in terms of detection effect, but there is still a possibility for improvement of inference speed. The easiest way is to reduce the size of the model. To this end, according to the characteristics of the SAR target, the backbone network, namely, NLCnet, was designed to be lightweight so as to better balance speed and accuracy.

NLCNet uses the deeply separable convolution mentioned by MobileNetV1 as the basic block. It is generally known that depthwise separable convolution is mainly divided into two processes, namely, depthwise convolution and pointwise convolution. Compared with conventional convolution, the number of parameters for depthwise separable convolution is about one-third of that for conventional convolution. Therefore, given the same number of parameters, the number of neural network layers using separable convolution can be deeper. Based on the LCNet network, a new network design is carried out. We reorganized and stacked these blocks to form a new backbone network which is mainly divided into six blocks. The stem part uses standard convolution, which is activated by the h-swish function. Block2 to block6 all use depthwise separable convolution. The main difference is that the number of superimposed depthwise separable convolutions is different, and in block5 and block6 5×5 convolution kernels are used in the depth-level convolution process. The NLCNet network achieved the highest precision with respect to recent work in the following two areas: (1) discarding of the squeeze-and-excitation networks (SE) module and (2) design of the lightweight convolution block. The structural details of NLCNet are shown in Figure 2.

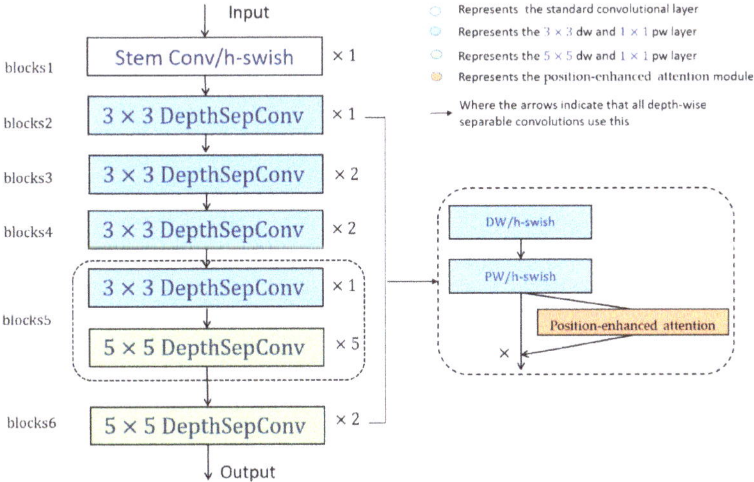

Figure 2. The details of the NLCNet backbone network.

3.2.1. Discarding of the Squeeze-and-Excitation Networks (SE) Module

The SE module [33] is widely used in many networks. It can help the model weight the channels in the network to obtain better features. However, we cannot blindly add the SE module to the model because not all SE modules will be more effective. Recently, through my own thinking and experiments, I found that the SE attention mechanism was added to the network, which resulted in a certain improvement in the classification task, but for target detection the effect is not obvious and sometimes it will affect the results, which may be similar to the network model. There is also a certain correlation. Considering this issue, we removed the SE module on the basis of LCnet in the experiments; the accuracy of the model was not reduced and the parameters of the model were relatively few.

3.2.2. Design of a Lightweight Convolution Block

Experiments showed that convolutional verification of different sizes would have a certain impact on network performance. The larger the convolution kernel, the larger the receptive field will be in the convolution process and the better it will be for constructing the global information of the target. In light of this, we chose to use a larger convolution kernel to balance speed and accuracy. It was found by YOLOX that placing the large convolution kernel at the tail of the network was the best choice because the performance achieved by these two methods was equivalent to replacing all layers of the network. Therefore, this substitution was only performed at the end of the network.

Through simple stacking and the use of corresponding technologies, the lightweight backbone used in this paper achieved a certain improvement in accuracy with respect to the SSDD dataset, while the number of parameters has also significantly decreased. Therefore, the advantages of NLCNet are obvious. The specific network structure is shown in Table 1.

Table 1. The details of NLCNet. PEA = position-enhanced attention.

Operator	Kernel Size	Stride	Input	Output	PEA
Conv2D	3×3	2	$640^2 \times 3$	$320^2 \times 16$	-
DepthSepConv	3×3	1	$320^2 \times 16$	$320^2 \times 32$	√
DepthSepConv	3×3	2	$320^2 \times 32$	$160^2 \times 64$	√
DepthSepConv	3×3	1	$160^2 \times 64$	$160^2 \times 64$	√
DepthSepConv	3×3	2	$160^2 \times 64$	$80^2 \times 128$	√
DepthSepConv	3×3	1	$80^2 \times 128$	$80^2 \times 128$	√
DepthSepConv	3×3	2	$80^2 \times 128$	$40^2 \times 256$	√
$5 \times$ DepthSepConv	5×5	1	$40^2 \times 256$	$40^2 \times 256$	√
DepthSepConv	5×5	2	$40^2 \times 256$	$20^2 \times 512$	√
DepthSepConv	5×5	1	$20^2 \times 512$	$20^2 \times 512$	√

3.3. Position-Enhanced Attention

Squeeze-and-excitation attention is a widely used attention mechanism that significantly enhances network performance and avoids many parameter calculations. Squeeze-and-excitation attention is widely used in various network models to highlight important channel information in features and is mainly used for the differential weighting of different channels through global pooling and a two-layer full connection layer without considering the influence of location information on features. Location information can further help the model to obtain target details in the image, thus improving model performance.

To highlight the key location information of features, we designed a new attention module in the network inspired by coordinate attention [17] called position-enhanced attention. It can embed the location information of the target in the image into the channel attention, which can better capture the interesting position information of the SAR target against a complex background and obtain a good global perception ability. At the same time, the computational cost of this process is relatively low. See Figure 3 for the position-enhanced attention architecture.

Since 2D global pooling does not contain location information, position-enhanced attention makes corresponding changes in 2D global pooling by splitting the original channel attention and forming two 1D global pooling operations. The specific process is that when the feature map is inputted, two 1D global pools are aggregated in a vertical and horizontal direction to form two independent feature maps with orientation awareness. The two generated feature maps with specific direction information are then encoded to form two attention maps. The two attention maps capture the independent and mutually dependent relationship of the input feature maps along a horizontal and vertical direction. From the above process, position information is obtained in the generated attention map and the two attention maps are applied to the input feature map, which can emphasize the target of interest in the image for better recognition.

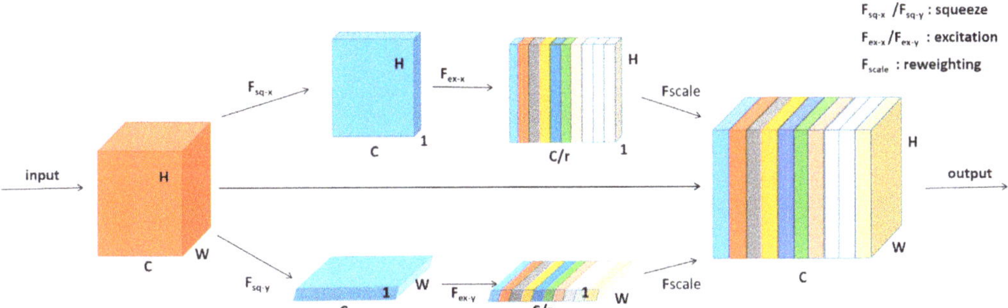

Figure 3. The details of the position-enhanced attention block (C, W, H, r represent the number of channels, width, height and reduction ratio, respectively).

For the accurate location information obtained, position-enhanced attention can be applied to coding channel relationships and remote dependencies. See Figure 4 for details of the position-enhanced attention architecture.

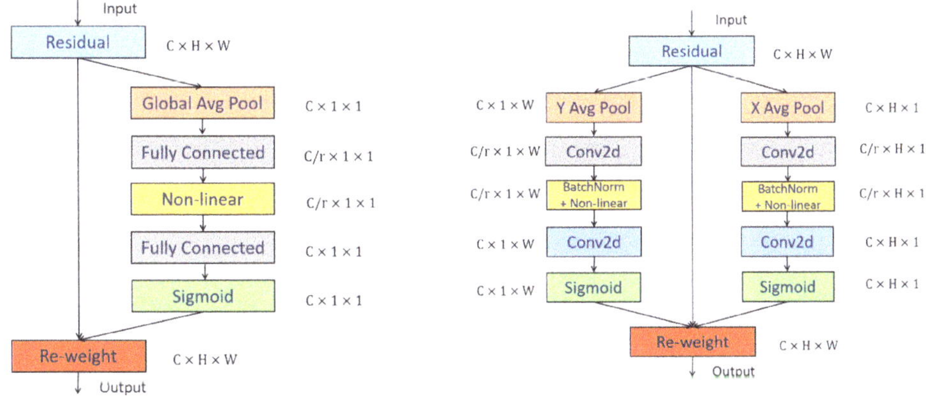

(**a**) Classic SE channel attention block   (**b**) Position-enhanced attention block

Figure 4. Structural contrast of the classic SE channel attention block and the position-enhanced attention block.

With channel attention, the spatial information in the image can usually establish the connection between channels through the global pooling operation, but it also causes the loss of position information, which is the result of the compression of the global information by the global pooling. In order to further utilize the location information of the target in the image, we split the 2D global pooling in the SE module to form two 1D global pooling operations. The 1D global pooling can extract the region of interest in the image in the horizontal and vertical directions so as to obtain better global perception ability and the two feature maps generated with specific directions save the position information of the target so the image target can be better identified and located. Specifically, given input X, two 1D global pooling operations are used to encode each channel in a horizontal and vertical direction and the size of the pooling kernel is (*H*, 1) or (1, *W*). Therefore, at height h, the output of channel c can be expressed as:

$$z_c^h(h) = \frac{1}{W} \sum_{0 \leq i \leq W} x_c(h, i) \tag{1}$$

At width w, the output of channel c can also be written as:

$$z_c^w(w) = \frac{1}{H} \sum_{0 \le j \le H} x_c(j, w) \tag{2}$$

Through the above transformation, we can aggregate the input features in two spatial directions and obtain two feature maps with directional perception characteristics. These two feature maps not only enable the corresponding attention module to save the remote dependency relationship between features but also to maintain accurate position information in the spatial direction, thereby helping the network to more accurately detect the target.

As mentioned above, through the extraction process of Equations (1) and (2), the attention branch channel can have a good global receptive field, can well retain global feature information and can encode precise location information.

Further, considering that the strong scattering characteristics of SAR targets against complex backgrounds cause their contours to be unclear and that the SAR target imaging angle changes greatly, we have carefully designed the follow-up attention processing flow. Previous studies have shown that 2D global pooling will lose position information. For this reason, Hou et al. [17] adopted two 1D pooling strategies and then performed channel concatenation. This method has difficulty handling the characteristics of SAR targets, mainly due to the following two problems: first, after the feature extraction and pooling operations of Equations (1) and (2), they are concatenated into a channel for subsequent processing because the feature correlation degrees of SAR targets in different spatial directions are very different, so this method loses the significant feature information of the two spatial directions, which is not conducive to characterizing the unique features of multi-oriented sparse SAR targets; second, this concatenation operation also increases the computational complexity of the channel.

To this end, we designed an attention strategy different from Hou et al.'s [17], namely, position-enhanced attention. Our starting point was to overcome the two problems of the above analysis, namely, directly designing two parallel branches to extract depth feature information in different spatial directions respectively. This operation can better extract salient feature information in two spatial directions and so can better characterize the characteristics of sparse SAR targets with different orientations; in addition, this parallel branch extraction can obtain a wider receptive field area so that better global awareness can be obtained.

Therefore, the aggregated feature maps in the two spatial directions were generated based on Equations (1) and (2). They respectively perform convolution operations along the spatial direction and the convolution function F is used for transformation, thereby generating:

$$f^h = \delta\left(Bn\left(F\left(z_c^h(h)\right)\right)\right) \tag{3}$$

$$f^w = \delta(Bn(F(z_c^w(w)))) \tag{4}$$

Among these:

$$h - swish(x) = x\frac{ReLU(x+3)}{6} \tag{5}$$

In the Equations (3) and (4), δ is the *h-swish* activation function and x is $Bn(F(\cdot))$. Bn is the batchnorm. f^h and f^w is the intermediate feature graph. f^h and f^w are transformed into tensors by using the other two 1×1 convolution transforms F_h and F_w.

$$g^h = \sigma\left(F_h\left(f^h\right)\right) \tag{6}$$

$$g^w = \sigma(F_w(f^w)) \tag{7}$$

where σ is the sigmoid function. Then, g^h and g^w are used in the position-enhanced attention block:

$$y_c(i,j) = x_c(i,j) \times g_c^h(i) \times g_c^w(j) \tag{8}$$

Position-enhanced attention considers the encoding of spatial information. As mentioned above, attention along both horizontal and vertical directions applies to the input tensor. This coding process allows position-enhanced attention to more accurately locate the target position in the image, thus helping the whole model to achieve better recognition. Experiments show that our method does achieve good results.

4. Experiments

To verify the proposed method, we conducted a series of related experiments to evaluate the model's detection performance. The content of this section includes details of some settings in the experiment and the main content of the SSDD dataset, followed by the evaluation indicators used in the experimental results, the influence of each module proposed in the ablation experiment on the model and a comparison with other target detection algorithms. Finally, LPEDet is compared with other recent SAR imaging methods.

4.1. Dataset and Experimental Settings

In our experiment, the datasets used were SSDD [34] and HRSID. For each ship, the detection algorithm predicts the frame of the ship target and gives the confidence of the ship target. The SSDD process is based on the PASCALVOC dataset and its data format is algorithmically compatible, making it easier to use with fewer code changes.

SSDD data are obtained by downloading public SAR images from the internet. Figure 5 shows part of the images in the dataset. The target area was cropped to approximately 500×500 pixels and the ship target location was manually marked. As long as there is a ship in the dataset, there are no requirements regarding ship type. The data in this dataset mainly include HH, HV, VV and VH polarization modes. There are 1160 images in the dataset and each image contains 2456 ships of different numbers and sizes. Although SSDD has few pictures, for the detection network, the number of targets that only recognize ships is sufficient. The corresponding relationship between the number of pictures and the number of ships in the dataset is shown in Table 2.

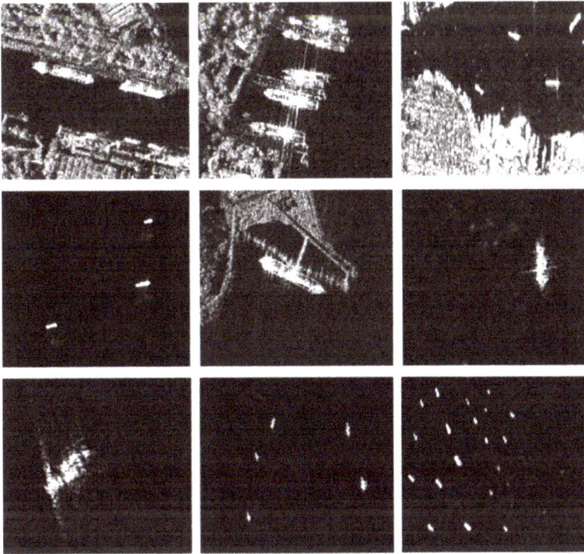

Figure 5. Illustration of the diversity of ship targets in the SSDD dataset.

Table 2. Correspondence between NoS and NoI in the SSDD dataset.

NoS	1	2	3	4	5	6	7	8	9	10	11	12	13	14
NoI	725	183	89	47	45	16	15	8	4	11	5	3	3	0

NoS = number of ships; NoI = number of images.

In addition, to verify the detection performance of our proposed method in different scenarios, we introduced another large-scale SAR target detection dataset, namely, the HRSID dataset. The images in this dataset are high-resolution SAR images, which are mainly used for ship detection, semantic segmentation and instance segmentation tasks. The dataset contains a total of 5604 high-resolution SAR images and 16,951 ship instances. The HRSID dataset borrows from the construction process of the Microsoft Common Objects in Context (COCO) dataset, including SAR images with different resolutions, polarizations, sea states, sea areas and coastal ports. For HRSID, the resolutions of the SAR images are: 0.5 m, 1 m and 3 m, respectively.

To make a fair comparison with previous work, we attempted to use the same settings that previous workers used. We randomly divided the original SSDD dataset according to the ratio of 8:2 commonly used in existing studies and 80% of the datasets were used for the training of all methods. The remaining 20% was used as a test set to evaluate the detection performance of all methods. The data in the training set and test set were not repeated at all among the methods to ensure the rigor and fairness of the experiment. Other parameters included a batch size of 8 and an image size for the input model of 640, RandomHorizontalFlip was adopted, ColorJitter and multiscale were used for data augmentation, and Mosaic and MixUp enhancement strategies were employed. Using the $lr \times batchsize/64$ learning rate, the cosine lr schedule and initial lr = 0.01 were employed. The weight decayed to 0.0005 and the SGD momentum was 0.9. A total of 600 epochs were trained. In the HRSID [35] dataset, we used a ratio of 6.5:3.5 to split the dataset, with 65% data for training and 35% for testing, the same as the original author split. The image size of the input model was 800. All experiments in this paper were carried out on an Ubuntu 18.04 operating system equipped with a GeForceRTX2060.

4.2. Evaluation Indicators

We used average precision (mAP) to analyze and verify the detection performance of our proposed method. Average accuracy can be derived from accuracy and recall.

Accuracy is the percentage of targets that are correctly identified in the test set. The percentage is defined by true positives (TPs) and false positives (FPs):

$$P = \frac{TP}{TP + FP} \tag{9}$$

TP means that the prediction of the classifier is positive and the prediction is correct; FP indicates that the prediction of the classifier is positive and the prediction is incorrect.

The recall rate is the probability that all positive samples in the test set are correctly identified, which is derived from true positives (TPs) and false negatives (FNs):

$$R = \frac{TP}{TP + FN} \tag{10}$$

FN indicates that the prediction of the classifier is negative and the prediction is incorrect.

Based on the accuracy and recall rate, an average accuracy value is also obtained. The graphical meaning can be clearly seen in the coordinate axis, that is, the area under the accuracy and recall rate curve, which is defined as follows:

$$mAP = \int_0^1 P(R)dR \tag{11}$$

4.3. Experimental Results and Analysis

4.3.1. Ablation Experiments on SSDD Datasets

To clearly compare the advantages of the added modules, we conducted the following ablation experiments. The first experiment ensured that other settings remained unchanged while replacing the backbone network Darknet-53 with a lightweight backbone NLCNet. Second, the attention module position-enhanced attention was added on the basis of the original network. This process did not change other settings and parameters.

It should be noted that the methods in the ablation experiment were reproduced according to the official open-source code of the comparison method and applied to the SSDD dataset for experimental comparison. The dataset used by the comparison method was exactly the same as that used by our proposed method; the hyperparameters of the comparison method were all set with standard default settings and the number of training epochs was also consistent with our method.

Influence of the NLCNet Backbone Network on the Experimental Results

The backbone Darknet-53 was replaced with our proposed NLCNet based on YOLOX, as previously shown in Figure 2. The mAP increased by 0.6% from 96.2% to 96.8% and the FLOPs dropped by 8.37 from 26.64 to 18.27. According to the data, our redesigned NLCNet showed advantages in feature extraction with respect to SAR image ship targets, not only improving accuracy but also reducing the number of parameters, making the model more lightweight and easier to transplant in industrial settings.

Influence of Position-Enhanced Attention on Experimental Results

To verify the effectiveness of our proposed position-enhanced attention and its advantages, we conducted ablation experiments with the original network without attention, the network with coordinate attention and the new network with our proposed position-enhanced attention in our dataset, respectively. The experimental results are shown in Table 3. The mAP of the network with our proposed position-enhanced attention was greatly improved compared to the network without attention, which increased from 96.8% to 97.4%. At the same time, the increase in FLOPs and params was negligible. The results show the effectiveness of our proposed position-enhanced attention. Compared with the original network with coordinate attention, the detection accuracy of our proposed model increased from 97.1% to 97.4% with the parameters and FLOPs unchanged. It should be noted that we kept two significant digits after the decimal point when we counted the experimental results. Therefore, it was calculated that the parameters of FLOPs and params of our position-enhanced attention model and the original coordinate attention model were the same size. Thus, the advantages of our designed positional attention are demonstrated by the results. The visualization results are shown in Figure 6.

Table 3. Results of ablation experiments. (mAP, FLOPs, Params and average inference time represent detection accuracy, computational complexity, parameter amount and average inference time, respectively).

Model	Backbone	Attention	mAP (%)	FLOPs (GMac)	Params (M)	Average Inference Time (ms)
YOLOX	Darknet-53	-	96.2	26.64	8.94	8.20
YOLOX	NLCNet	-	96.8	18.27	5.58	5.27
YOLOX	NLCNet	Coordinate attention	97.1	18.38	5.68	7.01
YOLOX	NLCNet	Position-enhanced attention	97.4	18.38	5.68	7.01

4.3.2. Comparison with the Latest Target Detection Methods Using SSDD Datasets

To further demonstrate the validity of our work, we compared LPEDet with some of the latest target detection methods, including the one-stage RetinaNet, SSD300, YOLOv3, YOLOX, YOLOv5, AFSar, two-stage Faster R-CNN, Cascade R-CNN, FPN and anchor-free CornerNet, CenterNet and FCOS methods. Among them, considering that our model mainly focuses on the lightweight design of the backbone network, for a fair comparison of performance, we cite the results of the backbone ablation experiments of AFSar [14]. The results are shown in Table 4. As seen from the table, our work not only outperformed other methods in terms of precision but also in terms of speed.

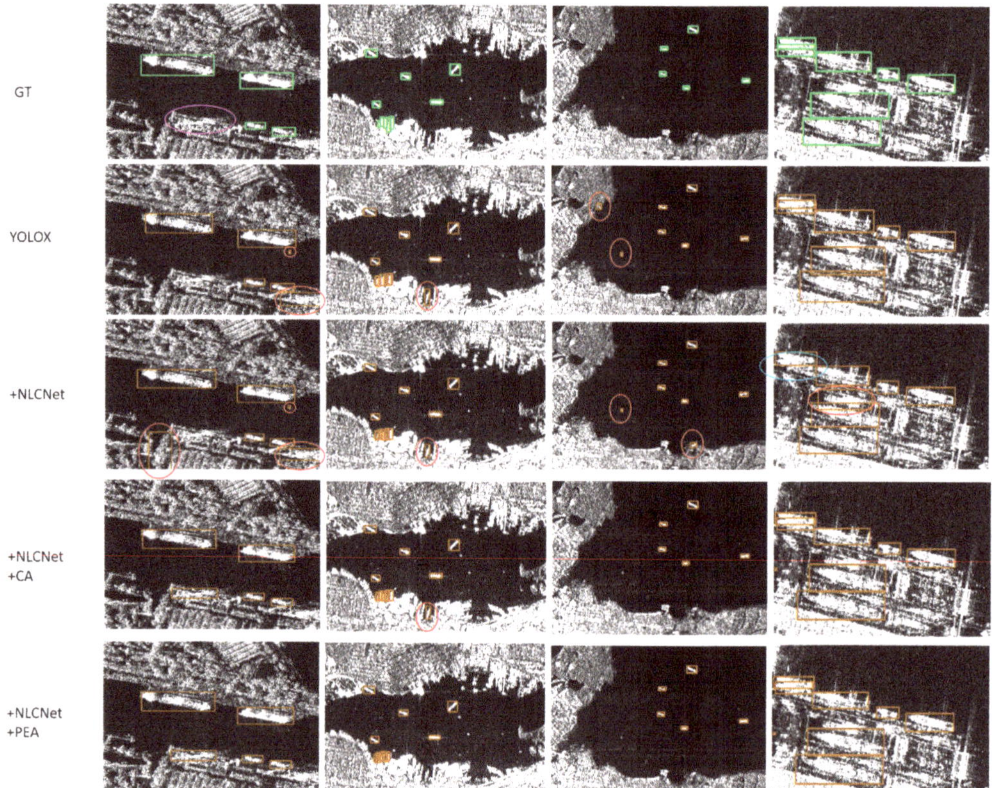

Figure 6. Visualization effect of the ablation experiment. The purple box is the target that was missed when marked, the red box is the target of misdetection and the blue box is a missed target. CA = coordinate attention; PEA = position-enhanced attention.

Table 4. Comparison with the latest target detection methods.

	Method	mAP (%)	FLOPs (GMac)	Params (M)	Average Inference Time (ms)
One-stage	RetinaNet [10]	91.2	81.69	36.1	44.10
	SSD300 [11]	93.1	154.45	34.31	53.62
	YOLOv3 [12]	96.2	77.54	61.52	45.81
	YOLOX	96.2	26.64	8.94	8.20
	YOLOv5	97.0	16.54	7.23	8.61
	AFSar [14]	96.7	8.66	-	-
Two-stage	Faster R-CNN [36]	96.4	91.41	41.12	45.36
	Cascade R-CNN [37]	96.8	119.05	69.17	65.56
	FPN [38]	96.5	71.65	63.56	78.30
Anchor-free	CornerNet [39]	94.7	707.75	201.04	95.61
	CenterNet [40]	95.1	20.4	14.21	31.54
	FCOS [41]	95.3	78.63	60.97	48.67
	LPEDet	97.4	18.38	5.68	7.01

It should be noted that, except for AFSar, the methods in the comparison experiments were reproduced according to the official open-source code of the comparison method and applied to the SSDD dataset for experimental comparison. The dataset used by the comparison method was exactly the same as that used by our proposed method; the hyperparameters of the comparison method were all set with standard default settings and the number of training epochs is also consistent with our method.

We also visualized the detection results of these methods. As shown in Figure 7, compared with the proposed LPEDet method, the detection rate of the above methods was significantly higher than that of the proposed LPEDet method. Our method has good performance with small target images, complex backgrounds and intensive target image detection. These findings show the effectiveness of our approach.

4.4. Comparison with the Latest SAR Ship Detection Methods Using SSDD Datasets

To further verify the performance of our method, we also compared it with the latest SAR ship detection methods, as shown in Table 5.

Table 5. Comparison with the latest SAR ship detection methods.

Methods	mAP (%)	FLOPs (GMac)	Params (M)	Average Inference Time (ms)
DCMSNN [30]	89.43	-	41.1	46.2
NNAM [31]	79.8	-	-	28
FBR-Net [42]	94.10	-	32.5	40.1
CenterNet++ [43]	95.1	-	-	33
EFGRNet [44]	91.1	-	-	33
Libra R-CNN [45]	88.7	-	-	57
DAPN [32]	89.8	-	-	41
LPEDet	97.4	18.38	5.68	7.01

The comparison methods and related experimental results in Table 5 need special explanation. Since none of these comparison methods have open-source original codes, it is difficult for us to completely reproduce the codes and parameter settings of the comparison methods. Therefore, to fairly compare the performance of different detection methods, we directly cite the highest detection results reported in the original reference of the comparison methods. Especially for the two indicators of FLOPs and params, most of the comparison methods have no relevant results. Therefore, we only cite the best experimental results for other indicators published in the references. In addition, the results of the comparison methods listed in Table 5 are mainly from references [42,43].

The results show that not only does our LPEDet achieve SOTA in accuracy but it also has a relatively faster inference speed, which shows the high efficiency of our method.

4.5. Comparison with the Latest SAR Ship Detection Methods Using HRSID Datasets

In order to fully verify the performance stability of the LPEDet method proposed in this paper with respect to different datasets, we introduce a new large-scale SAR target detection dataset, namely, HRSID, and compare a variety of state-of-the-art SAR target detection methods using this dataset. The specific results are shown in Table 6, below. By comparing the experimental results, it was found that, compared with the current latest SAR ship target detection methods, the LPEDet method proposed in this paper is superior in terms of accuracy, while the parametric and computational complexity of the model are also the lowest, which proves the stability of our method in relation to different datasets. As can be seen in Table 6, except for the fact that the data for CenterNet2 on AP, AP_{75}, AP_M and AP_L are slightly higher than for our model, the difference is not big, and the number of parameters in our model is almost 1/15 the number of its parameters, such that, considering accuracy and speed, our model still has better performance in comparison.

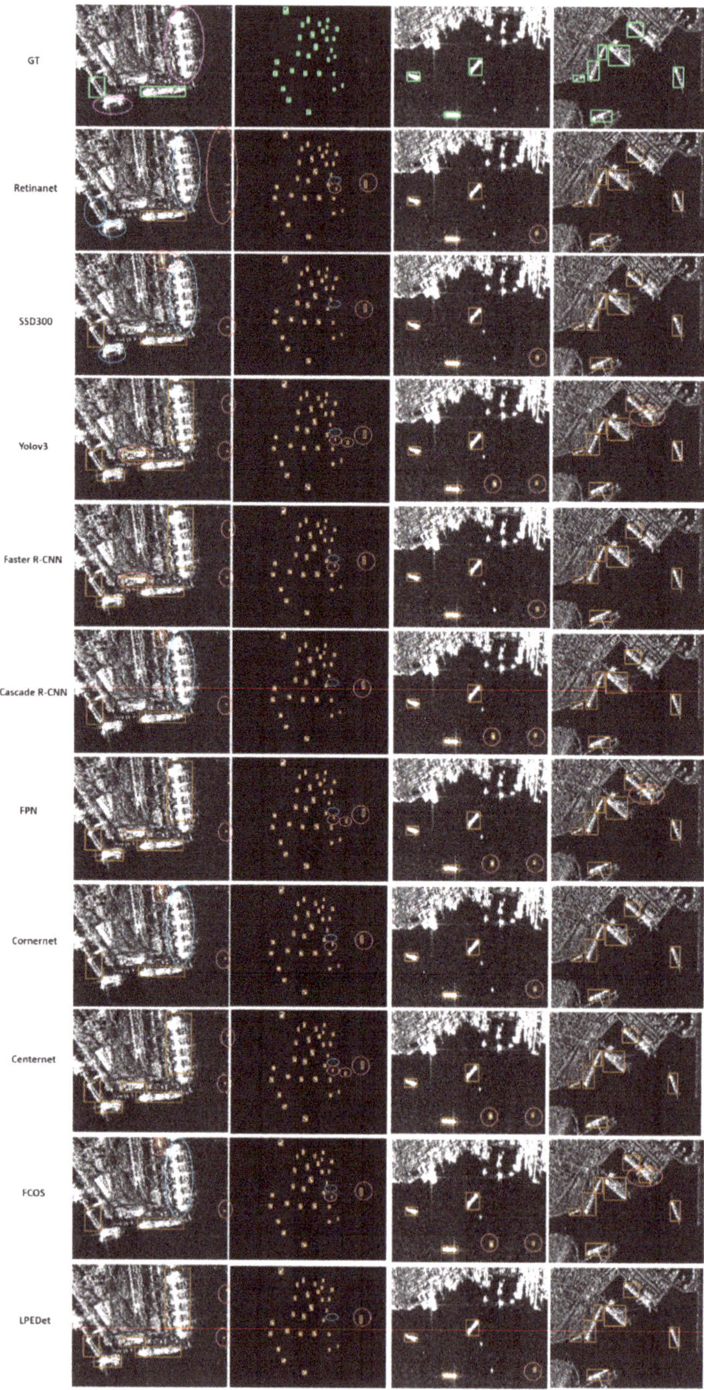

Figure 7. Visual detection results of the latest methods. The purple box is the target that was missed when marked, the red box is the target of misdetection and the blue box is a missed target.

Table 6. Comparison of the latest SAR target detection methods on HRSID.

Method	AP	AP$_{50}$	AP$_{75}$	AP$_S$	AP$_M$	AP$_L$	FLOPs (GMac)	Params (M)	Average Inference Time (ms)
YOLOv3 [12]	50.9	85.0	53.1	51.0	56.1	25.5	121.15	61.52	136
SAR-net [46]	-	84.7	-	-	-	-	104.2	42.6	118
CenterNet2 [47]	64.5	89.5	73.0	64.7	69.1	48.3	-	71.6	-
RetinaNet [10]	59.8	84.8	67.2	60.4	62.7	26.5	127.91	36.3	122
YOLOX [13]	61.4	87.2	68.9	63.0	57.0	21.8	26.64	8.94	8.20
LPEDet	64.4	89.7	71.8	65.8	63.4	24.2	18.38	5.68	7.01

4.6. The Effect of the Number of Training Sets on Detection Performance

In order to verify the robust performance of our proposed method under the conditions of different training data, we redivided the training and testing ratios of the dataset under the conditions of 33% and 66% of the training set data, respectively, to validate the performance of our model. The results are shown in Table 7 below. We combined the results of Table 4 in the paper for the analysis (the training data volume of all methods in Table 4 is 80% and above): when we only use 66% of the training data volume, mAP can still reaching 96.8%, the performance is still better than most state-of-the-art SAR target detection methods; and when we use only 33% of the training data volume, mAP can still reach 94.6%, outperforming RetinaNet and SSD300, such that, compared to other state-of-the-art SAR target detection methods, the results are not much different. The analysis of the above results shows that the proposed LPEDet method can still achieve better performance than the latest SAR target detection methods with fewer training data, that it still has good robustness and that it can greatly reduce the labor costs involved in manual labeling of data.

Table 7. Comparison of experimental results for different data volumes on SSDD.

Method	Datasets Rate	mAP (%)
LPEDet	33%	94.6
	66%	96.8
	80%	97.4

5. Conclusions

Multi-platform SAR earth observation equipment has accumulated massive amounts of high-resolution SAR target image data, and SAR image target detection has great engineering application value in military/civilian fields. Aimed at the problems of unclear target contour information, complex backgrounds, strong scattering and multiple scales in SAR images, a new anchor-free SAR ship detection algorithm, LPEDet, was proposed to improve the accuracy and speed of SAR ship detection in a balanced manner. First, YOLOX was used as the benchmark detection network; then, a new lightweight backbone, NLCNet, was designed. At the same time, to further improve localization accuracy, we designed a location-enhanced attention strategy. The experimental results based on the SSDD dataset showed that the mAP of our LPEDet reached 97.4%, achieving SOTA. Meanwhile, the average inference time for a single image is only 7.01ms when the input size is 640. With respect to the HRSID dataset, our model is also stable, with an AP$_{50}$ of 89.7%, which is superior to other state-of-the-art object detection methods, while the computational complexity, the number of parameters and the average inference time are lowest. In the future, based on the Hisea-1 SAR satellite that our research group participated in launching, our group independently constructed a larger-scale multi-type SAR target image dataset. We will verify the effectiveness of our proposed LPEDet algorithm on this large-scale dataset. Common SAR image artifacts such as speckle noise can affect SAR target detection results, and Mukherjee et al. [48] has demonstrated that their methods can respond to various types of image artifacts. Therefore, in the future, we will consider introducing image quality metrics to evaluate and correct the quality of the input SAR images so as to more comprehensively iterate and verify the robust performance of our designed SAR target detection method.

Author Contributions: Conceptualization, Y.F.; Methodology, Y.F.; Project administration, J.C. and Z.H.; Software, Y.F.; Supervision, J.C., Z.H., B.W., R.X., L.S. and M.X.; Validation, Y.F.; Visualization, Y.F.; Writing—original draft, Y.F.; Writing—review & editing, J.C. and H.W. All authors have read and agreed to the published version of the manuscript.

Funding: This work was supported in part by the National Natural Science Foundation of China under Grant 62001003, in part by the Natural Science Foundation of Anhui Province under Grant 2008085QF284 and in part by the China Postdoctoral Science Foundation under Grant 2020M671851.

Conflicts of Interest: The authors declare no conflict of interest.

References

1. Robey, F.C.; Fuhrmann, D.R.; Kelly, E.J. A CFAR adaptive matched filter detector. *IEEE Trans. Aerosp. Electron. Syst.* **1992**, *28*, 208–216. [CrossRef]
2. Huang, Y.; Liu, F. Detecting cars in VHR SAR images via semantic CFAR algorithm. *IEEE Geosci. Remote Sens. Lett.* **2016**, *13*, 801–805. [CrossRef]
3. Stein, G.W.; Zelnio, E.G.; Garber, F.D. Target detection using an improved fractal scheme. *Proc. SPIE—Int. Soc. Opt. Eng.* **2006**, *6237*, 19.
4. Zhang, T.; Zhang, X.; Shi, J.; Wei, S.; Wang, J.; Li, J.; Su, H.; Zhou, Y. Balance scene learning mechanism for offshore and inshore ship detection in SAR images. *IEEE Geosci. Remote Sens. Lett.* **2020**, *19*, 1–5. [CrossRef]
5. Zhang, T.; Zhang, X. ShipDeNet-20: An only 20 convolution layers and <1-MB lightweight SAR ship detector. *IEEE Geosci. Remote Sens. Lett.* **2020**, *18*, 1234–1238. [CrossRef]
6. Zhang, T.; Zhang, X.; Shi, J.; Wei, S. HyperLi-Net: A hyper-light deep learning network for high-accurate and high-speed ship detection from synthetic aperture radar imagery. *ISPRS J. Photogramm. Remote Sens.* **2020**, *167*, 123–153. [CrossRef]
7. Tz, A.; Xz, A.; Chang, L.B.; Js, A.; Sw, A.; Ia, C.; Xu, Z.A.; Yue, Z.D.; Dp, E.; Jl, F. Balance learning for ship detection from synthetic aperture radar remote sensing imagery. *ISPRS J. Photogramm. Remote Sens.* **2021**, *182*, 190–207.
8. Zhang, T.; Zhang, X.; Shi, J.; Wei, S. Depthwise separable convolution neural network for high-speed SAR ship detection. *Remote Sens.* **2019**, *11*, 2483. [CrossRef]
9. Zhang, T.; Zhang, X. High-speed ship detection in SAR images based on a grid convolutional neural network. *Remote Sens.* **2019**, *11*, 1206. [CrossRef]
10. Lin, T.-Y.; Goyal, P.; Girshick, R.; He, K.; Dollár, P. Focal loss for dense object detection. *IEEE Trans. Pattern Anal. Mach. Intell.* **2018**, *42*, 318–327. [CrossRef]
11. Liu, W.; Anguelov, D.; Erhan, D.; Szegedy, C.; Reed, S.; Fu, C.-Y.; Berg, A.C. Ssd: Single shot multibox detector. In *European Conference on Computer Vision*; Springer International Publishing: Cham, Switzerland, 2016; pp. 21–37.
12. Redmon, J.; Farhadi, A. Yolov3: An incremental improvement. *arXiv* **2018**, arXiv:1804.02767.
13. Ge, Z.; Liu, S.; Wang, F.; Li, Z.; Sun, J. Yolox: Exceeding yolo series in 2021. *arXiv* **2021**, arXiv:2107.08430.
14. Wan, H.Y.; Chen, J.; Huang, Z.X.; Xia, R.F.; Wu, B.C.; Sun, L.; Yao, B.D.; Liu, X.P.; Xing, M.D. AFSar: An anchor-free SAR target detection algorithm based on multiscale enhancement representation learning. *IEEE Trans. Geosci. Remote Sens.* **2021**, *60*, 1–14. [CrossRef]
15. Liu, S.; Huang, D. Receptive field block net for accurate and fast object detection. In Proceedings of the European Conference on Computer Vision (ECCV), Munich, Germany, 14 September 2018; pp. 385–400.
16. Cui, C.; Gao, T.; Wei, S.; Du, Y.; Guo, R.; Dong, S.; Lu, B.; Zhou, Y.; Lv, X.; Liu, Q. PP-LCNet: A lightweight CPU convolutional neural network. *arXiv* **2021**, arXiv:2109.15099.
17. Hou, Q.; Zhou, D.; Feng, J. Coordinate attention for efficient mobile network design. In Proceedings of the IEEE/CVF Conference on Computer Vision and Pattern Recognition, Nashville, TN, USA, 19–25 June 2021; pp. 13713–13722.
18. Ai, J.; Yang, X.; Yan, H. A local Cfar detector based on gray Intensity correlation in Sar imagery. In Proceedings of the IGARSS 2018—2018 IEEE International Geoscience and Remote Sensing Symposium, Valencia, Spain, 22–27 July 2018; pp. 697–700.
19. Kaplan, L.M. Improved SAR target detection via extended fractal features. *IEEE Trans. Aerosp. Electron. Syst.* **2001**, *37*, 436–451. [CrossRef]
20. Charalampidis, D.; Kasparis, T. Wavelet-based rotational invariant roughness features for texture classification and segmentation. *IEEE Trans. Image Process.* **2002**, *11*, 825–837. [CrossRef]
21. Simonyan, K.; Zisserman, A. Very deep convolutional networks for large-scale image recognition. *arXiv* **2014**, arXiv:1409.1556.

22. Szegedy, C.; Liu, W.; Jia, Y.; Sermanet, P.; Reed, S.; Anguelov, D.; Erhan, D.; Vanhoucke, V.; Rabinovich, A. Going deeper with convolutions. In Proceedings of the IEEE Conference on Computer Vision and Pattern Recognition, Boston, MA, USA, 7–12 June 2015; pp. 1–9.

23. Zhang, X.; Zhou, X.; Lin, M.; Sun, J. Shufflenet: An extremely efficient convolutional neural network for mobile devices. In Proceedings of the IEEE Conference on Computer Vision and Pattern Recognition, Salt Lake City, UT, USA, 18–22 June 2018; pp. 6848–6856.

24. Ma, N.; Zhang, X.; Zheng, H.-T.; Sun, J. Shufflenet v2: Practical guidelines for efficient cnn architecture design. In Proceedings of the IEEE Conference on Computer Vision and Pattern Recognition, Salt Lake City, UT, USA, 18–22 June 2018; pp. 116–131.

25. Han, K.; Wang, Y.; Tian, Q.; Guo, J.; Xu, C.; Xu, C. Ghostnet: More features from cheap operations. In Proceedings of the IEEE/CVF Conference on Computer Vision and Pattern Recognition, Seattle, WA, USA, 14–19 June 2020; pp. 1580–1589.

26. Howard, A.G.; Zhu, M.; Chen, B.; Kalenichenko, D.; Wang, W.; Weyand, T.; Andreetto, M.; Adam, H. Mobilenets: Efficient convolutional neural networks for mobile vision applications. *arXiv* **2017**, arXiv:1704.04861.

27. Sandler, M.; Howard, A.; Zhu, M.; Zhmoginov, A.; Chen, L.-C. Mobilenetv2: Inverted residuals and linear bottlenecks. In Proceedings of the IEEE Conference on Computer Vision and Pattern Recognition, Salt Lake City, UT, USA, 18–22 June 2018; pp. 4510–4520.

28. Sun, L.; Chen, J.; Feng, D.; Xing, M. The recognition framework of deep kernel learning for enclosed remote sensing objects. *IEEE Access* **2021**, *9*, 95585–95596. [CrossRef]

29. Sun, L.; Chen, J.; Feng, D.; Xing, M. Parallel ensemble deep learning for real-time remote sensing video multi-Target detection. *Remote Sens.* **2021**, *13*, 4377. [CrossRef]

30. Jiao, J.; Zhang, Y.; Sun, H.; Yang, X.; Gao, X.; Hong, W.; Fu, K.; Sun, X. A densely connected end-to-end neural network for multiscale and multiscene SAR ship detection. *IEEE Access* **2018**, *6*, 20881–20892. [CrossRef]

31. Chen, C.; He, C.; Hu, C.; Pei, H.; Jiao, L. A deep neural network based on an attention mechanism for SAR ship detection in multiscale and complex scenarios. *IEEE Access* **2019**, *7*, 104848–104863. [CrossRef]

32. Cui, Z.; Li, Q.; Cao, Z.; Liu, N. Dense attention pyramid networks for multi-scale ship detection in SAR images. *IEEE Trans. Geosci. Remote Sens.* **2019**, *57*, 8983–8997. [CrossRef]

33. Hu, J.; Shen, L.; Sun, G. Squeeze-and-excitation networks. In Proceedings of the IEEE Conference on Computer Vision and Pattern Recognition, Salt Lake City, UT, USA, 18–22 June 2018; pp. 7132–7141.

34. Zhang, T.; Zhang, X.; Li, J.; Xu, X.; Wang, B.; Zhan, X.; Xu, Y.; Ke, X.; Zeng, T.; Su, H. Sar ship detection dataset (ssdd): Official release and comprehensive data analysis. *Remote Sens.* **2021**, *13*, 3690. [CrossRef]

35. Wei, S.; Zeng, X.; Qu, Q.; Wang, M.; Su, H.; Shi, J. HRSID: A high-resolution SAR images dataset for ship detection and instance segmentation. *IEEE Access* **2020**, *8*, 120234–120254. [CrossRef]

36. Ren, S.; He, K.; Girshick, R.; Sun, J. Faster r-cnn: Towards real-time object detection with region proposal networks. *Adv. Neural Inf. Process. Syst.* **2016**, *39*, 1137–1149. [CrossRef]

37. Cai, Z.; Vasconcelos, N. Cascade r-cnn: Delving into high quality object detection. In Proceedings of the IEEE Conference on Computer Vision and Pattern Recognition, Salt Lake City, UT, USA, 18–22 June 2018; pp. 6154–6162.

38. Lin, T.-Y.; Dollár, P.; Girshick, R.; He, K.; Hariharan, B.; Belongie, S. Feature pyramid networks for object detection. In Proceedings of the IEEE Conference on Computer Vision and Pattern Recognition, Honolulu, HI, USA, 21–26 July 2017; pp. 2117–2125.

39. Law, H.; Deng, J. Cornernet: Detecting objects as paired keypoints. In Proceedings of the European Conference on Computer Vision (ECCV), Munich, Germany, 8–14 September 2018; pp. 734–750.

40. Duan, K.; Bai, S.; Xie, L.; Qi, H.; Huang, Q.; Tian, Q. Centernet: Keypoint triplets for object detection. In Proceedings of the IEEE/CVF International Conference on Computer Vision, Seoul, Korea, 27–28 October 2019; pp. 6569–6578.

41. Tian, Z.; Shen, C.; Chen, H.; He, T. Fcos: Fully convolutional one-stage object detection. In Proceedings of the IEEE/CVF International Conference on Computer Vision, Seoul, Korea, 27–28 October 2019; pp. 9627–9636.

42. Fu, J.; Sun, X.; Wang, Z.; Fu, K. An anchor-free method based on feature balancing and refinement network for multiscale ship detection in SAR images. *IEEE Trans. Geosci. Remote Sens.* **2020**, *59*, 1331–1344. [CrossRef]

43. Guo, H.; Yang, X.; Wang, N.; Gao, X. A CenterNet++ model for ship detection in SAR images. *Pattern Recognit.* **2021**, *112*, 107787. [CrossRef]

44. Nie, J.; Anwer, R.M.; Cholakkal, H.; Khan, F.S.; Pang, Y.; Shao, L. Enriched feature guided refinement network for object detection. In Proceedings of the IEEE/CVF International Conference on Computer Vision, Seoul, Korea, 27–28 October 2019; pp. 9537–9546.

45. Pang, J.; Chen, K.; Shi, J.; Feng, H.; Ouyang, W.; Lin, D. Libra r-cnn: Towards balanced learning for object detection. In Proceedings of the IEEE/CVF Conference on Computer Vision and Pattern Recognition, Long Beach, CA, USA, 16–20 June 2019; pp. 821–830.

46. Gao, S.; Liu, J.; Miao, Y.; He, Z. A High-Effective Implementation of Ship Detector for SAR Images. *IEEE Geosci. Remote Sens. Lett.* **2021**, *19*, 1–5. [CrossRef]

47. Zhou, X.; Koltun, V.; Krähenbühl, P. Probabilistic two-stage detection. *arXiv* **2021**, arXiv:2103.07461.

48. Mukherjee, S.; Valenzise, G.; Cheng, I. Potential of deep features for opinion-unaware, distortion-unaware, no-reference image quality assessment. In *International Conference on Smart Multimedia*; Springer International Publishing: Cham, Switzerland, 2019; pp. 87–95.

Article

A Low-Grade Road Extraction Method Using SDG-DenseNet Based on the Fusion of Optical and SAR Images at Decision Level

Jinglin Zhang [1], Yuxia Li [1,*], Yu Si [1], Bo Peng [1], Fanghong Xiao [1], Shiyu Luo [1] and Lei He [2]

1 School of Automation Engineering, University of Electronic Science and Technology of China, Chengdu 611731, China; 2018060507021@std.uestc.edu.cn (J.Z.); 201921060524@std.uestc.edu.cn (Y.S.); 2018060507023@std.uestc.edu.cn (B.P.); xiao_fanghong@std.uestc.edu.cn (F.X.); shiyuluo@uestc.edu.cn (S.L.)
2 School of Software Engineering, Chengdu University of Information Technology, Chengdu 610225, China; helei1978@cuit.edu.cn
* Correspondence: liyuxia@uestc.edu.cn

Abstract: Low-grade roads have complex features such as geometry, reflection spectrum, and spatial topology in remotely sensing optical images due to the different materials of those roads and also because they are easily obscured by vegetation or buildings, which leads to the low accuracy of low-grade road extraction from remote sensing images. To address this problem, this paper proposes a novel deep learning network referred to as SDG-DenseNet as well as a fusion method of optical and Synthetic Aperture Radar (SAR) data on decision level to extract low-grade roads. On one hand, in order to enlarge the receptive field and ensemble multi-scale features in commonly used deep learning networks, we develop SDG-DenseNet in terms of three modules: stem block, D-Dense block, and GIRM module, in which the Stem block applies two consecutive small-sized convolution kernels instead of the large-sized convolution kernel, the D-Dense block applies three consecutive dilated convolutions after the initial Dense block, and Global Information Recovery Module (GIRM) combines the ideas of dilated convolution and attention mechanism. On the other hand, considering the penetrating capacity and oblique observation of SAR, which can obtain information from those low-grade roads obscured by vegetation or buildings in optical images, we integrate the extracted road result from SAR images into that from optical images at decision level to enhance the extraction accuracy. The experimental result shows that the proposed SDG-DenseNet attains higher *IoU* and *F1* scores than other network models applied to low-grade road extraction from optical images. Furthermore, it verifies that the decision-level fusion of road binary maps from SAR and optical images can further significantly improve the *F1*, *COR*, and *COM* scores.

Keywords: low-grade road extraction; remote sensing; image segmentation; SAR image; deep learning

Citation: Zhang, J.; Li, Y.; Si, Y.; Peng, B.; Xiao, F.; Luo, S.; He, L. A Low-Grade Road Extraction Method Using SDG-DenseNet Based on the Fusion of Optical and SAR Images at Decision Level. *Remote Sens.* 2022, 14, 2870. https://doi.org/10.3390/rs14122870

Academic Editor: Giuseppe Scarpa

Received: 5 May 2022
Accepted: 14 June 2022
Published: 15 June 2022

1. Introduction

Research on extracting road information from remote sensing images has been carried out for many years. However, due to the different width and shape characteristics of different grades of roads, such as national roads, provincial roads, village roads, and mountain roads; roads with different materials have different color and texture characteristics, such as cement, asphalt, earth road, etc.; at the same time, the road area is blocked by buildings, trees, the central green belt of the road and many other factors, so the accurate extraction of road information is still the research frontier and poses a technical difficulty in the field of remote sensing information extraction.

Road extraction can be described as a pixel-level binary classification problem that distinguishes whether each pixel belongs to a road or not [1]. Recently, deep convolution neural networks (DCNNs) have been demonstrated to have significant improvements to typical computer vision tasks such as semantic segmentation [2]. Road semantic segmentation has applications in many fields, such as autonomous driving [3,4], traffic management [5], and smart city construction [6]. Semantic segmentation requires pixel-level classification [7–9], and it must combine pixel-level accuracy with multi-scale contextual reasoning [7–10]. In general, the simplest way to aggregate multi-scale context is inputting multi-scale information into the network for merging all scales of features. Some researchers have made much progress in the image processing fields. Farabet et al. [11] obtained different scale images by transforming the input image through a Laplacian pyramid. References [12,13] applied multi-scale inputs sequentially from coarse-to-fine. References [7,14,15] directly resized the input image for several scales. Meanwhile, another aggregating multi-scale context way is adopting an encoder-decoder structure, such as SegNet [16], U-Net [17], RefineNet [18], and other networks [19–21], which have demonstrated the effectiveness of models based on encoder-decoder structure. In addition, the context module is an effective way to aggregate multi-scale context information, such as merging DenseCRF [9] into DCNNs [22,23]. The spatial pyramid pool structure is also a common method to aggregate multi-scale context, such as Pyramid Scene Parsing Net (PSP) [24,25].

The larger receptive field is critical for networks because it can capture more global context information from the input images. For a standard convolution neural network (CNN), the traditional way to expand the receptive field is stacking more convolutional layers with a bigger convolutional kernel size, while the operation could result in the exponential expansion of the training parameters, which makes networks hard to train. The alternative way to expand the receptive field is stacking more pooling layers, which can expand the receptive field by reducing the dimension of the feature maps and maintaining the saliency characteristics. Although the pooling operations did not add the training parameters, much information would be lost because of the decrease in spatial resolution.

Reference [23] developed a convolutional network module, dilated convolution, which aggregates multi-scale contextual information without increasing the training parameters and decreasing resolution. Further, the module can also aggregate multi-scale contextual information with different expanding rates of dilated convolution kernel size. Besides, the module can be plugged into existing architectures for any resolution image, which is appropriate for dense prediction. Therefore, DeepLab v2 [26], DeepLab v3 [27], DeepLab v3+ [28], and D-LinkNet [1], which adopted dilated convolution for semantic segmentation, presented better performances.

Another effective strategy to increase the capture capabilities of global features is to introduce attention mechanism. Reference [29] first introduced an attention mechanism into computer vision tasks, which has been proven to be reliable. DANet [30] adopts a spatial and channel attention module to obtain more global context information. CBAM [31] introduced a lightweight spatial and channel attention module. DA-RoadNet [32] constructed a novel attention mechanism module to improve the network's ability to explore and integrate roads.

The network structure for semantic segmentation was divided into several parts, and those networks [1,26–28] only adopted dilated convolutions in one part. In fact, the encoder part and decoder part of existing architectures for semantic segmentation is built by stacking residual blocks or dense blocks. So, dilated convolution layers after each block have been added to capture more global context information. In the research, a new structure, D-Dense blocks, combined with traditional convolution layers and dilated convolution layers, has been proposed. Further, a network is built with D-Dense block and the center part of D-LinkNet for road extraction from satellite images. To increase the capabilities of capturing global features, the DA mechanism [30] is also introduced into the network. With the above design, the dilated convolution can run through the whole network and effectively integrate with the attention mechanism to obtain more global features and information. The presented context network was evaluated through controlled experiments with the Massachusetts Road dataset. The experiments demonstrate that the D-Dense block with attention mechanism architectures reliably increases the pixel-level accuracy for semantic segmentation.

Since SAR has the advantages of all-weather and strong penetration, using SAR images has irreplaceable advantages in remote sensing road extraction, which can further improve the accuracy of road information extraction. Many traditional road segmentation methods of SAR images have been proposed and proved effective. Methods based on human–computer interaction are called semi-automatic methods. Bentabet, L et al. [33] were the first to use the snake model for SAR image road extraction. The results of the experiments show that straight or curved roads could be accurately extracted by this model, but this model needs a large number of human–computer interactions [34]. Some automatic methods were also proven to be useful. Cheng Jianghua et al. [35] proposed a method based on the Markov random field (MRF). In order to maximize calculation efficiency, this method is developed on GPU-accelerated road extraction. Besides, there are also Deep-Learning methods of road extraction on SAR images. Wei X et al. [36] used Ordinal Regression and introduced Road-Topology Loss, which improves the baseline up to 11.98% in the *IoU* metric in their own dataset.

Focused on some problems of low-grade roads in remote sensing images, we study how to improve the accuracy of the road extraction in complex scenes using the powerful feature expression ability of deep learning and the penetrating feature of SAR images.

In this paper, we propose a novel deep learning network model called SDG-DenseNet to improve the accuracy of low-grade road extraction from optical remote sensing images. We fuse the extraction results from the SAR image into that of the optical image at the decision level, which improves the accuracy of low-grade road extraction in practical application scenarios. Therefore, the main contribution of this study can be summarized as:

(1) A novel SDG-DenseNet network for low-grade road extraction in optical images is proposed. The stem block is taken as the starting module to expand the receptive field and preserve image information, while the stem block also reduces the number of parameters. A novel D-dense block is introduced to construct the encoder and decoder of the network, which applies the dilated convolution in all parts from the encoder to the decoder to improve the receptive field of the network. Moreover, in order to make the dilated convolution run through the entire network, this paper introduces a GIRM module combining the dilated convolution and a double self-attention mechanism. The introduction of the GIRM module aims to enhance the network's ability to obtain global information. The segmentation effect of the novel network is better than that of many existing networks;

(2) A decision-level fusion method is proposed for the low-grade road extraction based on optical images and SAR images, which repairs some interrupted roads in the optical image extraction results. The extraction accuracy of decision-level fusion methods is higher than that of optical image-based deep learning methods in practical application scenarios.

2. Methods

In order to improve the image semantic segmentation accuracy, a novel SDG-Densenet network for low-grade road extraction in optical images is proposed. The construction of the novel SdG-Densenet for optical image semantic segmentation is composed of three parts: an encoder path, a decoder path, and the center part—the global information recovery module. The encoder path takes RGB images as input parameters and extracts features by stacking convolutional layers and pooling layers. The decoder path restores the detailed information and expands the spatial dimensions of the feature maps with deconvolutional layers. The center part is responsible for enlarging the receptive field, integrating multi-scale features, and maintaining the detailed information simultaneously. The skip connection encourages the reuse of the feature maps to help the decoder path recover spatially detailed information. Besides, a decision-level fusion method is also introduced in order to fuse the results optical image and SAR image, which mainly contains six steps: data preparation, pretreatment, image registration, road extraction, road segmented, and decision level integration. Figure 1 shows the overall structure of the proposed method.

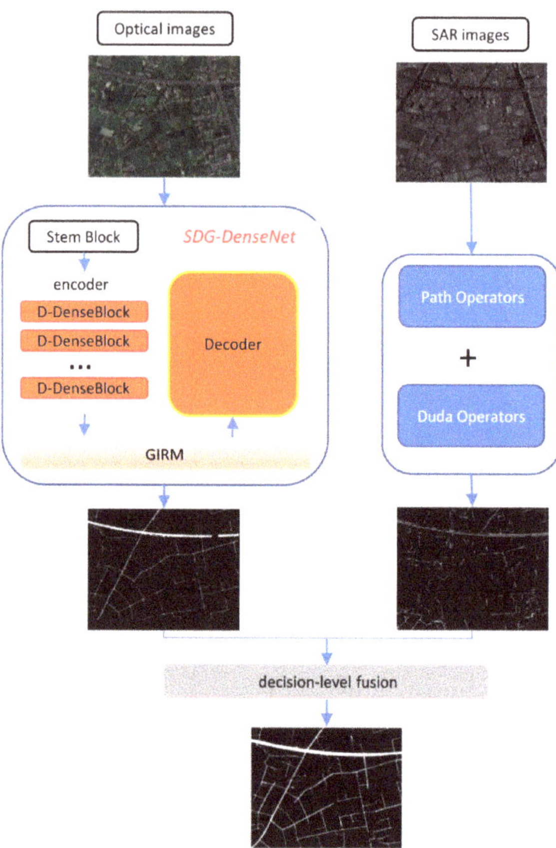

Figure 1. The overall technical flow of the proposed method.

2.1. Architecture of SDG-DenseNet Network

Because low-grade roads are easily blocked by vegetation or buildings, there are often problems of fracture and discontinuity in extracting low-grade roads in optical images. At the same time, due to the low construction standard of low-grade roads, their materials are often consistent with the surrounding environment, and they are often integrated into the background in the optical orthographic projection, resulting in a poor extraction effect. Based on the above problems, it is imperative to specialize in the novel network and improve the ability of global information extraction.

Based on D-LinkNet, the SDG-DenseNet was proposed. In order to improve the extraction ability of global information, the global information recovery module was introduced to the proposed new Network for semantic segmentation. Furthermore, the novel network took DenseNet as its backbone instead of ResNet and replaced the initial block with the stem block. Additionally, the Attention mechanism was introduced to improve the ability to obtain global information. The The overall structure of SDG-DenseNet network is shown in Figure 2.

2.2. Improved D-Dense Block and Stem Block

The construction of the D-Dense block is shown in Figure 3. In contrast to the original Dense block, we added three consecutive dilated convolution layers with different expanding rates after the original Dense block. The expanding rates of these three dilated convolutions are 2, 4, and 8, respectively. The structure of each dilated convolution could be set as BN-ReLU-Conv (1 × 1)-BN-ReLU-D_Conv (3 × 3, rate = 2, or 4, or 8). The same computation process with the original Dense block repeated (n + 3) times and makes the D-Dense block generate feature maps with (n + 3) × k channels.

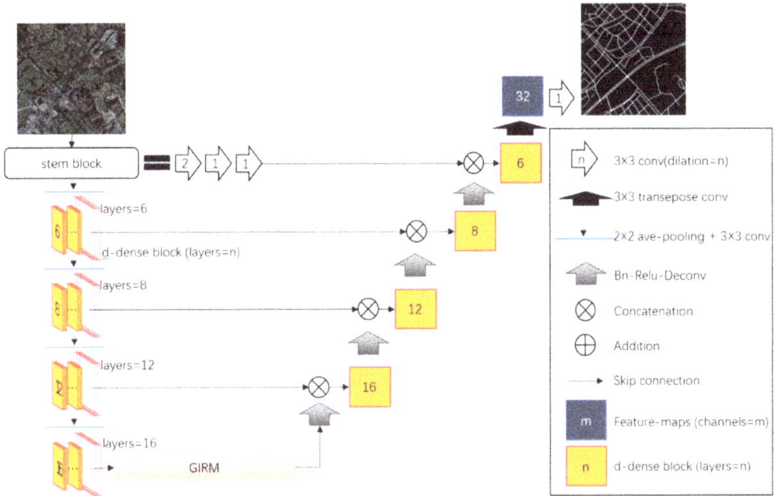

Figure 2. The construction of the SDG-DenseNet.

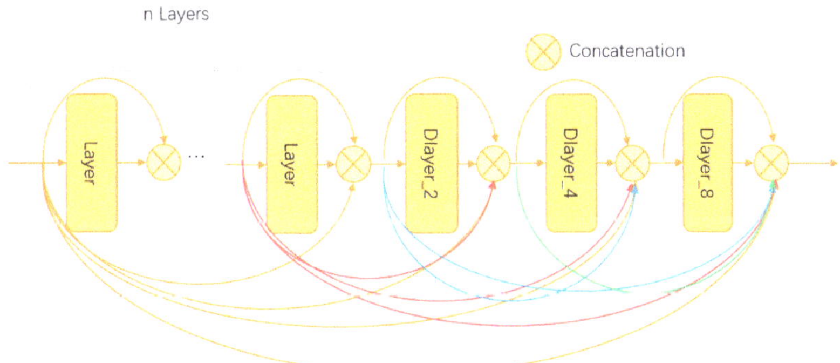

Figure 3. The construction of the D-Dense block.

The encoder starts with an initial block and performs convolution on the input image with a kernel of 7 × 7 size and a stride of 2 followed by a 3 × 3 max pooling. In addition, the output channels of the initial block are 64. Inspired by Inception v3 [37] and v4 [38], References [39,40] replaced the initial block [41] 7 × 7 convolution layer, stride = 2 followed by a 3 × 3 max pooling layer by the stem block. The Stem block is composed of three 3 × 3 convolution layers and one 2 × 2 mean pooling layer. The stride of the first convolution layer is 2 and the others are 1. In addition, the output channels for all the three convolution layers are 64. The experiment results in Reference [40] proved that the initial block applied would lose much information due to two consecutive down-sample operations, making it hard to recover the marginal information of the object in the decoder phase. The stem block is helpful for object detection, especially for small objects. So, the research also adopts the stem block at the beginning of the encoder phase.

2.3. Global Information Recovery Module (GIRM) Based on d-Blockplus and Attention Mechanism

In order to weaken and eliminate the problem of road fracture or low recall in low-grade road extraction, this paper proposes a global information recovery module, which is composed of a dual attention mechanism and d-blockplus. The global information extraction module aims to further improve the network's ability to obtain global information to ensure the integrity of the extraction results.

As shown in Figure 4, the global information extraction module is mainly composed of two parts. The dual attention mechanism mainly starts from the two directions of spatial attention and channel attention, extracts and integrates the global information of space and channel, and improves the attention to road targets. d-blockplus introduces multi-layer hole convolution to improve the receptive field, so as to improve the ability of the network to maintain the integrity of road extraction.

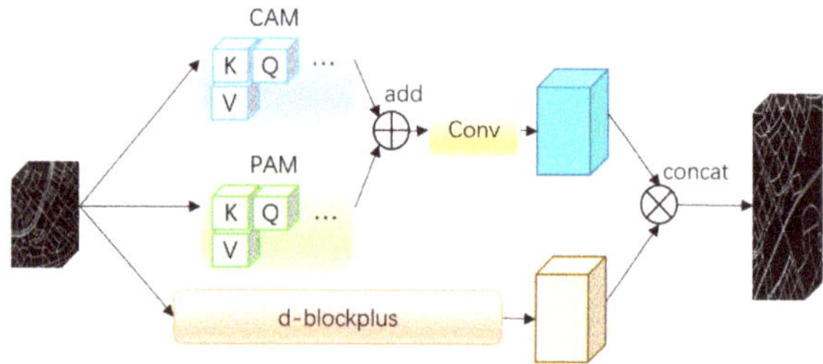

Figure 4. The construction of GIRM.

In the center part of the SDG-DenseNet, in addition to the d-blockplus, the position attention module (PAM) and the channel attention module (CAM) are also introduced. PAM and CAM are two reliable self-attention modules, which improve the ability of the network to obtain global information in the spatial dimension and channel dimension, respectively.

Figure 5 shows the structure of PAM. In PAM, the input feature maps go through two branches, and one of them will be used as Q and K to generate a $(H \times W) \times (H \times W)$ Attention probability map. In another branch, it is used as V. Where, V, Q, and K represent value features, query features, and key features, respectively; C, H, and W represent the channel, height, and weight of the characteristic graph, respectively. The overall structure of PAM is shown in Equation (1):

$$\begin{aligned} Att &= softmax(Q_{(C \times HW)} \cdot K_{(HW \times C)}) \\ F_{out} &= (V_{(C \times HW)} \cdot Att).reshape(C \times H \times W) + Input_{(C \times H \times W)} \end{aligned} \quad (1)$$

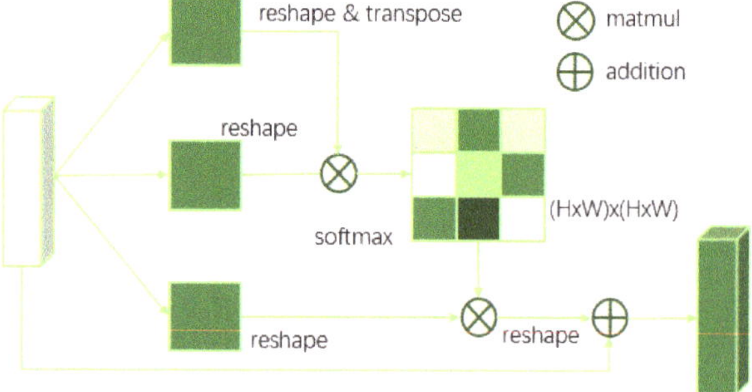

Figure 5. Architecture of the position attention module [30].

Figure 6 shows the structure of CAM. The structure of CAM is basically similar to that of PAM. CAM pays more attention to the information on the channel. In this network structure, the size of the probability map generated by CAM is ($C \times C$), which helps to boost feature discrimination. The overall structure of CAM is shown in Equation (2):

$$Att = softmax(Q_{(C \times HW)} \cdot K_{(HW \times C)})$$
$$F_{out} = (Att \cdot V_{(C \times HW)}).reshape(C \times H \times W) + Input_{(C \times H \times W)} \tag{2}$$

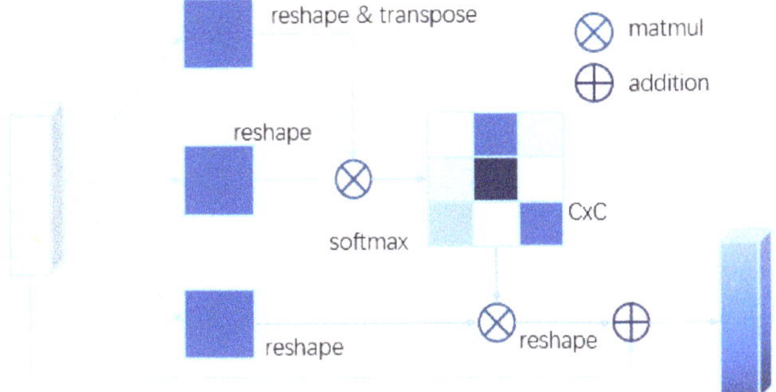

Figure 6. Architecture of the channel attention module [30].

D-block has four paths that contain dilated convolution in two cascade modes and two parallel modes, respectively. In each path, dilated convolutions are stacked with different expanding rates. Consequently, the receptive field of each path is different, and the network can aggregate multi-scale context information. Inspired by MobileNetV2 [42], to save network parameters and improve network performance, the bottleneck block is introduced into d-block to build d-blockplus. Figure 7 shows the structure of D-blockplus.

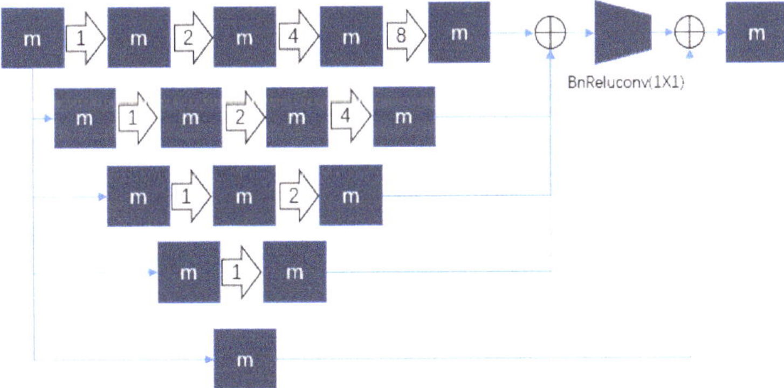

Figure 7. The construction of d-blockplus.

2.4. Decision-Level Fusion Algorithm for Low Grade Roads

In optical images, low-grade roads often show the problem where the roads are blocked by buildings, vegetation, shadows, and so on. However, the background of buildings and vegetation is often quite different from the road, and the blocked part is often not judged as a road in the process of deep learning, which directly leads to the phenomenon of fracture or undetected in the extraction results of low-grade roads. Figure 8 shows some examples of blocked roads. In these pictures, the roads in red boxes show fractures in the optical image because it is obscured by vegetation, buildings, or shadows.

Figure 8. Figures of blocked roads.

For the problems of the above complex scenes, the imaging mechanism of the optical image determines that the SDG-DenseNet network model cannot solve the problem of poor road continuity well. In this paper, the optical image extraction results based on the SDG-DenseNet network model and the SAR image extraction results based on Duda and path operators [43] realize decision-level fusion.

The Duda operator is a linear feature extraction operator that divides an $N \times N$ window into three parallel linear parts. The specific structure of the Duda operator is shown in Figure 9, where A, B, C, C1, and C2 represent the mean gray values of the three parts. What's more, the operator shown in Figure 9a has a relatively strong ability to extract roads in the horizontal direction, and the operator shown in Figure 9b has a relatively strong ability to extract roads with a certain inclination angle.

The other two types of Duda Operators are a 90-degree rotation of the above two. The function to determine the new value of a pixel can be expressed as follows:

$$H(x) = (1 - \frac{C}{A})(1 - \frac{C}{B})\frac{C1}{C2}. \tag{3}$$

Path operators refer to path openings and closings, which are morphological filters applied to analyze oriented linear structures in images. The morphological filter defines the adjacency graphs as structuring elements. Four different adjacency graphs are defined as horizontal lines, vertical lines, and two diagonal lines, respectively. Applying these four adjacency graphs to a binary image, the maximum path length of each pixel can be achieved. Then, the pixels, whose maximum path lengths are larger than the threshold Lmin, are retained in the image.

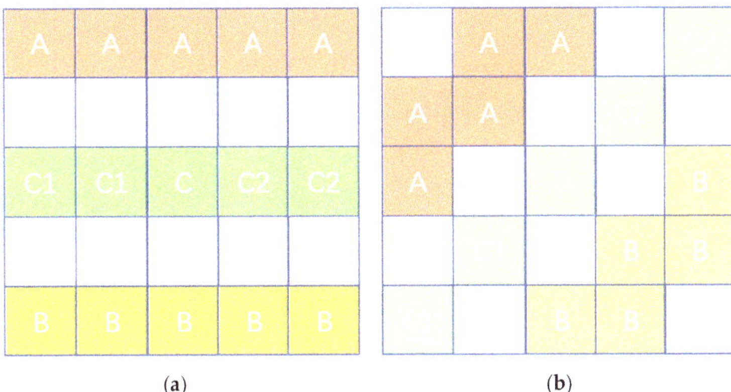

Figure 9. Window structure of two types of Duda operators.

The specific algorithm flow of the decision-level fusion method for low grade roads is shown in Figure 10.

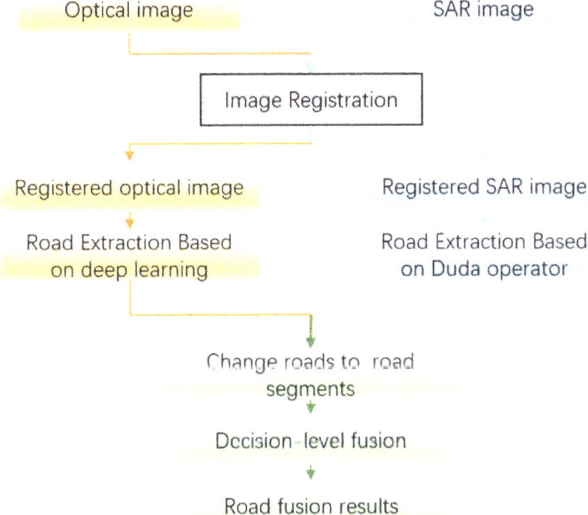

Figure 10. Low-grade road extraction algorithm for decision-level fusion of optical and SAR images.

Figure 10 shows the overall technical process of the road extraction algorithm based on the decision-level fusion of high-resolution optical and SAR remote sensing images. The specific steps of the algorithm are as follows.

Step 1: Data preparation. Obtain optical remote sensing images and SAR images in the same area, and their imaging time should be as close as possible;

Step 2: Pretreatment. The optical remote sensing image and SAR image are preprocessed, respectively, including radiometric correction, geometric correction, geocoding, and so on;

Step 3: Image registration. The optical remote sensing image and SAR image are matched and transformed into the same pixel coordinate system;

Step 4: Road extraction. Roads in optical remote sensing images are extracted by SDG-DenseNet and those in SAR images are extracted by the method in Reference [43], which is based on Duda and Path operator;

Step 5: Roads segmented. For the road extraction results of optical remote sensing image and SAR image, the road segments are obtained by segment method, and the attributes of each segment are recorded;

Step 6: Decision level fusion. Taking the line segment as the basic unit, the final road distribution map is obtained by decision-making level fusion of the roads extracted from the optical remote sensing image and SAR image.

3. Experiments

Our network experiments are performed on the Massachusetts Roads Dataset, and we test the fusion method in our own dataset that came from WorldView-2, WorldView-4, and TerraSAR-X. The TensorFlow platform was selected as the deep learning framework to train and test all networks. All models are trained on one NVIDIA GTX 2080 Ti GPU.

3.1. Dataset and Data Augmentation

Three sets of satellite images were applied to evaluate the Low-Grade road extraction method. To verify the effectiveness of the proposed SDG-DenseNet network on public datasets, we tested the SDG-DenseNet on the Massachusetts dataset. In addition, we conducted low-grade road extraction experiments on the self-built Chongzhou–Wuzhen dataset. Finally, we conducted decision-level fusion experiments on two sets of large-scale images from the Chongzhou and Wuzhen regions including optical and SAR images.

We trained and tested our SDG-DenseNet network model on the Massachusetts Roads Dataset [44], which consists of 1108 training images, 14 validation images, and 49 test images. The size of each image is 1500 × 1500. We cut each 1500 × 1500 image into four 1024 × 1024 images. Therefore, we obtained 4432 training images, 56 validation images, and 196 test images. Further, we performed data augmentation on the training set, including rotation, flipping, cropping, and color jittering, which could prevent the training set from overfitting. After data augmentation, we obtained 22,160 training images in total. Finally, we obtained 22,160 training images, 56 validation images, and 196 test images.

In order to test the proposed SDG-DenseNet network of low-grade road extraction, this paper also tests the SDG-DenseNet on the self-built dataset: The Chongzhou–Wuzhen dataset. Table 1 displays the three source images of the self-built dataset. We cut the three source images into 13,004 512 × 512 images. Therefore, we obtained 11,788 training images, 204 validation images, and 1012 test images. After the data augmentation of the training set, we got 47,152 training images. Finally, we obtained 47,152 training images, 204 validation images, and 1012 test images.

Table 1. The three source images of the self-built low-grade road dataset.

Data	Satellites	Resolution Ratio	Date	Scale	Area
1	WorldView-4	0.6 m	13 May 2018 (optical)	3469 × 4786	Chongzhou, Sichuan
2	WorldView-2	0.5 m	27 July 2018 (optical)	2800 × 3597	Wuzhen, Zhejiang
3	WorldView-2	0.5 m	27 July 2018 (optical)	2800 × 1798	Wuzhen, Zhejiang

We also test our decision-level fusion experiments on two sets of large-scale images from the Chongzhou and Wuzhen regions including optical and SAR images. The optical images came from WorldView-2 and WorldView-4, while we got the SAR images from TerraSAR-X. As shown in Table 2, in order to test the effect under application conditions, the decision-level fusion experiment is mainly tested on the two large-scale images.

Table 2. Two sets of large-scale images used in decision-level fusion experiments.

Data	Satellites	Resolution Ratio	Date	Scale	Area
1	WorldView-4	0.6 m	13 May 2018 (optical)	3469 × 4786	Chongzhou, Sichuan
	TerraSAR-X	0.8 m	20 September 2018 (SAR)		
2	WorldView-2	0.5 m	27 July 2018 (optical)	2800 × 3597	Wuzhen, Zhejiang
	TerraSAR-X	0.9 m	24 March 2019 (SAR)		

3.2. Hybrid Loss Function and Implementation Details

In previous work, most networks train their models only by using the cross-entropy loss [45], which is defined as Equation (3):

$$L_{ce} = -\frac{1}{N} \sum_{i=0}^{N} (y \log y' + (1-y) \log(1-y')), \tag{4}$$

where N indicates categories. y and y' mean the label and prediction vectors, respectively. Since an image consists of pixels, for road area segmentation, the imbalance of sample points (where the roads only cover a small part of the whole image) makes the direction of the gradient decrease toward the back corner (Figure 11a), which leads to a local optimum, especially in the early stage [46]. The Jaccard loss function is defined as:

$$L_{jaccard} = \frac{1}{N} \sum_{i=0}^{N} \frac{y_i y'_i}{y_i + y'_i - y_i y'_i} \tag{5}$$

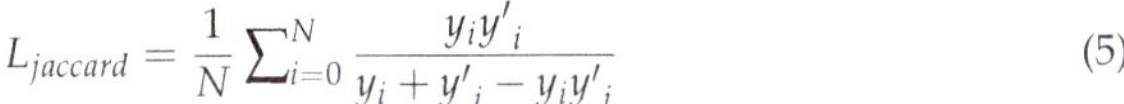

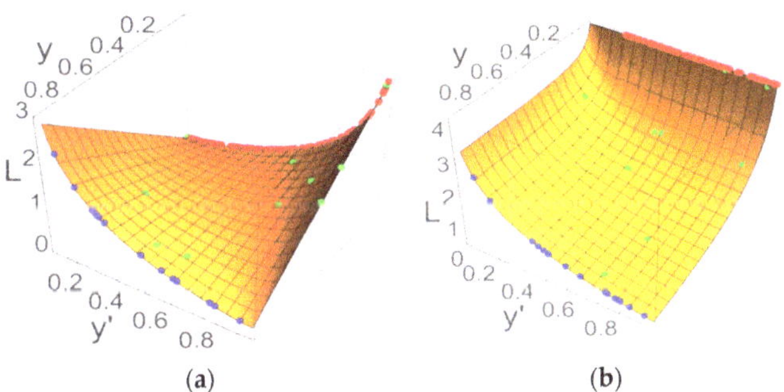

Figure 11. Different loss function surface. (**a**) Cross entropy surface; (**b**) Jaccard surface.

Its surface is shown in Figure 11b. As we can see, the Jaccard loss can address this problem if we sum the Jaccard loss and the cross-entropy loss together. So, the whole loss function is defined as:

$$L = L_{ce} - \lambda \log L_{jaccard}, \tag{6}$$

where λ is the weight of the Jaccard loss in the whole loss. Furthermore, the red, green, and blue points in Figure 11 represent the local maxima, saddle points, and local minima on the loss surface, respectively.

In the training phase, we chose Adam as our optimizer and originally set the learning rate to be 0.0001. We reduce the learning rate by 10 times while observing the loss value decreasing slowly. The loss weight λ is set to 1. The batch size during the training phase is set to 1.

3.3. Decision-Level Fusion Experiment

To verify the effect of every step in the decision-level fusion method for low-grade roads, we apply the fusion method to the road extraction results from the network and method based on the Duda operator and Path operator, using the large-scale images mentioned in Table 2 and the details of *Step 6*, where decision-level fusion is operated as in Figure 12. The detailed workflow of the Decision level fusion.

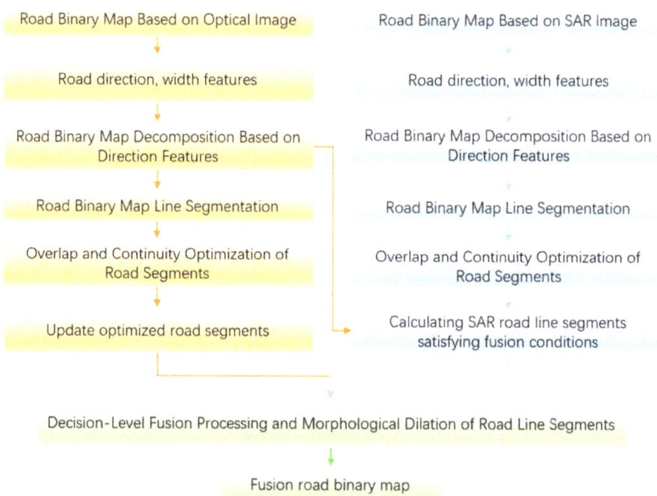

Figure 12. The detailed workflow of the decision level fusion.

As shown in Figure 12, the main process is divided into five steps:

Step 1: Road binary map extracted from input optical image and SAR image (not segmented);

Step 2: Segment the road binary map extracted from the SAR image, including extracting the road feature direction map, decomposing the binary map according to the direction feature, thinning the decomposed layer based on the curve fitting algorithm, and optimizing the line segment overlap, continuity and intersection to obtain the road segment set extracted from the SAR image;

Step 3: Segment the road binary map extracted from the optical image, optimize the overlap and continuity of segments, and record the updated segments of continuity optimization;

Step 4: For each road segment extracted from the SAR image, we judge whether it meets the fusion conditions with optical image road extraction results according to the overlap ratio in the corresponding optical extraction road binary layer, and record the qualified SAR road segments;

Step 5: After morphological expansion according to the width feature, the continuously optimized and updated line segments and the SAR road line segments meeting the fusion conditions are calculated with the original optically extracted road binary map according to pixels to obtain the fused Road Distribution binary map.

The specific method of searching line segments satisfying fusion conditions is shown in Figure 13. Assuming that A_m represents the road area on layer m after the decomposition of optical image extraction results, L_{mn} is the line segment n on layer m from SAR image road extraction results. They belong to the same layer m, that is, the road has similar directional features. We then count the number of pixels l_{n1} and l_{n2} where L_{mn} falls inside and outside the A_m region, and calculate the overlap rate $r = l_{n1}/(l_{n1} + l_{n2})$. If r is greater than the threshold T_r, L_{mn} is recorded as the road segment meeting the fusion conditions. In a practical application, the threshold tr takes an empirical value of 0.3. We traverse all

SAR extracted road segments until all SAR image-extracted road segments meeting the above fusion conditions are recorded.

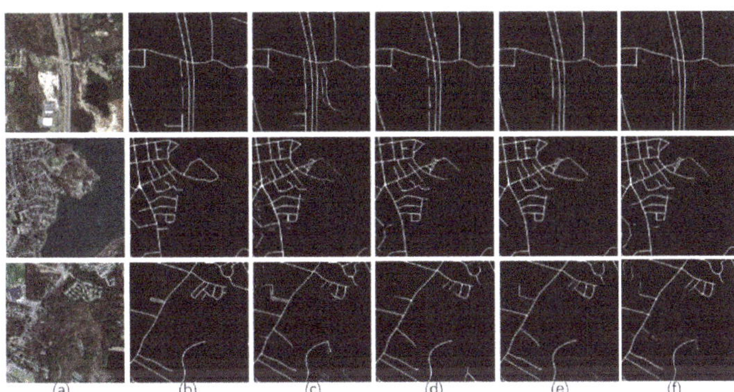

Figure 13. Schematic diagram of road fusion condition judgment for optical and SAR extraction.

3.4. Evaluation Metrics

In order to evaluate the performance of different road segmentation models, four evaluation metrics are used to evaluate the extraction results, including intersection-over-union (*IoU*), completeness (*COM*), correctness (*COR*), and F1-score [47], which are defined as:

$$IoU = \frac{TP}{TP+FN+FP} \qquad COR = \frac{TP}{TP+FP}$$

$$COM = \frac{TP}{TP+FN} \quad .... \quad F1 = \frac{2 \times COM \times COR}{COM+COR} \tag{7}$$

TP (True Positive) indicates that the extraction result is determined as a road, which is actually part of the road; *FP* (False Positive) indicates that the extraction result is determined as a road, but it is not actually part of the road; *FN* (False Negative) indicates that the extraction result is determined to be not a road, but it is actually part of the road. The *COM* scores of different models show the ability to maintain the completeness of the segmented roads. The higher the score, the better the road continuity extracted by the model. The *COR* scores of different models show the ability on reducing false detection of the segmented roads. The higher the score, the fewer areas will be falsely detected. The *IoU* and *F1* scores are the overall evaluation metrics that synthesize *COM* and *COR* scores and evaluate the overall quality of segmentation results.

Based on these evaluation metrics, we can obtain the performance of model road extraction results in different aspects from *COM* and *COR* scores, and obtain the overall performance judgment from *F1* and *IoU* scores.

4. Results and Discussion

4.1. Results of the Massachusetts Roads Dataset

In order to further verify the effectiveness of the proposed method, we evaluated our network with Massachusetts Roads Dataset. We divided the test images into two levels—general scene and complicated scene—according to the complexity of the image content scene. The sample results are shown in Figures 14–18.

Figure 14. Road extraction results in general scene images; (**a**) input image; (**b**) label image; (**c**) D-LinkNet; (**d**) S-DenseNet; (**e**) SD-DenseNet; (**f**) SDG-DenseNet.

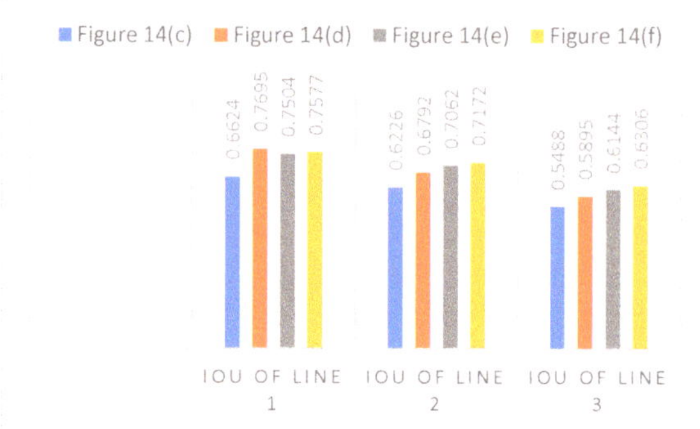

Figure 15. *IoU* scores of the methods in Figure 14.

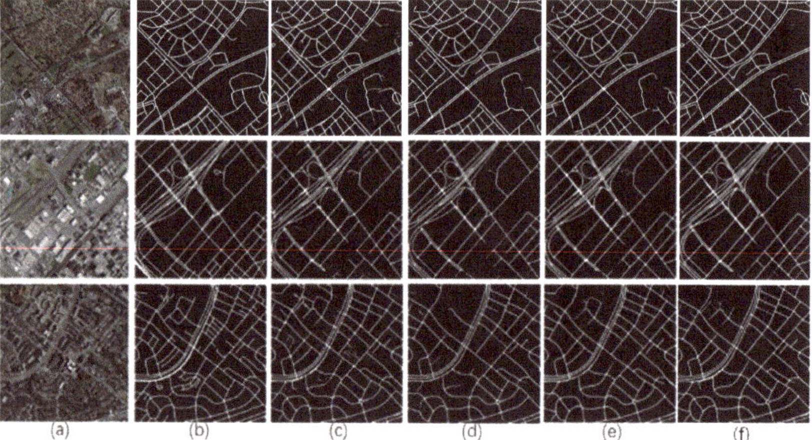

Figure 16. Road extraction results in complicated scene images (**a**) input image; (**b**) label image; (**c**) D-linkNet; (**d**) S-DenseNet; (**e**) SD-DenseNet; (**f**) SDG-DenseNet.

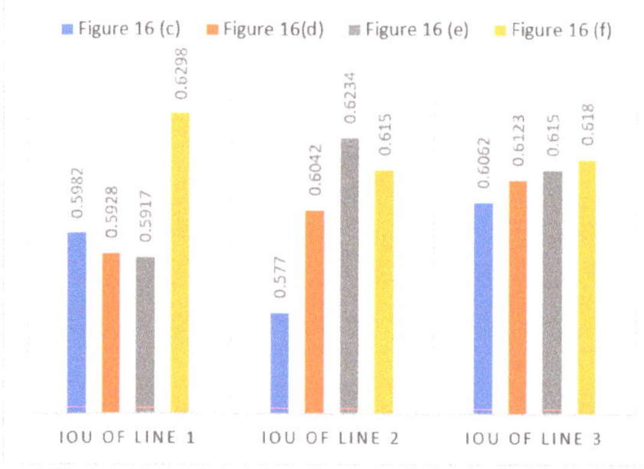

Figure 17. *IoU* scores of the methods in Figure 16.

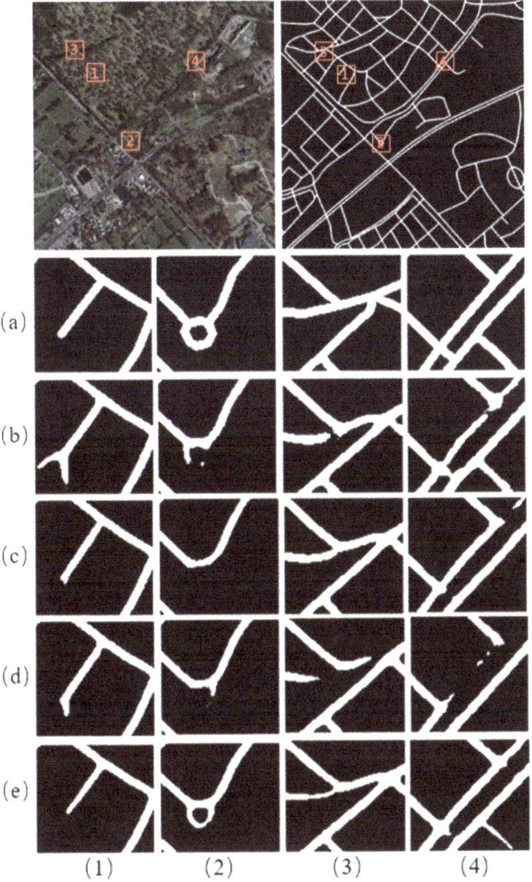

Figure 18. Four detailed areas of road extraction results in complicated scene images (**a**) area of label image; (**b**) D-LinkNet; (**c**) S-DenseNet; (**d**) SD-DenseNet; (**e**) SDG-DenseNet.

Figures 14 and 15 show the extraction results of general scene images. D-LinkNet shows the network built on residual blocks and the encoder part, DenseNet shows the network built on the Dense block, S-DenseNet shows the network built on Dense block and Stem block, SD-DenseNet represents that the network has also added dilated convolution on the basis of the previous networks, and SDG-DenseNet is built on the basis of the Stem block, D-Dense Blocks, and the GIRM module. The extraction results of the DLinkNet model contain some redundant information, and many independent patches are left in the image, which could affect the result of the overall accuracy. The parking lot areas, which are similar to roads, were successfully identified as backgrounds. However, some roads are not completely extracted. SDG-DenseNet has been further improved to make the completeness of roads better. The information extracted by the SDG-DenseNet network structure is more accurate.

Figure 15 shows the *IoU* scores for each image in each row in Figure 14. The proposed SDG-DenseNet achieves high *IoU* scores under all three optical images, which are 9.53%, 9.46%, and 8.18% higher than the baseline D-LinkNet, respectively.

Figure 16 shows the three extraction results of the complicated scene images from the 49 test images. Each road network includes more different level roads and flyover roads.

These complex situations seriously affect the road extraction results of every network model. However, the SDG-DenseNet can better extract every road including shadow obscured roads.

Figure 17 shows the *IoU* scores for each image in each row in Figure 16. Similar to Figure 15, in the three optical images, the proposed SDG-DenseNet achieves high *IoU* scores, which are 3.16%, 3.8%, and 1.18% higher than the baseline, respectively. At the same time, in order to improve the reliability of the results from a methodological point of view, the effects of different modules on different images are often different, for example, the *IoU* of Line 1 in Figure 17 shows that the two improved methods perform less well on this optical image. However, it is worth mentioning that the comprehensive results of statistics show that the average *IoU* score of the optimized model on the test set (196 test images) is higher than that of the baseline.

Figure 18 shows some detailed areas of the first image in Figure 16. The different results on Area 1 and Area 2 implied that SDG-DenseNet has a better ability on the correctness of segmentation. Area 1 shows the novel network's improvement in avoiding false segmentation, while in Area 2 it also emerges that SDG-DenseNet performs well in the recall ratio of extraction. Besides, Area 3 and Area 4 show that the novel network also performs perfectly when focusing on the completeness of the result of road extraction. In Figure 18, column (1), column (2), column (3), and column (4) correspond to area 1, area 2, area 3, and area 4 respectively.

Figures 14–18 show the semantic segmentation results of some randomly selected images in the Massachusetts test set. In order to further prove the effectiveness of the improved model on the test data set, this paper counts the evaluation indicators of the segmentation results of different models on the test set (196 test images). Through the training model and experiment, we get the D-LinkNet, DenseNet, S-DenseNet, SD-DenseNet, and SDG-DenseNet evaluation metrics index, as shown in Table 3. We found the *IoU* and *F1* scores of the network built on the Dense block or D-Dense block to be much higher than the network built with the residual block. Besides, the model based on DenseNet with the D-Dense block has higher *IoU* and *F1*-scores than that with Dense block.

Table 3. Results of the Massachusetts Roads Dataset of different models. The bold font indicates the optimal value under the current evaluation metrics.

Model's Description	*F1*	*IoU*	*COR*	*COM*
D-LinkNet	0.7688	0.6286	0.7712	0.7727
DenseNet	0.7786	0.6423	0.7780	**0.7854**
S-DenseNet	0.7810	0.6462	0.8153	0.7557
SD-DenseNet	0.7894	0.6562	**0.8190**	0.7667
SDG-DenseNet	**0.7963**	**0.6647**	0.8186	0.7767

In other words, compared with D-LinkNet, the novel network can extract roads more correctly and maintain good road completeness. Furthermore, when comparing the stem block with the initial block, we find that the network with the stem block is much better than the initial block in the correctness of road extraction. At the same time, stem block also improves the *IoU* and *F1* scores. The experiment results show the SDG-DenseNet could obtain better *IoU* and *F1* scores when performing well in the correctness of road extraction. It can also be seen from the table that the SDG-DenseNet is more balanced than other networks in its ability to maintain road completeness and correctness, while both *COR* and *COM* indices are kept at a relatively high level, thus achieving higher *F1* and *IoU* Scores.

4.2. Results on Massachusetts Roads Dataset of Different Methods

To evaluate our method performance, we compare its *IoU* scores with Residual Unet [46], Joint-Net [48], Dual Path Morph-Unet [49], and DA-RoadNet [32] which have been used in road extraction from satellite images.

Table 4 shows the scores obtained by different methods on the Massachusetts Roads Dataset. The SDG-DenseNet had the highest *F1* and *IoU* which proves the excellent performance of SDG-DenseNet in road extraction. Besides, as shown in Table 4, our new network achiever a higher *COM* score than other networks, while the *COR* score of the SDG-DenseNet is not much lower than other networks. In other words, our network achieves a good balance in maintaining the completeness and correctness of segmentation.

Table 4. Results of the Massachusetts Roads Dataset of different methods. The bold font indicates the optimal value under the current evaluation metrics.

Method	*F1*	*IoU*	*COR*	*COM*
Residual Unet [46]	*	0.6340	*	*
Joint-Net [48]	0.7805	0.6310	**0.8536**	0.7190
Dual Path Morph-Unet [49]	*	0.6440	*	*
DA-RoadNet [32]	0.7819	0.6419	0.8524	0.7124
SDG-DenseNet (ours)	**0.7963**	**0.6647**	0.8186	**0.7767**

('*' represents the metrics not mentioned in the cited papers).

4.3. Results of Low-Grade Roads on the Chongzhou–Wuzhen Dataset

In order to fully uncover the characteristics of the low-grade road extraction task and the performance of different networks on this task, this paper tests the low-grade road in the Chongzhou–Wuzhen dataset. In the test process, according to whether the low-grade roads are blocked, the complexity of the low-grade road structure, and the complexity of the background scene, the extraction difficulty is divided into four cases: simple, general, and complicated.

Figures 19–21 shows the extraction results of four different network models in simple scenes, general scenes, and complex scenes, respectively. Figure 19 shows the detection result in a simple scenario. The detection effect of D-LinkNet in a simple scenario is best, especially for the detection of the expressway; while its integrity is higher, there are fewer false detection parts. Figure 20 shows the detection effect in a general scenario. At this time, D-LinkNet has obvious road fracture and missing detection. SDG-DenseNet has the best detection effect. Compared with other networks, it extracts the most complete roads. Figure 21 shows the detection effect in complex scenes, and several networks show different degrees of missed detection and false detection. S-DenseNet shows the strongest ability to maintain integrity, but there are many false detection areas; SDG-DenseNet has a certain degree of road fracture, but there are few false detections.

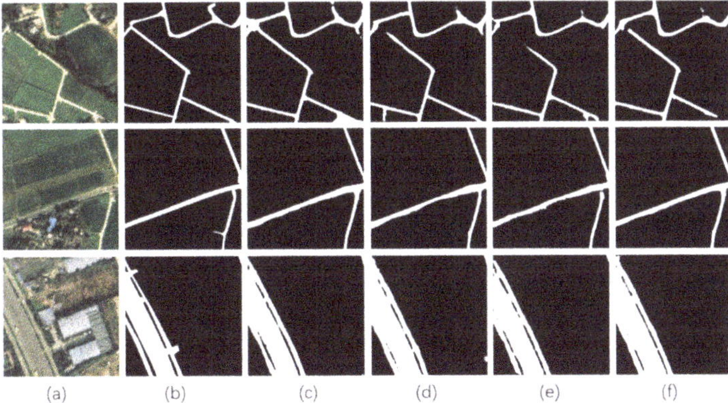

(a)   (b)   (c)   (d)   (e)   (f)

Figure 19. Low-grade road extraction in simple scene images (**a**) input image; (**b**) label image; (**c**) D-LinkNet; (**d**) S-DenseNet; (**e**) SD-DenseNet; (**f**) SDG-DenseNet.

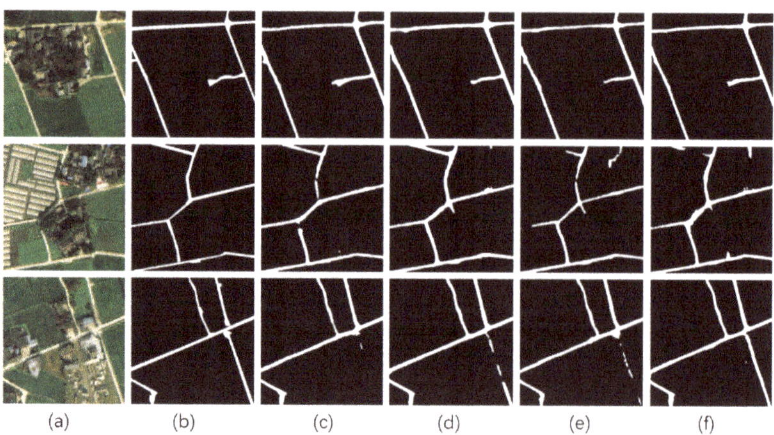

Figure 20. Low-grade road extraction in general scene images (**a**) input image; (**b**) label image; (**c**) D-LinkNet; (**d**) S-DenseNet; (**e**) SD-DenseNet; (**f**) SDG-DenseNet.

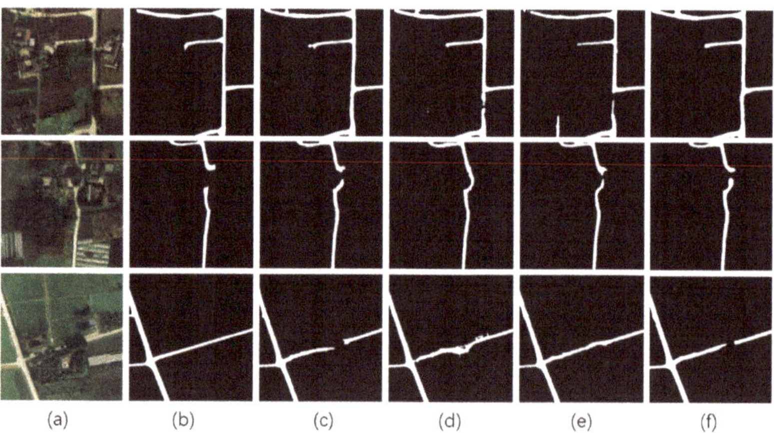

Figure 21. Low-grade road extraction in complicated scene images (**a**) input image; (**b**) label image; (**c**) D-linkNet; (**d**) S-DenseNet; (**e**) SD-DenseNet; (**f**) SDG-DenseNet.

Table 5 shows the *IoU* scores of the different models on the test of the Chongzhou–Wuzhen dataset. The result shows that SDG-DenseNet achieved the highest *IoU* scores while its model size is much less than D-LinkNet, which proves that the SDG-DenseNet has the best performance on low-grade road extraction tasks. S-DenseNet has the least parameters of the four networks, which is mainly due to the reduction of parameters by dense block.

Table 5. Results of the Chongzhou–Wuzhen test set on different models. The bold font indicates the optimal value under the current evaluation metrics.

Model's Description	*IoU*	Model Size
D-LinkNet	0.5236	0.98 GB
S-DenseNet	0.5558	**81.7 MB**
SD-DenseNet	0.5796	106 MB
SDG-DenseNet	**0.5901**	265 MB

4.4. Extraction Results of Low-Grade Roads on Large-Scale Images of the Fusion Method

In order to verify the feasibility and effect of the decision-level fusion method, and to test the overall effect of the process in the actual application scenario, we extracted the optical image based on SDG-DenseNet and the SAR image based on the Duda operator for the two large-scale images mentioned in Table 2, and then tested the effect of decision-level fusion.

In order to more intuitively reflect the effect of the fusion method, we compare the extracted roads with the roads in the label.

Figures 22 and 23 show the result of the decision-level fusion method tested on our own dataset in the Chongzhou area.

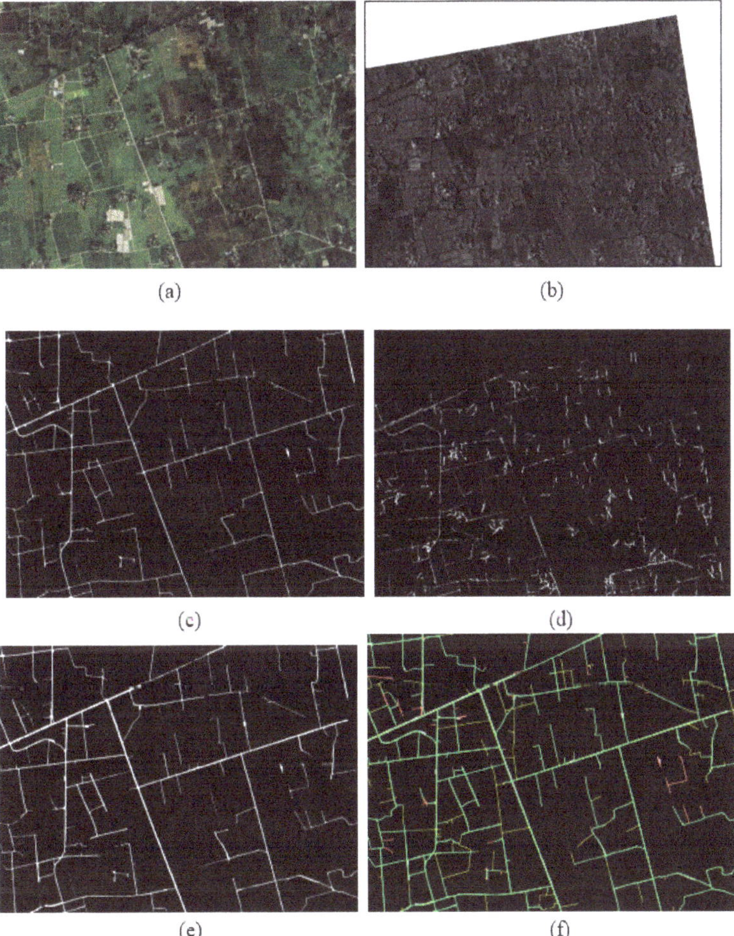

Figure 22. Tested data Area 1. Optical and SAR remote sensing images and road extraction results in the Chongzhou area. (**a**) Worldview-4 optical remote sensing image; (**b**) TerraSAR-X remote sensing image; (**c**) road extraction results of optical remote sensing image; (**d**) road extraction results from SAR remote sensing images; (**e**) road fusion extraction results of optical and SAR images; (**f**) ground truth and marking results of road fusion extraction results (green refers to the correctly extracted road, red refers to the incorrectly extracted road, and yellow refers to the omitted real road).

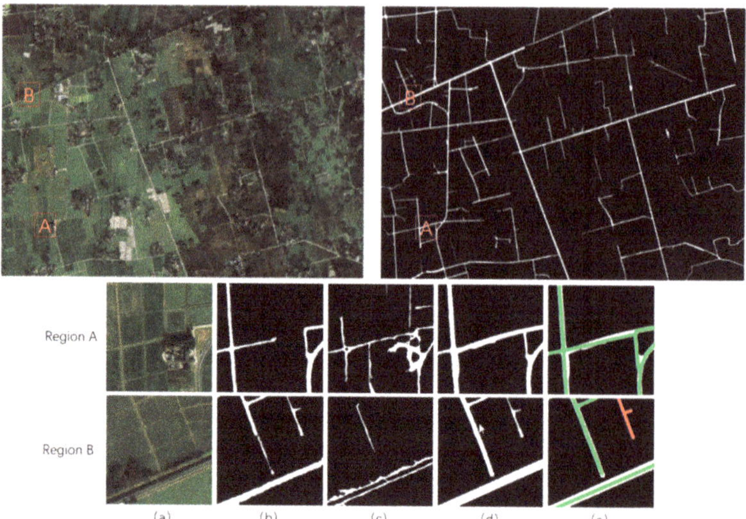

Figure 23. Tested data Area 1. Some details in optical and SAR remote sensing images and road extraction results in the Chongzhou area. (**a**) Worldview-4 optical remote sensing image; (**b**) road extraction results of optical remote sensing image; (**c**) road extraction results from the SAR remote sensing images; (**d**) road fusion extraction results of optical and SAR images; (**e**) ground truth and marking results of road fusion extraction results (green refers to the correctly extracted road, red refers to the incorrectly extracted road, and yellow refers to the omitted real road).

Figure 22 display the extraction effect of the optical image, SAR image, and decision-level fusion on low-grade roads in practical application scenarios. The road extracted from the optical image using SDG-DenseNet is more complete and continuous than the road extracted from the SAR image using the Duda and Path operators. However, the extraction results of the SAR images contain some information that are not found in optical image extraction results, such as some roads obscured by vegetation or buildings.

Figure 23 show some details from Figure 22. As shown in the figures, according to the extraction results of SAR images, decision-level fusion fixes some problems of road fracture and missing detection caused by vegetation or building occlusion in optical images.

We also tested the decision-level fusion method in the Wuzhen area of the self-made dataset, as shown in Figures 24 and 25. Figure 24 shows the extraction results in large-size images in practical application scenarios, which are similar to the results in Chongzhou. The extraction results of optical images are good in continuity, but there are also obvious problems of broken road extraction results and missing detection. After fusion with the SAR image, the partially occluded roads become continuous, and some roads that were missed in the optical image were detected.

Figure 25 shows some details of the detection results in Wuzhen. Figure 25 region A(b) shows the extracted road breaks due to bridge interference in the optical image, which can be seen in region A(d). After decision-level fusion, the broken extraction result is fixed. A similar situation occurs in region B due to the occlusion of vegetation, and the roads that are missed in the optical image are also repaired by the decision-level fusion method.

Table 6 shows the results of the two large-scale images mentioned in Table 2. We use the manual interpretation annotation method to evaluate and analyze the low-grade road extraction results, in other words, the matching degree between the extracted road network and the reference road is evaluated through completeness, correctness, and accuracy. As shown in Figures 22–25 and Table 6 through the decision-level fusion extraction of optical and SAR images, the $F1$-scores of road extraction can reach more than 0.85. The $F1$, COM, and *COR* scores are significantly higher than the results using only the optical image extraction method.

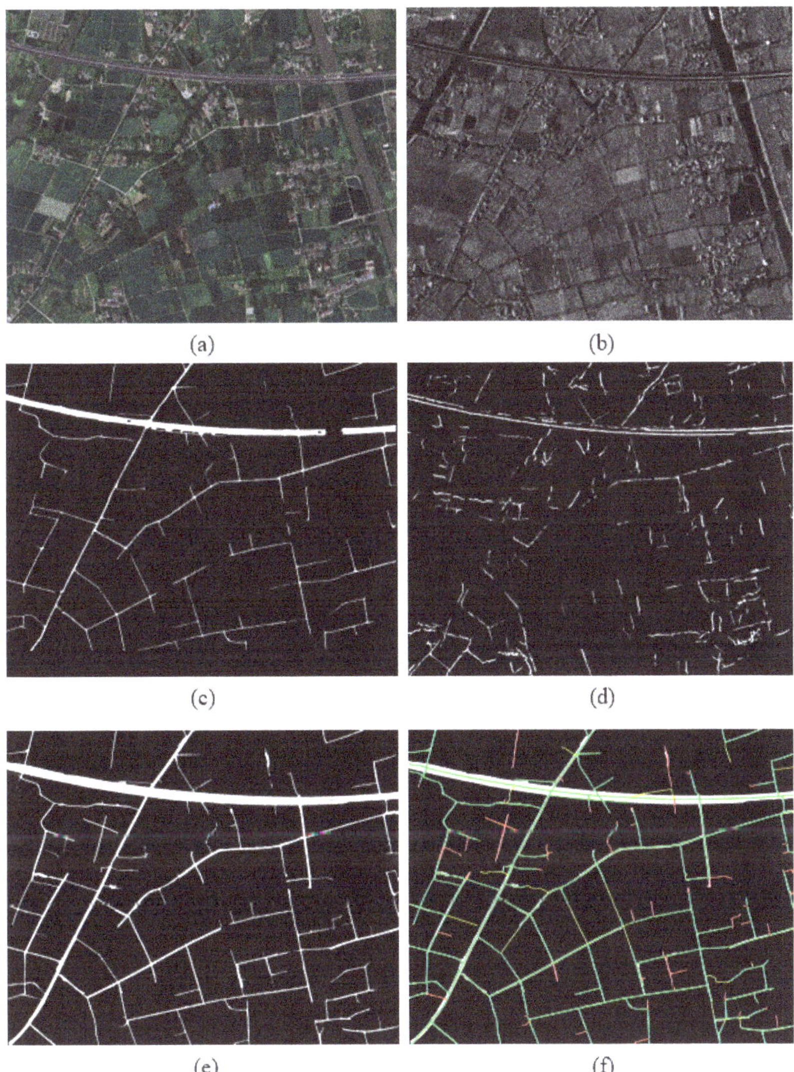

Figure 24. Tested data Area 2. Optical and SAR remote sensing images and road extraction results in the Chongzhou area. (**a**) Worldview-2 optical remote sensing image; (**b**) TerraSAR-X remote sensing image; (**c**) road extraction results of optical remote sensing image; (**d**) road extraction results from SAR remote sensing images; (**e**) road fusion extraction results of optical and SAR images; (**f**) ground truth and marking results of road fusion extraction results (green refers to the correctly extracted road, red refers to the incorrectly extracted road, and yellow refers to the omitted real road).

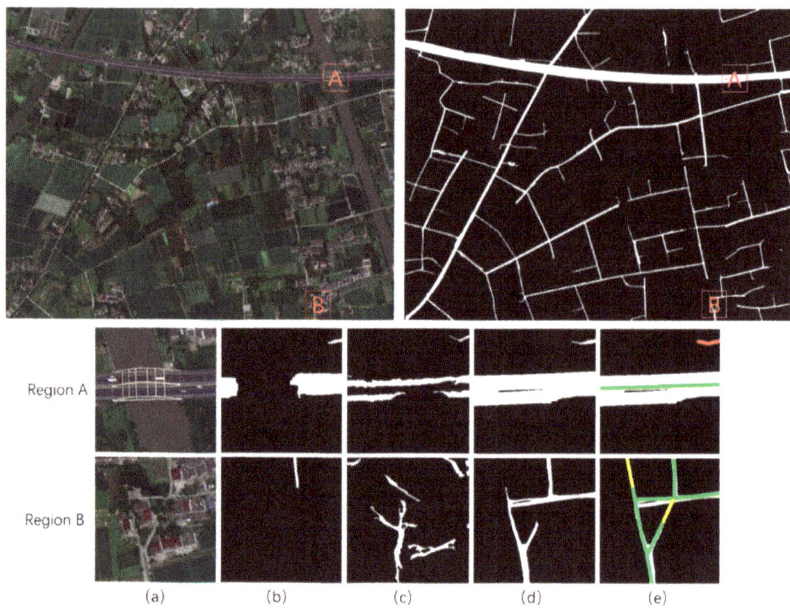

Figure 25. Tested data Area 2. Some details in optical and SAR remote sensing images and road extraction results in the Wuzhen area. (**a**) Worldview-4 optical remote sensing image; (**b**) road extraction results of optical remote sensing image; (**c**) road extraction results from SAR remote sensing images; (**d**) road fusion extraction results of optical and SAR images; (**e**) ground truth and marking results of road fusion extraction results (green refers to the correctly extracted road, red refers to the incorrectly extracted road, and yellow refers to the omitted real road).

Table 6. The results of two large scale images for the whole area.

Tested Data	Extraction Results Based on SDG-DenseNet (WorldView Optical Image)			Extraction Results Based on Decision-Level Fusion Method (SAR and Optical Image)		
Metrics	*F1*	*COR*	*COM*	*F1*	*COR*	*COM*
Area 1	0.7376	0.8567	0.6476	0.8528	0.9336	0.7849
Area 2	0.8047	0.7923	0.8176	0.8885	0.8680	0.9100

5. Conclusions

In this research, a D-Dense block module was proposed, which combined traditional convolution and dilated convolution based on a dense connection structure. Further, the new semantic segmentation network (SDG-DenseNet) was built with a D-Dense block, and it also adopted the center part of the D-LinkNet for high-resolution satellite imagery road extraction. Since the network also replaces the initial block with the stem block to hold more detailed information, it can be easier to recover the marginal information of the object in the decoder phase. In addition, the introduction of an attention mechanism also improves the ability of the network to obtain global information. Besides, to improve the accuracy of road extraction in large-scale images in practical application, a decision-level fusion method was proposed, which fused the information in optical images and SAR images.

Three sets of satellite images were applied to evaluate the network. The extraction results from the Massachusetts Roads dataset show that the SDG-DenseNet not only has the highest *IoU* and *F1* score but is also suitable to extract roads in complicated scenes.

Experiments showed that the *IoU* and *F*1 scores of SDG-DenseNet based on D-Dense block and GIRM modules were 3.61% and 2.75% higher, respectively, than the baseline D-LinkNet. The stem block is helpful to develop the accuracy for road extraction. Furthermore, the Chongzhou–Wuzhen dataset, based on three large-scale optical images, was applied to evaluate the models' extraction ability of the low-grade roads. The results show that the SDG-DenseNet performs best in four networks and its *IoU* score is 6.65% higher than that of D-LinkNet. At the same time, its model size is reduced by about 600 MB to D-LinkNet. Further, two pairs of large-scale optical and SAR images were applied to evaluate the decision-level fusion method. The results show that the fusion method performed well in accurately extracting the roads. After decision-level fusion of road binary map from SAR and optical image based on two tested data, the *F*1 is improved by about 8.4–11.5%, *COR* is about 7.4–7.7%, and *COM* is about 9.3–13.7%.

SDG-DenseNet improves d-block as d-blockplus and combines it with an attention mechanism, which not only ensures road completeness in the segmentation task but also greatly improves the correctness of the segmentation results. Therefore, the network maintains a perfect balance between correctness and completeness. In addition, the decision-level fusion method had been proposed to improve the extraction effect on the task of low-grade road extraction, and the presentation quality is better after the decision-level fusion. In future research, the contribution of each part of the network and every hyperparameter in the training phase should be taken into consideration.

Author Contributions: Methodology, J.Z.; software, F.X.; validation, Y.S.; formal analysis, B.P.; investigation, J.Z.; resources, Y.L.; data curation, B.P.; writing—original draft preparation, J.Z.; writing—review and editing, S.L.; supervision, Y.L.; funding acquisition, L.H. All authors have read and agreed to the published version of the manuscript.

Funding: This work was supported by the Key Projects of Global Change and Response of Ministry of Science and Technology of China under Grant 2020YFA0608203, in part by the Science and Technology Support Project of Sichuan Province under Grant 2021YFS0335, Grant 2020YFG0296, and Grant 2020YFS0338, in part by Fengyun Satellite Application Advance Plan under Grant FY-APP-2021.0304.

Data Availability Statement: The authors would like to thank the team of National Climate Center and University of Toronto for the data and experiments.

Conflicts of Interest: The authors declare no conflict of interest.

References

1. Zhou, L.; Zhang, C.; Wu, M. D-linknet: Linknet with pretrained encoder and dilated convolution for high resolution satellite imagery road extraction. In Proceedings of the IEEE Conference on Computer Vision and Pattern Recognition Workshops, Salt Lake City, UT, USA, 18–22 June 2018; pp. 182–186.
2. Long, J.; Shelhamer, E.; Darrell, T. Fully convolutional networks for semantic segmentation. In Proceedings of the IEEE Conference on Computer Vision and Pattern Recognition, Boston, MA, USA, 7–12 June 2015; pp. 3431–3440.
3. Wei, Y.; Zhang, K.; Ji, S. Simultaneous Road Surface and Centerline Extraction From Large-Scale Remote Sensing Images Using CNN-Based Segmentation and Tracing. *IEEE Trans. Geosci. Remote Sens.* **2020**, *58*, 8919–8931. [CrossRef]
4. Yang, F.; Wang, H.; Jin, Z. A fusion network for road detection via spatial propagation and spatial transformation. *Pattern Recognit.* **2020**, *100*, 107141. [CrossRef]
5. Zhou, M.; Sui, H.; Chen, S.; Wang, J.; Chen, X. BT-RoadNet: A boundary and topologically-aware neural network for road extraction from high-resolution remote sensing imagery. *ISPRS J. Photogramm. Remote Sens.* **2020**, *168*, 288–306. [CrossRef]
6. Chen, Z.; Wang, C.; Li, J.; Xie, N.; Han, Y.; Du, J. Reconstruction Bias U-Net for Road Extraction From Optical Remote Sensing Images. *IEEE J. Sel. Top. Appl. Earth Obs. Remote Sens.* **2021**, *14*, 2284–2294. [CrossRef]
7. He, X.; Zemel, R.S.; Carreira-Perpiñán, M.Á. Multiscale conditional random fields for image labeling. In Proceedings of the 2004 IEEE Computer Society Conference on Computer Vision and Pattern Recognition, Washington, DC, USA, 27 June–2 July 2004; Volume 2.
8. Shotton, J.; Winn, J.; Rother, C.; Criminisi, A. Textonboost for image understanding: Multi-class object recognition and segmentation by jointly modeling texture, layout, and context. *Int. J. Comput. Vis.* **2009**, *81*, 2–23. [CrossRef]
9. Krähenbühl, P.; Koltun, V. Efficient inference in fully connected crfs with gaussian edge potentials. *Adv. Neural Inf. Process. Syst.* **2011**, *24*, 109–117.
10. Galleguillos, C.; Belongie, S. Context based object categorization: A critical survey. *Comput. Vis. Image Underst.* **2010**, *114*, 712–722. [CrossRef]

11. Farabet, C.; Couprie, C.; Najman, L.; Lecun, Y. Learning hierarchical features for scene labeling. *IEEE Trans. Pattern Anal. Mach. Intell.* **2013**, *35*, 1915–1929. [CrossRef]

12. Eigen, D.; Fergus, R. Predicting depth, surface normal and semantic labels with a common multi-scale convolutional architecture. In Proceedings of the IEEE International Conference on Computer Vision, Santiago, Chile, 7–13 December 2015; pp. 2650–2658.

13. Pinheiro PH, O.; Collobert, R. Recurrent convolutional neural networks for scene labeling. In Proceedings of the 31st International Conference on Machine Learning, Beijing, China, 21–26 June 2014.

14. Chen, L.C.; Yang, Y.; Wang, J.; Xu, W.; Yuille, A.L. Attention to scale: Scale-aware semantic image segmentation. In Proceedings of the IEEE Conference on Computer Vision and Pattern Recognition, Las Vegas, NV, USA, 27–30 June 2016; pp. 3640–3649.

15. Lin, G.; Shen, C.; Van Den Hengel, A.; Reid, I. Efficient piecewise training of deep structured models for semantic segmentation. In Proceedings of the IEEE Conference on Computer Vision and Pattern Recognition, Las Vegas, NV, USA, 27–30 June 2016; pp. 3194–3203.

16. Badrinarayanan, V.; Kendall, A.; Cipolla, R. Segnet: A deep convolutional encoder-decoder architecture for image segmentation. *IEEE Trans. Pattern Anal. Mach. Intell.* **2017**, *39*, 2481–2495. [CrossRef]

17. Ronneberger, O.; Fischer, P.; Brox, T. U-net: Convolutional networks for biomedical image segmentation. In Proceedings of the International Conference on Medical Image Computing and Computer-Assisted Intervention, Munich, Germany, 5–9 October 2015; Springer: Cham, Switzerland, 2015; pp. 234–241. [CrossRef]

18. Lin, G.; Milan, A.; Shen, C.; Reid, I. RefineNet: Multi-path refinement networks with identity mappings for high-resolution semantic segmentation. *arXiv* **2016**, arXiv:1611.06612.

19. Pohlen, T.; Hermans, A.; Mathias, M.; Leibe, B. Full-resolution residual networks for semantic segmentation in street scenes. In Proceedings of the IEEE Conference on Computer Vision and Pattern Recognition, Honolulu, HI, USA, 21–26 July 2017; pp. 4151–4160.

20. Peng, C.; Zhang, X.; Yu, G.; Sun, J. Large Kernel Matters—Improve Semantic Segmentation by Global Convolutional Network. In Proceedings of the IEEE Conference on Computer Vision and Pattern Recognition, Honolulu, HI, USA, 21–26 July 2017; pp. 4353–4361.

21. Amirul Islam, M.; Rochan, M.; Bruce ND, B.; Wang, Y. Gated feedback refinement network for dense image labeling. In Proceedings of the IEEE Conference on Computer Vision and Pattern Recognition, Honolulu, HI, USA, 21–26 July 2017; pp. 3751–3759.

22. Chen, L.C.; Papandreou, G.; Kokkinos, I.; Murphy, K.; Yuille, A.L. Semantic image segmentation with deep convolutional nets and fully connected crfs. *arXiv* **2014**, arXiv:1412.7062.

23. Yu, F.; Koltun, V. Multi-scale context aggregation by dilated convolutions. *arXiv* **2015**, arXiv:1511.07122.

24. Zhao, H.; Shi, J.; Qi, X.; Wang, X.; Jia, J. Pyramid scene parsing network. In Proceedings of the IEEE Conference on Computer Vision and Pattern Recognition, Honolulu, HI, USA, 21–26 July 2017; pp. 2881–2890.

25. Hu, C.; Bai, X.; Qi, L.; Chen, P.; Xue, G.; Mei, L. Vehicle color recognition with spatial pyramid deep learning. *IEEE Trans. Intell. Transp. Syst.* **2015**, *16*, 2925–2934. [CrossRef]

26. Chen, L.C.; Papandreou, G.; Kokkinos, I.; Murphy, K.; Yuille, A.L. Deeplab: Semantic image segmentation with deep convolutional nets, atrous convolution, and fully connected crfs. *IEEE Trans. Pattern Anal. Mach. Intell.* **2018**, *40*, 834–848. [CrossRef]

27. Chen, L.C.; Papandreou, G.; Schroff, F.; Adam, H. Rethinking atrous convolution for semantic image segmentation. *arXiv* **2017**, arXiv:1706.05587.

28. Chen, L.C.; Zhu, Y.; Papandreou, G.; Schroff, F.; Adam, H. Encoder-decoder with atrous separable convolution for semantic image segmentation. In Proceedings of the European Conference on Computer Vision (ECCV), Munich, Germany, 8–14 September 2018; pp. 801–818.

29. Wang, X.; Girshick, R.; Gupta, A.; He, K. Non-local neural networks. In Proceedings of the 2018 IEEE/CVF Conference on Computer Vision and Pattern Recognition, Salt Lake City, UT, USA, 18–23 June 2018; pp. 7794–7803.

30. Fu, J.; Liu, J.; Tian, H.; Li, Y.; Bao, Y.; Fang, Z.; Lu, H. Dual attention network for scene segmentation. In Proceedings of the IEEE/CVF Conference on Computer Vision and Pattern Recognition, Long Beach, CA, USA, 16–17 June 2019; pp. 3146–3154.

31. Woo, S.; Park, J.; Lee, J.Y.; Kweon, I.S. CBAM: Convolutional Block Attention Module. In *Computer Vision—ECCV 2018. ECCV 2018*; Ferrari, V., Hebert, M., Sminchisescu, C., Weiss, Y., Eds.; Lecture Notes in Computer Science; Springer: Cham, Switzerland, 2018; Volume 11211.

32. Wan, J.; Xie, Z.; Xu, Y.; Chen, S.; Qiu, Q. DA-RoadNet: A Dual-Attention Network for Road Extraction From High Resolution Satellite Imagery. *IEEE J. Sel. Top. Appl. Earth Obs. Remote Sens.* **2021**, *14*, 6302–6315. [CrossRef]

33. Bentabet, L.; Jodouin, S.; Ziou, D.; Vaillancourt, J. Road vectors update using SAR imagery: A snake-based method. *IEEE Trans. Geosci. Remote Sens.* **2003**, *41*, 1785–1803. [CrossRef]

34. Sun, Z.; Geng, H.; Lu, Z.; Scherer, R.; Woźniak, M. Review of Road Segmentation for SAR Images. *Remote Sens.* **2021**, *13*, 1011. [CrossRef]

35. Jiang, Y.H.; Pi, Y.J. SAR image road detection based on Hough transform and genetic algorithm. *Radar Sci. Technol.* **2005**, *3*, 156–162.

36. Wei, X.; Lv, X.; Zhang, K. Road Extraction in SAR Images Using Ordinal Regression and Road-Topology Loss. *Remote Sens.* **2021**, *13*, 2080. [CrossRef]

37. Szegedy, C.; Ioffe, S.; Vanhoucke, V.; Alemi, A.A. Inception-v4, inception-resnet and the impact of residual connections on learning. In Proceedings of the Thirty-First AAAI Conference on Artificial Intelligence, San Francisco, CA, USA, 2–4 February 2017.
38. Rahman, M.A.; Wang, Y. Optimizing intersection-over-union in deep neural networks for image segmentation. In Proceedings of the International Symposium on Visual Computing, Las Vegas, NV, USA, 12–14 December 2016; pp. 234–244.
39. Jégou, S.; Drozdzal, M.; Vazquez, D.; Romero, A.; Bengio, Y. The one hundred layers tiramisu: Fully convolutional densenets for semantic segmentation. In Proceedings of the IEEE Conference on Computer Vision and Pattern Recognition Workshops, Honolulu, HI, USA, 21–26 July 2017; pp. 11–19.
40. LeCun, Y.A.; Bottou, L.; Orr, G.B.; Orr, G.B.; Muller, K.R. *Efficient Backprop in Neural Networks: Tricks of the Trade*; Springer: Berlin/Heidelberg, Germany, 2012; pp. 9–48.
41. Huang, G.; Liu, Z.; Van Der Maaten, L.; Weinberger, K.Q. Densely connected convolutional networks. In Proceedings of the IEEE Conference on Computer Vision and Pattern Recognition, Honolulu, HI, USA, 21–26 July 2017; pp. 4700–4708.
42. Sandler, M.; Howard, A.; Zhu, M.; Zhmoginov, A.; Chen, L.-C. MobileNetV2: Inverted Residuals and Linear Bottlenecks. In Proceedings of the 2018 IEEE/CVF Conference on Computer Vision and Pattern Recognition, Salt Lake City, UT, USA, 18–23 June 2018; pp. 4510–4520. [CrossRef]
43. Xiao, F.; Chen, Y.; Tong, L.; He, L.; Tan, L.; Wu, B. Road detection in high-resolution SAR images using Duda and path operators. In Proceedings of the 2016 IEEE International Geoscience and Remote Sensing Symposium (IGARSS), Beijing, China, 10–15 July 2016; pp. 1266–1269. [CrossRef]
44. Mnih, V.; Hinton, G.E. Learning to Detect Roads in High-Resolution Aerial Images. In Proceedings of the Computer Vision—ECCV 2010—11th European Conference on Computer Vision, Heraklion, Crete, Greece, 5–11 September 2010; Proceedings, Part VI; Springer: Berlin/Heidelberg, Germany, 2010.
45. Sun, T.; Chen, Z.; Yang, W.; Wang, Y. Stacked u-nets with multi-output for road extraction. In Proceedings of the IEEE/CVF Conference on Computer Vision and Pattern Recognition Workshops (CVPRW), Salt Lake City, UT, USA, 18–22 June 2018; pp. 202–206.
46. Zhang, Z.; Liu, Q.; Wang, Y. Road extraction by deep residual U-Net. *IEEE Geosci. Remote Sens. Lett.* **2018**, *15*, 749–753. [CrossRef]
47. Zhang, L.; Lan, M.; Zhang, J.; Tao, D. Stagewise Unsupervised Domain Adaptation with Adversarial Self-Training for Road Segmentation of Remote-Sensing Images. *IEEE Trans. Geosci. Remote Sens.* **2021**, *60*, 5609413. [CrossRef]
48. Zhang, Z.; Wang, Y. JointNet: A common neural network for road and building extraction. *Remote Sens.* **2019**, *11*, 696. [CrossRef]
49. Dey, M.S.; Chaudhuri, U.; Banerjee, B.; Bhattacharya, A. Dual-Path Morph-UNet for Road and Building Segmentation From Satellite Images. *IEEE Geosci. Remote Sens. Lett.* **2022**, *19*, 3004005. [CrossRef]

 *remote sensing*

Article

A Lightweight Self-Supervised Representation Learning Algorithm for Scene Classification in Spaceborne SAR and Optical Images

Xiao Xiao, Changjian Li and Yinjie Lei *

College of Electronics and Information Engineering, Sichuan University, Chengdu 610064, China;
xiaoxiaox@stu.scu.edu.cn (X.X.); li_changjian@stu.scu.edu.cn (C.L.)
* Correspondence: yinjie@scu.edu.cn

Abstract: Despite the increasing amount of spaceborne synthetic aperture radar (SAR) images and optical images, only a few annotated data can be used directly for scene classification tasks based on convolution neural networks (CNNs). For this situation, self-supervised learning methods can improve scene classification accuracy through learning representations from extensive unlabeled data. However, existing self-supervised scene classification algorithms are hard to deploy on satellites, due to the high computation consumption. To address this challenge, we propose a simple, yet effective, self-supervised representation learning (Lite-SRL) algorithm for the scene classification task. First, we design a lightweight contrastive learning structure for Lite-SRL, we apply a stochastic augmentation strategy to obtain augmented views from unlabeled spaceborne images, and Lite-SRL maximizes the similarity of augmented views to learn valuable representations. Then, we adopt the stop-gradient operation to make Lite-SRL's training process not rely on large queues or negative samples, which can reduce the computation consumption. Furthermore, in order to deploy Lite-SRL on low-power on-board computing platforms, we propose a distributed hybrid parallelism (DHP) framework and a computation workload balancing (CWB) module for Lite-SRL. Experiments on representative datasets including OpenSARUrban, WHU-SAR6, NWPU-Resisc45, and AID dataset demonstrate that Lite-SRL can improve the scene classification accuracy under limited annotated data, and it is generalizable to both SAR and optical images. Meanwhile, compared with six state-of-the-art self-supervised algorithms, Lite-SRL has clear advantages in overall accuracy, number of parameters, memory consumption, and training latency. Eventually, to evaluate the proposed work's on-board operational capability, we transplant Lite-SRL to the low-power computing platform NVIDIA Jetson TX2.

Keywords: synthetic aperture radar; optical images; scene classification; on-board; lightweight self-supervised algorithm

Citation: Xiao, X.; Li, C.; Lei, Y. A Lightweight Self-Supervised Representation Learning Algorithm for Scene Classification in Spaceborne SAR and Optical Images. *Remote Sens.* **2022**, *14*, 2956. https://doi.org/10.3390/rs14132956

Academic Editors: Tianwen Zhang, Tianjiao Zeng and Xiaoling Zhang

Received: 24 May 2022
Accepted: 15 June 2022
Published: 21 June 2022

1. Introduction

The remote sensing scene classification (RSSC) task aims to classify scene regions into different semantic categories [1–5], which plays an essential role in various Earth observation applications, i.e., land resource exploration, forest inventory, urban-area monitoring [6–8]. In recent years, Landsat, Sentinel, and other missions have provided an increasing number of spaceborne images for scene classification task, including synthetic aperture radar (SAR) images and optical images. With more available data, scene classification methods based on convolution neural networks (CNN) have undergone rapid growth [7,9].

However, the amount of annotated scene data available for supervised CNN training remains limited. Taking SAR data as an example, SAR images are affected by speckle noise due to the imaging mechanism, resulting in poor image quality [10,11]. In addition, the random fluctuation of pixels makes it difficult to distinguish between scene categories [12]. Therefore, the annotation of SAR images requires experienced experts and is a time-consuming task [13]. The same problem of high annotation costs exists for optical images. This leads to the total images number of RSSC datasets, i.e., OpenSARUrban [14], WHU-SAR6 [11], NWPU-Resisc45 [3], and AID [15], compared with natural image datasets, i.e., ImageNet [16], being much smaller; the specific images number for each datasets is shown Figure A1. With limited annotated samples, CNN tends to be overfitted after training [17], leading to poor generalization performance in RSSC task. Therefore, exploring methods to reduce RSSC task's reliance on annotated data is appealing.

Recently, self-supervised learning (SSL) has emerged as an attractive candidate for solving the problem of labeled data shortage [18]. SSL methods can learn valuable representations from unlabeled images through solving pretext tasks [19]; the network trained in a self-supervised fashion can be used as a pre-trained model to enable higher accuracy with fewer training samples [20]. To this end, an increasing number of RSSC studies have concentrated on SSL. In practice, remote sensing images (RSIs) differ significantly from natural images in the acquisition and transmission stage—RSIs suffer from noise impact and high transmission costs [21]. Performing self-supervised training on satellites can solve these issues; however, existing SSL algorithms are hard to deploy on satellites due to the high computation consumption. The method based on self-supervised instance discrimination [22] was first applied in RSSC task; soon after, the SSL algorithm represented by contrastive multiview coding [20] showed good performance in RSSC tasks. These methods relies on a large batch of negative samples, while the training process needs to maintain large queues, which can consume much computation resources. Other self-supervised methods utilize images with the same geographic coordinate regions from different times and introduce loss functions based on geographic coordinate with complex feature extraction modules [23], which also consume a lot of resources during training. Therefore, we need to reduce the computation consumption during self-supervised training.

As mentioned above, we attempt to deploy a self-supervised algorithm on satellites. A lightweight network is necessary, while a practical on-board training approach can also provide support. Since it is impracticable to carry high-power GPUs on satellites, current trend is to use edge devices, i.e., NVIDIA Jetson TX2 [24] as on-board computing devices [25]. Latest radiation characterized on-board computing modules, such as the S-A1760 Venus [26], utilizes TX2 inside the product to help spacecraft achieve high performance AI computing. Accordingly, we also use TX2 as the deployment platform. As under resource-limited scenarios (limited memory, i.e., memory size of 8 G, limited computation resources, i.e., bandwidth of 59.7 GB/S), distributed strategies are typically applied to train the network; thus, for on-board training a flexible distributed training framework is required. However, the approaches adopted by deep learning frameworks, i.e., PyTorch [27], TensorFlow [28], and Caffe [29], for distributed training remain primitive. Existing dedicated distributed training frameworks, such as Mesh-TensorFlow [30] and Nemesyst [31], are likewise incapable for on-board scenarios, because they fail to consider the case of limited on-board computation resources.

Based on the above observation, we need (i) a self-supervised learning algorithm that satisfies guaranteed accuracy and low computation consumption simultaneously; and (ii) an effective distributed strategy for on-board self-supervised training deployment. To address these challenges, we propose a lightweight On-board Self-supervised Representation Learning (Lite-SRL) algorithm for RSSC task. Lite-SRL uses a contrastive learning structure that contains lightweight modules, by maximizing the similarity of RSIs' augmented views to capture distinguishable feature from unlabeled images. The augmentation strategies we used to obtain contrast views differ slightly between SAR and optical images. Meanwhile, inspired by self-supervised learning algorithm BYOL [32] and SimSiam [33], we use the stop-gradient operation making the training process not rely on large batch size, queues, or negative sample pairs, which greatly reduces the computation workload with guaranteed accuracy. Moreover, the structure of Lite-SRL is adapted to distributed training for deployment. Experiments on representative scene classification datasets including OpenSARUrban, WHU-SAR6, NWPU-Resisc45, and AID dataset demonstrate that

Lite-SRL can improve the scene classification accuracy with limited annotated data; it also demonstrates that Lite-SRL is generalizable to both SAR and optical images in RSSC task. Meanwhile, experiments with six state-of-the-art self-supervised algorithms demonstrate that Lite-SRL has clear advantages in overall accuracy, number of parameters, memory consumption, and training latency.

In order to deploy Lite-SRL algorithm to the low-power computing platform Jetson TX2, we propose a distributed hybrid parallelism (DHP) training framework along with a generic training computation workload balancing module (CWB). Since a single TX2 node cannot complete the whole network training, CWB automatically partitions the network according to the workload balancing principle (View Algorithm 2 for details) and assigns each part to DHP to realize distributed hybrid parallelism training. The integration of CWB and DHP enables training neural networks under limited on-board resources. Eventually, we transplant Lite-SRL algorithm to the on-board computing platform through the distributed training modules.

The main contributions of this article are as follows:

1. To improve the scene classification accuracy under insufficient annotated data, we proposed a simple yet effective self-supervised representation learning algorithm called Lite-SRL. To reduce computation consumption, we design a lightweight contrastive learning structure in Lite-SRL and adopt the stop-gradient operation;
2. To realize on-board deployment of Lite-SRL algorithm, we proposed a training framework called DHP and a generic computation workload balancing module CWB. As far as we know, we represent the first work to combine self-supervised learning with on-board data processing;
3. Extensive experiments on four representative datasets demonstrated that Lite-SRL could improve the scene classification accuracy under limited annotated data, and it is generalizable to SAR and optical images. Compared with six state-of-the-art methods, Lite-SRL had clear advantages in overall accuracy, number of parameters, memory consumption, and training latency;
4. Eventually, to evaluate the proposed work's on-board operational capability, we transplant Lite-SRL to the low-power computing platform NVIDIA Jetson TX2.

The remainder of this paper is organized as follows: Section 2 covers research works related to this article. Section 3 presents the detailed research steps. Section 4 presents the experimental setups. In Section 5 detailed experimental results are presented and summarized. Section 6 provides detailed records for the deployment process. Section 7 provides conclusions. Appendix A lists all the abbreviations in this article and their corresponding full names.

2. Related Works

In this section, we provide a brief review of existing related works. We present solutions of related RSSC works under limited labeled samples, among which, the methods based on self-supervised contrastive learning show excellent results, and we further present the development of self-supervised contrastive learning. We also offer the existing related studies on distributed training.

2.1. RSSC under Limited Annotated Samples

Recently, self-supervised learning (SSL) has attracted considerable interest in the study of RSSC for solving the problem of labeled data shortage. SCL_MLNet [34] introduced an end-to-end self-supervised contrastive learning-based metric network for few-shot RSSC task. Li et al. [35] proposed Meta-FSEO model to improve the performance of few-shot RSSC task in varying urban scenes. These few-shot learning tasks validate that SSL enables RSSC models to achieve well generalization performance from only a few annotated data. Meanwhile, studies [20,36] proved that using the same domain images for SSL training in RSSC task can help to overcome classical transfer learning problems, which further demonstrates the effectiveness of using SSL as a pre-training process in RSSC. The authors of [20,36,37] explored the effectiveness of several SSL networks in RSSC task, among which the contrastive learning-based [22,23] SSL algorithm performed best in the RSSC task. Moreover, Jung et al. [38] presented self-supervised contrastive learning solution with smoothed representation for RSSC based on the SimCLR [22] framework. Zhao et al. [39] introduced a self-supervised contrastive learning algorithm to achieve hyperspectral image classification for problems with few labeled samples. It has been proved by the above works that self-supervised contrastive learning provides a great improvement for RSSC task; thus, our work also adopts the self-supervised contrastive learning method for RSSC task.

2.2. Self-Supervised Contrastive Learning

Through solving pretext tasks, self-supervised methods utilize unlabeled data to learn representations that can be transferred to downstream tasks. In self-supervised learning methods, relative position predicting [19,40], image inpainting [41], and instance-wise contrastive learning [22] are three common pretext tasks. As mentioned above, the validity of contrastive learning is superior to image in-painting and relative position predicting in RSSC tasks. Current state-of-the-art contrastive learning methods differ in detail. SimCLR [22] and MoCo [42] benefit from a large queue of negative samples. Based on earlier versions, MoCo-v2 [43] adds the same nonlinear layer as SimCLR to the encoder representation. MoCo-v2 and SimCLR perform well when maintaining a larger batch. SwAV [44] is another type of clustering-based idea that combines clusters into contrastive learning networks. SwAV computes assignments separately from the two augmented views to perform unsupervised clustering. The clustering-based approach likewise requires large queues or memory banks to supply sufficient samples for clustering. BYOL [32] is characterized by not requiring negative sample pairs and, thus, can eliminate the need to maintain a very large batch of negative sample queues. With no reliance on negative samples, BYOL is more robust to the choice of data enhancement methods. SimSiam [33] is similar to BYOL but with no momentum encoder; meanwhile, it directly shares weights between two branches. SimSiam's experiments demonstrate that without using any of the negative sample pairs, large batch, and momentum encoders, contrastive learning structures can still learn valuable representations. We applied these methods to RSSC tasks and synthetically compared them with our proposed algorithm.

2.3. Distributed Training under Limited Resources

Distributed training assigns the training process to multiple computing devices for collaborative execution [45]. Current mainstream distributed training methods can be divided into data parallelism [46,47], model parallelism [47,48], and hybrid parallelism [49,50]. In data parallelization, each node trains a duplicate of the model using different mini-batches of data. All nodes contain a complete copy of the model and compute the gradients individually [47], after training the parameters of the final model can be updated through the server. In model parallelization, the network layers are divided into multiple partitions and distributed over multiple nodes for parallel training [47,49]. During model parallelization training, each node has different parameters and is responsible for the computation of different partition layers, and each node updates only the weights of assigned partitions. Hybrid parallelism is a combination of data parallelism and model parallelism, which is the development trend of distributed training. Mesh-TensorFlow [30] and Nemesyst [31] are two end-to-end hybrid parallel training frameworks, both using small independent batches of data for training. Based on Mesh-TensorFlow, Moreno-Alvarez et al. [51] proposed a static load balancing approach for the model parallelism scheme. Akintoye et al. [49] proposed a generalized hybrid parallelization approach to optimize partition allocation on available GPUs. FlexFlow [47] framework applied a simulator to predict optimal parallelization strategy in order to improve training efficiency on GPU clusters. However, the above distributed training frameworks failed to consider resource-limited scenarios; in addition, they do not perform computation workload balancing for training process.

3. Methods

3.1. Overview of the Proposed Framework

The overview of the proposed work is shown in Figure 1; our work consists of two main parts: (i) we propose a self-supervised algorithm Lite-SRL for RSSC task, the algorithm satisfies guaranteed accuracy and low computation consumption simultaneously. (ii) We use a low-power computing platform for deployment and we propose a set of distributed training modules to satisfy the requirements.

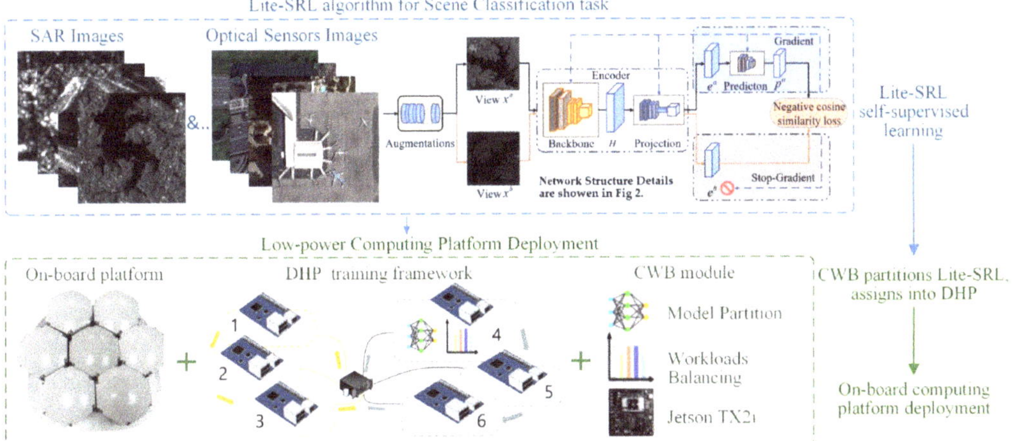

Figure 1. Overview of the proposed work. Lite-SRL: on-board self-supervised representation learning algorithm for RSSC task; CWB: computation workload balancing module; DHP: on-board distributed hybrid parallelism training framework.

To improve the scene classification accuracy with limited annotated data, Lite-SRL learns valuable representations from unlabeled RS images. During the algorithm deployment, CWB automatically partitions the training process of Lite-SRL according to the workload balancing principle (View Algorithm 2 for details) and assigns each partition into the DHP training framework, achieving high efficiency on-board self-supervised training.

3.2. Lite-SRL Self-Supervised Representation Learning Network

3.2.1. Network Structure

We propose an On-board Self-supervised Representation Learning (Lite-SRL) network for RSSC tasks. Since SimSiam [33] and BYOL [32] have excelled as effective self-supervised contrastive learning methods for many downstream tasks, we use a similar structure as the pretext task for self-supervised contrastive learning and make the training process less resource-intensive. Based on SimSiam's experiment results, our Lite-SRL directly maximizes the similarity of two augmented views of an image without using either negative pairs or momentum encoders, and, thus, the training process does not rely on large batches or queues. Lite-SRL adopts lightweight structures as detailed in Figure 2, allowing us to achieve high accuracy with fewer parameters and training resource usage.

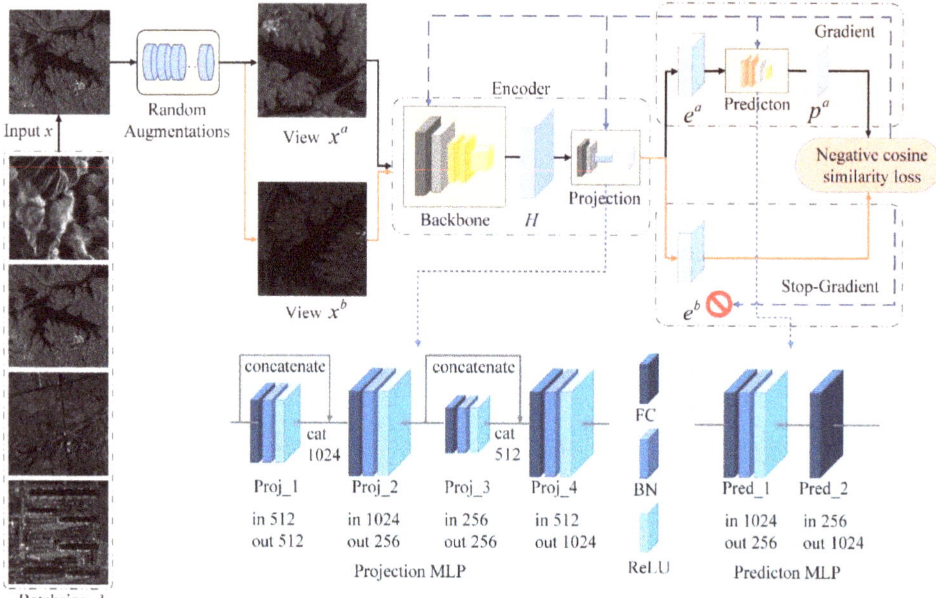

Figure 2. Network structure.

The structure of Lite-SRL is shown in Figure 2, where two randomly augmented views x^a and x^b are obtained from the training batch $\{x_1, x_2, \cdots x_k\}$ as inputs, with the top and bottom paths sharing the parameters of Encoder. These two views are processed separately by encoder E, which consists of Backbone and Projection. Prediction is denoted as P, it converts the output of one view after encoder and matches it with the other view. Express the output vectors of x^a and x^b are expressed as $p^a = P(E(x^a))$ and $e^b = E(x^b)$. Again, perform the above procedure in reverse order for x^a and x^b, the output vectors are p^b and e^a. Vectors' negative cosine similarity is expressed as follows:

$$N\left(p^a, e^b\right) = -\frac{p^a}{\parallel p^a \parallel_2} \cdot \frac{e^b}{\parallel e^b \parallel_2} \tag{1}$$

here $\parallel \Delta \parallel_2$ is *l*2-norm, $\parallel x \parallel = \sqrt{\sum_i^n (x_i)^2}$. The symmetrization loss is expressed as follows:

$$L\left(x^a, x^b\right) = -\frac{1}{2}\frac{P(E(x^a))}{\parallel P(E(x^a)) \parallel_2} \cdot \frac{E\left(x^b\right)}{\parallel E\left(x^b\right) \parallel_2} - \frac{1}{2}\frac{P\left(E\left(x^b\right)\right)}{\parallel P\left(E\left(x^b\right)\right) \parallel_2} \cdot \frac{E(x^a)}{\parallel E(x^a) \parallel_2} \tag{2}$$

using Equation (1) to simplify the symmetrization loss calculation, Equation (2) yields the following equation:

$$L = \frac{1}{2}N\left(p^a, e^b\right) + \frac{1}{2}N\left(p^b, e^a\right) \tag{3}$$

The overall loss during training is the average of all images in the batch. The study of SimSiam and BYOL demonstrated that the stop gradient operation is the key to avoid collapse during training. More importantly, stop gradient operation allows the training process to not rely on large batch size, queues, or negative sample pairs, which greatly reduces the computation workload. We also use the Stop-Grad operation, as shown in

Figure 2, for the way that does not go through P we apply stop gradient operation to it when performing back propagation, modifying (1) as follows:

$$N\left(p^a, stop\_gradient\left(e^b\right)\right) \tag{4}$$

which means e^b is considered as a constant in this term. By adding Stop-Grad operation, the form in Equation (3) is realized as:

$$L = \frac{1}{2}N\left(p^a, stop\_gradient\left(e^b\right)\right) + \frac{1}{2}N\left(p^b, stop\_gradient(e^a)\right) \tag{5}$$

The encoder of x^b in the first term of Equation (5) does not receive the gradient from e^b, instead receives the gradient from p^b in the second term, and the operation performed on the gradient of x^a is opposite to that of x^b. After obtaining the contrastive loss, we use the stochastic gradient descent (SGD) optimizer to perform back propagation and update the network parameters. The learning procedure is formally presented in Algorithm 1. The structure of the Projection and Prediction multi-layer perceptron (MLP) modules in Lite-SRL are shown in Figure 2. We use lightweight MLP modules, each fully connected layer in Projection MLP is connected to batch normalization (BN) layer and rectified linear unit (ReLU), we incorporated two concatenate layers in the structure. Prediction MLP uses a bottleneck structure, as detailed in Figure 2. Neither BN nor ReLU is used in the last output layer, and such a structure prevents training collapse [39,44].

Algorithm 1. Learning Procedure of Lite-SRL

E: Encoder with Backbone and Projection MLP
P: Prediction MLP
Aug: random image augmentation
θ: parameters of E and P
Stop: stop-gradient operation
Input: Training samples $\{x_1, x_2, \cdots x_k\}$
Output: negative cosine similarity loss
1: **for** number of training epochs do
2: Training samples $\{x_1, x_2, \cdots x_k\}$ in a minibatch form
3: Do augmentation $\left(\{x_1^a, x_2^a, \cdots x_k^a\}, \{x_1^b, x_2^b, \cdots x_k^b\}\right) - Aug(\{x_1, x_2, \quad x_k\})$
4: In Lite-SRL 2-way do
$\{e_1^a, e_2^a, \cdots e_k^a\} = E(\{x_1^a, x_2^a, \cdots x_k^a\})$ and $\{p_1^a, p_2^a, \cdots p_k^a\} = P(\{e_1^a, e_2^a, \cdots e_k^a\})$;
$\{e_1^b, e_2^b, \cdots e_k^b\} = E\left(\{x_1^b, x_2^b, \cdots x_k^b\}\right)$ and $\{p_1^b, p_2^b, \cdots p_k^b\} = P\left(\{e_1^b, e_2^b, \cdots e_k^b\}\right)$
5: Calculate negative cosine similarity with stop-gradient operation
$Loss = \dfrac{N(\{p_1^a, p_2^a, \cdots p_k^a\}, Stop(\{e_1^b, e_2^b, \cdots e_k^b\}))}{2} + \dfrac{N(\{p_1^b, p_2^b, \cdots p_k^b\}, Stop(\{e_1^a, e_2^a, \cdots e_k^a\}))}{2}$
6: Do backwards propagation with SGD optimizer
7: Update weights θ
8: **end for**
9: After training, use pre-trained model for downstream Remote Sensing Scene Classification

---

Lite-SRL uses a simple, yet effective, network structure, which has significant advantages over existing self-supervised algorithms in (i) network parameters, (ii) memory consumption, and (iii) the average training latency. Detailed experimental results are shown in Section 5.1.

3.2.2. Lite-SRL Network Partition

In order to deploy the algorithm on low-power computing platforms, the training process of Lite-SRL is adapted to a sequential structure as shown in Figure 3a. For two views, x^a and x^b, of an augmented image, first perform concatenate operation and send them to Encoder together, get the combined output of e^a and e^b, keep the values of e^a and e^b and do not preserve the gradient information. Then send them to Prediction part and get p^a and p^b, use the retained e^a and e^b when calculating the contrastive loss with p^a and p^b, considering e^a or e^b as constant values when applying the stop-gradient operation. This allows the two contrastive losses to be calculated simultaneously.

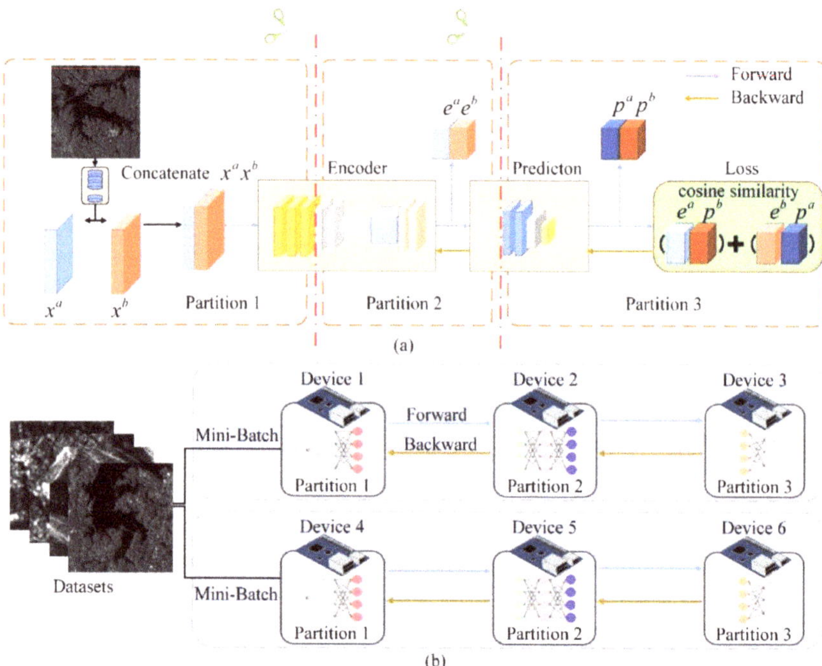

Figure 3. (**a**) We design the training process of Lite-SRL as a sequential structure to adapt model parallelization. (**b**) Schematic of the proposed distributed hybrid parallel (DHP) training baseline.

Each convolution layer within a CNN structure can be used as a single partition to achieve highly efficient model parallelism training capabilities. Given a network M that consists of layers $\{L_1, L_2 \cdots, L_q\}$. Divide network M into n partitions $\{P_1, P_2 \cdots, P_n\}$ where $P_i = \{L_j, L_{j+1} \cdots, L_{j+K}\}$, $\{L_j, L_{j+1} \cdots, L_{j+K}\}$ denotes partition P_i is start from layer L_j and contains k layers. Except for this, the calculation of all partitions is sequential, partition P_i transmit its output feature to its next partition P_{i+1}, while the gradient calculated by partition P_i is transmitted to the front partition P_{i-1}. At iteration t, during forward propagation send the input $A_{L_{i-1}}^t$ from partition P_{i-1} to partition P_i and delivers activation $A_{L_i}^t$. Identically, during backward propagation of iteration t, the $G_{L_{i+1}}^t$ indicates the gradient calculated by partition P_{i+1}. With each layer $L_i \leq L_x \leq L_q$, we denote the weight parameter of layer L_x as w_x, the gradient is given as:

$$\hat{G}_{w_x}^{t-i} = \frac{\delta A_{L_x}^{t-i}}{\delta w_x^{t-i-1}} \cdot G_{L_{x+1}}^{t-i} \tag{6}$$

We denote the learning rate as γ_{t-i}, Equation (6) is updated by the following equation:

$$w_x^{t-i} = w_x^{t-i-1} - \gamma_{t-i} \cdot \hat{G}_{w_x}^{t-i} \tag{7}$$

For layers in non-sequential CNN, parallel paths are not partitioned; instead the parallel zone is considered as a block. After the network partitioning, the network can be trained in model parallelism mode.

In Figure 3b, the feature and gradient are transferred between devices. Take Device 1 and Device 2, for example; Device 1 is in charge of Partition 1's training and Device 2 is in charge of Partition 2's training. In iteration t, during forward propagation, the last layer of Partition 1 in Device 1 transmits feature value $A^t_{L_1}$ to the first layer of Partition 2 in Device 2. During backward propagation, Device 2 transmits the gradient value $G^t_{L_2}$ to Device 1, the gradient of Partition 1's last layer is $\frac{\delta A^t_{L_1}}{\delta w^{t-1}_1} \cdot G^t_{L_2}$, where w^{t-1}_1 is the weight parameter of Partition 1's last layer obtained from iteration $t-1$. Device 1 updates the weight parameters according to Equations (6) and (7): $w^t_1 = w^{t-1}_1 - \gamma_t \frac{\delta A^t_{L_1}}{\delta w^{t-1}_1} \cdot G^t_{L_2}$, where γ_{t-i} is learning rate.

3.3. Distributed Training Strategy

We use a combination of six TX2 nodes and one high-speed switch to form the low-power computing platform as shown in Figure 15.

With multiple nodes, different amounts of nodes can be flexibly scheduled to participate in the training according to the computing requirements. We propose a distributed hybrid parallelized DHP training framework based on the PyTorch framework, the schematic of DHP is shown in Figure 3b. DHP framework uses TCP communication protocol, and the data transmitted between nodes mainly include the output feature of each layer in the forward propagation, the gradient values obtained from each layer in backward propagation, and the parameters of layers aggregated by each node after reaching the number of iterations. Meanwhile, we propose a generic computation workload balancing module CWB, which can perform model partitioning and workload balancing for a given network structure and working conditions. CWB is the core that enables training CNN under limited computing power. Furthermore, based on our DHP framework, we propose a dynamic chain system that can promote the training speed without sacrificing training accuracy.

3.3.1. Computation Workload Balancing Module

Under model parallelism, each node has different parameters and is responsible for the computation of different model layers respectively, updating only the weights of the assigned model layers. Setting appropriate network partitioning points for network partitioning can improve the efficiency of distributed training. TX2 uses Jetson series SOC, with CPU and GPU sharing 8 GB memory and the memory requirements during Lite-SRL training process are larger than the computing capacity of a single TX2; thus, network partitioning and workload balancing are required.

We propose a gen*eric Computation workload Balancing module, CWB; it works as follows. For a given network structure and specified batch of input data, take Lite-SRL as an example. Lite-SRL contains a total of q layers of networks $\{L_1, L_2 \cdots, L_q\}$; CWB first collects the forward inference and back propagation time of each layer running on TX2, where the forward inference time of each layer is denoted as $\{T_{f1}, T_{f2} \cdots, T\}$, and the back propagation is denoted as $\{T_{b1}, T_{b2} \cdots, T_{bq}\}$. CWB then calculates the memory size occupied by the model parameters of each layers $\{M_{w1}, M_{w2} \cdots, M_{wq}\}$, and the memory size occupied by the output of the intermediate layers $\{M_{l1}, L_{l2} \cdots, L_{lq}\}$. CWB partitions Lite-SRL into n partitions $\{P_1, P_2 \cdots, P_n\}$ and assigns them to n TX2 $\{TX2_1, TX2_2 \cdots, TX2_n\}$, where $P_i = \{L_j, L_{j+1} \cdots, L_{j+K}\}$, then between P_i and P_{i+1}, that is, between $TX2_i$ and $TX2_{i+1}$ need to transmit the feature data from layer L_{j+K} to layer L_{j+K+1}, and the gradient value needs to be transmitted back during back propagation. Record the ratio of the file size to the transmission rate between TX2 as the theoretical transmission latency T_t, CWB calculates the transmission latency $\{T_{t1}, T_{t2} \cdots, T_{tq}\}$ for all candidate partition points.

The training time T_{all} for a mini-batch is:

$$T_{all} = \sum_0^q T_{fi} + \sum_0^q T_{bi} + T_{ta} + T_{tb} \qquad (8)$$

The equation for calculating the equipment utilization index is as follows:

$$E = -ln\frac{\sum\left(\frac{T_{tx2}^n}{T_{all}} - \frac{1}{n}\right)^2}{n} \qquad (9)$$

The process of CWB searching for the best partition point is formally presented in Algorithm 2. After network partitioning, each partition is assigned to the DHP system for distributed training. The detailed implementation of CWB is recorded in Figure 13.

Algorithm 2. CWB search for the best partition point

Step1:CWB performs memory workload balancing
Step2:CWB performs time equalization
1: Assign $\{M_{w1}, M_{w2} \cdots, M_{wq}\}$ and $\{M_{l1}, L_{l2} \cdots, L_{lq}\}$ to $\{TX2_1, TX2_2 \cdots, TX2_n\}$
2: Assume 3 TX2s can satisfy memory allocation, then 2 sets of candidate partition point that satisfy memory workload balancing are recorded as $[[a, a+1\cdots], [b, b+1\cdots]]$
3: **for** a in $[a, a+1\cdots]$ do
4: **for** b in $[b, b+1\cdots]$ do
5: Partition point 1 adopts a, partition point 2 adopts b
6: Denote the running time of $TX2_1$ as $T_{tx2}^1 = \sum_0^a T_{fi} + \sum_0^a T_{bi}$
7: Denote the running time of $TX2_2$ as $T_{tx2}^2 = \sum_a^b T_{fi} + \sum_a^b T_{bi}$
8: Denote the running time of $TX2_3$ as $T_{tx2}^3 = \sum_b^q T_{fi} + \sum_b^q T_{bi}$
9: The training time T_{all} for a mini-batch is
$T_{all} = \sum_0^q T_{fi} + \sum_0^q T_{bi} + T_{ta} + T_{tb}$
10: The partition point use $[a, b]$, the ratio of running time to waiting time of $TX2_n$ is
$\frac{T_{tx2}^n}{T_{all}}$
11: Calculate the equipment utilization indices E using Equation (9)
$E = -ln\frac{\sum\left(\frac{T_{tx2}^n}{T_{all}} - \frac{1}{n}\right)^2}{n}$
12: **end for**
13: **end for**
14: The partition point of E with the highest score is the best partition point

---

3.3.2. Dynamic Chain System

Figure 4a shows our hybrid parallel distributed training baseline schematic, where each node in the chain is fixedly linked to its front and back nodes, and the later nodes in the chain have to wait for the front nodes to finish forward and backward propagation. Overlap network computation time with transmission time is a common method to improve efficiency in distributed training [52], which can improve training efficiency. Our modified dynamic chain is shown in Figure 4b, where three nodes are responsible for the computation of partition 1, two for the computation of partition 2, and one for the computation of partition 3 of the model. We add a communication scheduler module to our distributed training framework, enabling the node that first completes the computation to search for the available nodes in the next layer. Each mini-batch will form a dynamic chain that performs forward and backward propagation, and after each node completes its current backward propagation, it will automatically leave the current chain and construct a new chain with the node that is waiting. Dynamic chain system has higher training efficiency than baseline, improving node utilization without reducing training accuracy. The dynamic chain system can be well generalized for different training demands, and we have conducted additional experiments for different training computations as detailed in Section 6.2.

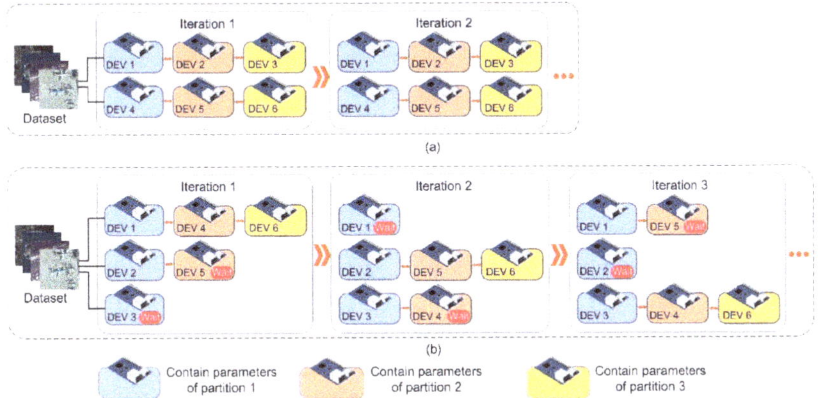

Figure 4. Illustration of our proposed distributed hybrid parallel training baseline and dynamic chain system. (**a**) Distributed hybrid parallel training baseline. (**b**) Dynamic chain system. In iteration 1, Devices 1, 4, and 6 forms a computation chain, while Devices 3 and 5 are in a waiting state. During this time, Device 5 completes the forward computation from Device 2. At the end of iteration 1, Device 6 disconnects from Device 4, automatically links to Device 5, and immediately performs the third part of the training, Device 4 links to Device 3 and waits to link with Device 6. In iteration 3, Device 4 links to Device 6, and the rest nodes also link to available nodes. Subsequent iterations follow the same procedure.

4. Experimental Setups

4.1. Datasets Description

For SAR images, we use the OpenSARUrban [14] dataset and the WHU-SAR6 [11] dataset for experiments. We use a small number of training samples to predict a large number of test samples in our experiments, the training proportions for OpenSARUrban dataset are 10% and 20%, and for WHU-SAR6 dataset we set training proportions as 10% and 20%.

- The OpenSARUrban [14] dataset consists of 10 categories of urban scene images collected from Sentinel-1; its scene images cover 21 major cities in China. Each category contains about 40 to 2000 images with a size of 100×100 pixels, and the resolution of the images is about 20 m;
- The WHU-SAR6 [11] dataset consists of six categories of scene images collected form Sentinel-1 and GF-3. Each category contains about 250 to 420 images with ranging in size from 500 to 600 pixels. Since the total number of WHU-SAR6 images is relatively small, to increase the dataset volume we crop the images into small patches of 256×256 pixels without destroying the scene semantic information.

For optical images, we use the NWPU-RESISC45 [3] dataset and the Aerial Image dataset (AID) [15]. The training proportions for NWPU-RESISC45 dataset are 10% and 20%, which are more challenging since they both require using a small number of training samples to predict labels for many test data. For the AID dataset, we set training proportions as 10%, 20%, and 50%. Detailed information is shown in Table 1.

- The NWPU-RESISC45 [3] dataset is the current largest open benchmark dataset for scene classification task, consisting of 45 categories of scene images. Each category contains 700 images with a size of 256×256 pixels, and the spatial resolution of the images is about 0.2 to 30 m.
- The AID [15] dataset consists of 30 categories of scene images; each category containing about 200 to 400 images, for a total of 10,000 samples, each with a size of 600×600 pixels.

Table 1. Datasets description and training proportions.

Datasets	Images Number	Categories Number	Training Proportions
OpenSARUrban [1] [14]	16679	10	10%, 20%
WHU-SAR6 [2] [11]	17590	6	10%, 20%
NWPU-RESISC45 [3]	31500	45	10%, 20%
AID [15]	10000	30	10%, 20%, 50%

[1] OpenSARUrban dataset has VH and VV polarizations, we used the VH data. [2] For WHU-SAR6 dataset, we cropped the images into small patches of 256 × 256 pixels to increase the dataset volume.

4.2. Data Augmentation

By performing random crop and resize on target image, the receptive field of the network can achieve both global and local prediction, which is crucial for RSSC task. We perform spatial transformations such as random crop, flip, rotate, and resize to enable the model to learn rotation invariants and scaling invariants simultaneously. Further, we simulate temporal transformations with Gaussian blur, color jitter, and random grayscale. The augmentation strategies differ between SAR images and optical images. Detailed data augmentation result is shown in Figure 5.

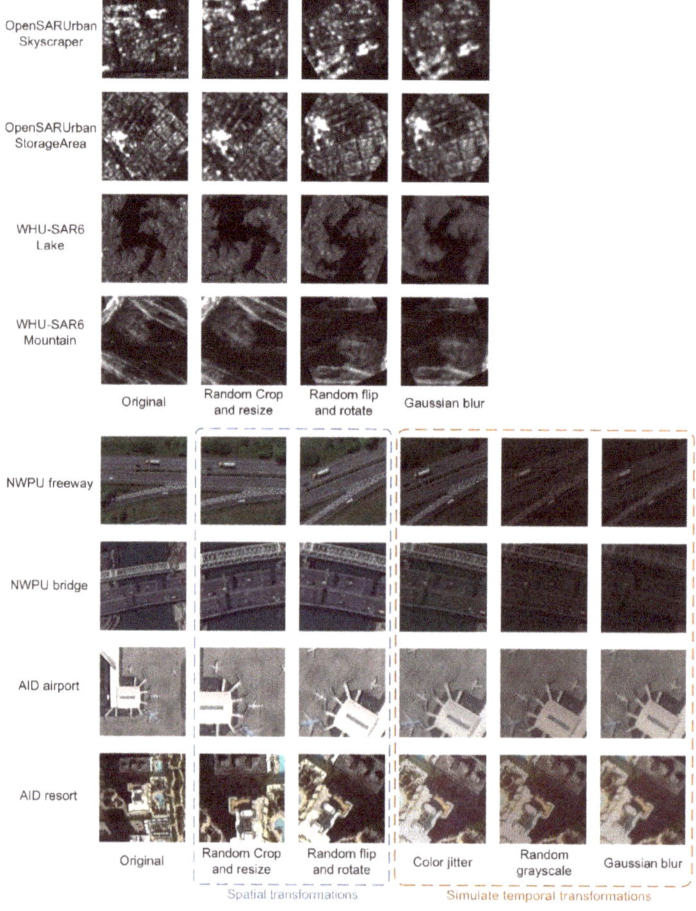

Figure 5. Illustration of data augmentations.

4.3. Implementation Details

The input images were normalized to 224×224 and used the data augmentation settings shown in Figure 5. The batch size was set to 64, and all methods were trained for 400 epochs. For all competitive algorithms, we used ResNet-18 [53] as the backbone and removed the fully connected layer after Advpooling in ResNet-18. Since the loss functions and optimizers of these competitive methods are different, the experimental results are obtained under the individual methods' respective optimal hyperparameter settings. All competitive algorithms were implemented using PyTorch 1.7, Python3.7. The proposed Lite-SRL method used an SGD optimizer with a momentum of 0.9 and a weight decay of 1×10^{-4}, the initial learning rate was 0.05, and the learning rate decreased using the cosine decay.

The experimental section consists of two parts.

- Experiments of self-supervised learning. In this part we use workstations to compare the proposed Lite-SRL with other advanced self-supervised methods comprehensively. The two workstations are identically configured with NVIDIA RTX 3090GPU, Intel Xeon CPU E5-1650, and 64 G RAM.
- Experiments for on-board deployment of Lite-SRL algorithm. We used the proposed distributed training modules and provided detailed records for the deployment process. The experimental on-board computing platform consists of NVIDIA Jetson TX2 nodes and a high-speed switch.

5. Experimental Results

The flowchart of self-supervised learning experiments is shown in Figure 6. Encoders obtained by self-supervised training are used both as (i) frozen feature extractors (Freeze experiment), and as (ii) initial fine-tune model (Fine-tune experiment). For both the freeze and fine-tune experiments, we connected a linear classifier after the encoder and used an Adam optimizer with a batch size of 64, the learning rate reduced in a cosine manner within 200 epochs.

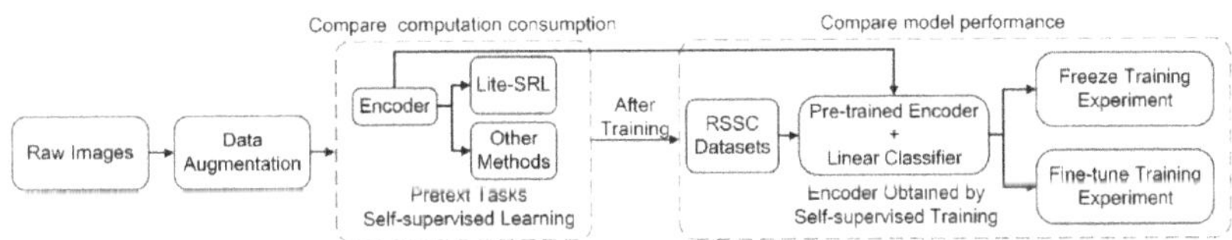

Figure 6. The flowchart of self-supervised learning experiments.

5.1. Guaranteed Accuracy with Less Computation

We compare (i) overall accuracy, (ii) number of parameters, (iii) memory consumption, and (iv) average training latency with competitive self-supervised algorithms. The memory consumption during network training consists of the following elements. The memory occupied by the model: including the consumption of parameters, gradients and optimizer momentum. The memory occupied by the network intermediate layers' outputs: including the inputs and outputs of each layer.

Considering that on-board scenario is highly sensitive to the computation workload, the algorithm is required to achieve higher accuracy and less computation simultaneously. Experiments show that Lite-SRL can achieve optimal classification accuracy with minimum computation. As shown in Figure 7, Lite-SRL shows the best accuracy in the RSSC task, while Lite-SRL has a clear advantage in terms of computation consumption. Thus Lite-SRL provides a lightweight yet effective solution for on-board self-supervised representation learning.

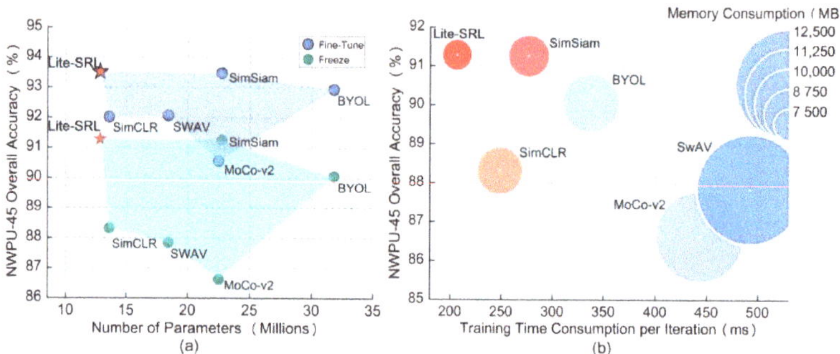

Figure 7. Guaranteed accuracy with less computation. (**a**) Fine-tune and freeze experiment results on NWPU-45 dataset with training proportion of 20%, the horizontal axis compares the number of parameters. (**b**) Freeze experiment results on NWPU-45 dataset with training proportion of 20%; horizontal axis compares the training time consumption per iteration, and the diameter of the bubble is proportional to the memory consumption during network training.

5.2. Self-Supervised Representation Extractor

In freeze training experiment, we use the encoders obtained from each method as feature extractors to evaluate their performance in scene classification. To visualize the effectiveness of Lite-SRL, we fed the test set images to the pre-trained model learned from Lite-SRL, and applied t-SNE [54] to map the output features to a 2-dimensional space. As shown in Figure 8, features from different classes can be well distributed by our self-supervised method, with significantly better results than ImageNet supervised pre-trained model. This demonstrates that by utilizing unlabeled RSI data, our proposed representation learning strategy enables the model to produce a valuable feature representation for downstream RSSC task.

In Figure 8c, we marked the samples from the OpenSARUrabn dataset that Lite-SRL failed to distinguish. We found that these SAR samples contain confusable features. For instance, the six different scene categories in Figure 8c all contain a river flowing through the city. Since we do not use any labels during self-supervised learning, Lite-SRL may extract the wrong features for these confusing scene images.

Table 2 shows the results of freeze training. Experimental results show that in RSSC task, these self-supervised models get better results than supervised models pre-trained on ImageNet, despite the fact that the datasets used for self-supervised pre-training are much smaller than the ImageNet dataset (OpenSARUrban dataset has 16,670 images, WHU-SAR6 dataset has 17,590 images, NWPU-RESISC45 dataset has a total of 34,500 images, the ImageNet pre-trained model used approximately 1.5 million images). Lite-SRL achieved the highest classification accuracy, while higher accuracy can be achieved using the fine-tuning method, as detailed in Table 3.

5.3. Improving the Scene Classification Accuracy with Limited Annotated Data

The proposed self-supervised learning method can solve the problem of annotated data shortage in scene classification task, as high accuracy is achieved in the test set using a small number of training samples.

The fine-tune results of the competitive self-supervised methods are shown in Table 3. Note that due to the differences in these methods, our experimental results record the best results of each method with different learning rates. All of the self-supervised methods showed significant improvements over the randomly initialized models, and at the same time, all of the methods outperformed the models pre-trained in a supervised manner on ImageNet. In 10% training proportion experiments, we used a small number of training samples to predict a large number of test samples. Even so, we achieved high classification accuracy with a simple classification network structure by using the self-supervised pre-trained model as the start point for fine-tuning, proving the effectiveness of the proposed self-supervised learning method.

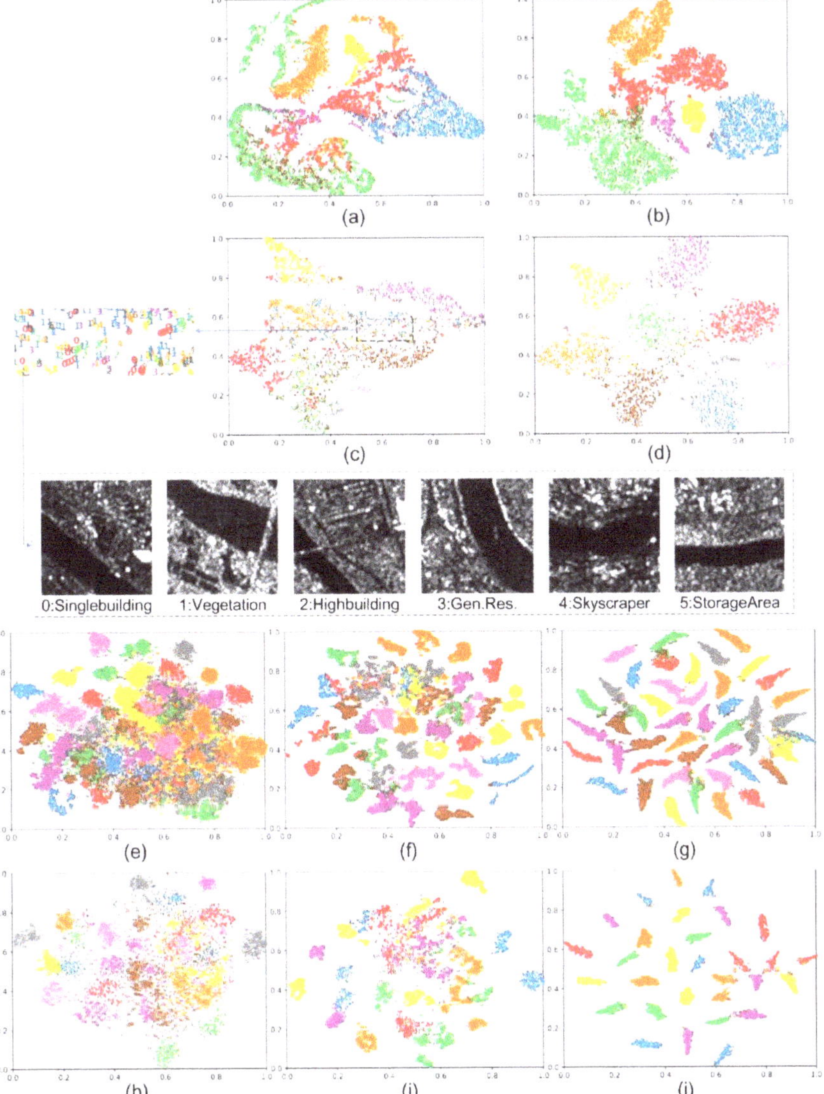

Figure 8. The t-SNE visualization of feature distributions on different datasets. (**a**) Lite-SRL model on WHU-SAR6 dataset; (**b**) fine-tuned Lite-SRL model on WHU-SAR6 dataset; (**c**) Lite-SRL model on OpenSARUrban dataset; (**d**) fine-tuned Lite-SRL model on OpenSARUrban dataset; for SAR dataset due to the imaging mechanism, we did not use ImageNet's pre-trained model. (**e**) ImageNet pre-trained model on NWPU-45 dataset; (**f**) Lite-SRL model on NWPU-45 dataset; (**g**) fine-tuned Lite-SRL model on NWPU-45 dataset; (**h**) ImageNet pre-trained model on AID dataset; (**i**) Lite-SRL model on AID dataset; (**j**) fine-tuned Lite-SRL model on NWPU-45 dataset.

Table 2. Results of freeze experiment in terms of overall accuracy (%).

Method: Freeze	Parameters (Millions)	Overall Accuracy (%)							
		WHU-SAR6		OpenSARUrban		NWPU-45		AID	
		10%	20%	10%	20%	10%	20%	10%	20%
ImageNet [1] (Supervised) [16]	-	-	-	-	-	73.17	77.08	79.40	80.45
SimCLR [22]	13.57	83.40	86.73	67.87	68.33	86.45	88.32	85.52	87.23
MoCo-v2 [43]	22.48	82.39	85.07	65.52	66.07	83.37	86.63	84.56	86.05
SWAV [44]	18.45	83.04	86.30	65.98	67.28	84.16	87.85	84.85	86.59
BYOL [32]	31.81	86.11	87.75	68.36	69.73	88.63	90.06	87.24	88.32
SimSiam [33]	22.73	87.59	**88.64**	70.20	70.86	91.19	91.26	89.15	90.49
Lite-SRL (ours)	**12.82**	**87.71**	88.56	**70.23**	**71.09**	**91.22**	**91.28**	**89.27**	**90.67**

[1] The ImageNet is the encoder obtained by supervised pre-training on ImageNet dataset.

Table 3. Results of fine-tune experiment in terms of overall accuracy (%).

Method: Fine-Tune	Parameters (Millions)	Overall Accuracy (%)							
		WHU-SAR6		OpenSARUrban		NWPU-45		AID	
		10%	20%	10%	20%	10%	20%	10%	20%
Randomly initialized	-	-	-	-	-	77.16	82.87	80.63	83.47
ImageNet (Supervised)	-	-	-	-	-	84.74	89.93	89.81	90.54
SimCLR	13.57	91.85	93.74	80.21	83.87	90.35	92.02	91.32	93.54
MoCo-v2	22.48	90.59	92.70	79.07	82.75	88.71	90.56	89.96	91.47
SWAV	18.45	91.58	93.37	79.85	83.63	89.26	92.07	91.53	92.84
BYOL	31.81	93.21	94.86	80.62	84.88	90.57	92.94	91.95	93.68
SimSiam	22.73	**94.77**	95.69	81.49	85.29	92.68	93.48	92.38	94.63
Lite-SRL (ours)	**12.82**	94.57	**95.83**	**81.76**	**85.43**	**92.77**	**93.51**	**92.55**	**94.82**

Note that our method exhibited higher accuracy with small training batch, while with large training batch, methods such as SimCLR, MoCo-v2, and SWAV, which need to maintain large queues or negative sample pairs, would have improved accuracy.

In Table 4 we illustrate the classification performance of some state-of-the-art methods.

Table 4. Compare with some SOTA methods, in terms of overall accuracy (%).

Method	Overall Accuracy (%)			
	NWPU 10%	NWPU 20%	AID 20%	AID50%
D-CNN with GoogLeNet [55]	86.89	90.49	86.89	90.49
RTN [56]	89.90	92.71	92.44	-
MG-CAP(Sqrt-E) [57]	90.83	92.95	93.34	96.12
ResNet-101 [53]	89.41	92.51	93.31	96.34
ResNet-101+MTL [58]	91.61	93.93	93.67	96.61
ResNet-18+Lite-SRL (ours)	92.77	93.51	94.82	95.78
ResNet-101+Lite-SRL (ours)	**93.41**	**94.43**	**95.29**	**96.82**

Including multi-granularity canonical appearance pools (MG-CAP) [57], recurrent transformer networks (RTN) [56], and MTL [58] using a self-supervised approach. Our Lite-SRL produces an accuracy close to ResNet-101 when using ResNet-18 as the encoder. Further, we set up experiments using ResNet-101 as the encoder in Lite-SRL and produced a top accuracy of 94.43%, which is in pair with the ResNet-101+MTL [58] approach, representing the state-of-the-art performance. The promising performance of Lite-SRL further validates the effectiveness of self-supervised learning in RSSC task.

5.4. Confusion Matrix Analysis

As can be seen from the OpenSARUrban (20%) confusion matrix shown in Figure 9a, the accuracy of the entire test set is 85.43%. High Building category reported the lowest recognition accuracy, and 6.2% were incorrectly identified as Single Building. Urban building areas including Gen.Res, High Building, Single Building, and Denselow showed high misclassifications, as these urban functional areas have similar characteristics. Due to the imbalance of each category, Railway only had 20 test samples with three incorrect classifications.

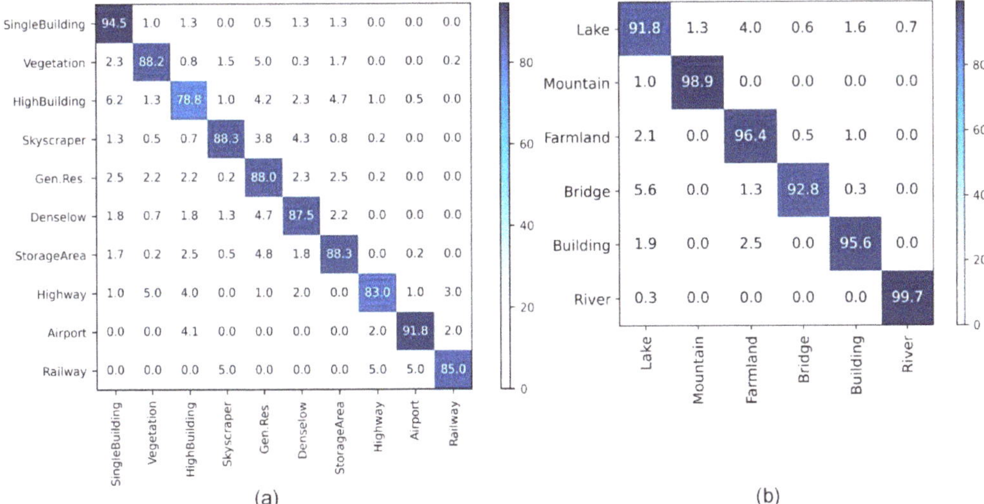

Figure 9. Confusion matrix of fine-tuned results: (**a**) on OpenSARUrban with 20% training proportion; (**b**) on WHU-SAR6 with 20% training proportion.

Figure 9b shows the confusion matrix of fine-tuned results on WHU-SAR6 (20%), the accuracy of the entire test set is 95.83%, with four of the six categories achieving 95% or higher accuracy. Lake and Bridge are the two classes with the highest confusion rates because these two categories both contain water areas.

As can be seen from the NWPU-45 (20%) confusion matrix shown in Figure 10, With 38 of the 45 categories achieving 90% or higher accuracy, the accuracy of the entire test set is 93.51%. Churches and palaces are the two classes with the highest confusion rates because the buildings have similar distribution and appearance in these two groups.

Figure 11 shows the confusion matrix of fine-tuned results on the AID (50%) dataset. With 26 of the 30 categories reaching 90% or higher accuracy and 23 categories achieving higher than 95%, the accuracy of the entire test set is 95.78%. Resort and park, and center and square are the categories with the highest confusion rate because the images of resorts and parks have a similar distribution of greenery, while center and square are urban scenes with similar characteristics.

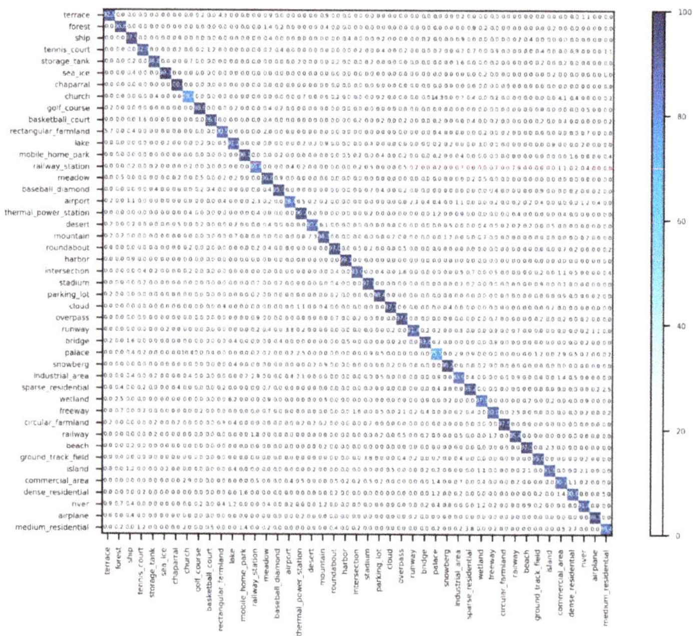

Figure 10. Confusion matrix of fine-tuned results on NWPU-45 20% training proportion.

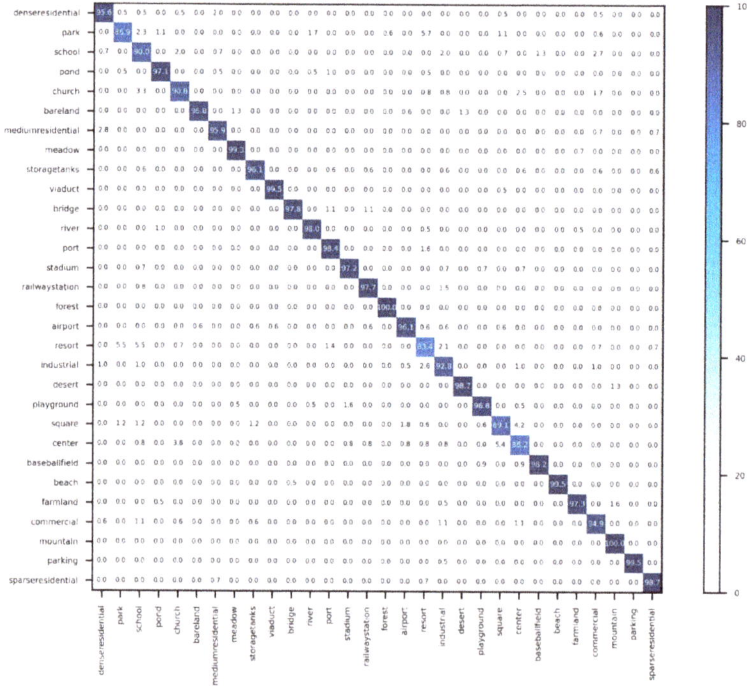

Figure 11. Confusion matrix of fine-tuned results on AID 50% training proportion.

6. Deployment of Lite-SRL

We applied the Lite-SRL self-supervised algorithm to the proposed DHP distributed training system. The flowchart of deployment is shown in Figure 12.

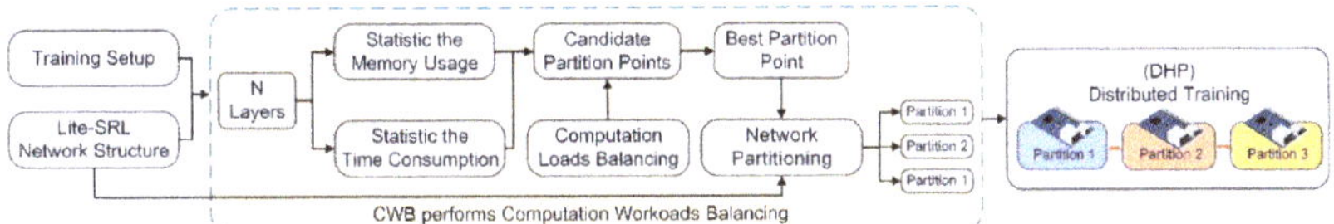

Figure 12. The flowchart of Lite-SRL's deployment corresponds to the content in the following. N-Layers corresponds to Figure 13a; Statistic the Memory Usage corresponds to Figure 13b; Statistic the Time Consumption corresponds to Figure 13c; Candidate Partition Points corresponds to Figure 14a; Best Partition Point corresponds to Figure 14b.

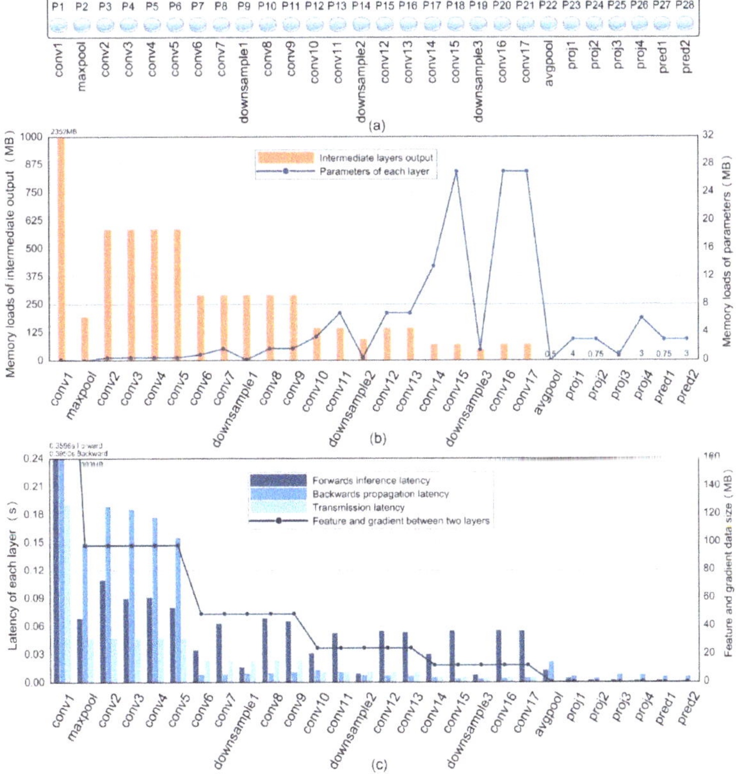

Figure 13. Data collected by CWB. (**a**) Partitionable layers contained in the Lite-SRL network structure, corresponding to 28 partitionable points $\{p_1, p_2, \cdots p_{28}\}$. (**b**) CWB calculated the memory workload occupied by each network layer during the training process, including the intermediate variables and network parameters for each layer. (**c**) CWB measured time consumption, including inference latency and backward propagation latency of each layer when trained on TX2, together with the data transmission latency between TX2. The transmission latency was derived from the gradient data size between two layers and the inter-device transfer rate.

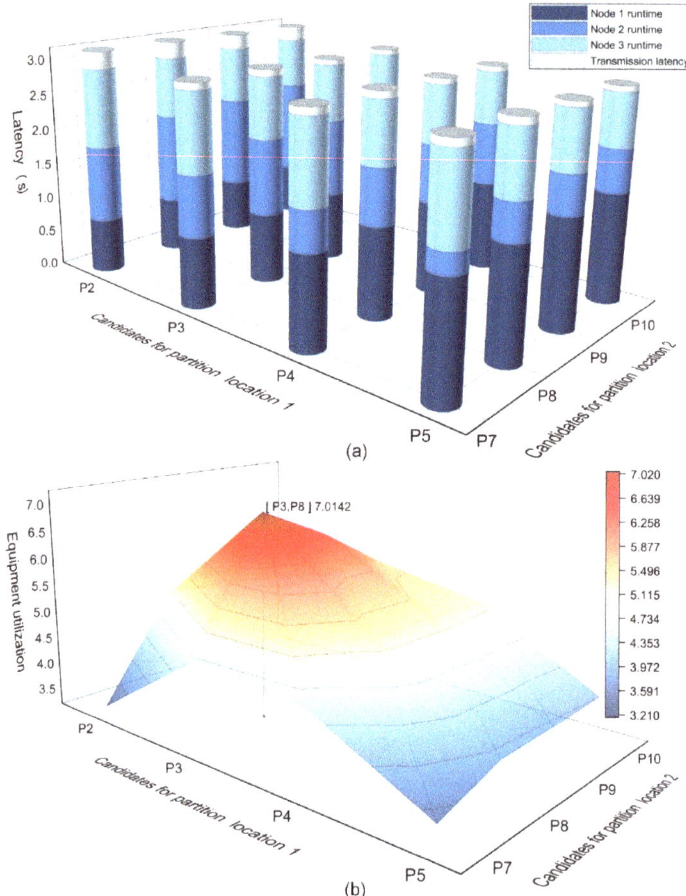

Figure 14. CWB calculates the optimal partition points. Two sets of candidate partition points are $\{p_2, p_3, p_4, p_5\}$ and $\{p_7, p_8, p_9, p_{10}\}$, the rest of the partition points have been screened out as they cannot satisfy the memory allocation requirements. (**a**) Runtime proportion of each node under candidate partition points. (**b**) Using Equation (9) to calculate equipment utilization evaluation indices under candidate partition points.

6.1. Computation Workload Balancing

As shown in Figure 13a, CWB figured out all partitionable points over the given Lite-SRL Network structure. CWB requires the following training setup information: (i) the training batch size, (ii) the type of optimizer being used, and (iii) the data exchange rate between TX2 devices to calculate the figures required for workload balancing. In the experiment, the batch is set to 64 and used SGD optimizer with momentum for training. The rate of data transmission between TX2 is simulated by the Linux traffic control tool. According to the partition points and the above setup information, CWB statistic the memory usage and the time consumption of each layer when training on TX2.

In experiments we uniformly use the float32 format data type, one data occupies 4 bytes of memory. CWB first performed memory workload balancing and computed the candidate partition points using the data shown in Figure 13b, the theoretical memory usage for all intermediate values during training is 7599.7 MB. Based on experience, each TX2 can achieve a preferable working state when allocating about 3 GB memory of computation, so it requires 3 TX2s to collaborate the training process of one mini-batch. The two sets of candidate partition points calculated by CWB are $\{p_2, p_3, p_4, p_5\}$ and $\{p_7, p_8, p_9, p_{10}\}$, satisfying that the training of each partition can be carried out on a single TX2, the rest of the partition points have been screened out. CWB then performed time equalization utilizing the data in shown Figure 13c, by accumulating the forward and backward latency of individual layers under different candidate partition points, to obtain the running time of each TX2 node during a training batch.

As shown in Figure 14, CWB used the equipment utilization index to find the optimal partition point among the candidate partition points. Figure 14a shows the runtime proportion of each node under candidate partition points, the transmission latency varies depending on different combinations of partition points. In Figure 14b CWB found the partition point combination with the highest equipment utilization $[p_3, p_8]$, representing the optimal partition points. For the given Lite-SRL network structure and training settings, CWB partitioned it into the following three parts: partition 1 $\{p_1, p_2\}$, partition 2 $\{p_3, p_4, p_5, p_6, p_7\}$, partition 3 $\{p_8, \cdots p_{28}\}$.

6.2. Distributed Training with Higher Efficiency

We used six TX2 nodes to compose an on-board computation platform and tested the proposed distributed training baseline along with the improved dynamic chain system. In our on-board distributed training baseline experiments, six nodes formed a two-chain hybrid parallel training according to Figure 4b. After the workload balancing, we allocated the training of three network partitions to TX2 nodes on each chain. In dynamic chain system experiments, three nodes are responsible for the computation of partition 1, two for the computation of partition 2, and one for the computation of partition 3 of the model, as shown in Figure 4b. The two distributed training methods were performed 1000 iterations each, the distributed training system performed parameter aggregation every 100 iterations and updated the model parameters in each node using the aggregated parameters. Here one iteration referred to the completion of one mini-batch's forward and backward propagation.

As shown in Tables 5 and 6, the average runtime of executing one iteration in the baseline is 3572 s, while the average time in dynamic chain system is 2750 s.

Table 5. Distributed training baseline.

Average Running Time of One Iteration (ms)		Partition 1		Partition 2		Partition 3	
		Node 1	Node 2	Node 3	Node 4	Node 5	Node 6
		Average runtime of each node in one iteration (ms)					
Baseline	3572	1035	1039	1145	1139	921	923
		Running iterations					
		500	500	500	500	500	500

Table 6. Improved dynamic chain system.

Average Running Time of One Iteration (ms)		Partition 1		Partition 2		Partition 3	
		Node 1	Node 2	Node 3	Node 4	Node 5	Node 6
		Average runtime of each node in one iteration (ms)					
Dynamic	2750	1036	1038	1037	1140	1142	922
		Running iterations					
		329	331	340	406	594	1000

The baseline used 6 nodes to complete 1000 iterations of training in **3572 s**, and the 2 chains had each been running for 500 iterations. In comparison, the dynamic chain system used the same nodes to complete 1000 iterations of training in **2750 s**. With the scheduling of the communication module, the system can be viewed as containing 3 chains; the nodes responsible for the first and second partitions end up with different running iterations, and node 6 completes 1000 iterations. Dynamic chain system improved training efficiency by 23.01% over the baseline without compromising training accuracy.

The experimental platform is shown in Figure 15. Our distributed system consists of six nodes, thus allowing flexible distributed training. More chains can be constituted when the calculation demand is low, and more nodes can be invoked to join the training when the calculation demand is high. The proposed Lite-SRL with ResNet-18 as backbone can be completed using 3 TX2 nodes, the baseline and dynamic chain are shown in Figure 15a. When more complex networks need to be trained, it can be achieved by invoking more nodes to join the training, which manifests the advantage of distributed multi-nodes. To this end, we conducted additional experiments. We use the Lite-SRL algorithm, replacing more complex backbone structure as detailed in the following.

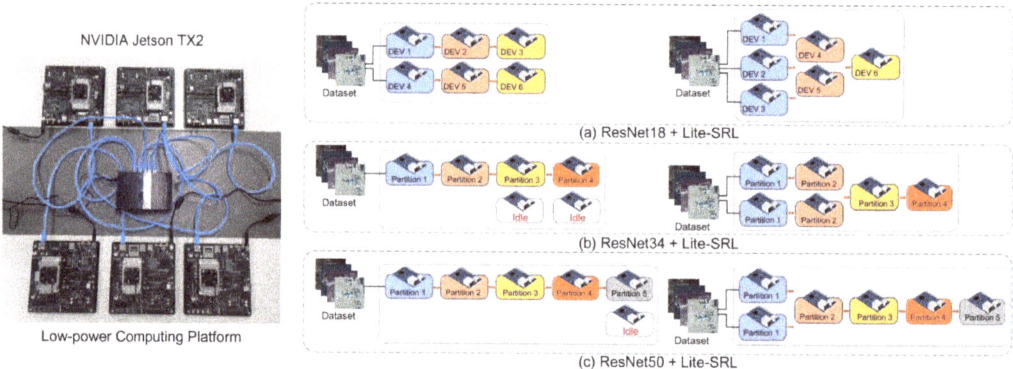

Figure 15. The left side is the illustration of baseline and the right side is the illustration of dynamic chain system. (**a**) Lite-SRL with ResNet-18 as encoder; three nodes are required to complete the training of a mini-batch, baseline uses six nodes to form two chains, and dynamic can form three chains. (**b**) Lite-SRL with ResNet-34 as encoder; four nodes are required to complete the training of a mini-batch. Baseline forms one chain with two nodes idle, while the dynamic chain system can schedule all nodes for training. (**c**) Lite-SRL with ResNet-50 as encoder, five nodes are required to complete the training of a mini-batch. Baseline forms one chain with 1 node idle, while the dynamic chain system can schedule all nodes for training.

The time consumption of DHP under different training computations is shown in Table 7. The dynamic link system can avoid node idleness and, thus, improve the efficiency of training. Furthermore, through this experiment, we demonstrate the potential of our distributed training system, which can be applied to a wider range of neural network training tasks.

Table 7. Distributed Training Time Consumption, corresponding to the illustration in Figure 15.

Method	Memory Consumption (MB)	Distributed Training Time Consumption			Accuray [2] (%)	
		Baseline (ms)	Dynamic (ms)	Improvement	Baseline	Dynamic
ResNet18 + Lite-SRL	7599.7	3572	2750	23.0%	91.31	91.27
ResNet34 + Lite-SRL	10,185.9	4895	3984	18.6%	91.75	91.78
ResNet50 [1] + Lite-SRL	13,039.3	6473	5962	6.9%	92.11	92.09

[1] For ResNet50 the training batch size is 32, the rest of training settings remain unchanged. [2] To compare the accuracy, we use the NWPU-45 dataset and the accuracy test method is the same as the freeze experiment above.

7. Conclusions

In this article, we propose a self-supervised algorithm Lite-SRL for the scene classification task. Our algorithm has clear advantages in terms of overall accuracy, number of parameters, memory consumption, and training latency. We demonstrate that self-supervised algorithms can effectively alleviate the shortage of remote sensing labeled data. Taking the experimental results on NWPU-45 dataset as an example, with training proportions of 10% and 20%, which require few labeled data to predict a large number of test samples, we achieve 92.77% and 93.51% accuracy with a simple network structure after self-supervised pre-training. Previous RSSC studies usually require more complex structures and multiple tricks to achieve such classification accuracies. Meanwhile, our algorithm has far better performance than other methods under 10% training proportion, proving that Lite-SRL's self-supervised training provides an effective feature extractor.

We exploit the advantage of self-supervised learning by training on satellites. The integration of CWB and DHP enables training neural networks under limited on-board resources. In addition, we add a communication scheduler module to the DHP framework to improve the training speed on top of the baseline. On the experimental computing platform, we successfully transplant Lite-SRL and verify the effectiveness of proposed on-board distributed training modules.

We believe that on-board self-supervised distributed training can facilitate the development of on-board data processing techniques. Not only for RSSC task, but also other tasks in remote sensing such as remote sensing image segmentation [59], target detection [10], etc., can utilize this working paradigm. Our proposed distributed training modules provide strong adaptability, other types of deep learning algorithms can also be deployed in the distributed training framework, making it possible to enhance the intelligence in remote sensing applications.

The next step of our work will be as follows:

- We will design a dedicated lightweight feature extractor in the self-supervised structure to further reduce the memory computation;
- We will explore techniques such as gradient compression, network pruning, etc., to further improve distributed training efficiency;
- We will explore hardware acceleration solutions for onboard distributed training;
- We expect to add more remote sensing observation missions to on-board distributed self-supervised training applications.

Author Contributions: Conceptualization, X.X. and Y.L.; methodology, X.X. and C.L.; software, X.X. and C.L.; investigation, X.X.; resources, X.X.; data curation, X.X. and C.L.; writing—original draft preparation, X.X. and Y.L.; writing—review and editing, X.X. and Y.L.; visualization, X.X.; funding acquisition, Y.L. All authors have read and agreed to the published version of the manuscript.

Funding: This research received no external funding.

Data Availability Statement: No new data were created or analyzed in this study. Data sharing is not applicable to this article.

Conflicts of Interest: The authors declare no conflict of interest.

Appendix A

As shown in Table A1, we list all the abbreviations and their corresponding full names in this article.

Table A1. The abbreviations and corresponding full names, organized in alphabetical order.

Abbreviation	Full Name
AID	Aerial Image Dataset
BN	Batch Normalization
BYOL	Bootstrap Your Own Latent
CNN	Convolution Neural Network
CWB	Computation workload Balancing module
MG-CAP	Multi-Granularity Canonical Appearance Pools
MLP	Multi-Layer Perceptron
MoCo	Momentum Contrast for Visual Representation Learning
MTL	Multitask Learning
NWPU-45	NWPU-Resisc45 Dataset
DHP	Distributed Hybrid Parallelism Training Framework
Lite-SRL	Lightweight Self-supervised Representation Learning algorithm
ReLU	Rectified Linear Unit
RSIs	Remote Sensing Images
RSSC	Remote Sensing Scene Classification
SGD	Stochastic Gradient Descent
SimCLR	Simple Framework For Contrastive Learning
Simsiam	Simple Siamese Representation Learning
SwAV	Unsupervised Learning By Contrasting Cluster Assignments
t-SNE	T-Distributed Stochastic Neighbor Embedding

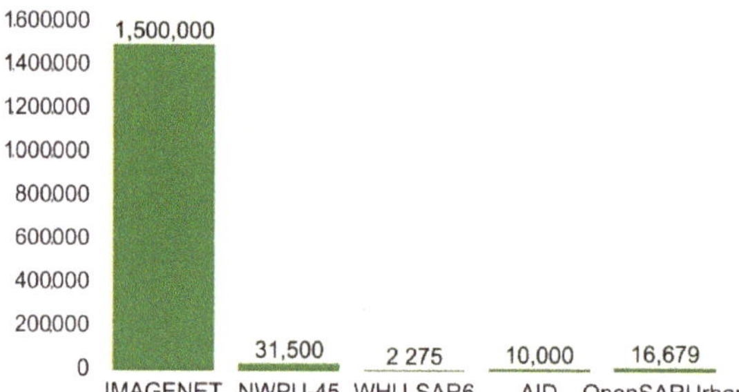

Figure A1. The total images number of RSSC datasets, i.e., OpenSARUrban [14], WHU-SAR6 [11], NWPU-Resisc45 [3], AID [15] compared with natural image datasets, i.e., ImageNet.

References

1. Hu, F.; Xia, G.-S.; Hu, J.; Zhang, L. Transferring Deep Convolutional Neural Networks for the Scene Classification of High-Resolution Remote Sensing Imagery. *Remote Sens.* **2015**, *7*, 14680–14707. [CrossRef]
2. Ni, K.; Liu, P.; Wang, P. Compact Global-Local Convolutional Network with Multifeature Fusion and Learning for Scene Classification in Synthetic Aperture Radar Imagery. *IEEE J. Sel. Top. Appl. Earth Obs. Remote Sens.* **2021**, *14*, 7284–7296. [CrossRef]
3. Cheng, G.; Han, J.; Lu, X. Remote Sensing Image Scene Classification: Benchmark and State of the Art. *Proc. IEEE* **2017**, *105*, 1865–1883. [CrossRef]
4. Xu, X.; Zhang, X.; Zhang, T. Multi-Scale SAR Ship Classification with Convolutional Neural Network. In Proceedings of the IEEE International Geoscience and Remote Sensing Symposium (IGARSS), Online Event, 11–16 July 2021; pp. 4284–4287.
5. Lu, X.; Sun, X.; Diao, W.; Feng, Y.; Wang, P.; Fu, K. LIL: Lightweight Incremental Learning Approach through Feature Transfer for Remote Sensing Image Scene Classification. *IEEE Trans. Geosci. Remote Sens.* **2022**, *60*, 5611320. [CrossRef]
6. Zhang, T.; Zhang, X. Squeeze-And-Excitation Laplacian Pyramid Network with Dual-Polarization Feature Fusion for Ship Classification in SAR Images. *IEEE Geosci. Remote Sens. Lett.* **2022**, *19*, 4019905. [CrossRef]
7. Gu, Y.; Wang, Y.; Li, Y. A Survey on Deep Learning-Driven Remote Sensing Image Scene Understanding: Scene Classification, Scene Retrieval and Scene-Guided Object Detection. *Appl. Sci.* **2019**, *9*, 2110. [CrossRef]

8. Zhang, T.; Zhang, X.; Ke, X.; Liu, C.; Xu, X. HOG-ShipCLSNet: A Novel Deep Learning Network with HOG Feature Fusion for SAR Ship Classification. *IEEE Trans. Geosci. Remote Sens.* **2022**, *60*, 5210322. [CrossRef]
9. Liao, N.; Datcu, M.; Zhang, Z.; Guo, W.; Zhao, J.; Yu, W. Analyzing the Separability of SAR Classification Dataset in Open Set Conditions. *IEEE J. Sel. Top. Appl. Earth Obs. Remote Sens.* **2021**, *14*, 7895–7910. [CrossRef]
10. Zhang, T.; Zhang, X.; Shi, J.; Wei, S. HyperLi-Net: A Hyper-Light Deep Learning Network for High-Accurate and High-Speed Ship Detection from Synthetic Aperture Radar Imagery. *ISPRS J. Photogramm. Remote Sens.* **2020**, *167*, 123–153. [CrossRef]
11. Su, B.; Liu, J.; Su, X.; Luo, B.; Wang, Q. CFCANet: A Complete Frequency Channel Attention Network for SAR Image Scene Classification. *IEEE J. Sel. Top. Appl. Earth Obs. Remote Sens.* **2021**, *14*, 11750–11763. [CrossRef]
12. Zhang, T.; Zhang, X. A Polarization Fusion Network with Geometric Feature Embedding for SAR Ship Classification. *Pattern Recognit.* **2022**, *123*, 108365. [CrossRef]
13. Dumitru, C.O.; Schwarz, G.; Datcu, M. SAR Image Land Cover Datasets for Classification Benchmarking of Temporal Changes. *IEEE J. Sel. Top. Appl. Earth Obs. Remote Sens.* **2018**, *11*, 1571–1592. [CrossRef]
14. Zhao, J.; Zhang, Z.; Yao, W.; Datcu, M.; Xiong, H.; Yu, W. OpenSARUrban: A Sentinel-1 SAR Image Dataset for Urban Interpretation. *IEEE J. Sel. Top. Appl. Earth Obs. Remote Sens.* **2020**, *13*, 187–203. [CrossRef]
15. Xia, G.-S.; Hu, J.; Hu, F.; Shi, B.; Bai, X.; Zhong, Y.; Zhang, L.; Lu, X. AID: A Benchmark Data Set for Performance Evaluation of Aerial Scene Classification. *IEEE Trans. Geosci. Remote Sens.* **2017**, *55*, 3965–3981. [CrossRef]
16. Deng, J.; Dong, W.; Socher, R.; Li, L.-J.; Li, K.; Fei-Fei, L. ImageNet: A large-scale hierarchical image database. In Proceedings of the 2009 IEEE Conference on Computer Vision and Pattern Recognition (CVPR), Miami, FL, USA, 20–25 June 2009; pp. 248–255.
17. Zhang, T.; Zhang, X. A Full-Level Context Squeeze-And-Excitation ROI Extractor for SAR Ship Instance Segmentation. *IEEE Geosci. Remote Sens. Lett.* **2022**, *19*, 4506705. [CrossRef]
18. Kolesnikov, A.; Zhai, X.; Beyer, L. Revisiting self-supervised visual representation learning. In Proceedings of the IEEE/CVF Conference on Computer Vision and Pattern Recognition (CVPR), Long Beach, CA, USA, 15–20 June 2019; pp. 1920–1929.
19. Noroozi, M.; Favaro, P. Unsupervised learning of visual representations by solving jigsaw puzzles. In *European Conference on Computer Vision*; Springer: Cham, Switzerland, 2016; pp. 69–84.
20. Stojnic, V.; Risojevic, V. Self-supervised learning of remote sensing scene representations using contrastive multiview coding. In Proceedings of the IEEE/CVF Conference on Computer Vision and Pattern Recognition (CVPR), Nashville, TN, USA, 19–25 June 2021; pp. 1182–1191.
21. Zhang, T.; Zhang, X.; Shi, J.; Wei, S.; Wang, J.; Li, J.; Su, H.; Zhou, Y. Balance Scene Learning Mechanism for Offshore and Inshore Ship Detection in SAR Images. *IEEE Geosci. Remote Sens. Lett.* **2022**, *19*, 4004905. [CrossRef]
22. Chen, T.; Kornblith, S.; Norouzi, M.; Hinton, G. A simple framework for contrastive learning of visual representations. In Proceedings of the International Conference on Machine Learning, Vienna, Austria, 12–18 July 2020; Volume 119, pp. 1597–1607.
23. Ayush, K.; Uzkent, B.; Meng, C.; Tanmay, K.; Burke, M.; Lobell, D.; Ermon, S. Geography-aware self-supervised learning. In Proceedings of the IEEE/CVF International Conference on Computer Vision (ICCV), Montreal, QC, Canada, 10–17 October 2021; pp. 10181–10190.
24. Franklin, D. NVIDIA Developer Blog: NVIDIA Jetson TX2 Delivers Twice the Intelligence to the Edge. Available online: https://devblogs.nvidia.com/jetson-tx2-delivers-twice-intelligence-edge/ (accessed on 13 April 2022).
25. Xu, X.; Zhang, X.; Zhang, T. Lite-YOLOv5: A Lightweight Deep Learning Detector for On-Board Ship Detection in Large-Scene Sentinel-1 SAR Images. *Remote Sens.* **2022**, *14*, 1018. [CrossRef]
26. Aitech's S-A1760 Venus™ Brings NVIDIA-Based AI Supercomputing to Next Generation Space Applications: Radiation-CharActerized COTS System Qualified for Use in Small Sat Clusters and Short-Duration Spaceflights. Available online: https://aitechsystems.com/aitechs-s-a1760-venus-brings-nvidia-based-ai-supercomputing-to-next-generation-space-applications/ (accessed on 13 April 2022).
27. Paszke, A.; Gross, S.; Massa, F.; Lerer, A.; Bradbury, J.; Chanan, G.; Killeen, T.; Lin, Z.; Gimelshein, N.; Antiga, L. Pytorch: An imperative style, high-performance deep learning library. *Adv. Neural Inf. Processing Syst.* **2019**, *32*, 8026–8037.
28. Abadi, M.; Barham, P.; Chen, J.; Chen, Z.; Davis, A.; Dean, J.; Devin, M.; Ghemawat, S.; Irving, G.; Isard, M. Tensorflow: A system for large-scale machine learning. In Proceedings of the 12th {USENIX} Symposium on Operating Systems Design and Implementation ({OSDI} 16), Savannah, GA, USA, 2–4 November 2016; pp. 265–283.
29. Jia, Y.; Shelhamer, E.; Donahue, J.; Karayev, S.; Long, J.; Girshick, R.; Guadarrama, S.; Darrell, T. Caffe: Convolutional architecture for fast feature embedding. In Proceedings of the 22nd ACM International Conference on Multimedia, Orlando, FL, USA, 3–7 November 2014; pp. 675–678.
30. Shazeer, N.; Cheng, Y.; Parmar, N.; Tran, D.; Vaswani, A.; Koanantakool, P.; Hawkins, P.; Lee, H.; Hong, M.; Young, C.; et al. Mesh-tensorflow: Deep learning for supercomputers. *arXiv* **2018**, arXiv:1811.02084.
31. Onoufriou, G.; Bickerton, R.; Pearson, S.; Leontidis, G. Nemesyst: A hybrid parallelism deep learning-based framework applied for internet of things enabled food retailing refrigeration systems. *Comput. Ind.* **2019**, *113*, 103133. [CrossRef]
32. Grill, J.-B.; Strub, F.; Altché, F.; Tallec, C.; Richemond, P.H.; Buchatskaya, E.; Doersch, C.; Pires, B.A.; Guo, Z.D.; Azar, M.G. Bootstrap your own latent: A new approach to self-supervised learning. *arXiv* **2020**, arXiv:2006.07733.
33. Chen, X.; He, K. Exploring simple siamese representation learning. In Proceedings of the IEEE/CVF Conference on Computer Vision and Pattern Recognition (CVPR), Nashville, TN, USA, 19–25 June 2021; pp. 15750–15758.

34. Li, X.; Shi, D.; Diao, X.; Xu, H. SCL-MLNet: Boosting Few-Shot Remote Sensing Scene Classification via Self-Supervised Contrastive Learning. *IEEE Trans. Geosci. Remote Sens.* **2022**, *60*, 5801112. [CrossRef]

35. Li, Y.; Shao, Z.; Huang, X.; Cai, B.; Peng, S. Meta-FSEO: A Meta-Learning Fast Adaptation with Self-Supervised Embedding Optimization for Few-Shot Remote Sensing Scene Classification. *Remote Sens.* **2021**, *13*, 2776. [CrossRef]

36. Tao, C.; Qi, J.; Lu, W.; Wang, H.; Li, H. Remote Sensing Image Scene Classification With Self-Supervised Paradigm Under Limited Labeled Samples. *IEEE Geosci. Remote Sens. Lett.* **2022**, *19*, 8004005. [CrossRef]

37. Kang, J.; Fernandez-Beltran, R.; Duan, P.; Liu, S.; Plaza, A.J. Deep Unsupervised Embedding for Remotely Sensed Images Based on Spatially Augmented Momentum Contrast. *IEEE Trans. Geosci. Remote Sens.* **2021**, *59*, 2598–2610. [CrossRef]

38. Jung, H.; Oh, Y.; Jeong, S.; Lee, C.; Jeon, T. Contrastive Self-Supervised Learning with Smoothed Representation for Remote Sensing. *IEEE Geosci. Remote Sens. Lett.* **2021**, *19*, 8010105. [CrossRef]

39. Zhao, L.; Luo, W.; Liao, Q.; Chen, S.; Wu, J. Hyperspectral Image Classification with Contrastive Self-Supervised Learning under Limited Labeled Samples. *IEEE Geosci. Remote Sens. Lett.* **2022**, *19*, 6008205. [CrossRef]

40. Doersch, C.; Gupta, A.; Efros, A.A. Unsupervised visual representation learning by context prediction. In Proceedings of the IEEE International Conference on Computer Vision (ICCV), Santiago, Chile, 7–13 December 2015; pp. 1422–1430.

41. Pathak, D.; Krahenbuhl, P.; Donahue, J.; Darrell, T.; Efros, A.A. Context encoders: Feature learning by inpainting. In Proceedings of the IEEE/CVF Conference on Computer Vision and Pattern Recognition (CVPR), Las Vegas, NV, USA, 27–30 June 2016; pp. 2536–2544.

42. He, K.; Fan, H.; Wu, Y.; Xie, S.; Girshick, R. Momentum contrast for unsupervised visual representation learning. In Proceedings of the IEEE/CVF Conference on Computer Vision and Pattern Recognition (CVPR), Seattle, WA, USA, 13–19 June 2020; pp. 9729–9738.

43. Chen, X.; Fan, H.; Girshick, R.; He, K. Improved baselines with momentum contrastive learning. *arXiv* **2020**, arXiv:2003.04297.

44. Caron, M.; Misra, I.; Mairal, J.; Goyal, P.; Bojanowski, P.; Joulin, A. Unsupervised learning of visual features by contrasting cluster assignments. *arXiv* **2020**, arXiv:2006.09882.

45. Krizhevsky, A.; Sutskever, I.; Hinton, G.E. Imagenet classification with deep convolutional neural networks. *Adv. Neural Inf. Process. Syst.* **2012**, *25*, 1097–1105. [CrossRef]

46. Kim, S.; Yu, G.-I.; Park, H.; Cho, S.; Jeong, E.; Ha, H.; Lee, S.; Jeong, J.S.; Chun, B.-G. Parallax: Sparsity-aware data parallel training of deep neural networks. In Proceedings of the Fourteenth EuroSys Conference, Dresden, Germany, 25–28 March 2019; pp. 1–15.

47. Jia, Z.; Zaharia, M.; Aiken, A. Beyond data and model parallelism for deep neural networks. *Proc. Mach. Learn. Syst.* **2019**, *1*, 1–13.

48. Lee, S.; Kim, J.K.; Zheng, X.; Ho, Q.; Gibson, G.; Xing, P. *On Model Parallelization and Scheduling Strategies for Distributed Machine Learning*; Carnegie Mellon University: Pittsburgh, PA, USA, 2014; pp. 2834–2842.

49. Akintoye, S.B.; Han, L.; Zhang, X.; Chen, H.; Zhang, D. A hybrid parallelization approach for distributed and scalable deep learning. *arXiv* **2021**, arXiv:2104.05035. [CrossRef]

50. Demirci, G.V.; Ferhatosmanoglu, H. Partitioning sparse deep neural networks for scalable training and inference. In Proceedings of the ACM International Conference on Supercomputing, Virtual Event, 14–17 June 2021; pp. 254–265.

51. Moreno-Alvarez, S.; Haut, J.M.; Paoletti, M.E.; Rico-Gallego, J.A. Heterogeneous model parallelism for deep neural networks. *Neuro Comput.* **2021**, *441*, 1–12. [CrossRef]

52. Das, D.; Avancha, S.; Mudigere, D.; Vaidynathan, K.; Sridharan, S.; Kalamkar, D.; Kaul, B.; Dubey, P. Distributed deep learning using synchronous stochastic gradient descent. *arXiv* **2016**, arXiv:1602.06709.

53. He, K.; Zhang, X.; Ren, S.; Sun, J. Deep residual learning for image recognition. In Proceedings of the IEEE/CVF Conference on Computer Vision and Pattern Recognition (CVPR), Las Vegas, NV, USA, 27–30 June 2016; pp. 770–778.

54. Van Hinton, G. Visualizing data using t-SNE. *J. Mach. Learn. Res.* **2008**, *9*, 2579–2605.

55. Cheng, G.; Yang, C.; Yao, X.; Guo, L.; Han, J. When Deep Learning Meets Metric Learning: Remote Sensing Image Scene Classification via Learning Discriminative CNNs. *IEEE Trans. Geosci. Remote Sens.* **2018**, *56*, 2811–2821. [CrossRef]

56. Chen, Z.; Wang, S.; Hou, X.; Shao, L.; Dhabi, A. Recurrent transformer network for remote sensing scene categorisation. In Proceedings of the 2018 British Machine Vision Conference, Newcastle, UK, 3–6 September 2018; Volume 266, p. 0987.

57. Wang, S.; Guan, Y.; Shao, L. Multi-Granularity Canonical Appearance Pooling for Remote Sensing Scene Classification. *IEEE Trans. Image Proces.* **2020**, *29*, 5396–5407. [CrossRef]

58. Zhao, Z.; Luo, Z.; Li, J.; Chen, C.; Piao, Y. When Self-Supervised Learning Meets Scene Classification: Remote Sensing Scene Classification Based on a Multitask Learning Framework. *Remote Sens.* **2020**, *12*, 3276. [CrossRef]

59. Zhang, T.; Zhang, X. HTC+ for SAR Ship Instance Segmentation. *Remote Sens.* **2022**, *14*, 2395. [CrossRef]

Article

Triangle Distance IoU Loss, Attention-Weighted Feature Pyramid Network, and Rotated-SARShip Dataset for Arbitrary-Oriented SAR Ship Detection

Zhijing Xu [1], Rui Gao [1,*], Kan Huang [1] and Qihui Xu [2]

[1] College of Information Engineering, Shanghai Maritime University, Shanghai 201306, China
[2] School of Communication & Information Engineering, Shanghai University, Shanghai 200444, China
* Correspondence: 202030310004@stu.shmtu.edu.cn; Tel.: +86-198-2173-5586

Abstract: In synthetic aperture radar (SAR) images, ship targets are characterized by varying scales, large aspect ratios, dense arrangements, and arbitrary orientations. Current horizontal and rotation detectors fail to accurately recognize and locate ships due to the limitations of loss function, network structure, and training data. To overcome the challenge, we propose a unified framework combining triangle distance IoU loss (TDIoU loss), an attention-weighted feature pyramid network (AW-FPN), and a Rotated-SARShip dataset (RSSD) for arbitrary-oriented SAR ship detection. First, we propose a TDIoU loss as an effective solution to the loss-metric inconsistency and boundary discontinuity in rotated bounding box regression. Unlike recently released approximate rotational IoU losses, we derive a differentiable rotational IoU algorithm to enable back-propagation of the IoU loss layer, and we design a novel penalty term based on triangle distance to generate a more precise bounding box while accelerating convergence. Secondly, considering the shortage of feature fusion networks in connection pathways and fusion methods, AW-FPN combines multiple skip-scale connections and attention-weighted feature fusion (AWF) mechanism, enabling high-quality semantic interactions and soft feature selections between features of different resolutions and scales. Finally, to address the limitations of existing SAR ship datasets, such as insufficient samples, small image sizes, and improper annotations, we construct a challenging RSSD to facilitate research on rotated ship detection in complex SAR scenes. As a plug-and-play scheme, our TDIoU loss and AW-FPN can be easily embedded into existing rotation detectors with stable performance improvements. Experiments show that our approach achieves 89.18% and 95.16% AP on two SAR image datasets, RSSD and SSDD, respectively, and 90.71% AP on the aerial image dataset, HRSC2016, significantly outperforming the state-of-the-art methods.

Keywords: synthetic aperture radar (SAR) image; arbitrary-oriented ship detection; differentiable rotational IoU algorithm; triangle distance IoU loss; attention-weighted feature pyramid network; multiple skip-scale connections; attention-weighted feature fusion; Rotated-SARShip dataset (RSSD)

Citation: Xu, Z.; Gao, R.; Huang, K.; Xu, Q. Triangle Distance IoU Loss, Attention-Weighted Feature Pyramid Network, and Rotated-SARShip Dataset for Arbitrary-Oriented SAR Ship Detection. *Remote Sens.* **2022**, *14*, 4676. https://doi.org/10.3390/rs14184676

Academic Editor: Domenico Velotto

Received: 10 August 2022
Accepted: 13 September 2022
Published: 19 September 2022

Publisher's Note: MDPI stays neutral with regard to jurisdictional claims in published maps and institutional affiliations.

1. Introduction

As an active microwave sensor, synthetic aperture radar (SAR) enables all-day, all-weather, and long-distance space-to-Earth observation without being limited by light and climate conditions [1]. With the development of spaceborne SAR high-resolution imaging technology, ship detection in SAR images has become a current research hotspot [2–8].

In recent years, with the breakthrough of convolutional neural networks (CNNs) [9] in computer vision, CNN-based methods have been introduced into SAR ship detection [10–15]. Though these works have promoted the development of this field to some extent, most of them simply apply the horizontal bounding box (HBB)-based methods used in natural scenes to SAR scenes, which still encounter severe challenges, stated as follows:

1. **Complexity of SAR scenes**—since SAR images are taken from a bird's eye perspectives, they contain diverse and intricate spatial patterns. As shown in Figure 1a, instances of small ships tend to be overwhelmed by complex inshore scenes, which inevitably interferes with the recognition of foreground objects, making it difficult for HBB-based methods to accurately distinguish ships from other background components;

2. **Diversity of ship distribution**—in SAR images, ship targets are characterized by varying scales, large aspect ratios, dense arrangements, and arbitrary orientations. In Figure 1b, the HBBs of ships with tilt angles and large aspect ratios contain considerable redundant areas, which introduce background clutter. Moreover, two HBBs of densely arranged ships have a high intersection-over-union (IoU), which is not conducive to non-maximum suppression (NMS), leading to missed detection [16].

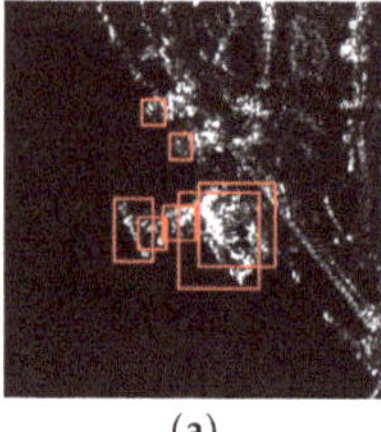 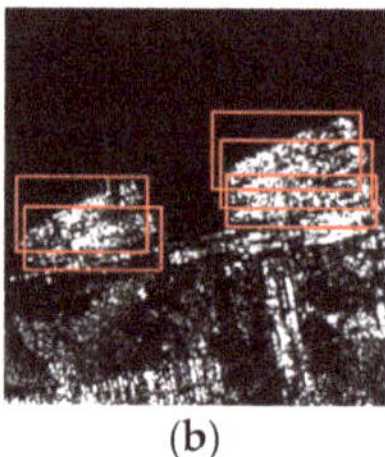 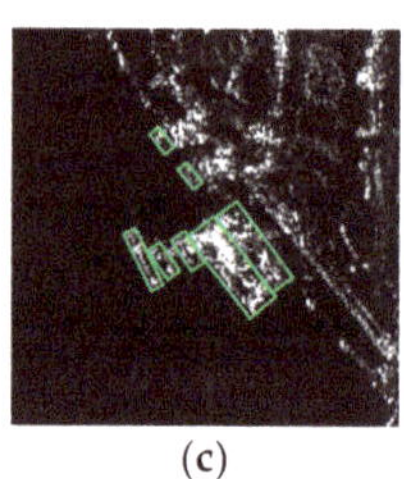 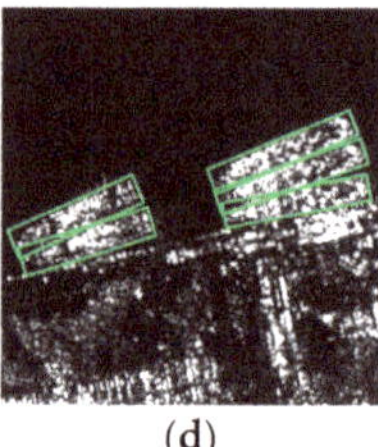

(a)　　　　　(b)　　　　　(c)　　　　　(d)

Figure 1. Densely arranged ships in complex inshore scenes. Here, (**a**,**b**) show the detecting of ship targets using the HBB-based RetinaNet [17]; (**c**,**d**) show the detecting of ship targets using the OBB-based RetinaNet with the proposed TDIoU loss and AW-FPN. The red and green boxes denote the detection results.

To eliminate the defects of HBB-based methods in detecting ships in SAR scenes, oriented bounding box (OBB)-based methods have emerged [18–22]. As shown in Figure 1c,d, OBBs can effectively avoid overlap and attenuate the influence of background clutter, enabling more precise prediction of the location and orientation of ships.

However, OBB-based methods still have the following limitations in SAR scenes:

1. **Problems of rotation detectors based on angle regression**—most rotation detectors adopt l_n-norms as the regression loss in the training phase and intersection-over-union (IoU) as the evaluation metric in the test phase, which will lead to loss-metric inconsistency. In addition, due to the periodicity of the angle parameter, regression-based rotation detectors usually suffer from angular boundary discontinuity [23];

2. **Constraints of multi-scale feature fusion**—due to the large variation in the shapes and scales of ship targets in SAR images, the conventional feature fusion networks [24–27], which are limited by their connection pathways and fusion methods, are not effective in detecting ships with large aspect ratios or small sizes;

3. **Deficiencies of existing SAR ship datasets**—the vast majority of SAR ship detection datasets [28–33] are still annotated by horizontal bounding boxes. Meanwhile, with potential drawbacks, such as insufficient samples, small image sizes, and relatively simple scenes, in these datasets, relevant research is hindered.

To overcome these bottlenecks, we propose a unified framework for rotated SAR ship detection. Inspired by IoU-based losses in horizontal detection, we develop a triangle distance IoU loss (TDIoU loss) and implement the forward and backward processes to ensure its trainability. Thanks to its well-designed penalty term, TDIoU loss not only solves the problems caused by angle regression but also dramatically improves convergence speed and simplifies computation. Second, to enables more effective multi-scale feature fusion for detecting ships with large aspect ratios and varying scales in complex SAR scenes, an attention-weighted feature pyramid network (AW-FPN) combining multiple skip-scale connections and the attention-weighted feature fusion (AWF) mechanism is proposed.

Finally, to promote further research in this field, a novel dataset, the rotated-SARShip dataset (RSSD), is released to provide a challenging benchmark for arbitrary-oriented ship detection in SAR images. Extensive experiments and visual analysis on three datasets prove that our approach achieves better detection accuracy than other advanced methods.

To sum up, the main contributions of this paper are summarized as follows:

1. To the best of our knowledge, TDIoU loss is the first IoU loss specifically for rotated bounding box regression. To solve the non-differentiable problem of rotational IoU, we derive an algorithm based on the Shoelace formula and implement back-propagation for it. The TDIoU loss aligns the training target with the evaluation metric and is immune to boundary discontinuity by measuring the sampling point distance and the triangle distance between OBBs without directly introducing the angle parameter. Furthermore, it is still informative for learning even when there is no overlap between two OBBs or they are in an inclusion relationship, a common occurrence in small ship detection;

2. Our AW-FPN outperforms previous methods in both connection pathways and fusion methods. Skip-scale connections inject more abundant semantic and location information into multi-scale features, facilitating the recognition and localization of ships. The AWF mechanism generates non-linear fusion weights of the same size as the input feature via a multi-scale channel attention module (MCAM) and multi-scale spatial attention module (MSAM), enabling soft feature selections in an element-wise manner, which is critical for detecting ships with large aspect ratios or small sizes;

3. We construct a large-scale RSSD for detecting ships with arbitrary orientations and large aspect ratios in SAR images. To ensure data diversity, we collect original images from three SAR satellites and select different imaging areas. With the help of the automatic identification system (AIS) and Google Earth, 8013 SAR images, including 21,479 ships, are precisely annotated by rotated ground truths. Moreover, we conduct comprehensive statistical analysis and provide results of 15 baseline methods on our dataset. Notably, RSSD is the largest current dataset for rotated SAR ship detection;

4. We embed TDIoU loss and AW-FPN as plug-ins into baseline models and conduct comparative experiments with a dozen popular rotation detectors on two SAR image datasets, the RSSD and the SSDD, and one aerial image dataset, HRSC2016. The results prove that our approach not only achieves state-of-the-art performance in SAR scenes, but also that it shows excellent generalization ability in optical remote sensing scenes.

The rest of the paper is organized as follows: Section 2 reviews related works. Section 3 describes the problems in angle regression and conventional IoU-based losses. Section 4 introduces the proposed TDIoU loss and the AW-FPN for rotated SAR ship detection. Section 5 presents details of the proposed RSSD. Extensive experiments and comprehensive discussions are provided in Section 6. Section 7 summarizes the whole work.

2. Related Work

In this section, we first review CNN-based SAR ship detection methods, then discuss the related works dealing with the problems caused by angle regression and multi-scale feature fusion, and finally analyze several existing publicly available SAR ship datasets.

2.1. SAR Ship Detection Methods Based on Convolutional Neural Networks

In the field of object detection, convolutional neural networks have become the mainstream algorithm. In recently years, CNN-based methods have made significant progress in SAR ship detection. As a pioneering work, Li et al. [10] discussed the defects of Faster R-CNN [34] in SAR ship detection and proposed an improved framework based on feature fusion and hard negative mining. Zhang et al. [11] proposed a novel concept of balance learning (BL) for high-quality SAR ship detection. Zhang et al. [12] proposed a grid convolutional network with depthwise separable convolution that accelerates ship detection by griding the input image. To enhance the detailed features of ships, Liang et al. [13] proposed a visual attention mechanism. Furthermore, the means dichotomy method and speed block kernel density estimation method were used for adaptive hierarchical ship detection. Gao et al. [14] achieved better ship detection accuracy by using the anchor-free CenterNet [35] based on an attention mechanism and feature reuse strategy. Zhang et al. [15] designed a quad feature pyramid network consisting of four unique FPNs and verified its effectiveness on five SAR datasets.

However, the above methods fail to take into account the large aspect ratio and multi-angle characteristics of ships, leading to missed and false detection. Therefore, in recent years, there has been some research on rotated ship detection. For instance, Wang et al. [18] added the angle regression and semantic aggregation method to SSD. The attention module was used to adaptively select meaningful features of ships. Chen et al. [19] presented a feature-guided alignment module and a lightweight non-local attention module to balance the detection accuracy and inference speed of single-stage rotation detectors. Pan et al. [16] constructed a multi-stage rotational region-based network that generates rotated anchors through a rotation-angle-dependent strategy. To reduce the false alarm rate, Yang et al. [20] devised a novel loss to balance the loss contribution of various negative samples. To enhance the detection of small ships, An et al. [21] proposed an anchor-free rotation detector with a flexible frame. Sun et al. [22] applied the bi-directional feature fusion module and angle classification technique to a YOLO-based rotated ship detector.

2.2. Loss-Metric Inconsistency and Angular Boundary Discontinuity

To eliminate the gap between the bounding box regression loss and the evaluation metric, IoU-based losses have been introduced in horizontal detectors [36–40]. Unfortunately, they cannot be simply applied to rotation detection, as the general rotational IoU algorithm is non-differentiable for back-propagation. In addition, unlike other bounding box parameters, the angle parameter is periodic in nature, which will lead to a surge in loss value at the boundary of the angle definition range when using l_n-norm losses.

Some studies have attempted to address part of the above issues from two perspectives. One idea is to design differentiable approximate IoU losses for angle regression. To control the loss value by the amplitude of IoU, Yang et al. [41] added an extra IoU factor into the smooth L1 loss. Furthermore, PIoU [42] estimated the intersection area of two rotated bounding boxes by roughly counting the number of pixels. Aiming to address the uncertainty of convex shapes, Zheng et al. [43] presented an affine transformation to estimate the intersection area. The GWD [23] converted the oriented bounding box to two-dimensional Gaussian distribution, using the Gaussian–Wasserstein distance to approximate the rotational IoU loss. Although these improved regression losses alleviate the problems to some extent, their gradient directions are still not dominated by IoU, and they cannot accurately guide training.

Another idea is to treat the angle prediction as a discrete classification task so as to properly constrain the prediction results. Yang et al. [44] developed a circular smooth label (CSL) technique that directly uses the angle parameter as the category label to tackle the periodicity of the angle and improve the tolerance of adjacent angles. The DCL [45] analyzed the problems of over-thick prediction heads in sparse coded labels and converted the angle categories into dense codes, such as the binary codes and gray codes, to further improve the detection efficiency. Although angle classification techniques avoid angular boundary discontinuity, they are still limited by angular discretization granularity, which inevitably leads to theoretical errors in high-precision angle prediction.

As of now, no full-fledged method exists to address all the above issues. In a sense, the proposed differentiable rotational IoU algorithm opens up the possibility of using the IoU-based loss for rotated bounding box regression, and the newly designed TDIoU loss fundamentally eliminates all these problems in an ingenious manner.

2.3. Multi-Scale Feature Fusion

In CNNs, high-level features contain richer semantic information and broader receptive fields, making them beneficial for detecting large ship targets. Low-level features are of high resolution and contain abundant shallow information, which is conducive to locating small ship targets. One of the difficulties in SAR ship detection is how to effectively fuse multi-scale features. Figure 2 displays several mainstream feature fusion networks [24–27]. Analysis shows that they still suffer from the following limitations in SAR scenes:

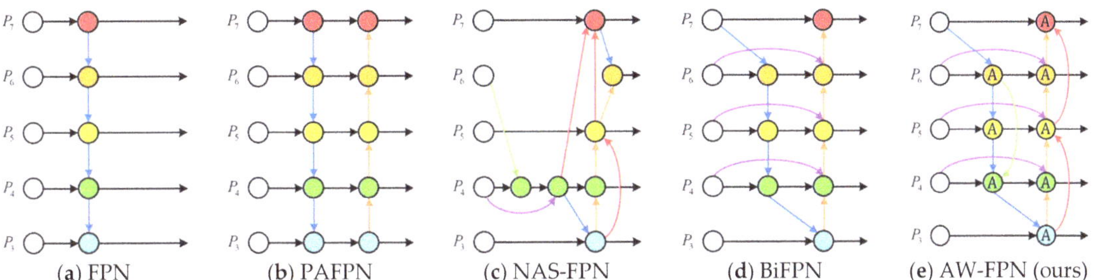

Figure 2. Feature fusion networks. Here, P_i indicates the feature pyramid level i. (**a**) The FPN proposes a top-down pathway to fuse multi-scale features from P_3 to P_7 ; (**b**) PANet builds up an extra bottom-up pathway; (**c**) NAS-FPN designs the network topology by the neural architecturesearch; (**d**) BiFPN adds transverse skip-scale connections and learnable scalar fusion weights; (**e**) our AW-FPN with multiple skip-scale connections and attention-weighted feature fusion (AWF) mechanism.

1. **Restricted connection pathway**—the conventional feature pyramid network (FPN) [24] is inherently limited by a single top-down information flow. Therefore, in PANet [25], an extra bottom-up path aggregation network is added. The above two methods only consider adjacent-level feature fusion. To solve this problem, BiFPN [27] added transverse skip-scale connections from input nodes to output nodes. However, such single same-level feature reuse ignores semantic interactions between cross-level features. Due to the relatively long pathways between high-level features and low-level features, semantics are likely be weakened during layer-to-layer transmission, which is not conducive to the detection of ships with extreme shapes and scales;

2. **Inappropriate fusion method**—most works on feature fusion focus only on designing complicated connection pathways. The fusion method, usually realized by simple addition, is rarely mentioned. Due to the different resolutions of different feature levels, their contributions to the output should also be unequal. The BiFPN added learnable scalar weights to the input features of each node. However, such a rough weighting method, which makes no distinction between all feature points, is still a linear combination of feature maps. Since ship targets in the same image usually have significant differences in scale, simple linear aggregation might not be the best choice.

In recent years, several investigations on visual attention have begun to focus on the fusion method. In SKNet [46] and ResNest [47], the global channel attention mechanism [48] is used to conduct dynamic weighted averaging of features from multiple kernels or groups. Although these attention-based approaches achieve non-linear feature fusion, they only show solicitude for the feature selections in the same layer, leaving no solution for fusing cross-level features of inconsistent semantics and scales. Furthermore, global channel attention only generates a scalar fusion weight for each channel of the feature map, which is obviously not appropriate for scenes with large variations in target scale. Generally speaking, multi-scale networks need to learn diverse feature representations, and a single global channel interaction will weaken the context information of small targets. Recently, aiming to provide a paradigm for cross-level feature fusion, Dai et al. [49] proposed an attentional feature fusion (AFF) mechanism. Regrettably, as with previous approaches, AFF only tends to focus on the salience representations of features in the channel dimension, which might result in the loss of multi-scale spatial contexts.

Our AW-FPN has improved on both of the above. To enrich the semantic and location information in feature maps, both transverse and longitudinal skip-scale connections are used. To generate high-quality fusion weights, a novel AWF mechanism is proposed. The MCAM and MSAM in AWF aggregate both multi-scale channel and spatial contexts, so as to emphasize the region around real ship targets and suppress background clutter.

2.4. SAR Image Datasets for Ship Detection

Due to the limitations of SAR imaging conditions, the datasets of SAR scenes are not as diverse as those of natural scenes. Recent research has been committed to constructing larger and more comprehensive SAR ship detection datasets. Table 1 shows the statistics of six existing datasets [28–33]. However, they still suffer from the following defects:

1. **Insufficient training samples**—the existing SAR ship datasets, such as SSDD [28], DSSDD [30], and AIR-SARShip [31], have a relatively small number of image samples and, therefore, require a large amount of data augmentation before training, which is not conducive to training a high-precision ship detection network;

2. **Small image sizes and relatively simple scenes**—in the SAR-Ship-Dataset [29], ship slices are only 256 × 256 pixels in size. As a matter of fact, small ship slices are more suitable for ship classification since they contain simpler scene information and less inshore scattering. As a result, detectors trained on these ship slices may have difficulty in locating ships near highly reflective objects in large-scale scenes [32];

3. **Inappropriate annotations**—most existing datasets in this field, which fail to consider the large aspect ratio and multi-angle characteristics of ships, are still annotated by HBBs without shape and orientation information. In contrast, OBBs can better fit the approximate shape of ships and mitigate the effect of background clutter. Notably, HRSID [32] and SSDD adopt the polygon annotation for ship instance segmentation. Semantic segmentation divides each pixel of an image into a semantically interpretable class and highlights instances of the same class with the same color. On this basis, instance segmentation employs the results of object detection to perform an instance-level segmentation on different targets of the same class. Although segmented polygons generated by pixel-wise masks enable more accurate contour detection, they are costly in both annotation and detection. For ships in SAR images, we prefer to learn about their general shapes, such as aspect ratio and orientation. On balance, the OBB annotation is a relatively suitable choice. So far, only SSDD provides OBB annotations. However, it contains only 1160 images with 2587 ships, which is far from meeting the demands of ship detection in complex SAR scenes. Hence, it is necessary to construct a large-scale dataset specifically for arbitrary-oriented SAR ship detection.

Table 1. Statistics of the six SAR ship detection datasets released in references [28–33] and our proposed RSSD.

Datasets	Satellite	Polarization	Resolution (m)	Image Size (Pixel)	Image Number	Ship Number	Annotations
SSDD [28]	RadarSat-2, TerraSAR-X, Sentinel-1	HH, HV, VV, VH	1~15	(214~653) × (190~526)	1160	2587	HBB, OBB, Polygon
SAR-Ship-Dataset [29]	Gaofen-3, Sentinel-1	Single, Dual, Ful	3, 5, 8, 10, 25, etc.	256 × 256	43,819	59,535	HBB
DSSDD [30]	RadarSat-2, TerraSAR-X, Sentinel-1	–	1~5	416 × 416	1174	–	HBB
AIR-SARShip [31]	Gaofen-3	Single, VV	1, 3	1000 × 1000, 3000 × 3000	331	–	HBB
HRSID [32]	Sentinel-1, TerraSAR-X	HH, HV, VV	0.5, 1, 3	800 × 800	5604	16,951	HBB, Polygon
LS-SSDD-v1.0 [33]	Sentinel-1	VV, VH	5 × 20	about 24,000 × 16,000	15	6015	HBB
RSSD (ours)	**Sentinel-1, TerraSAR-X, Gaofen-3**	**Single, HH, HV, VV**	**0.5, 1, 3, 5 × 20**	**800 × 800**	**8013**	**21,479**	**OBB**

Our proposed RSSD acquires data from three SAR satellites with different resolutions, polarizations, and imaging modes. The imaging areas are selected in ports and canals with busy trade. All images have been meticulously pre-processed and split into 8013 ship slices of 800×800 pixels. With the help of professional tools, 21,479 ships are precisely annotated by OBBs. All these treatments contribute to the complexity and diversity of our dataset.

3. Analysis of Angle Regression Problems and Conventional IoU-Based Losses

In this section, we first discuss two major problems in the existing rotation detectors mainly caused by angle regression. Then, we review the conventional IoU-based losses and analyze the limitations they may encounter in rotated bounding box regression. Finally, we summarize several requirements that should be met for the rotational IoU loss.

3.1. Problems of Rotation Detectors Based on Angle Regression

Figure 3 demonstrates two generic parametric definitions of oriented bounding boxes (i.e., OpenCV definition and long-edge definition). According to the above two definitions, any two-dimensional bounding box can be represented as a group of five parameters (cx, cy, w, h, and θ), where (cx, cy) represents the centroid coordinate of the oriented bounding box, w and h indicate the width and height, respectively, and θ denotes the rotation angle. To predict the angle θ of the bounding box, most rotation detectors directly introduce an additional output channel into the regression subnet and use l_n-norms as the regression loss during the training phase. However, in the testing stage, the performance is evaluated by IoU. Obviously, such a mismatch may present some problems, which we will now summarize.

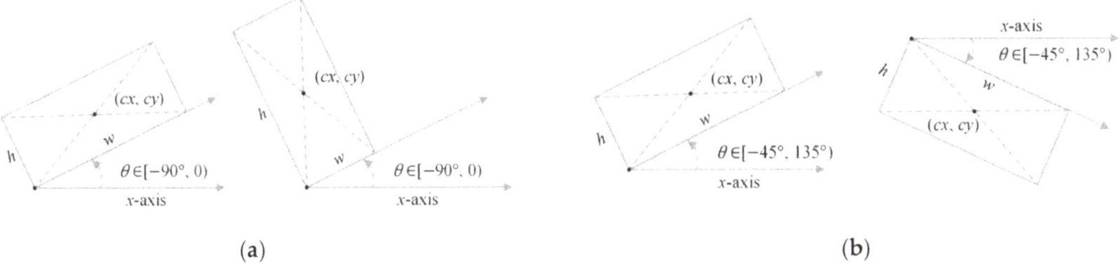

(a)   (b)

Figure 3. Two generic parametric definitions of oriented bounding boxes. (**a**) OpenCV definition, where θ indicates the acute or right angle between the width w and the x-axis; (**b**) Long-edge definition, where w and h signify the long side and short side of a bounding box, respectively. Here, θ denotes the angle from the x-axis to the direction of the width w.

3.1.1. Loss-Metric Inconsistency

In Figure 4a, we compare the relationships between different regression losses and angle differences. Despite the fact that they are all monotonic, only the IoU loss (the light blue curve) and our TDIoU loss (the navy blue curve) are concave, indicating that the gradient directions of l_n-norms are inconsistent with that of IoU. Figure 4b displays the relationship between the rotational IoU and angle differences under different aspect ratios. For a target with a large aspect ratio, a slight angle difference will also lead to a rapid drop in the IoU value. Figure 4c displays the relationships between different regression losses and aspect ratios. All l_n-norm losses remain constant regardless of aspect ratio variations, while the IoU-based losses vary dramatically. The loss-metric inconsistency leads to the conclusion that even a small training loss cannot guarantee high detection performance.

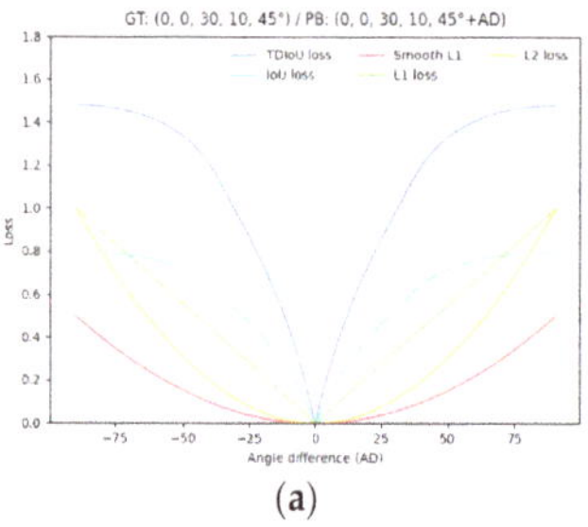 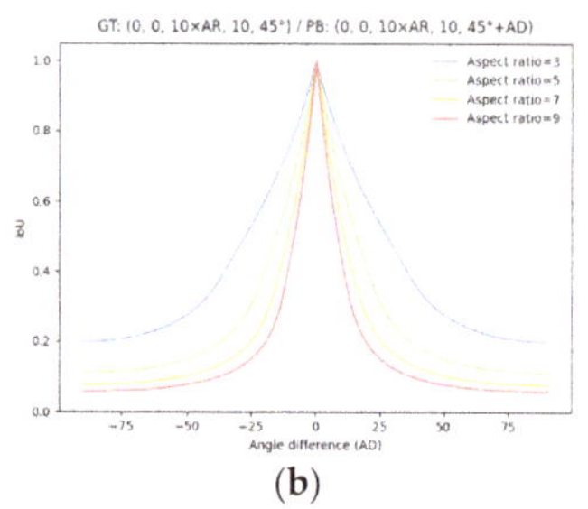 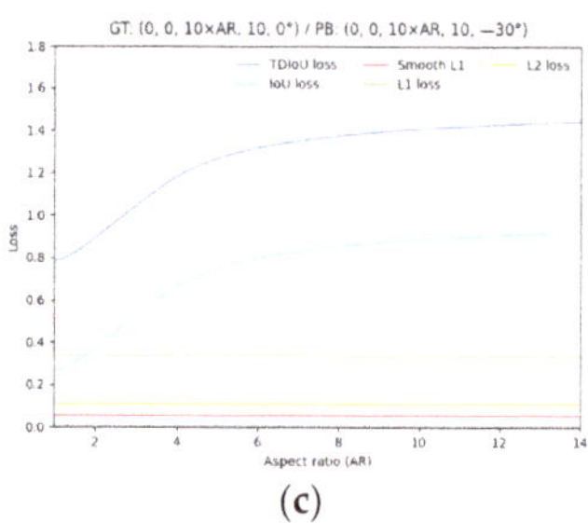

(**a**)   (**b**)   (**c**)

Figure 4. Loss-metric inconsistency. All ground truths (GT) and predicted boxes (PB) are represented as (cx, cy, w, h, and θ) under the long-edge definition. (**a**) Regression loss variations versus angle differences (AD); (**b**) rotational IoU variations versus angle differences under different aspect ratios (AR); (**c**) regression loss variations versus aspect ratios.

3.1.2. Angular Boundary Discontinuity

The angular boundary discontinuity refers to the surge in loss at the boundary of the angle definition range due to the periodicity of the angle (PoA) and the exchangeability of edges (EoE) [23]. Figure 5a shows the boundary problem under the OpenCV definition. Suppose there is a blue anchor/proposal and a green ground truth. The angle of the anchor/proposal is exactly around the maximum or minimum of the defined range. The ideal regression form is to rotate the anchor/proposal counterclockwise by a small angle to the position of the red box. However, due to the angle periodicity, the angle of the predicted box exceeds the defined range $[-90°, 0)$, and the width and height are interchanged relative to the ground truth, leading to a large smooth L1 loss. At this point, the anchor/proposal has to be regressed in a more complex way. For example, it should be rotated clockwise by a larger angle, and its width and height should be scaled at the same time. A similar phenomenon also occurs under the long-edge definition, as shown in Figure 5b.

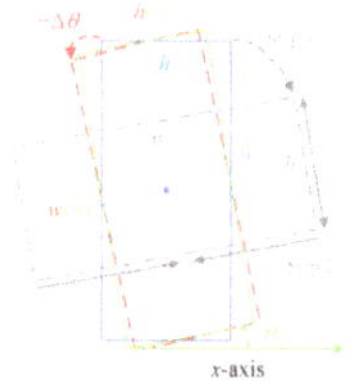 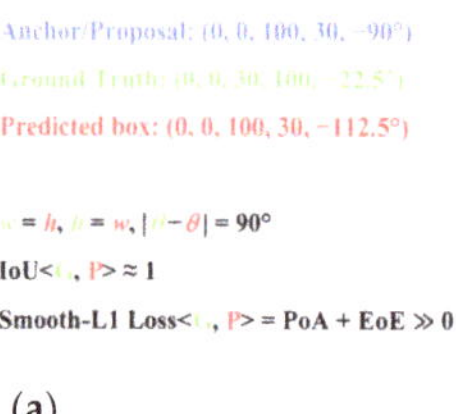

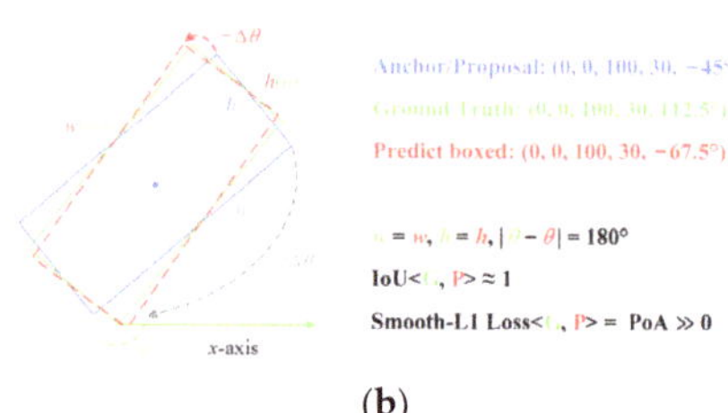

(**a**)   (**b**)

Figure 5. Angular boundary discontinuity under (**a**) the OpenCV definition and (**b**) the long-edge definition.

In essence, angular boundary discontinuity is a kind of manifestation of loss-metric inconsistency. In the boundary case, even if the IoU between the predicted box and the ground truth is very high, a considerable loss will be incurred. Based on the above analysis, we can conclude that the l_n-norms are inapplicable to rotated bounding box regression.

3.2. Limitations of Conventional IoU-Based Losses

It has been demonstrated in horizontal detection methods that the IoU-based losses [36–40] can ensure that the training target remains consistent with the evaluation metric. In theory, they should also work in the rotation case, as the only difference is that the IoU computation for oriented bounding boxes is more complex than that for horizontal ones.

Compared to l_n-norms, the IoU loss has several merits. Firstly, the IoU computation involves all of the geometric properties of bounding boxes, including location, orientation, shape, etc. Secondly, instead of treating the parameters as independent variables as in the case of l_n-norms, IoU implicitly encodes the relationship between each parameter by area calculation. Finally, IoU is scale-invariant, making it ideal for solving scale and range disparities between individual parameters. The original IoU loss is defined as follows [37]:

$$L_{IoU} = 1 - \text{IoU} \tag{1}$$

Here, L_{IoU} is valid only when two bounding boxes have overlap and would not offer any moving gradient for non-overlapping cases. Moreover, it cannot reflect the manner in which the boxes intersect. In Figure 6, the relative positions between the predicted box and the ground truth are obviously different, while the evaluation results of L_{IoU} remain constant.

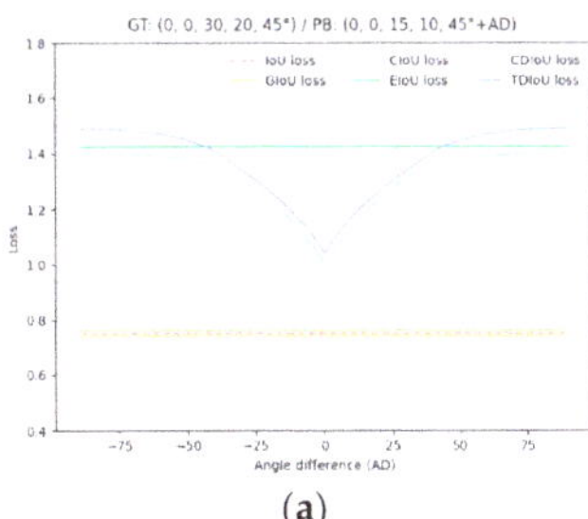

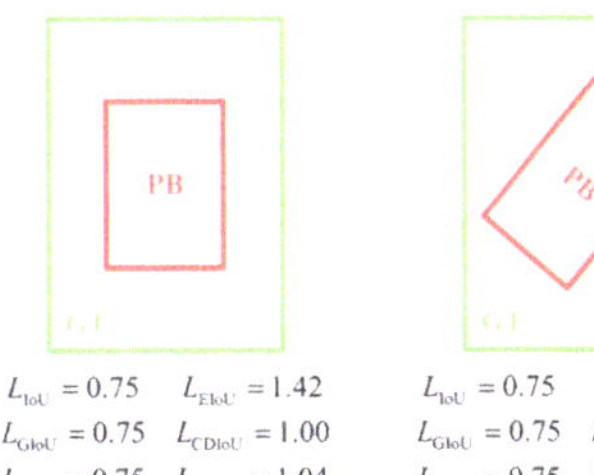

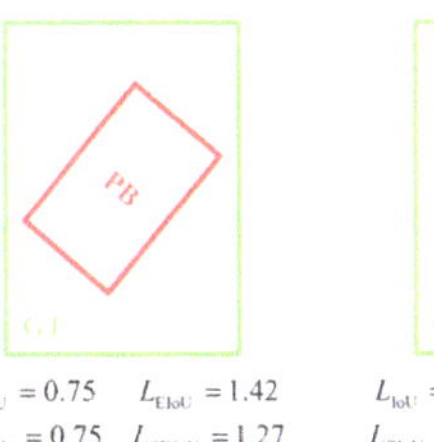

(a)   (b)

Figure 6. Comparison between different IoU-based losses. (**a**) Different IoU-based loss curves versus angle differences; (**b**) some examples from (**a**). When B^{pb} and B^{gt} with coincident centroids are in a containment relationship and their widths and heights are constant, GIoU loss, CIoU loss, and EIoU loss all degenerate into the original IoU loss. In contrast, our TDIoU loss (the navy blue curve) is still able to stably reflect the angle difference and is informative for learning.

The GIoU loss [37] alleviates the issue of gradient disappearance in the non-overlapping case by adding an additional penalty term, which is expressed as follows:

$$L_{GIoU} = 1 - \text{IoU} + \frac{\left| C - B^{pb} \cup B^{gt} \right|}{|C|} \tag{2}$$

where B^{pb} and B^{gt} are the predicted box and the ground truth, and C denotes the smallest enclosing box covering B^{pb} and B^{gt}. Research shows that GIoU first tries to increase the size of B^{pb} to overlap B^{gt} and then uses the IoU term to maximize the intersection area of the bounding boxes [40]. Moreover, GIoU loss requires more iterations to converge.

When designing the penalty term, CIoU loss [38] takes into account the centroid distance and the aspect ratio of the bounding boxes, which is defined as follows:

$$L_{CIoU} = 1 - \text{IoU} + \frac{\rho^2\left(b^{pb},\, b^{gt} \right)}{c^2} + \alpha v \tag{3}$$

$$v = \frac{4}{\pi^2}\left(\arctan\frac{w^{gt}}{h^{gt}} - \arctan\frac{w^{pb}}{h^{pb}} \right), \ \alpha = \frac{v}{(1 - \text{IoU}) + v} \tag{4}$$

where b^{pb} and b^{gt} represent the centroids of B^{pb} and B^{gt}, respectively; $\rho(\cdot)$ indicates the Euclidean distance; c denotes the diagonal length of the smallest enclosing box; w^{pb} and h^{pb} signify the width and height of B^{pb}, respectively; w^{gt} and h^{gt} signify the width and height of B^{gt}, respectively. In CIoU loss, v only reflects the difference in the aspect ratio, rather than the actual difference between w^{pb} and w^{gt} (or h^{pb} and h^{gt}).

To solve this problem, EIoU loss [39] proposes a more efficient form of penalty term:

$$L_{EIoU} = 1 - \text{IoU} + \frac{\rho^2\left(b^{pb}, b^{gt}\right)}{c^2} + \frac{\rho^2\left(w^{pb}, w^{gt}\right)}{c_w^2} + \frac{\rho^2\left(h^{pb}, h^{gt}\right)}{c_h^2} \tag{5}$$

where c_w and c_h indicate the width and height of the smallest enclosing box, respectively. The EIoU loss directly minimizes the difference in the width and height between B^{pb} and B^{gt}, leading to faster convergence and more accurate bounding box regression.

Recently, a new form of penalty term was released in CDIoU loss [40], which narrows the difference between B^{pb} and B^{gt} by minimizing the distance between their vertices, as follows:

$$L_{CDIoU} = 1 - \text{IoU} + \frac{B^{pb} - B^{gt}_{\,2}}{c^2} \tag{6}$$

where $B^{pb} - B^{gt}_{\,2}$ is the distance between the corresponding vertices of B^{pb} and B^{gt}.

However, the above IoU-based losses are all designed for horizontal detection. Due to the introduction of the angle parameter, applying them to oriented bounding box regression will bring some problems. As shown in Figure 6a,b, when B^{pb} and B^{gt} with coincident centroids are in a containment relationship and their widths and heights are constant, the values of GIoU loss, CIoU loss, and EIoU loss remain the same regardless of changes in the angle of B^{pb}. At this point, they completely degenerate into the original IoU loss, making the regression more difficult and the convergence slower. In other words, general parameter-based penalty terms cannot effectively measure the angle difference between B^{pb} and B^{gt}. A natural idea is to introduce the angle parameter into the penalty term. Nevertheless, such a treatment will reintroduce the angular boundary discontinuity, which goes against our original intention. In addition, we also find that the penalty term of CDIoU loss based on the vertex distance is sensitive to the angle parameter. Unfortunately, the denominator of its penalty term involves computing the smallest enclosing box covering B^{pb} and B^{gt}, an extremely tricky task for two rotated boxes. Since the shape of the convex hull formed by the vertices of B^{pb} and B^{gt} is not fixed, the oriented minimum bounding box algorithm [50] requires exhaustive enumeration to obtain the final result, which will consume a lot of computing time and delay the whole training process.

To sum up, a qualified rotational IoU loss should at least meet the following four requirements:

1. **Requirement 1**—it should be differentiable for back-propagation;
2. **Requirement 2**—it should be continuous at the boundary of the angle definition range;
3. **Requirement 3**—it should stably reflect the angle difference between bounding boxes;
4. **Requirement 4**—the computation of the penalty term should be as simple as possible.

4. The Proposed Method

This section elaborates on our proposed unified framework for detecting arbitrary-oriented ships in SAR images, including the differentiable rotational IoU algorithm based on the Shoelace formula, the triangle distance IoU loss (TDIoU loss), and the attention-weighted feature pyramid network (AW-FPN) combining multiple skip-scale connections and the attention-weighted feature fusion (AWF) mechanism.

4.1. Differentiable Rotational IoU Algorithm Based on the Shoelace Formula

Figure 7 visualizes the computation of the intersection-over-union (IoU) for horizontal and oriented bounding boxes. For two-dimensional object detection, the IoU between the ground truth B^{gt} and the predicted box B^{pb} is defined as follows [51]:

$$\text{IoU}\left(B^{gt},\, B^{pb}\right) = \frac{\left|B^{gt} \cap B^{pb}\right|}{\left|B^{gt} \cup B^{pb}\right|} = \frac{Area_{intersect}}{Area_{union}} = \frac{Area_{intersect}}{Area_{gt} + Area_{pb} - Area_{intersect}} \tag{7}$$

where $\left| B^{gt} \cap B^{pb} \right|$ and $Area_{intersect}$ signify the area of the intersection area, and $\left| B^{gt} \cup B^{pb} \right|$ and $Area_{union}$ imply the area of the union area. $Area_{gt}$ and $Area_{pb}$ denote the area of B^{gt} and B^{pb}, respectively. It can be found that how to calculate $Area_{intersect}$ is the core issue. However, as shown in Figure 7b, the IoU computation for OBBs is more complex than that for HBBs, since the shape of the intersection area in the rotation case could be any polygon with fewer than eight edges. In addition, the general rotational IoU algorithm [52] is non-differentiable, as it uses triangulation to calculate $Area_{intersect}$. To address the above issue, we derive a differentiable rotational IoU algorithm based on the Shoelace formula [53], whose pseudo code is provided in Algorithm 1 (Pseudo code of the proposed rotational IoU algorithm based on the Shoelace formula). To further apply it to the IoU loss layer, we implement its forward and backward computation, as illustrated in Figure 8.

Algorithm 1: IoU computation for oriented bounding boxes

Input: Vertex coordinates of B^{gt} and B^{pb}
output: IoU value

1:     Compute the area of B^{gt} and B^{pb}: $Area_{gt} \leftarrow RectArea(B^{gt})$; $Area_{pb} \leftarrow RectArea(B^{pb})$;
2:     Get the edges of B^{gt} and B^{pb}: $Edge_{gt} \leftarrow GetEdge(B^{gt})$; $Edge_{pb} \leftarrow GetEdge(B^{pb})$;
3:     Initialize $A \leftarrow 0$ and the vertices of the intersection area $V \leftarrow EmptySet$;
4:     **for** $i \leftarrow 1$ **to 4 do**
5:     Get the vertices of B^{gt} inside B^{pb}: $V \leftarrow V.add\left(DotProduct\left(B^{gt}(i), B^{pb} \right) \right)$;
6:     Get the vertices of B^{pb} inside B^{gt}: $V \leftarrow V.add\left(DotProduct\left(B^{gt}, B^{pb}(i) \right) \right)$;
7:     **for** $j \leftarrow 1$ **to 4 do**
8:     Get the intersection of edges: $V \leftarrow V.add\left(Bezier\left(Edge_{gt}(i), Edge_{pb}(j) \right) \right)$;
9:     **end for**
10:    **end for**
11:    Sort the vertices of the intersection area: $Indices \leftarrow SortVertex(V)$;
12:    Gather the sorted vertex coordinates according to indices: $V' \leftarrow Gather(V, Indices)$;
13:    **for** $n \leftarrow 1$ **to** $len(V)$ **do**
14:    Shoelace Formula: $A \leftarrow A + V'(n, 1) \times V'(n+1, 2) - V'(n, 2) \times V'(n+1, 1)$
15:    **end for**
16:    Compute the area of the intersection area: $Area_{intersect} \leftarrow A / 2$;
17:    **return** IoU $\leftarrow Area_{intersect} / (Area_{gt} + Area_{pb} - Area_{intersect})$

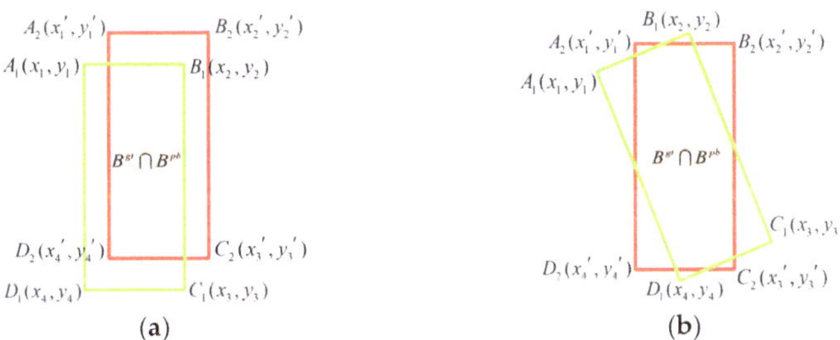

Figure 7. IoU computation for (**a**) horizontal and (**b**) oriented bounding boxes. Red and green boxes represent the predicted box and the ground truth, and the intersection area is highlighted in orange.

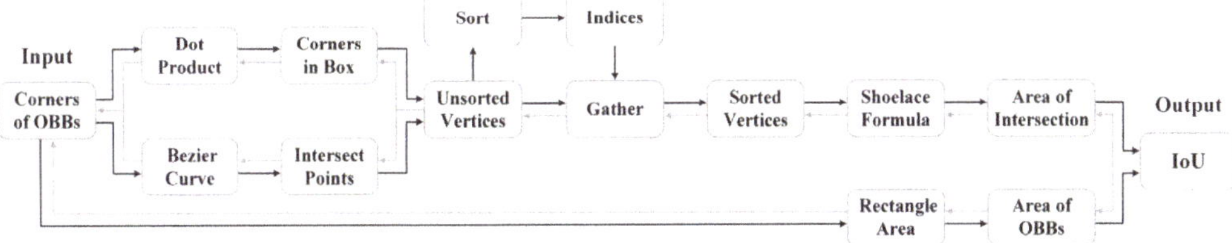

Figure 8. The forward and backward computation of the proposed rotational IoU algorithm. Green and purple boxes signify tensors and operators, respectively. Black and grey arrows indicate forward and backward processes, respectively.

4.1.1. Forward Process

On the basis of Algorithm 1 and Figure 8, the forward process is as follows:

Step 1—convert the ground truth B^{gt} and the predicted box B^{pb} into vertex coordinate representations and calculate their areas (i.e., $Area_{gt}$ and $Area_{pb}$, respectively);

Step 2—find the vertices of the intersection area of B^{gt} and B^{pb}. These are located on the basis of two cases, as follows: (1) from the vertex of B^{gt} and B^{pb}, which falls just inside the other box, and (2) from the intersection point between the edges of two rotated boxes.

In the former case, we use the dot product to calculate the projection of each vertex of B^{gt} and B^{pb} onto two adjacent edges of the other box, respectively, and then determine whether the vertex falls inside the other box, by judging whether the projection exceeds the extent of the corresponding edge. In the latter case, since each edge of rotated boxes is a line segment defined by two vertices, the problem is transformed into locating the intersection point between two line segments in two-dimensional space [54].

Suppose L_1 is an edge of B^{gt}, defined by two vertices (x_1, y_1) and (x_2, y_2), and L_2 is an edge of B^{pb}, defined by two vertices (x_3, y_3) and (x_4, y_4). The line segments L_1 and L_2 can be defined in terms of first-degree Bezier parameters, as follows [55]:

$$L_1 = \begin{bmatrix} x_1 \\ y_1 \end{bmatrix} + t \begin{bmatrix} x_2 - x_1 \\ y_2 - y_1 \end{bmatrix}, \; L_2 = \begin{bmatrix} x_3 \\ y_4 \end{bmatrix} + u \begin{bmatrix} x_4 - x_3 \\ y_4 - y_3 \end{bmatrix} \tag{8}$$

where both t and u are real numbers, and can be expressed as follows:

$$t = \frac{\det \begin{bmatrix} x_1 - x_3 & x_3 - x_4 \\ y_1 - y_3 & y_3 - y_4 \end{bmatrix}}{\det \begin{bmatrix} x_1 - x_2 & x_3 - x_4 \\ y_1 - y_2 & y_3 - y_4 \end{bmatrix}}, \; u = \frac{\det \begin{bmatrix} x_1 - x_3 & x_1 - x_2 \\ y_1 - y_3 & y_1 - y_2 \end{bmatrix}}{\det \begin{bmatrix} x_1 - x_2 & x_3 - x_4 \\ y_1 - y_2 & y_3 - y_4 \end{bmatrix}} \tag{9}$$

where $\det[\,\cdot\,]$ represents the determinant computation. If, and only if, $0 \le t \le 1$ and $0 \le u \le 1$, an intersection point (P_x, P_y) exists as follows:

$$(P_x, P_y) = (x_1 + t(x_2 - x_1), \; y_1 + t(y_2 - y_1)) = (x_3 + u(x_4 - x_3), \; y_3 + u(y_4 - y_3)) \tag{10}$$

In particular, when L_1 and L_2 are collinear (parallel or coincident), they do not intersect. By traversing each edge of B^{gt} and B^{pb}, we obtain all the intersection points.

By computing the above two cases, we finally determine the vertices of the intersection area. If the vertex does not exist, the IoU value is zero;

Step 3—sort the vertices of the intersection area. In general, the vertices of the intersection area form a convex hull. To compute its area, we need to sort its vertices. First, calculate the mean value of the abscissa and the ordinate of these vertices, and note it as the centroid of the polygon. Second, compute the vectors from the centroid to each vertex and normalize them to simplify the sort operation. Finally, scan all the vertices in counterclockwise order from the positive direction of the x-axis to obtain the sorted vertex indices.

Step 4—perform the gather operation to successively fetch the actual coordinate values of the sorted vertices from the unsorted vertex tensor according to the indices;

Step 5—compute the area of the intersection polygon using the Shoelace formula, as follows [56]:

$$Area_{intersect} = \frac{1}{2} \left| \sum_{i=1}^{n} x_i(y_{i+1} - y_{i-1}) \right| = \frac{1}{2} \left| \sum_{i=1}^{n} y_i(x_{i+1} - x_{i-1}) \right| = \frac{1}{2} \left| \sum_{i=1}^{n} \det \begin{bmatrix} x_i & x_{i+1} \\ y_i & y_{i+1} \end{bmatrix} \right| \tag{11}$$

where n represents the number of edges of the intersection polygon; (x_i, y_i) indicate the sorted vertices of the polygon, where i $= 1, 2, \cdots, n$. Note that $x_{n+1} = x_1$ and $y_{n+1} = y_1$;

Step 6—compute the rotational IoU value of B^{gt} and B^{pb} according to Equation (7).

4.1.2. Backward Process

During the forward process, the sort operation returns the indices of sorted vertices in counterclockwise order. Since the return value is discrete (an integer number) rather than continuous (a float number), it is non-differentiable and, therefore, cannot participate in the backward process. However, the computation part of the rotational IoU is still differentiable. This is because we only use the gather operation to obtain the coordinate values of the sorted vertices on the basis of the indices returned by the sort operation, and then adopt the Shoelace formula to compute the area of the intersection area. Throughout the process, the sort operation is not really involved in the area calculation. In most existing deep learning frameworks, the gather function is defined to gather values from the input tensor along a specified dimension and according to a specified index. As its return value is continuous by definition, it is differentiable. Furthermore, the computing process of IoU, including the dot product, the line–line intersection algorithm, and the Shoelace formula, only comprises some essential additive and multiplicative operations, ensuring that the process is robust to the rotational case and feasible for back-propagation.

4.2. Triangle Distance IoU Loss

The proposed rotational IoU algorithm enables back-propagation of the IoU loss layer and, thus, meets **Requirement 1**. In this part, we aim to design a rotational IoU-based loss, which fulfills **Requirements 2, 3, and 4** by constructing a proper penalty term.

Similarly to [37], we define the IoU-based loss as follows:

$$L = 1 - \text{IoU} + \mathcal{R}\left(B^{pb}, B^{gt}\right) \tag{12}$$

where $\mathcal{R}\left(B^{pb}, B^{gt}\right)$ is the penalty term for the predicted box B^{pb} and the ground truth B^{gt}.

Inspired by CDIoU, we apply the distance between corresponding sampling points (i.e., centroids and vertices) of B^{gt} and B^{pb} to the penalty term to measure the overall similarity between them, while avoiding the angular boundary discontinuity caused by the direct introduction of the angle parameter. To reduce the computing complexity, a novel reference term, namely triangle distance, is devised as the denominator of the penalty term to replace the diagonal length of the smallest enclosing box. Following this idea, we design a triangle distance IoU loss (TDIoU loss), which is defined as follows:

$$L_{TDIoU} = 1 - \text{IoU} + \mathcal{R}_{TDIoU} \tag{13}$$

According to Figure 9a, the penalty term of TDIoU loss is defined as follows:

$$\mathcal{R}_{TDIoU} = \frac{|AE| + |BF| + |CG| + |DH| + |PQ|}{\left|\Delta_{AEQ}^{AQ,EQ}\right| + \left|\Delta_{BFQ}^{BQ,FQ}\right| + \left|\Delta_{CGQ}^{CQ,GQ}\right| + \left|\Delta_{DHQ}^{DQ,HQ}\right| + \left|\Delta_{APQ}^{AP,AQ}\right|} \tag{14}$$

where *ABCD* and *EFGH* indicate the corresponding vertices of the predicted box B^{pb} and the ground truth B^{gt}. Here, *P* and *Q* represent the centroids of B^{pb} and B^{gt}, respectively. Furthermore, $|\cdot|$ refers to the Euclidean distance between two sampling points, while $\left|\Delta_{AEQ}^{AQ,EQ}\right|$ indicates the sum of the two edges *AQ* and *EQ* of ΔAEQ (the same applies for other similar terms).

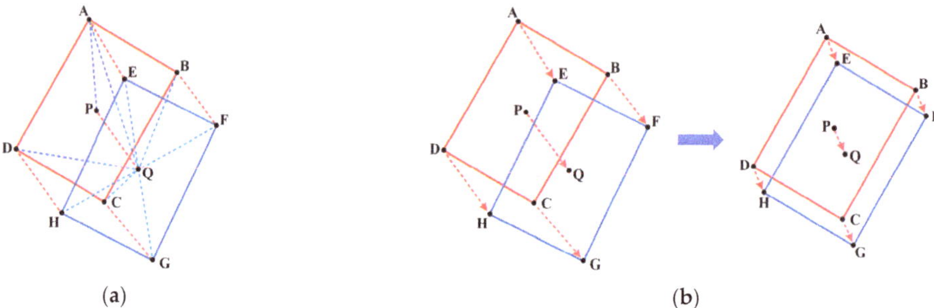

(a)   (b)

Figure 9. The schematic diagram of the TDIoU loss. (**a**) The computation of $\mathcal{R}_{TDIoU}$. The red and blue boxes indicate the predicted box B^{pb} and the ground truth B^{gt}, respectively. The red and blue lines denote the distance between sampling points; (**b**) the process of bounding box regression guided by TDIoU loss. After back-propagation, the model tends to pull the centroids and vertices of the anchor/proposal toward the corresponding points of the ground truth until they overlap.

Note that each group of corresponding sampling points is exploited to construct independent triangles in $\mathcal{R}_{TDIoU}$. To illustrate this process, here we use the vertices A and E. As shown in Figure 9a, we use A, E, and the centroid of B^{gt}, Q, to construct $\triangle AEQ$, which obviously satisfies $|AE| < |AQ| + |EQ|$. Then, $|AE|$ is put into the numerator of $\mathcal{R}_{TDIoU}$ to directly measure the distance between the vertices A and E, while $|AQ|$ and $|EQ|$ are introduced into the denominator of $\mathcal{R}_{TDIoU}$ as part of the reference term. In this way, we finally establish the entire reference term by traversing each group of sampling points, specifically as follows:

$$
\begin{aligned}
|AE| &< |AQ| + |EQ| \\
|BF| &< |BQ| + |FQ| \\
|CG| &< |CQ| + |GQ| \\
|DH| &< |DQ| + |HQ| \\
|PQ| &< |AP| + |AQ|
\end{aligned}
\tag{15}
$$

In the denominator reference term of $\mathcal{R}_{TDIoU}$, the triangle distance plays a similar role to the diagonal length of the smallest enclosing box, ensuring that the value of the penalty term is limited to $[0, 1)$. The difference is that the computing process of the triangle distance is much simpler than that of the latter as it only involves the computation of the distance between two points, which is able to save more training resources and time.

Overall, our TDIoU loss is a unified solution to all the above requirements. Compared to other bounding box regression losses, it has several advantages in rotation detection, as follows:

1. The TDIoU loss inherits all the virtues of existing IoU-based losses. As shown in Figure 4c, though the width w and the height h are not directly used in $\mathcal{R}_{TDIoU}$, TDIoU loss can reflect the overall difference between B^{pb} and B^{gt}, and is sensitive to aspect ratio changes. As an improvement to CDIoU, the centroid distance is introduced in $\mathcal{R}_{TDIoU}$ to speed up bounding box alignment. As illustrated in Figure 9b, the bounding box regression guided by TDIoU loss tends to pull the centroids and vertices of the anchor/proposal toward the corresponding points of the ground truth until they overlap. This process steadily matches the location, shape, and orientation of B^{pb} and B^{gt};

2. By measuring the sampling point distance, $\mathcal{R}_{TDIoU}$ realizes the implicit encoding of the relationship between each parameter. As shown in Figure 6, even when B^{pb} and B^{gt} are in a containment relationship, TDIoU loss is able to reflect the angle difference without directly introducing the angle θ, thus, fundamentally immunizing the angular boundary discontinuity. Hence, our TDIoU loss fulfills **Requirements 2 and 3**;

3. The penalty term of TDIoU loss takes into account the computing complexity by using triangles formed by each group of sampling points to construct the denominator, which significantly reduces the training time and satisfies **Requirement 4.**

Additionally, as a novel training metric, TDIoU loss has the following properties:

1. Here, $0 \leq \mathcal{R}_{TDIoU} < 1$. The lower the value of $\mathcal{R}_{TDIoU}$, the higher the similarity between two boxes; the higher the value of $\mathcal{R}_{TDIoU}$, the higher the difference between two boxes.
2. Here, $0 \leq L_{TDIoU} < 2$. When two bounding boxes are completely coincident, $L_{TDIoU} = 0$. When two bounding boxes are far apart, $L_{TDIoU} \rightarrow 2$.

4.3. Attention-Weighted Feature Pyramid Network

In this part, we introduce the main idea of the proposed attention-weighted feature pyramid network (AW-FPN), which improves the conventional feature fusion networks from the following two aspects: the connection pathway and the fusion method.

4.3.1. Skip-Scale Connections

First used as the identity mapping shortcut in residual blocks [57–59], the skip connection has been a significant component in convolutional networks. In BiFPN, same-level features at different scales are fused via transverse skip-scale connections. However, this single same-level feature reuse neglects the semantic interactions between cross-level features and fails to avoid the semantic loss during layer-to-layer transmission. To search for better network topology, NAS-FPN uses the neural architecture search (NAS) technique. Although it has a haphazard structure that is difficult to interpret, it can guide us in designing a more preferable feature network. As shown in Figure 2c, NAS-FPN contains not only transverse skip-scale connections but also longitudinal skip-scale connections.

Motivated by the above analysis, we devise a more effective feature pyramid network structure, as demonstrated in Figure 10. First, we retain transverse skip-scale connections used in BiFPN for the same-level feature reuse while avoiding adding much cost. Second, to enhance semantic interactions between features of different resolutions, two types of longitudinal skip-scale connections are added in the bi-directional pathways, as follows:

1. **Top-down skip-scale connections**, which integrate higher-level semantic information into lower-level features to improve the classification performance;
2. **Bottom-up skip-scale connections**, which incorporate shallow positioning information into higher-level features to locate small ship targets more accurately.

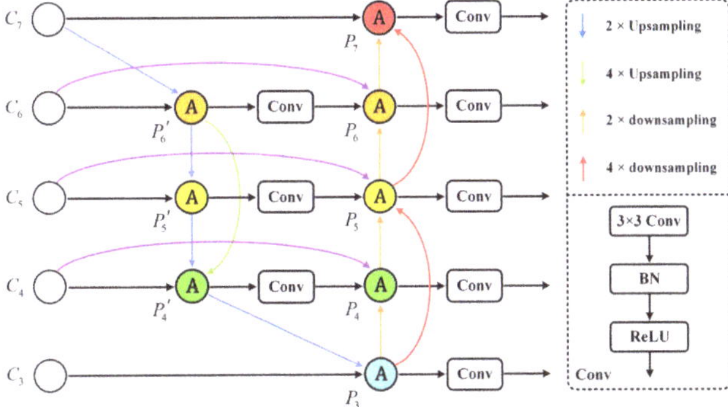

Figure 10. Architecture of AW-FPN. Where **A** denotes the attention-weighted feature fusion (AWF).

4.3.2. Attention-Weighted Feature Fusion (AWF)

When fusing features of inconsistent semantics and scales, a common approach is to directly add them together. The BiFPN assigns a learnable scalar weight for each connection pathway. Nevertheless, in the case of considerable variations in target scales, these linear fusion methods still face obstacles. The AFF [49] provides a non-linear attentional feature fusion scheme. To some extent, our proposed attention-weighted feature fusion (AWF) mechanism can be regarded as its follow-up work, but differs in at least three aspects, as follows:

1. While AFF focuses only on the channel attention, neglecting the spatial context aggregation, our AWF gathers global and local feature contexts in both a multi-scale channel attention module (MCAM) and multi-scale spatial attention module (MSAM);
2. The attentional feature fusion strategy in AFF is restricted to two cross-level features, while our AWF extends it to circumstance of multiple input features;
3. To extract the global channel descriptor, AFF employs only average-pooled features. However, a single average-pooling squeeze may result in the loss of specific spatial information. Hence, to capture finer grained global descriptors, the proposed MCAM and MSAM adopt both average-pooling and max-pooling operations.

Figure 11 describes the process of implementing the AWF. The given N input features from different pyramid levels $F_n \in \mathbb{R}^{C \times H_n \times W_n}$ ($n = 1, 2, \cdots, N$). As they are of different widths and heights, we resize them to the same resolution in advance, as follows:

$$Resize: F_n \rightarrow F'_n \in \mathbb{R}^{C \times H \times W} \tag{16}$$

where *Resize* is an upsampling or downsampling operation. To integrate the information flows of different scales from multiple inputs, we first combine them to construct a fully context aware initial integration $U \in \mathbb{R}^{C \times H \times W}$, where $\oplus$ is an element-wise summation, as follows:

$$U = F'_1 \oplus F'_2 \oplus \cdots \oplus F'_n \tag{17}$$

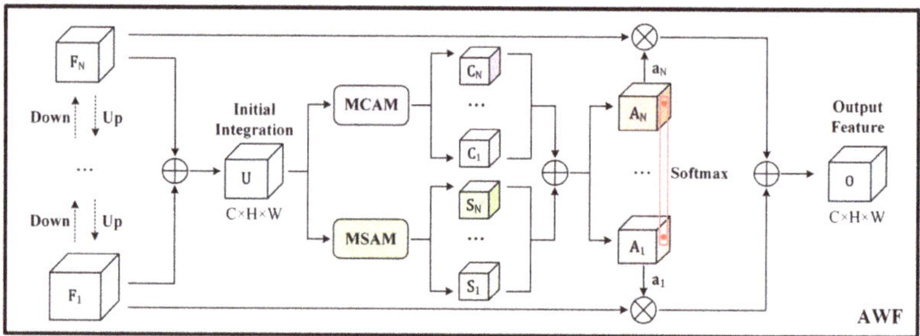

Figure 11. Diagram of the AWF. To generate the non-linear fusion weight a_n, an element-wise softmax is performed on the integrated attention descriptors A_n, fused by multi-scale channel attention C_n and multi-scale spatial attention S_n.

Then, to aggregate global and local feature contexts, the initial integration **U** is transmitted to two parallel multi-scale attention modules MCAM and MSAM, as shown in Figure 12.

MCAM—to polymerize global spatial information for each channel, we employ both average-pooling and max-pooling operations to squeeze the spatial dimension of **U**, so as to generate two distinct channel-wise statistics. Next, we merge them via an element-wise summation to obtain a refined global channel descriptor. Meanwhile, we follow the idea of AFF to aggregate local channel contexts by altering the pooling size. A simple approach is to directly use **U** as the local channel descriptor. After that, the global and local channel descriptors are fed into two independent excitation branches. As the fully connected layer used in [46,48] cannot be directly performed on the three-dimensional tensor, we adopt the convolution operation with a kernel size of 1×1, which only uses point-wise channel interactions at each spatial location to learn the non-linear association between channels.

The global channel context $C_g(U)$ and the local channel context $C_L(U)$ are defined as follows:

$$C_g(U) = \mathcal{B}\left(Conv_2^{1\times1}\left(\delta\left(\mathcal{B}\left(Conv_1^{1\times1}(AvgPool(U) \oplus MaxPool(U))\right)\right)\right)\right) \tag{18}$$

$$C_L(U) = \mathcal{B}\left(Conv_2^{1\times1}\left(\delta\left(\mathcal{B}\left(Conv_1^{1\times1}(U)\right)\right)\right)\right) \tag{19}$$

where $C_g(U) \in \mathbb{R}^{NC\times1\times1}$ and $C_L(U) \in \mathbb{R}^{NC\times H\times W}$. Here, $\mathcal{B}$ denotes the batch normalization [60]. Additionally, δ is the ReLU function, and $Conv^{1\times1}$ is the 1×1 convolution. To simplify computation, the first convolution of each branch is used for channel reduction, while the second is used to restore the channel dimension. Hence, the numbers of filters of $Conv_1^{1\times1}$ and $Conv_2^{1\times1}$ are set to C/r and NC, where r is the channel reduction ratio. Then, $C_g(U)$ and $C_l(U)$ are fused via the broadcasting mechanism to construct the multi-scale channel context $C(U)$. This can be seen as follows:

$$C(U) = C_g(U) \oplus C_l(U) \tag{20}$$

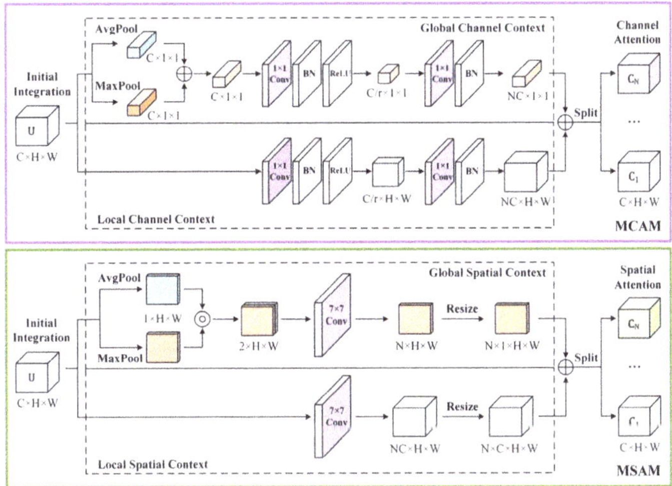

Figure 12. Diagram of each attention sub-module. Multi-scale feature contexts are aggregated in both MCAM and MSAM.

Since $C(U) \in \mathbb{R}^{NC\times H\times W}$ is a channel context aggregation of N input features, it is subsequently split into $C_n \in \mathbb{R}^{C\times H\times W}$ as the multi-scale channel attention for each input.

MSAM—similarly to MCAM, to learn the global and local cross-spatial relationships of the initial integration **U**, we use two parallel branches. First, to obtain a refined global spatial descriptor, we perform the average-pooling and max-pooling operations along the channel dimension and concatenate them, while the initial integration is simply treated as the local spatial descriptor. Then, the convolution layer with a kernel size of 7×7, which has a broader receptive field, is selected as the spatial context aggregator to encode emphasized or suppressed positions of spatial descriptors. On this basis, the global spatial context $S_g(U)$ and the local spatial context $S_l(U)$ can be defined as follows:

$$S_g(U) = Resize\left(Conv_1^{7\times7}(AvgPool(U) \odot MaxPool(U))\right) \tag{21}$$

$$S_l(U) = Resize\left(Conv_2^{7\times7}(U)\right) \tag{22}$$

where $S_g(U) \in \mathbb{R}^{n \times 1 \times H \times W}$ and $S_l(U) \in \mathbb{R}^{n \times C \times H \times W}$. Here, $\odot$ indicates a concatenate operation. The numbers of filters of $Conv_1^{7 \times 7}$ and $Conv_2^{7 \times 7}$ are set to N and NC. As the convolution outputs of the two branches cannot be added directly, we resize them and then fuse them via the broadcast mechanism to obtain the multi-scale spatial context $S(U) \in \mathbb{R}^{n \times C \times H \times W}$.

$$S(U) = S_g(U) \oplus S_l(U) \tag{23}$$

We split $S(U)$ into $S_n \in \mathbb{R}^{C \times H \times W}$ as the multi-scale spatial attention, and the integrated attention descriptor $A_n \in \mathbb{R}^{C \times H \times W}$ can be computed by $A_n = C_n \oplus S_n$. Next, to generate the non-linear fusion weight a_n for each input feature, a softmax operation is executed on each group of corresponding elements of all attention descriptors A_n.

$$a_n = \frac{e^{A_n}}{e^{A_1} \oplus e^{A_2} \oplus \cdots \oplus e^{A_n}} \tag{24}$$

Each element $a_n^{x,\,y,\,z}$ of a_n is a real number between 0 and 1 and fulfills $\sum_{m=1}^{n} a_n^{x,\,y,\,z} = 1$. As $a_n \in \mathbb{R}^{C \times H \times W}$ have the same size as resized features F_n', they preserve and emphasize the subtle details in all inputs, enabling high-quality soft feature selections between F_n'.

$$O = \left(a_1 \otimes F_1'\right) \oplus \left(a_2 \otimes F_2'\right) \oplus \cdots \oplus \left(a_n \otimes F_n'\right) \tag{25}$$

Here, O signifies the final fused feature and $\otimes$ implies an element-wise multiplication.

4.3.3. The Forward Process of the AW-FPN

Our ultimate AW-FPN combines both multiple skip-scale connections and attention-weighted feature fusion. As shown in Figure 10, it takes level 3–7 features extracted by the backbone network as the input $C^{in} = \{C_3,\ C_4,\ C_5,\ C_6,\ C_7\}$, where C_i denotes a feature level with a resolution of $1/2^i$ of the input image. The top-down and bottom-up aggregation pathways are constructed successively. Here, we take level 5 as an example to illustrate the forward process. On the top-down pathway, the intermediate feature of level 6 (P_6') is upsampled $2 \times$ and then fused with C_5 via AWF, followed by a 3×3 convolution to generate the intermediate feature P_5'. On the bottom-up pathway, the outputs of levels 3 and 4 (P_4 and P_3) are subjected to $4 \times$ and $2 \times$ downsampling operations, respectively, and then fused with C_5 and P_5'. The final output P_5 is generated by the 3×3 convolution, as follows:

$$P_5' = Conv\left(AWF\left(C_5,\ Resize\left(P_6'\right)\right)\right) \tag{26}$$

$$P_5 = Conv\left(AWF\left(C_5,\ P_5',\ Resize(P_4),\ Resize(P_3)\right)\right) \tag{27}$$

where $Conv$ implies the 3×3 convolution, which is followed by a batch normalization operation and a ReLU function. All other feature levels are constructed in a similar way.

5. Rotated-SARShip Dataset

In this section, we introduce the collection process and data statistics of our proposed rotated-SARShip dataset (RSSD) for arbitrary-oriented ship detection in SAR images.

5.1. Original SAR Image Acquisition

Table 2 provides detailed information of the original SAR imageries used to establish our RSSD. First, from the Copernicus Open Access Hub [61], we downloaded three raw Sentinel-1 images with a resolution of 5 m $\times$ 20 m, characterized by large scales and wide coverages (25,340 $\times$ 17,634 pixels on average). As shown in Figure 13, the imagery acquisition areas are selected in the Dalian Port, Panama Canal, and the Tokyo Port (these ports have huge cargo throughputs, and the canal has busy trade). In general, the polarization, imaging mode, and the incident angle of sensors tend to influence the imaging condition of SAR images to a certain extent. For the Sentinel-1 images, the basic polarimetric combination is VV and VH. The imaging mode is interferometric wide swath (IW), which is the primary sensor mode for data acquisition in marine surveillance zones. Furthermore, to minimize redundant interferences, such as foreshortening, layover, and shadowing of vessels, we choose an incident angle of 27.6~34.8° [32].

Table 2. Detailed information of the original SAR imageries used to establish our RSSD.

No.	Data Source	Polarization	Imaging Mode	Incident Angle (°)	Resolution (m)	Image Size (Pixel)	Location	Date and Time
1	Sentinel-1	VV, VH	IW	27.6~34.8	5 × 20	25,313 × 16,704	Dalian Port	5 October 2021, 09:48:20
2	Sentinel-1	VV, VH	IW	27.6~34.8	5 × 20	25,136 × 19,488	Panama Canal	30 September 2021, 11:06:41
3	Sentinel-1	VV, VH	IW	27.6~34.8	5 × 20	25,480 × 16,709	Tokyo Port	1 October 2021, 08:41:23
4~256	Gaofen-3 (AIR-SARShip)	Single, VV	SpotLight, SM	–	1, 3	1000 × 1000, 3000 × 3000	–	–
257~5792	Sentinel-1, TerraSAR-X (HRSID)	HH, HV, VV	S3-SM, ST, HS	27.6~34.8, 20~45, 20~60, 20~55	0.5, 1, 3	800 × 800	Barcelona, Sao Paulo, Houston, etc.	–

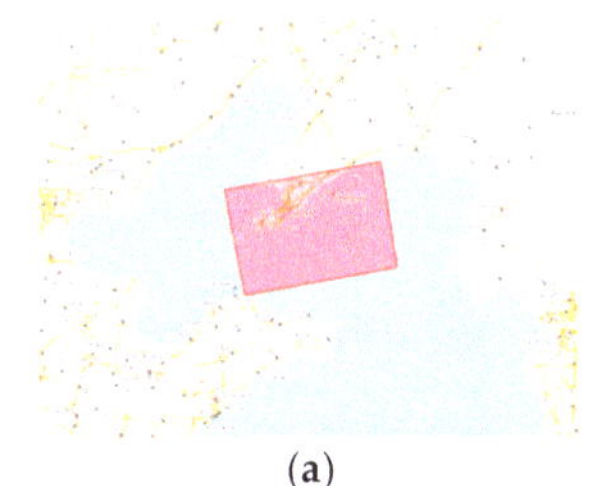 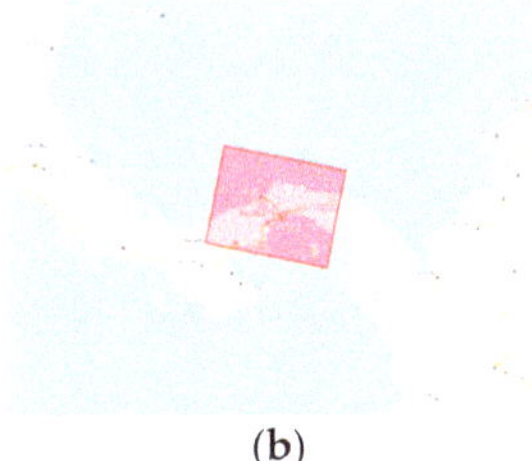 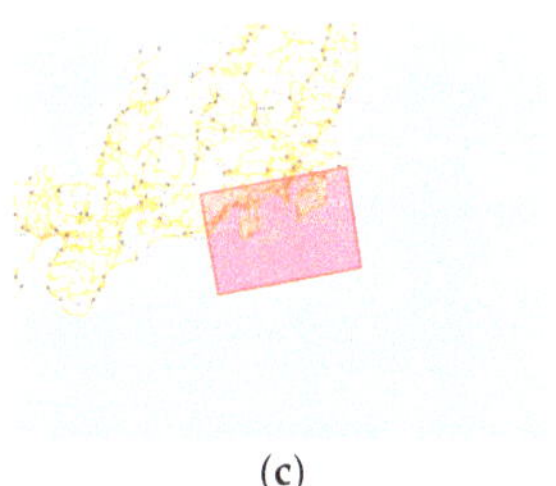

(a)   (b)   (c)

Figure 13. Coverage areas of No. 1–3 Sentinel-1 images. (**a**) The Dalian Port; (**b**) the Panama Canal; (**c**) the Tokyo Port.

To ensure complex and diverse image scenes, we also screen 252 and 5535 SAR images from AIR-SARShip [31] and HRSID [32], respectively. As shown in Table 2, the HRSID images shot by Sentinel-1 and TerraSAR-X have resolutions of 0.5 m, 1 m, and 3 m. The polarizations are HH, HV, and VV, and the imaging modes are S3-StripMap (S3-SM), Staring SpotLight (ST), and High-Resolution SpotLight (HS). The AIR-SARShip images from Gaofen-3 have resolutions of 1 m and 3 m, polarizations of single and VV, and imaging modes of SpotLight and StripMap (SM). Since these images have different resolutions and imaging conditions, ships in them usually appear in different forms. Notably, images with a resolution of less than 3 m can retain the detailed features of ships, while images with a resolution of 5 m × 20 m can increase the number of small ship targets.

5.2. SAR Image Pre-Processing and Splitting

The above original SAR imageries still need to be pre-processed before annotation. To display recognizable features of ships, we first apply the Sentinel-1 toolbox [62] to convert the raw Sentinel-1 data into grayscale images in the 16-bit tag image file format (TIFF), followed by geometrical rectification and radiometric calibration operations. Since images selected from AIR-SARShip and HRSID have already undergone the above processing, we directly perform the de-speckling operation on all the original images to suppress the influence of background noise. Finally, we transform all images into portable network graphics (PNG) files in the same format as the DOTA dataset [63].

Due to the side-scan imaging mechanism of SAR satellites, the original SAR imagery generally has a huge size and should be split into ship slices to fit the input size of CNN-based detectors. First, to avoid duplicate splitting, the offshore areas with a relatively dense ship distribution are separated from the images in advance [32]. After that, a sliding window of 800 × 800 pixels is used to shift over the whole image with a stride of 600 pixels in width and height (25% overlap rate) to preserve the relatively intact features of ships. Since the images screened from HRSID have already been cropped to the expected size matching the network input, the splitting operation is performed only on the images from Sentinel-1 and AIR-SARShip. Furthermore, we reserve the complex inshore scenes containing ships and artificial facilities and remove the negative samples with only pure background.

5.3. Dataset Annotation

With the assistance of the official document and the Sentinel-1 toolbox, we can easily acquire the precise imaging time and geographic location of each Sentinel-1 image, which will help the automatic identification system (AIS) and Google Earth to provide support for the annotation work. As shown in Figure 14, we first identify the approximate location of the imaging area of each Sentinel-1 image in AIS and Google Earth. Since AIS provides the movement trajectories of most ships around the time the images were shot, it is possible to grasp the approximate distribution of ships and estimate their possible positions in the imaging area. Subsequently, we match the AIS message with each Sentinel-1 image and determine the topographical features and marine conditions of the coverage area with the help of Google Earth. On this basis, we adopt RoLabelImg [64] to annotate the oriented bounding boxes of ships, obtaining relatively accurate ground truths. To ensure that the annotations meet the requirements of most rotation detectors, we convert them to the DOTA format, using four ordered vertices to represent ship ground truths, as shown in Figure 15. In fact, there are still some islands and reefs incorrectly labeled as ships. Thus, we employ Google Earth for further in-depth inspection and correction to ensure the accuracy of the annotations. Note that since the specific shooting information of the images from AIR-SARShip and HRSID cannot be acquired directly, we first refer to their original horizontal ground truths and carefully check whether there are errors and omissions. Then, we annotate them with more elaborate oriented bounding boxes.

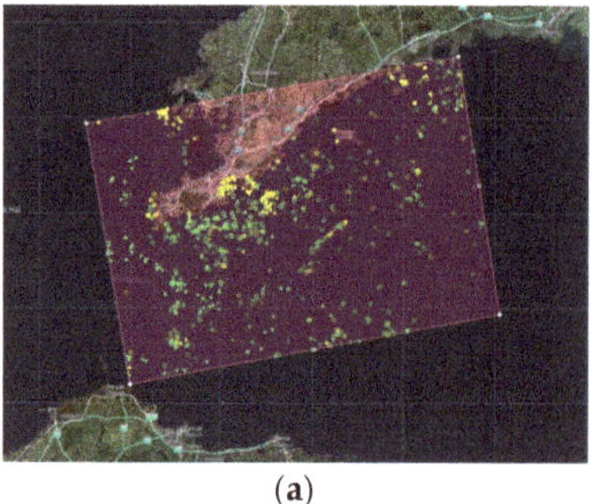

(a)

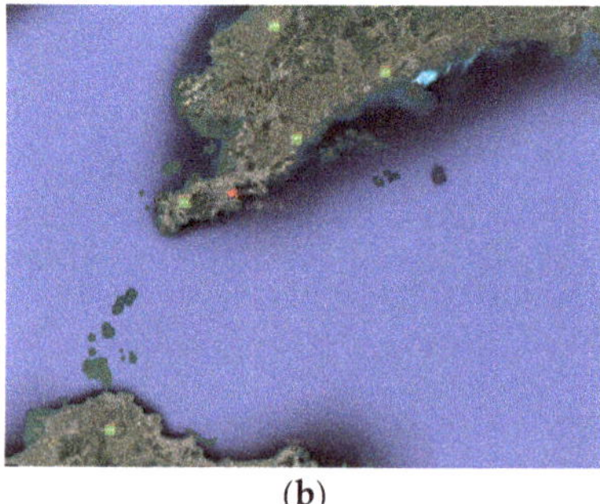

(b)

Figure 14. AIS query and Google Earth correction (take No.1 Sentinel-1 image as an example). (**a**) AIS information of the coverage. Marks of different shapes and colors represent different types of ships; (**b**) corresponding Google Earth image.

So far, we have established the RSSD, and there have been 8013 SAR images with corresponding annotation files, including 21,479 ship targets annotated by rotated ground truths. Figure 16 displays ship ground truth annotations of diverse SAR images in RSSD.

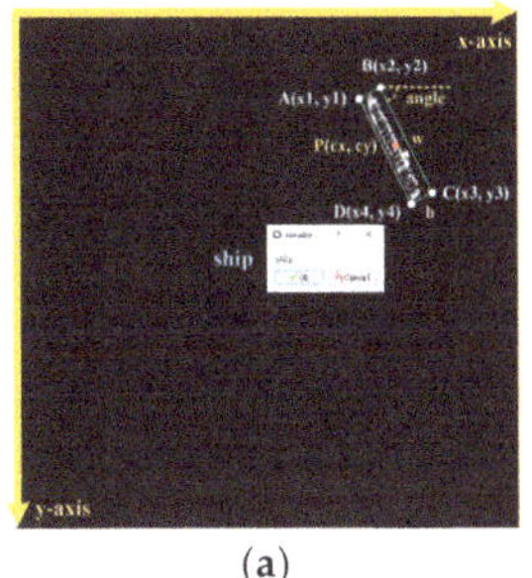
(a)

(b)

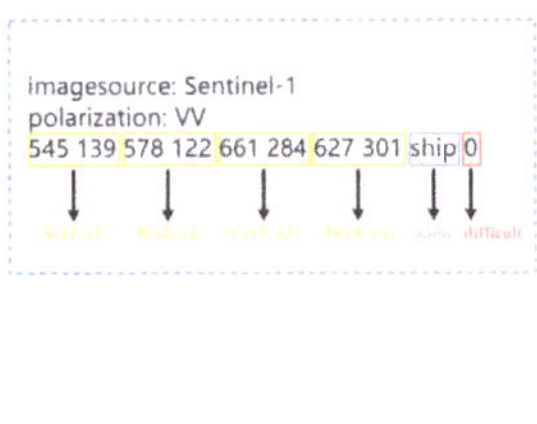
(c)

Figure 15. The ship annotation. (**a**) The OBB label in a SAR image; (**b**) the xml label file annotated by RoLabelImg. Each ship target is represented by an oriented bounding box as its ground truth. Where (cx, cy) is the centroid, w and h denote the width and height, and θ indicates the rotation angle; (**c**) the txt label file in DOTA format. Each ship is represented by four ordered vertices. Note that the top-left vertex is taken as the starting point, and the four vertices are arranged in clockwise order.

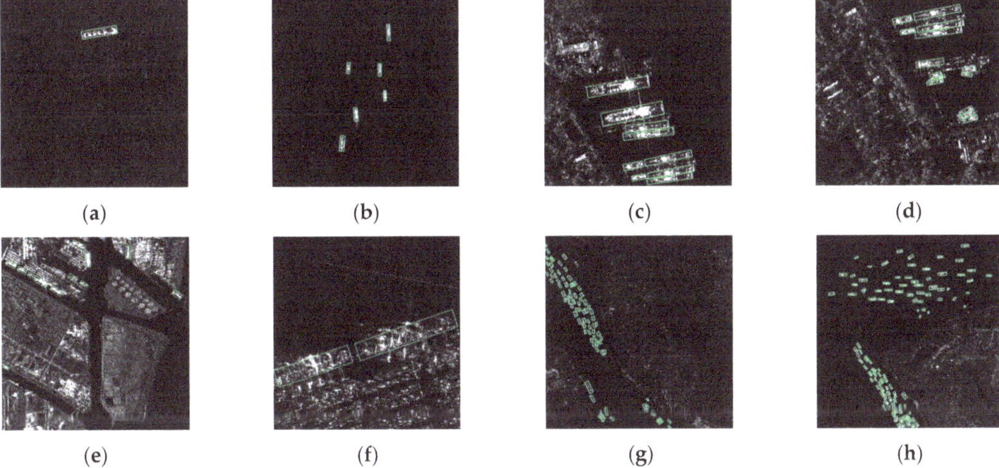
(a) (b) (c) (d)
(e) (f) (g) (h)

Figure 16. Ship ground truth annotations of diverse SAR images in RSSD. Real ships are accurately marked in green OBBs. (**a**) Offshore single ship; (**b**) offshore multiple ships; (**c**,**d**) densely arranged ships; (**e**) ships lying off the port; (**f**) ships with large aspect ratios; (**g**,**h**) small ships in the canal.

5.4. Statistical Analysis on the RSSD

Figure 17 visualizes the comprehensive statistical comparison between our RSSD and SSDD, both of which adopt OBB annotations. As Table 3 shows, 70% of the RSSD images are randomly selected as the training set, and 30% are selected as the test set. For the SSDD, we divide all images in the ratio of 8:2 according to [28]. As shown in Table 1, SSDD contains 1160 SAR images with 2587 annotated ships, indicating that each image contains only 2.2 ships on average, while in our dataset, each image contains about 2.7 ships. Figure 17a,e display the width and height distribution of ship ground truths. Compared to the extreme funnel-like distribution of SSDD, our RSSD features a more uniform ship size distribution and more prominent multi-scale characteristics. As per Figure 17b,f, the aspect ratio of ship ground truths in the SSDD is generally below 3, whereas it is concentrated in the range of 2~5 in the RSSD, indicating that most instances in our dataset are with relatively high aspect ratios. Since the difficulty in detecting ships typically increases with the aspect ratio, our RSSD is more challenging compared to other datasets. As per Figure 17c,g, according to the MS COCO evaluation metric [65], the numbers of small ships ($Area_{OBBs} < 1024$ pixels), medium ships ($1024 < Area_{OBBs} < 9216$ pixels), and large ships ($Area_{OBBs} > 9216$ pixels) in the RSSD are 13,369, 7741, and 369, respectively, (62.24%, 36.04%, and 1.72% of all ships, respectively), while in the SSDD, the proportions are 71.12%, 28.30%, and 0.58%, respectively. Ships in both datasets are relatively small in size but have large variations in scale. As shown in Figure 17d,h, the angle distribution of ship ground truths in the RSSD is more balanced than that in the SSDD. This ensures that rotation detectors learn the multi-angle features better.

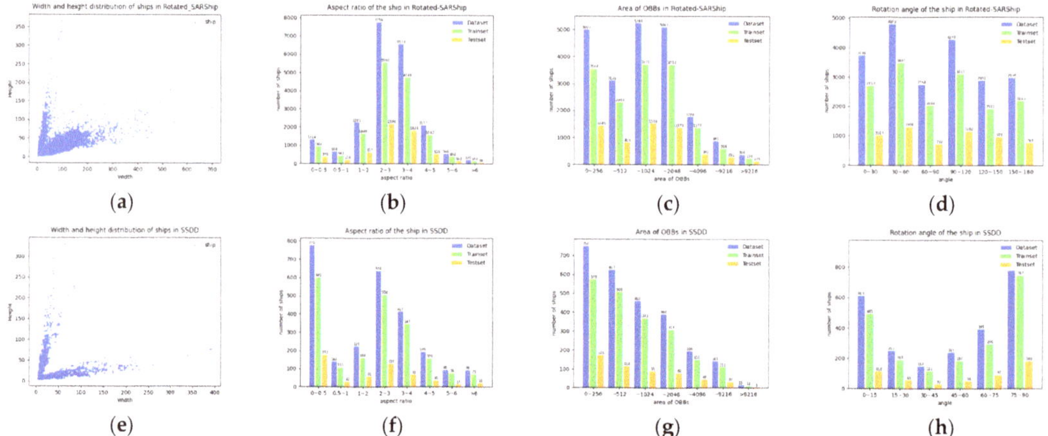

Figure 17. Statistical comparison between the proposed RSSD and the SSDD. Here, (**a**) and (**e**) show the width and height distribution of ship ground truths in RSSD and SSDD, respectively; (**b**) and (**f**) display the aspect ratio distribution of ship OBBs; (**c**) and (**g**) indicate the area distribution of ship OBBs; (**d**) and (**h**) show the rotation angle distribution of ship OBBs.

Table 3. Details of dataset division.

Dataset	Train	Test	All	Inshore (Test)	Offshore (Test)
RSSD (ours)	5692	2321	8013	479	1842
SSDD	928	232	1160	46	186
HRSC2016	617	444	1061	–	–

Based on the above analysis, it is obvious that the ship targets in our RSSD not only differ significantly in orientation degrees but also have multi-scale characteristics, which provides a challenging benchmark for arbitrary-oriented ship detection in SAR images.

6. Experiments and Discussion

In this section, we first present the benchmark datasets, implementation details, and evaluation metrics. Then, extensive comparative experiments with existing methods are carried out to verify the superiority and robustness of our approach. Meanwhile, comprehensive discussions are provided to analyze and interpret the experimental results.

6.1. Benchmark Datasets and Implementation Details

The proposed rotated-SARShip dataset (RSSD) and the public SAR ship detection dataset (SSDD), specific information about which is provided in Section 5, are used to evaluate the performance of the proposed method. In our experiments, all SSDD images are resized to 512×512 pixels, with padding operation to avoid distortion, while the RSSD images of 800×800 pixels are directly used as the network input. The ratio of training set to test set for the RSSD is set to 7:3, while that for the SSDD is set to 8:2. To better assess the performance of our approach in different SAR scenes, we further divide the test sets into inshore and offshore scenes. Figure 18 and Table 3 show the details of dataset division.

Figure 18. The proportion of inshore and offshore scenes in the test sets of (**a**) RSSD and (**b**) SSDD.

Furthermore, a public benchmark for OBB-based ship detection in optical remote sensing images, the HRSC2016 dataset [66], is used to verify the generalization ability of the proposed method across different scenarios. It contains 1061 high-resolution aerial images, including 2976 different types of ships annotated by oriented bounding boxes, with the image size ranging from 300×300 to 1500×900 pixels. We employ the training (436 images) and validation (181 images) sets for training, and the test set (444 images) for testing. All images are resized to 800×512 pixels without altering the original aspect ratio.

The experiments are conducted on the platform with Ubuntu 18.04 OS, 32 GB of RAM, and a NVIDIA GTX 1080Ti GPU. For all datasets, we train the models in 72 epochs. The SGD optimizer is adopted with a batch size of 2 and an initial learning rate of 0.0025. The momentum and weight decay are 0.9 and 0.0001, respectively. As for the learning schedule, we apply the warmup strategy for 500 iterations, and the learning rate is dropped 10-fold at each decay step. If not specified, ResNet50 [57] is employed as the default backbone network. Its parameters are initialized by ImageNet pretrained weights. For fair comparisons with other methods and to avoid over-fitting, we only use random flipping and rotation for data augmentation in the training phase. If not specified, no extra tricks are used.

6.2. Evaluation Metrics

To qualitatively and quantitatively evaluate the detection performance of different methods in our experiments, two normative metrics, the precision–recall curve (P–R curve) and average precision (AP), are leveraged. Specifically, the precision and the recall can be expressed as follows:

$$Precision = \frac{TP}{TP + FP} \tag{28}$$

$$Recall = \frac{TP}{TP + FN} \tag{29}$$

where TP (true positives), FP (false positives), and FN (false negatives) represent the number of correctly detected ships, false alarms, and undetected ships, respectively. The P–R curve, with precision as the y-axis and recall as the x-axis, reveals the relationship between these two metrics. The AP is defined as the area under the P–R curve, as follows:

$$AP = \int_0^1 P(R)dR \tag{30}$$

where P and R indicate the precision and recall, respectively. The AP evaluates the overall performance of detectors under different IoU thresholds (0.5 by default) and, the larger the value, the better the performance. Furthermore, we use the total training time as a metric to evaluate the computing complexity and training efficiency of different losses.

6.3. Ablation Study

In this part, we first introduce two robust rotation detectors as baselines. On this basis, a series of component-wise experiments on the RSSD, the SSDD, and the HRSC2016 are carried out to validate the effectiveness of the proposed TDIoU loss and AW-FPN.

6.3.1. Baseline Rotation Detectors

Two rotation detectors, RetinaNet [17] and CS^{2}A-Net [67], are selected as baselines in our experiments. As a typical single-stage detector, RetinaNet consists of a backbone network, a feature pyramid network, and detection heads. It uses a ResNet [57] to generate a multi-scale feature pyramid and attaches a detection head to each pyramid level (P_3 to P_7). Each detection head is made up of a classification sub-network and a regression sub-network. To implement a RetinaNet-based rotation detector (RetinaNet-R), we modify the regression output to an OBB (cx, cy, w, h, and θ) under the long-edge definition, where (cx, cy), w, h, and θ denote the centroid, the width, the height, and the angle, respectively, and $\theta \in [-45°, 135°)$. Accordingly, the angle θ is taken into consideration in the anchor generation. At each pyramid level, we set anchors in three aspect ratios, {1:2, 1:1, and 2:1}, three scales, {1, $2^{1/3}$, and $2^{2/3}$}, and six angles, {$-45°$, $-15°$, $15°$, $45°$, $75°$, and $105°$}. The proposed TDIoU loss and AW-FPN can be easily embedded into RetinaNet-R, as shown in Figure 19a.

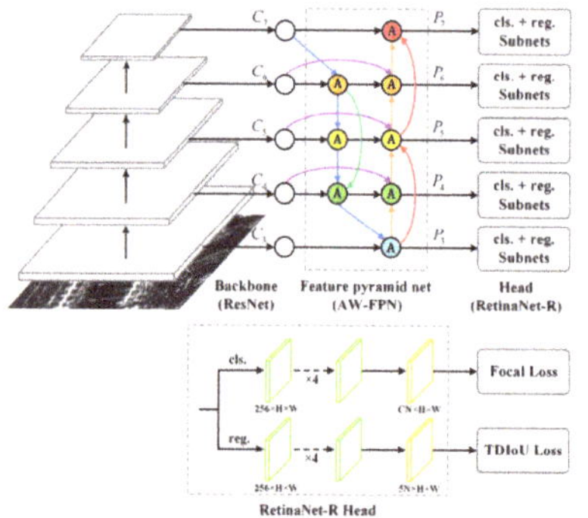
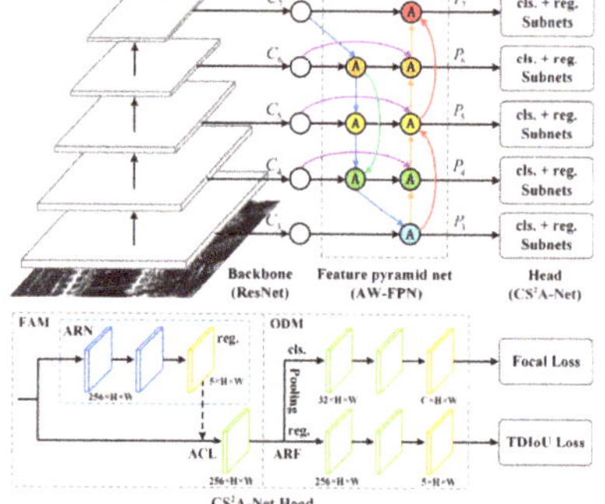

(a) RetinaNet-R with TDIoU loss and AW-FPN

(b) CS^{2}A-Net with TDIoU loss and AW-FPN

Figure 19. Architectures of two baselines. As a plug-and-play scheme, TDIoU loss and AW-FPN can be easily embedded into the above rotation detectors. (**a**) The regression output of RetinaNet is modified to an OBB under long-edge definition. Here, 'C' denotes the number of categories, and 'N' denotes the number of anchors on each feature point; (**b**) the CS^{2}A-Net head consisting of the FAM and ODM can be cascaded to improve accuracy. The number of cascade heads is set to 2 by default.

The CS^{2}A-Net is an advanced rotation detector based on the RetinaNet architecture. Its detection head consists of a feature alignment module (FAM) and an oriented detection module (ODM), which can be cascaded to improve accuracy. The FAM uses an anchor refinement network (ARN) to generate refined rotated anchors, and then sends refined anchors and input features to an alignment convolution layer (ACL) to learn aligned features. In ODM, the active rotating filter (ARF) learns orientation-sensitive features, and then a pooling operation extracts the orientation-invariant features for classification and regression. Our TDIoU loss and AW-FPN can also be integrated into CS^{2}A-Net, as shown in Figure 19b.

The multi-task loss function of two baseline detectors is defined as follows:

$$L = \frac{1}{N} \sum_{N=1}^{n} S_n L_{\text{reg}}\left(B_n^{pb}, B_n^{gt}\right) + \frac{\lambda_2}{N} \sum_{N=1}^{n} L_{\text{cls}}\left(p_n^{pb}, p_n^{gt}\right) \tag{31}$$

where λ_1 and λ_2 indicate the loss balance hyper-parameter and are set to 1 by default, N denotes the number of anchors in a mini-batch, and S_n is a binary value ($S_n = 1$ for positive anchors and $S_n = 0$ for negative anchors). The vectors B_n^{pb} and B_n^{gt} denote the locations of the n-th predicted box and the corresponding ground truth, respectively. The values p_n^{pb} and p_n^{gt} indicate the predicted classification score and the true label of the nth object, respectively. In our experiments, the regression loss L_{reg} is set as the smooth L1 loss, the TDIoU loss, etc. The classification loss L_{cls} is set as the focal loss [17], as follows:

$$L_{focal}(p_t) = -\alpha_t(1 - p_t)^{\gamma}\log(p_t) \tag{32}$$

where $(1 - p_t)^{\gamma}$ and α_t are two modulation factors that satisfy the following conditions:

$$p_t = \begin{cases} p_n^{pb}, & p_n^{gt} = 1 \\ 1 - p_n^{pb}, & \text{otherwise} \end{cases} \quad \text{and} \quad \alpha_t = \begin{cases} \alpha, & p_n^{gt} = 1 \\ 1 - \alpha, & \text{otherwise} \end{cases} \tag{33}$$

where α and γ are two hyper-parameters, which are set to 0.25 and 2, respectively, by default.

6.3.2. Effectiveness of the TDIoU Loss

We evaluate the TDIoU loss with two baseline detectors on three datasets, as shown in Tables 4–6. Both detectors adopt ResNet50 and the original FPN. To ensure the objectivity and richness of the ablation study, we implement two approximate IoU losses (IoU-smooth L1 and GWD loss) and five IoU-based losses (IoU, GIoU, CIoU, EIoU, and CDIoU loss) to compare the performance of different regression losses. Only the regression loss is modified, and all other settings remain intact for fair comparisons.

Table 4. Comparison of different regression losses on RSSD. Here, R-50-FPN denotes ResNet50 with FPN, **LMI** and **ABD** denote the loss-metric inconsistency and angular boundary discontinuity, respectively, and ✓ indicates that the method has corresponding issue. **Training** represents the total training time (in hours) for 72 epochs with a single GPU and a batch size of 2. Bold items are the best result of each column.

Detector	Regression Loss	LMI	ABD	Inshore AP	Offshore AP	Test AP	Training (h)
RetinaNet-R (R-50-FPN)	Smooth L1 (baseline)	✓	✓	44.30	91.38	72.13	**10.2**
	IoU-smooth L1 [41]	✓		45.49 (+1.19)	92.37 (+0.99)	73.22 (+1.09)	12.9
	GWD [23]	✓		48.28 (+3.98)	93.36 (+1.98)	74.92 (+2.79)	12.1
	IoU [36]			47.37 (+3.07)	93.11 (+1.73)	74.36 (+2.23)	12.6
	GIoU [37]			47.43 (+3.13)	93.16 (+1.78)	74.43 (+2.30)	26.2
	CIoU [38]			47.76 (+3.46)	93.25 (+1.87)	74.65 (+2.52)	26.8
	EIoU [39]			48.01 (+3.71)	93.30 (+1.92)	74.77 (+2.64)	26.6
	CDIoU [40]			48.54 (+4.24)	93.46 (+2.08)	75.05 (+2.92)	26.5
	AIoU		✓	NAN	NAN	NAN	–
	TDIoU			**49.68 (+5.38)**	**94.09 (+2.71)**	**75.93 (+3.80)**	13.0
CS^{2}A-Net (R-50-FPN)	Smooth L1 (baseline)	✓	✓	70.99	96.13	85.95	**11.7**
	TDIoU			75.17 (+4.18)	96.56 (+0.43)	87.65 (+1.70)	14.8

Table 4 shows results on our RSSD. Compared with smooth L1, RetinaNet-R based on approximate IoU losses improves the AP of inshore scenes, offshore scenes, and the entire test set by 1.19~3.98%, 0.99~1.98%, and 1.09~2.79%, respectively. Conventional IoU-based losses improve the AP by 3.07~4.24%, 1.73~2.08%, and 2.23~2.92%, respectively. The proposed TDIoU loss improves the AP by 5.38%, 2.71%, and 3.80%, respectively. Even with the advanced CS^{2}A-Net, TDIoU loss still improves the inshore AP, offshore AP, and test AP by 4.18%, 0.43%, and 1.70%, respectively, indicating that our method dramatically improves ship detection performance, especially in the complex inshore scenes. Similar experimental conclusions are also reflected in the other two datasets. Table 5 shows results on the SSDD. The TDIoU-based RetinaNet-R is improved by 8.50%, 1.01%, and 3.42% on inshore AP, offshore AP, and test AP, respectively, compared to the approximate IoU losses (2.24~5.88%, 0.27~0.61%, and 0.90~2.25%) and the traditional IoU-based losses (2.94~6.75%, 0.39~0.67%, and 1.17~2.61%). When CS^2A-Net is used as the base detector, our TDIoU loss further improves the AP by 3.47%, 0.73%, and 1.51%. Table 6 shows results on the HRSC2016. The RetinaNet-R achieves the best accuracy by using the TDIoU loss (i.e., improvement by 3.49% and 4.39% in terms of the 2007 and 2012 evaluation metrics, respectively). Similarly, our TDIoU loss achieves considerable improvement on CS^2A-Net, with an increase of 0.32% and 2.48%, respectively.

Table 5. Comparison of different regression losses on SSDD.

Detector	Regression Loss	LMI	ABD	Inshore AP	Offshore AP	Test AP	Training (h)
RetinaNet-R (R-50-FPN)	Smooth L1 (baseline)	✓	✓	59.35	97.10	86.14	1.1
	IoU-Smooth L1 [41]	✓		61.59 (+2.24)	97.37 (+0.27)	87.04 (+0.90)	1.5
	GWD [23]	✓		65.23 (+5.88)	97.71 (+0.61)	88.39 (+2.25)	1.3
	IoU [36]			62.29 (+2.94)	97.49 (+0.39)	87.31 (+1.17)	1.4
	GIoU [37]			63.36 (+4.01)	97.57 (+0.47)	87.68 (+1.54)	2.9
	CIoU [38]			64.13 (+4.78)	97.63 (+0.53)	87.98 (+1.84)	3.1
	EIoU [39]			64.24 (+4.89)	97.65 (+0.55)	88.02 (+1.88)	3.0
	CDIoU [40]			66.10 (+6.75)	97.77 (+0.67)	88.75 (+2.61)	2.9
	TDIoU		✓	**67.85 (+8.50)**	**98.11 (+1.01)**	**89.56 (+3.42)**	1.6
CS^2A-Net (R-50-FPN)	Smooth L1 (baseline)			75.79	98.79	92.08	1.2
	TDIoU	✓	✓	**79.26 (+3.47)**	**99.52 (+0.73)**	**93.59 (+1.51)**	1.7

Table 6. Comparison of different regression losses on HRSC2016. Here, AP_{07} and AP_{12} indicate the PASCAL VOC 2007 and 2012 metrics, respectively.

Detector	Regression Loss	LMI	ABD	Test AP_{07}	Test AP_{12}	Training (h)
RetinaNet-R (R-50-FPN)	Smooth L1 (baseline)	✓	✓	81.63	84.82	**1.1**
	IoU-Smooth L1 [41]	✓		82.64 (+1.01)	85.84 (+1.02)	1.4
	GWD [23]	✓		83.94 (+2.31)	87.78 (+2.96)	1.2
	IoU [36]			83.07 (+1.44)	86.64 (+1.82)	1.3
	GIoU [37]			83.22 (+1.59)	86.83 (+2.01)	2.8
	CIoU [38]			83.62 (+1.99)	87.33 (+2.51)	3.1
	EIoU [39]			83.78 (+2.15)	87.55 (+2.73)	3.0
	CDIoU [40]			84.13 (+2.50)	88.06 (+3.24)	2.9
	TDIoU		✓	**85.12 (+3.49)**	**89.21 (+4.39)**	1.5
CS^2A-Net (R-50-FPN)	Smooth L1 (baseline)			89.94	94.91	**1.1**
	TDIoU	✓	✓	**90.26 (+0.32)**	**97.39 (+2.48)**	1.6

Figure 20 shows P–R curves of RetinaNet-R using different regression losses on three datasets. The area under the P–R curve of TDIoU loss is always larger than that of the other losses, indicating that the overall performance of our method is better. The possible causes are summarized as follows: (1) Compared to the approximate IoU losses, we fundamentally eliminate the loss-metric inconsistency by introducing the differentiable rotational IoU algorithm. (2) In contrast to the parameter-based IoU losses, the TDIoU penalty term effectively reflects the overall difference between OBBs by measuring the distance between sampling points. In Table 4, to further investigate the effect of the angle parameter, we directly introduce it into the EIoU penalty term, which is named AIoU loss. However, AIoU loss is prone to non-convergence in the training phase, which is probably because the direct introduction of angle parameter will bring back the boundary discontinuity. On the contrary, the distance-based penalty term can reflect angle differences without directly employing the angle parameter. (3) Compared to the CDIoU loss, the introduction of the centroid distance is able to speed up bounding box alignment. Figure 21 displays different regression loss curves in the training phase. The TDIoU loss directly minimizes the distance between corresponding centroids and vertices of two boxes and, thus, converges much faster than other losses. Moreover, since we use the triangle distance rather than the diagonal length of the smallest enclosing box to construct the denominator of penalty term, TDIoU loss reduces the training time by nearly half compared with other IoU-based losses, indicating that the computing complexity of our method is greatly reduced. All in all, the proposed TDIoU loss is more applicable to rotated bounding box regression.

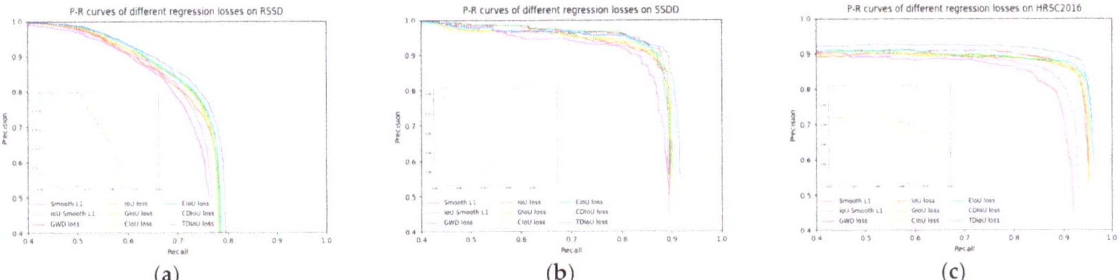

Figure 20. P–R curves of RetinaNet-R based on different regression losses on (**a**) RSSD, (**b**) SSDD, and (**c**) HRSC2016.

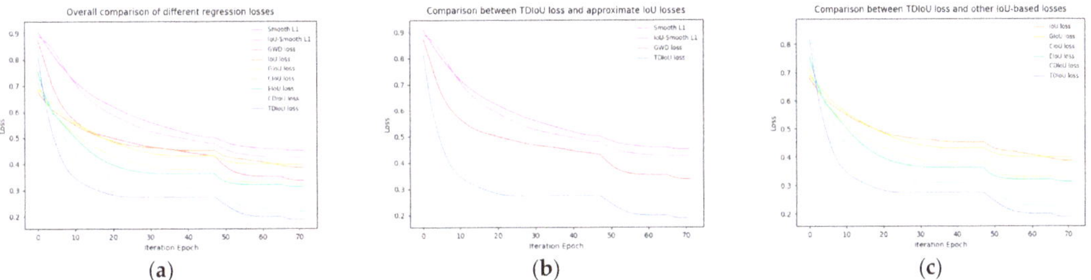

Figure 21. Regression loss curves of RetinaNet-R on RSSD. (**a**) Overall comparison of different regression losses; (**b**) comparison between TDIoU loss and approximate IoU losses; (**c**) comparison between TDIoU loss and other IoU-based losses.

6.3.3. Effectiveness of the AW-FPN

Since the proposed AW-FPN combines both multiple skip-scale connections and the attention-weighted feature fusion (AWF) strategy, we want to understand their respective contributions to accuracy improvement. Hence, we implement seven feature fusion networks with different connection pathways and fusion methods to verify the effectiveness of the AW-FPN, as shown in Tables 7–9. Notably, to eliminate the effect of irrelevant factors, the structure of all feature fusion networks is used only once.

Table 7 shows the results on our RSSD. The comparison between different connection pathways shows that the traditional FPN is inevitably limited by a single top-down information flow and achieves the lowest accuracy. The PANet with an extra bottom-up pathway improves by 0.68%, 0.59%, and 0.63% on inshore AP, offshore AP, and test AP, respectively. The BiFPN with single transverse skip-scale connections and the linear weighted fusion (LWF) strategy improves the AP by 1.87%, 1.13%, and 1.35%, respectively. For the AW-FPN with both transverse and longitudinal skip-scale connections, even the simplest additive fusion method can achieve performance similar to that of BiFPN. When using the same LWF method as BiFPN, the AW-FPN improves the AP by 2.38%, 1.34%, and 1.58%, indicating that longitudinal skip-scale connections are also crucial for feature fusion. For comparisons between different fusion methods, AW-FPN improves by 2.91%, 1.63%, and 1.97% when using AFF (channel attention only) and by 4.09%, 2.06%, and 2.84% when using the proposed AWF (both channel and spatial attention), indicating that the attention-based fusion methods outperform the linear fusion methods and, that to generate non-linear fusion weights, it is better to use both channel and spatial attention rather than only using single channel attention. When CS^2A-Net is used as the base detector, our ultimate AW-FPN further improves the AP by 3.96%, 0.38%, and 1.50%. Similar experimental results are obtained on the other two datasets. From Tables 8 and 9, for the SSDD and HRSC2016, the proposed AW-FPN achieves the most outstanding performance on RetinaNet-R and a considerable improvement on the advanced CS^2A-Net, which proves the effectiveness of our approach.

Table 7. Comparison of different feature fusion networks on RSSD. The structure of all feature fusion networks is used only once in our experiment. Here, ADD represents the direct addition of feature maps, while LWF indicates the linear weighted fusion method in BiFPN [27].

Detector	Fusion Network	Fusion Method	Fusion Type	Inshore AP	Offshore AP	Test AP
RetinaNet-R (R-50)	FPN (baseline)	ADD	Linear	44.30	91.38	72.13
	PANet [25]	ADD	Linear	44.98 (+0.68)	91.97 (+0.59)	72.76 (+0.63)
	BiFPN [27]	LWF	Linear	46.17 (+1.87)	92.51 (+1.13)	73.48 (+1.35)
	AW-FPN	ADD	Linear	45.72 (+1.42)	92.42 (+1.04)	73.31 (+1.18)
	AW-FPN	LWF	Linear	46.68 (+2.38)	92.72 (+1.34)	73.71 (+1.58)
	AW-FPN	AFF (channel)	Soft Selection	47.21 (+2.91)	93.01 (+1.63)	74.10 (+1.97)
	AW-FPN	AWF (channel + spatial)	Soft Selection	**48.39 (+4.09)**	**93.44 (+2.06)**	**74.97 (+2.84)**
CS^2A-Net (R-50)	FPN (baseline)	ADD	Linear	70.99	96.13	85.95
	AW-FPN	AWF (channel + spatial)	Soft Selection	**74.95 (+3.96)**	**96.51 (+0.38)**	**87.45 (+1.50)**

Table 8. Comparison of different feature fusion networks on SSDD.

Detector	Fusion Network	Fusion Method	Fusion Type	Inshore AP	Offshore AP	Test AP
RetinaNet-R (R-50)	FPN (baseline)	ADD	Linear	59.35	97.10	86.14
	PANet [25]	ADD	Linear	60.98 (+1.63)	97.30 (+0.20)	86.80 (+0.66)
	BiFPN [27]	LWF	Linear	61.83 (+2.48)	97.42 (+0.32)	87.13 (+0.99)
	AW-FPN	ADD	Linear	61.62 (+2.27)	97.38 (+0.28)	87.05 (+0.91)
	AW-FPN	LWF	Linear	62.02 (+2.67)	97.44 (+0.34)	87.20 (+1.06)
	AW-FPN	AFF (channel)	Soft Selection	63.11 (+3.76)	97.54 (+0.44)	87.58 (+1.44)
	AW-FPN	AWF (channel + spatial)	Soft Selection	**65.71 (+6.36)**	**97.91 (+0.81)**	**88.63 (+2.49)**
CS^2A-Net (R-50)	FPN (baseline)	ADD	Linear	75.79	98.79	92.08
	AW-FPN	AWF (channel + spatial)	Soft Selection	**78.83 (+3.04)**	**99.32 (+0.53)**	**93.38 (+1.30)**

Table 9. Comparison of different feature fusion networks on HRSC2016.

Detector	Fusion Network	Fusion Method	Fusion Type	Test AP_{07}	Test AP_{12}
RetinaNet-R (R-50)	FPN (baseline)	ADD	Linear	81.63	84.82
	PANet [25]	ADD	Linear	82.44 (+0.81)	85.62 (+0.80)
	BiFPN [27]	LWF	Linear	82.74 (+1.11)	85.92 (+1.10)
	AW-FPN	ADD	Linear	82.50 (+0.87)	85.68 (+0.86)
	AW-FPN	LWF	Linear	82.86 (+1.23)	86.30 (+1.48)
	AW-FPN	AFF (channel)	Soft Selection	83.27 (+1.64)	86.89 (+2.07)
	AW-FPN	AWF (channel + spatial)	Soft Selection	**84.10 (+2.47)**	**88.02 (+3.20)**
CS^2A-Net (R-50)	FPN (baseline)	ADD	Linear	89.94	94.91
	AW-FPN	AWF (channel + spatial)	Soft Selection	**90.21 (+0.27)**	**97.22 (+2.31)**

Figure 22 shows the P–R curves of RetinaNet-R with different feature fusion networks. The P–R curve of AW-FPN is always higher than that of other methods. This may be because multiple skip-scale connections enhance semantic interactions between features of different resolutions and scales, which contributes to the complement of context information. In addition, in contrast to other linear fusion methods and the AFF using only channel attention, the proposed AWF aggregates global and local feature contexts in both the multi-scale channel attention module (MCAM) and the multi-scale spatial attention module (MSAM) to generate higher quality fusion weights. Figure 23 shows the feature visualization of different feature fusion networks. The region of interest (ROI) is highlighted in the feature heat map. The ROI in the feature maps generated by other methods is usually overlarge and contains considerable background clutter. In contrast, the contour and location of ships in the feature map generated by our AW-FPN is more distinct and accurate, which helps the detectors to focus more on the real ship targets rather than background clutter, and to learn more useful context information.

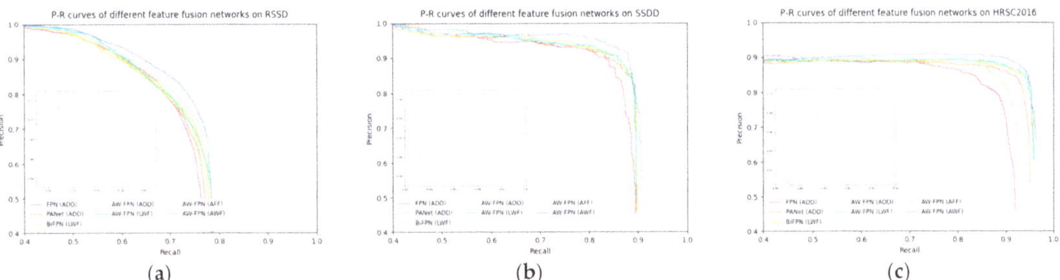

Figure 22. P–R curves of RetinaNet-R with different feature fusion networks on (**a**) RSSD, (**b**) SSDD, and (**c**) HRSC2016.

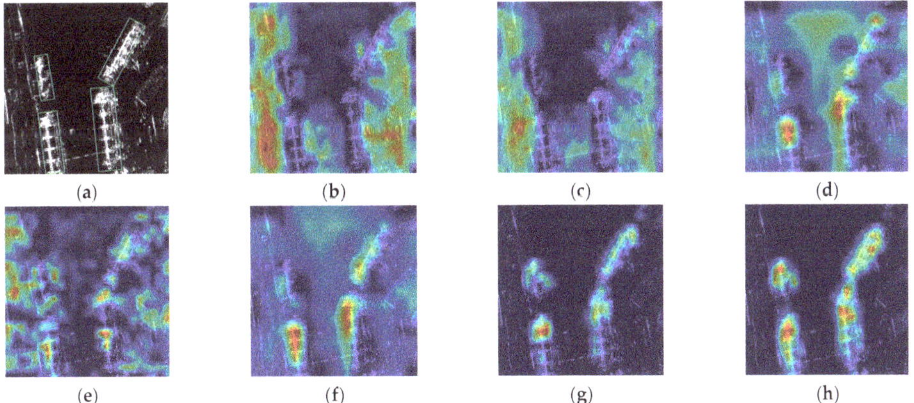

Figure 23. Feature visualization of different feature fusion networks (take P_5 as an example). (**a**) Input; (**b**) FPN (ADD); (**c**) PANet (ADD); (**d**) BiFPN (LWF); (**e**) AW-FPN (ADD); (**f**) AW-FPN (LWF); (**g**) AW-FPN (AFF); (**h**) AW-FPN (AWF).

6.4. Comparison with the State-of-the-Art

We embed the proposed AW-FPN into CS^2A-Net and train it with our TDIoU loss. and then compare our approach with the state-of-the-art methods on three datasets.

6.4.1. Results on the RSSD

Table 10 provides a quantitative comparison of different methods on RSSD. As can be seen, the latest two-stage detection methods, such as CSL, SCRDet++, and ReDet, generally achieve outstanding performance. However, they always adopt complex structures in exchange for improved accuracy at the expense of detection efficiency. Lately, some single-stage detection methods, such as R^3Det, GWD, and CS^2A-Net, have been presented, which show competitive performance and efficiency on RSSD. Our method can further improve the accuracy of these rotation detectors and has a minimal impact on detection efficiency. As per Table 10, the proposed approach achieves 75.41%, 96.62%, and 87.87% accuracy in terms of inshore AP, offshore AP, and test AP on CS^2A-Net, respectively, without using multi-scale training and testing, which is already extremely close to the performance of the advanced ReDet and GWD. When employing a stronger backbone (i.e., ResNet101) and multi-scale training and testing, our approach achieves state-of-the-art performance, with the AP of 77.65%, 97.35%, and 89.18%, respectively, which is 1.98%, 0.65%, and 1.11% higher than that of the suboptimal method (i.e., GWD). Furthermore, the inference speed of our method reaches 12.1 fps, which is 11.1 fps and 2.5 fps faster than that of ReDet and GWD, respectively. Compared to the original CS^2A-Net, our approach trades off a speed loss of only 1.1 fps for significant gains, of 3.48%, 0.87%, and 1.85%, in accuracy.

Table 10. Comparison with state-of-the-art methods on RSSD. Here, **MS** indicates the multi-scale training and testing, **FPS** is obtained by calculating the overall inference time and the number of images, TDIoU + AW-FPN represents the CS^2A-Net detector based on TDIoU loss and AW-FPN, R-50 refers to ResNet50 (likewise R-101, R-152), and ReR-50 and H-104 denote ReResNet50 [68] and Hourglass104, respectively [69].

Method	Backbone	Stage	MS	Inshore AP	Offshore AP	Test AP	FPS
SCRDet [41]	R-101	Two	✓	65.47	95.53	83.65	5.0
RSDet [70]	R-152	Two		68.48	95.88	84.85	–
Gliding Vertex [71]	R-101	Two		70.83	96.11	85.80	–
CSL [44]	R-152	Two	✓	71.45	96.18	86.15	4.0
SCRDet++ [72]	R-101	Two	✓	71.66	96.21	86.24	5.0
ReDet [68]	ReR-50	Two	✓	75.57	96.65	88.03	<1.0
RetinaNet-R [17]	R-50	Single		44.30	91.38	72.13	17.5
DRN [73]	H-104	Single	✓	67.95	95.68	84.40	–
R^3Det [74]	R-152	Single	✓	71.47	96.25	86.21	9.6
DCL [45]	R-101	Single	✓	71.92	96.23	86.36	12.0
GWD [23]	R-152	Single	✓	75.67	96.70	88.07	9.6
CS^2A-Net [67]	R-50	Single		70.99	96.13	85.95	16.5
CS^2A-Net [67]	R-101	Single	✓	74.17	96.48	87.33	13.2
TDIoU+AW-FPN (ours)	R-50	Single		75.41	96.62	87.87	15.1
TDIoU + AW-FPN (ours)	R-101	Single	✓	77.65	97.35	89.18	12.1

Figure 24 shows qualitative results of different methods on RSSD. As per the results of the offshore scene containing multiple ships (the first row), the other four methods detect islands and reefs incorrectly as ships, while our method is more robust in distinguishing small ships from background components. For the complex inshore scenes (the second row to the fourth row), the results of other methods include false alarms and leave some vessels undetected. In contrast, our method succeeds in detecting all ships and locating them more precisely, especially for densely arranged ships close to man-made facilities.

6.4.2. Results on the SSDD

Table 11 shows experimental results of different methods on the SSDD. Since SSDD contains few SAR images and the scenes are relatively simple, the ship detection accuracy is generally high. As shown in Table 11, based on CS2A-Net (R-50), our approach achieves 80.75%, 99.64%, and 94.05% of inshore AP, offshore AP, and test AP, respectively. When using ResNet101 as the backbone network, the AP of our approach reaches 84.34%, 99.71%, and 95.16%, compared to the state-of-the-art detectors ReDet (82.80%, 99.18%, and 94.27%) and GWD (81.99%, 99.66%, and 94.35%). Moreover, our approach improves the overall accuracy by 1.26% and 0.44% compared to BiFA-YOLO and R2FA-Det, respectively, and the inference speed by 5.5 fps compared to the suboptimal R2FA-Det, indicating that the proposed method achieves the best performance and satisfies high detection efficiency.

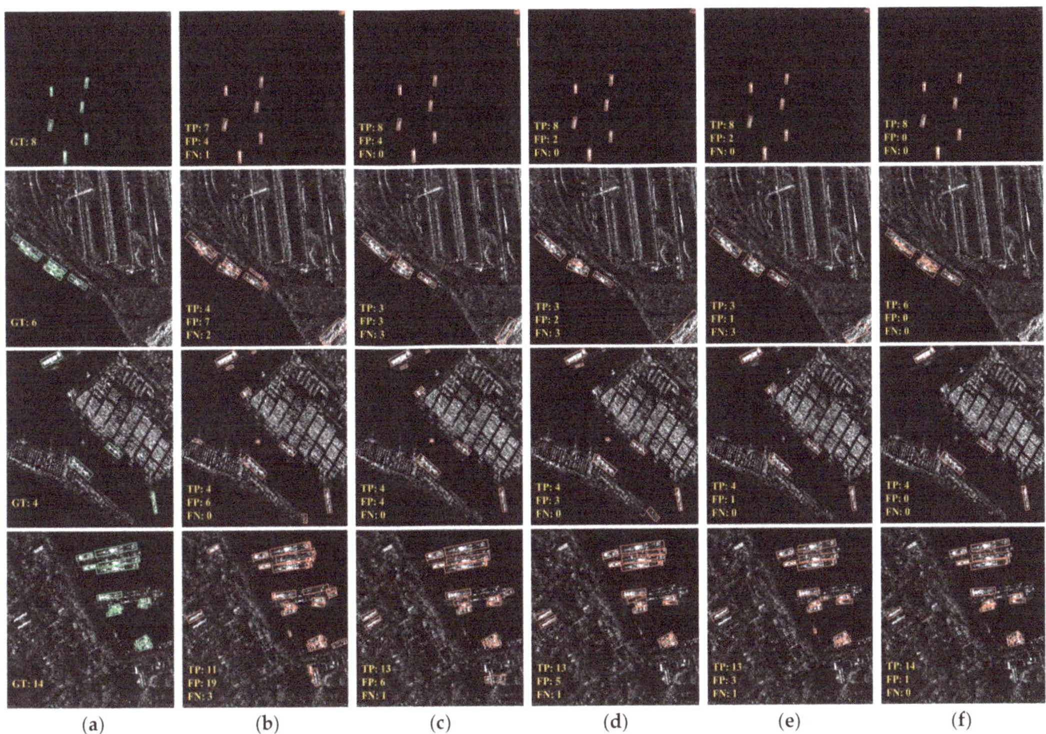

Figure 24. Detection results of different methods on RSSD. (**a**) Ground truth (GT); (**b**) RetinaNet-R; (**c**) CS^{2}A-Net; (**d**) ReDet; (**e**) GWD; (**f**) TDIoU + AW-FPN (ours). Green and red boxes represent real ship targets and detection results, respectively.

Table 11. Comparison with state-of-the-art methods on SSDD. Here, V-16 and C-53 denote VGG16 [75] and CSPDarknet53 [76]. The method with * indicates that its results are from the corresponding paper. Here, (<800) indicates that the long side of images is less than 800 pixels.

Method	Backbone	Stage	Image Size	Inshore AP	Offshore AP	Test AP	FPS
Cascade RCNN * [19]	R-50	Multiple	300×300	–	–	88.45	2.8
MSR2N * [16]	R-50	Two	$(<800) \times 350$	–	–	90.11	9.7
Gliding Vertex [71]	R-101	Two	512×512	75.23	98.35	91.88	–
CSL [44]	R-152	Two	512×512	76.15	98.87	92.16	7.0
SCRDet + + [72]	R-101	Two	512×512	77.17	99.16	92.56	8.8
ReDet [68]	ReR-50	Two	512×512	82.80	99.18	94.27	<1.0
RetinaNet-R [17]	R-50	Single	512×512	59.35	97.10	86.14	**30.6**
R^{3}Det [74]	R-152	Single	512×512	76.92	99.09	92.29	16.9
DRBox-v2 * [77]	V-16	Single	300×300	–	–	92.81	18.1
BiFA-YOLO * [22]	C-53	Single	512×512	–	–	93.90	–
GWD [23]	R-152	Single	512×512	81.99	99.66	94.35	16.9
R^{2}FA-Det * [19]	R-101	Single	300×300	–	–	94.72	15.8
CS^{2}A-Net [67]	R-50	Single	512×512	75.79	98.79	92.08	29.0
CS^{2}A-Net [67]	R-101	Single	512×512	79.01	99.41	93.47	23.2
TDIoU + AW-FPN (ours)	R-50	Single	512×512	80.75	99.64	94.05	26.6
TDIoU + AW-FPN (ours)	R-101	Single	512×512	84.34	99.71	95.16	21.3

Figure 25 visualizes some detection results of different methods on the SSSD. In the complex inshore scenes, the other three methods suffer from missed and false detection under background clutter interference. In contrast, our approach is highly robust and displays superiority in detecting densely distributed small ships.

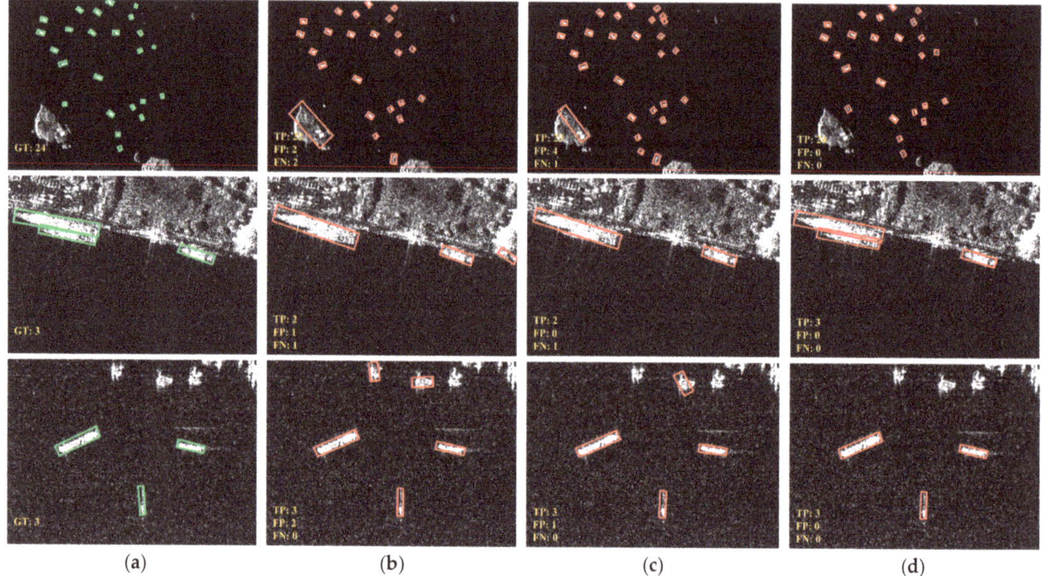

Figure 25. Detection results of different methods on SSDD. (**a**) GT; (**b**) CS^{2}A-Net; (**c**) GWD; (**d**) TDIoU + AW-FPN (ours).

6.4.3. Results on the HRSC2016

To verify the effectiveness and robustness of our approach in optical remote sensing scenarios, we conduct experiments with state-of-the-art methods on the HRSC2016, which contains a great number of ships with large aspect ratios and arbitrary orientations. As shown in Table 12, our approach achieves 90.71% and 98.65% accuracy on the metrics AP_{07} and AP_{12}, respectively, outperforming other comparison methods. Compared with the suboptimal approach (i.e., ReDet), the proposed method improves the accuracy by 0.25% and 1.02%. In addition, the inference speed of our method is 16.9 fps, which is much faster than that of the two-stage method ReDet (<1.0 fps). As per the above results, our method shows excellent generalization ability in other rotation detection scenarios.

Table 12. Comparison with state-of-the-art methods on HRSC2016. The method with * indicates that its results are from the corresponding paper.

Method	Backbone	Stage	Image Size	Test AP_{07}	Test AP_{12}	FPS
RoI-Transformer * [78]	R-101	Two	800 × 512	86.20	–	6.0
RSDet * [70]	R-50	Two	800 × 800	86.50	–	–
Gliding Vertex * [71]	R-101	Two	–	88.20	–	–
CenterMap-Net * [79]	R-50	Two	–	–	92.80	–
CSL [44]	R-101	Two	800 × 800	89.62	96.10	5.0
ReDet [68]	ReR-50	Two	800 × 512	90.46	97.63	<1.0
RetinaNet-R [17]	R-50	Single	800 × 512	81.63	84.82	**24.4**
DRN * [73]	H-104	Single	–	–	92.70	–
R^{3}Det [74]	R-101	Single	800 × 800	89.26	96.01	12.0
DCL [45]	R-101	Single	800 × 800	89.46	96.41	12.0
GWD [23]	R-101	Single	800 × 800	89.85	97.37	12.0
CS^{2}A-Net [67]	R-50	Single	800 × 512	89.94	94.91	23.0
CS^{2}A-Net [67]	R-101	Single	800 × 512	90.17	95.01	18.4
TDIoU + AW-FPN (ours)	R-50	Single	800 × 512	90.35	97.54	21.1
TDIoU + AW-FPN (ours)	R-101	Single	800 × 512	90.71	98.65	16.9

To evaluate the capability of our method to detect ships with extreme aspect ratios, we choose three images containing ships with large aspect ratios. As shown in Figure 26, our approach has fewer false alarms than any other methods. In addition, the position and orientation of the predicted box generated by our method are much closer to those of the ground truth.

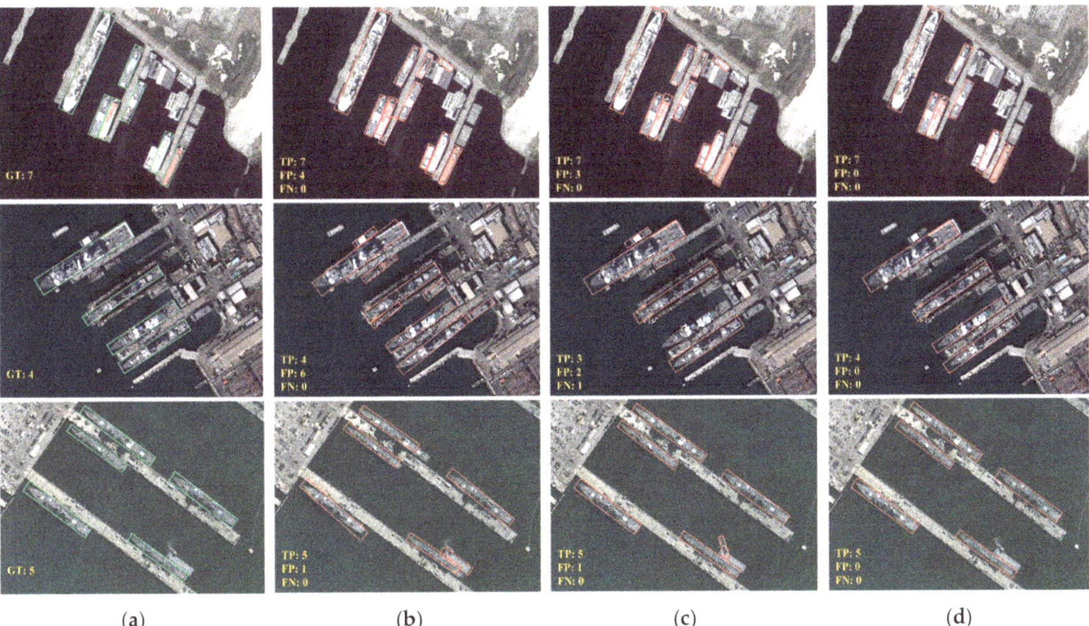

Figure 26. Detection results of different methods on HRSC2016. (**a**) GT; (**b**) CS^{2}A-Net; (**c**) GWD; (**d**) TDIoU + AW-FPN (ours).

Figure 27 displays P–R curves of different methods on RSSD, SSDD, and HRSC2016. It can be found that the P–R curve of our method is almost always higher than those of the other methods. Through all the above experiments and discussions, we can draw the conclusion that the proposed TDIoU loss and AW-FPN can improve the detection accuracy of arbitrary-oriented ships in both SAR scenes and optical remote sensing scenes, especially in the case of extreme scale and aspect ratio variations. This may be attributed to the fact that TDIoU loss fundamentally eliminates the loss-metric inconsistency and angular boundary discontinuity, so as to guide the rotation detector to achieve more accurate boundary box regression. Furthermore, the proposed AW-FPN is improved in terms of both the connection pathway and the fusion method, enabling high-quality semantic interactions and soft feature selections between features of inconsistent resolutions and scales.

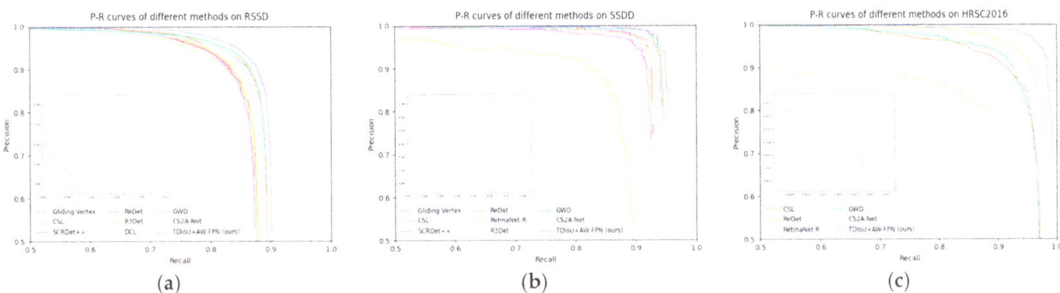

Figure 27. P–R curves of different methods on (**a**) RSSD, (**b**) SSDD and (**c**) HRSC2016.

7. Conclusions

In this paper, a unified framework combining TDIoU loss, AW-FPN, and RSSD is proposed to improve the capability of rotation detectors in recognizing and locating ships in SAR images. (1) The rotational IoU algorithm based on the Shoelace formula opens up the possibility of using IoU-based loss for rotated bounding box regression. On this basis, an effective TDIoU penalty term is designed to overcome the defects of existing IoU-based losses and solve the problems caused by angle regression. (2) Here, AW-FPN improves previous methods from connection pathways and fusion methods. Skip-scale connections enhance semantic interactions between multi-scale features. The AWF generates attention fusion weights via MCAM and MSAM to encode emphasized and suppressed positions in feature maps, making detectors focus more on real ship targets. (3) We construct a challenging benchmark, namely RSSD, for arbitrary-oriented SAR ship detection. Ships in RSSD not only differ significantly in orientations but also features multi-scale characteristics. In addition, 15 baseline results are provided for research. (4) Extensive experiments are conducted on three datasets. When using TDIoU loss and AW-FPN, even the advanced CS^{2}A-Net is able to improve upon the AP by 1.85%, 1.69%, and 0.54% on RSSD, SSDD, and HRSC2016, respectively, fully demonstrating the effectiveness and robustness of our approach.

Our future work is summarized as follows:

1. Though numerous innovative methods have emerged in SAR ship detection, due to the limitation of datasets, most of them are still based on HBBs. Therefore, we will further improve our TDIoU loss and AW-FPN, and try to combine them with more advanced rotation detection methods to improve the detection accuracy of arbitrary-oriented ships, especially in complex inshore scenes;
2. We will keep maintaining and updating RSSD to v2.0 or higher. Specifically, this will involve increasing the number of ship slices, incorporating more diverse SAR scenarios, building more standardized baselines, providing more accurate polygon annotations, etc. In the near future, it will be publicly available to facilitate further research in this field.
3. We will explore the possibility of multi-classification of ship targets in SAR images, which is an emerging research topic. With the development of high-resolution SAR image generation technology, the category information will be integrated into ship detection, which is beneficial for the progress of SAR intelligent interpretation technology.

Author Contributions: Conceptualization, R.G.; methodology, R.G.; software, R.G.; validation, R.G.; formal analysis, R.G.; investigation, R.G. and Z.X.; resources, R.G. and Z.X.; data curation, R.G., Z.X. and Q.X.; writing—original draft preparation, R.G.; writing—review and editing, R.G., Z.X., K.H. and Q.X.; visualization, R.G.; supervision, Z.X. and K.H.; project administration, R.G.; funding acquisition, Z.X. and K.H. All authors have read and agreed to the published version of the manuscript.

Funding: This research was funded by National Key Research and Development Program, grant number 2019YFB1600605; The Youth Fund from National Natural Science Foundation of China, grant number 62101316; Shanghai Sailing Program, grant number 20YF1416700.

Data Availability Statement: Not applicable.

Conflicts of Interest: The authors declare no conflict of interest.

References

1. Moreira, A.; Prats-Iraola, P.; Younis, M.; Krieger, G.; Hajnsek, I.; Papathanassiou, K.P. A tutorial on synthetic aperture radar. *IEEE Geosci. Remote Sens. Mag.* **2013**, *1*, 6–43. [CrossRef]
2. Zhang, T.; Zhang, X. HTC+ for SAR Ship Instance Segmentation. *Remote Sens.* **2022**, *14*, 2395. [CrossRef]
3. Wang, X.; Chen, C.; Pan, Z.; Pan, Z. Fast and Automatic Ship Detection for SAR Imagery Based on Multiscale Contrast Measure. *IEEE Geosci. Remote Sens. Lett.* **2019**, *16*, 1834–1838. [CrossRef]
4. Zhang, T.; Zhang, X. A polarization fusion network with geometric feature embedding for SAR ship classification. *Pattern Recognition.* **2022**, *123*, 108365. [CrossRef]
5. Ao, W.; Xu, F.; Li, Y.; Wang, H. Detection and Discrimination of Ship Targets in Complex Background from Spaceborne ALOS-2 SAR Images. *IEEE J. Sel. Top. Appl. Earth Obs. Remote Sens.* **2018**, *11*, 536–550. [CrossRef]

6. Zhang, T.; Zhang, X. A mask attention interaction and scale enhancement network for SAR ship instance segmentation. *IEEE Geosci. Remote Sens. Lett.* **2022**, *19*, 1–5. [CrossRef]

7. He, C.; Tu, M.; Liu, X.; Xiong, D.; Liao, M. Mixture Statistical Distribution Based Multiple Component Model for Target Detection in High Resolution SAR Imagery. *ISPRS Int. J. Geo-Inf.* **2017**, *6*, 336. [CrossRef]

8. Zhang, T.; Zhang, X.; Shi, J.; Wei, S. Depthwise Separable Convolution Neural Network for High-Speed SAR Ship Detection. *Remote Sens.* **2019**, *11*, 2483. [CrossRef]

9. Krizhevsky, A.; Sutskever, I.; Hinton, G.E. Imagenet classification with deep convolutional neural networks. In Proceedings of the Advances in Neural Information Processing Systems, Lake Tahoe, NV, USA, 3–8 December 2012; pp. 1097–1105.

10. Li, J.; Qu, C.; Shao, J. Ship detection in SAR images based on an improved faster R-CNN. In Proceedings of the 2017 SAR in Big Data Era: Models, Methods and Applications (BIGSARDATA), Beijing, China, 13–14 November 2017; pp. 1–6.

11. Zhang, T.; Zhang, X.; Liu, C. Balance learning for ship detection from synthetic aperture radar remote sensing imagery. *ISPRS J. Photogramm. Remote Sens.* **2021**, *182*, 190–207. [CrossRef]

12. Zhang, T.; Zhang, X. High-Speed Ship Detection in SAR Images Based on a Grid Convolutional Neural Network. *Remote Sens.* **2019**, *11*, 1206. [CrossRef]

13. Liang, Y.; Sun, K.; Zeng, Y.; Li, G.; Xing, M. An Adaptive Hierarchical Detection Method for Ship Targets in High-Resolution SAR Images. *Remote Sens.* **2020**, *12*, 303. [CrossRef]

14. Gao, F.; He, Y.; Wang, J.; Hussain, A.; Zhou, H. Anchor-free Convolutional Network with Dense Attention Feature Aggregation for Ship Detection in SAR Images. *Remote Sens.* **2020**, *12*, 2619. [CrossRef]

15. Zhang, T.; Zhang, X.; Ke, X. Quad-FPN: A Novel Quad Feature Pyramid Network for SAR Ship Detection. *Remote Sens.* **2021**, *13*, 2771. [CrossRef]

16. Pan, Z.; Yang, R.; Zhang, Z. MSR2N: Multi-Stage Rotational Region Based Network for Arbitrary-Oriented Ship Detection in SAR Images. *Sensors* **2020**, *20*, 2340. [CrossRef] [PubMed]

17. Lin, T.-Y.; Goyal, P.; Girshick, R.; He, K.; Dollar, P. Focal loss for dense object detection. In Proceedings of the IEEE International Conference on Computer Vision, Venice, Italy, 22–29 October 2017; pp. 2999–3007.

18. Wang, J.; Lu, C.; Jiang, W. Simultaneous Ship Detection and Orientation Estimation in SAR Images Based on Attention Module and Angle Regression. *Sensors* **2018**, *18*, 2851. [CrossRef]

19. Chen, S.; Zhang, J.; Zhan, R. R^{2}FA-Det: Delving into High-Quality Rotatable Boxes for Ship Detection in SAR Images. *Remote Sens.* **2020**, *12*, 2031. [CrossRef]

20. Yang, R.; Wang, G.; Pan, Z.; Lu, H.; Zhang, H.; Jia, X. A Novel False Alarm Suppression Method for CNN-Based SAR Ship Detector. *IEEE Geosci. Remote Sens. Lett.* **2020**, *18*, 1401–1405. [CrossRef]

21. An, Q.; Pan, Z.; You, H.; Hu, Y. Transitive Transfer Learning-Based Anchor Free Rotatable Detector for SAR Target Detection with Few Samples. *IEEE Access* **2021**, *9*, 24011–24025. [CrossRef]

22. Sun, Z.; Leng, X.; Lei, Y.; Xiong, B.; Ji, K.; Kuang, G. BiFA-YOLO: A Novel YOLO-Based Method for Arbitrary-Oriented Ship Detection in High-Resolution SAR Images. *Remote Sens.* **2021**, *13*, 4209. [CrossRef]

23. Yang, X.; Yan, J.; Ming, Q.; Wang, W.; Zhang, X.; Tian, Q. Rethinking Rotated Object Detection with Gaussian Wasserstein Distance Loss. *arXiv* **2021**, arXiv:2101.11952. [CrossRef]

24. Lin, T.Y.; Dollár, P.; Girshick, R.; He, K.; Hariharan, B.; Belongie, S. Feature pyramid networks for object detection. In Proceedings of the IEEE Conference on Computer Vision and Pattern Recognition, Honolulu, HI, USA, 21–26 July 2017; pp. 2117–2125.

25. Liu, S.; Qi, L.; Qin, H.; Shi, J.; Jia, J. Path aggregation network for instance segmentation. In Proceedings of the IEEE Conference on Computer Vision and Pattern Recognition, Salt Lake City, UT, USA, 18–23 June 2018; pp. 8759–8768.

26. Ghiasi, G.; Lin, T.-Y.; Le, Q.V. NAS-FPN: Learning Scalable Feature Pyramid Architecture for Object Detection. In Proceedings of the IEEE/CVF Conference on Computer Vision and Pattern Recognition, Long Beach, CA, USA, 15–20 June 2019; pp. 7029–7038. [CrossRef]

27. Tan, M.; Pang, R.; Le, Q.V. EfficientDet: Scalable and efficient object detection. In Proceedings of the IEEE/CVF Conference on Computer Vision and Pattern Recognition, Seattle, WA, USA, 13–19 June 2020; pp. 10778–10787. [CrossRef]

28. Zhang, T.; Zhang, X.; Li, J.; Xu, X.; Wang, B.; Zhan, X.; Xu, Y.; Ke, X.; Zeng, T.; Su, H.; et al. SAR Ship Detection Dataset (SSDD): Official Release and Comprehensive Data Analysis. *Remote Sens.* **2021**, *13*, 3690. [CrossRef]

29. Wang, Y.; Wang, C.; Zhang, H.; Dong, Y.; Wei, S. A SAR Dataset of Ship Detection for Deep Learning under Complex Backgrounds. *Remote Sens.* **2019**, *11*, 765. [CrossRef]

30. Chang, Y.-L.; Anagaw, A.; Chang, L.; Wang, Y.C.; Hsiao, C.-Y.; Lee, W.-H. Ship Detection Based on YOLOv2 for SAR Imagery. *Remote Sens.* **2019**, *11*, 786. [CrossRef]

31. Sun, X.; Wang, Z.; Sun, Y.; Diao, W.; Zhang, Y.; Kun, F. AIR-SARShip-1.0: High-resolution SAR Ship Detection Dataset. *J. Radars* **2019**, *8*, 852.

32. Wei, S.; Zeng, X.; Qu, Q.; Wang, M.; Su, H.; Shi, J. HRSID: A High-Resolution SAR Images Dataset for Ship Detection and Instance Segmentation. *IEEE Access* **2020**, *8*, 120234–120254. [CrossRef]

33. Zhang, T.; Zhang, X.; Ke, X.; Zhan, X.; Shi, J.; Wei, S.; Pan, D.; Li, J.; Su, H.; Zhou, Y.; et al. LS-SSDD-v1.0: A Deep Learning Dataset Dedicated to Small Ship Detection from Large-Scale Sentinel-1 SAR Images. *Remote Sens.* **2020**, *12*, 2997. [CrossRef]

34. Ren, S.; He, K.; Girshick, R.; Sun, J. Faster r-cnn: Towards real-time object detection with region proposal networks. In Proceedings of the 28th International Conference on Neural Information Processing Systems, Montreal, QC, Canada, 8–13 December 2015; pp. 91–99.

35. Zhou, X.; Wang, D.; Krähenbühl, P. Objects as points. *arXiv* **2019**, arXiv:1904.07850. [CrossRef]

36. Yu, J.; Jiang, Y.; Wang, Z.; Cao, Z.; Huang, T. UnitBox: An Advanced Object Detection Network. In Proceedings of the 24th ACM International Conference on Multimedia, New York, NY, USA, 15–19 October 2016; pp. 516–520.

37. Rezatofighi, H.; Tsoi, N.; Gwak, J.; Sadeghian, A.; Reid, I.; Savarese, S. Generalized Intersection Over union: A metric and a Loss for Bounding Box Regression. In Proceedings of the IEEE Conference on Computer Vision and Pattern Recognition 2019, Long Beach, CA, USA, 16–20 June 2019; pp. 658–666. [CrossRef]

38. Zheng, Z.; Wang, P.; Liu, W.; Li, J.; Ye, R.; Ren, D. Distance-IoU Loss: Faster and Better Learning for Bounding Box Regression. In Proceedings of the Proceedings of the AAAI Conference on Artificial Intelligence, New York, NY, USA, 7 February 2020; Volume 34, pp. 12993–13000.

39. Zhang, Y.-F.; Ren, W.; Zhang, Z.; Jia, Z.; Wang, L.; Tan, T. Focal and efficient IOU loss for accurate bounding box regression. *Neurocomputing* **2022**, *506*, 146–157. [CrossRef]

40. Chen, D.; Miao, D. Control Distance IoU and Control Distance IoU Loss Function for Better Bounding Box Regression. *arXiv* **2021**, arXiv:2103.11696. [CrossRef]

41. Yang, X.; Yang, J.; Yan, J.; Zhang, Y.; Zhang, T.; Guo, Z.; Sun, X.; Fu, K. SCRDet: Towards More Robust Detection for Small, Cluttered and Rotated Objects. In Proceedings of the 2019 IEEE/CVF International Conference on Computer Vision (ICCV), Seoul, South Korea, 27 October–2 November 2019; pp. 8231–8240. [CrossRef]

42. Chen, Z.; Chen, K.; Lin, W.; See, J.; Yu, H.; Ke, Y.; Yang, C. Piou loss: Towards accurate oriented object detection in complex environments. In Proceedings of the European Conference on Computer Vision (ECCV), Glasgow, UK, 23–28 August 2020.

43. Zheng, Y.; Zhang, D.; Xie, S.; Lu, J.; Zhou, J. Rotation-robust intersection over union for 3D object detection. In Proceedings of the European Conference on Computer Vision, Glasgow, UK, 23–28 August 2020; pp. 464–480.

44. Yang, X.; Yan, J. Arbitrary-Oriented Object Detection with Circular Smooth Label. In Proceedings of the European Conference on Computer Vision, Glasgow, UK, 23–28 August 2020; pp. 677–694.

45. Yang, X.; Hou, L.; Zhou, Y.; Wang, W.; Yan, J. Dense Label Encoding for Boundary Discontinuity Free Rotation Detection. In Proceedings of the IEEE/CVF Conference on Computer Vision and Pattern Recognition (CVPR), Nashville, TN, USA, 20–25 June 2021; pp. 15814–15824. [CrossRef]

46. Li, X.; Wang, W.; Hu, X.; Yang, J. Selective Kernel Networks. In Proceedings of the IEEE/CVF Conference on Computer Vision and Pattern Recognition, Long Beach, CA, USA, 16–20 June 2019; pp. 510–519.

47. Zhang, H.; Wu, C.; Zhang, Z.; Zhu, Y.; Lin, H.; Zhang, Z.; Sun, Y.; He, T.; Mueller, J.; Manmatha, R.; et al. ResNeSt: Split-Attention Networks. In Proceedings of the IEEE/CVF Conference on Computer Vision and Pattern Recognition, New Orleans, LA, USA, 19–20 June 2022. [CrossRef]

48. Hu, J.; Shen, L.; Sun, G. Squeeze-and-Excitation Networks. In Proceedings of the IEEE Conference on Computer Vision and Pattern Recognition, Salt Lake City, UT, USA, 18–23 June 2018; pp. 7132–7141.

49. Dai, Y.; Gieseke, F.; Oehmcke, S.; Wu, Y.; Barnard, K. Attentional Feature Fusion. In Proceedings of the IEEE/CVF Winter Conference on Applications of Computer Vision, Virtual, 5–9 January 2021; pp. 3560–3569.

50. Minimum Bounding Box Algorithms. Available online: https://en.wikipedia.org/wiki/Minimum_bounding_box_algorithms (accessed on 29 June 2022).

51. Zhou, D.; Fang, J.; Song, X.; Guan, C.; Yin, J.; Dai, Y.; Yang, R. IoU Loss for 2D/3D Object Detection. In Proceedings of the 2019 International Conference on 3D Vision (3DV), Quebec City, QC, Canada, 16–19 September 2019; pp. 85–94. [CrossRef]

52. Ma, J.; Shao, W.; Ye, H.; Wang, L.; Wang, H.; Zheng, Y.; Xue, X. Arbitrary-Oriented Scene Text Detection via Rotation Proposals. *IEEE Trans. Multimedia* **2018**, *20*, 3111–3122. [CrossRef]

53. Rotated IoU. Available online: https://github.com/lilanxiao/Rotated_IoU (accessed on 29 June 2022).

54. Line–Line Intersection. Available online: https://en.wikipedia.org/wiki/Line-line_intersection (accessed on 29 June 2022).

55. Bézier Curve. Available online: https://en.wikipedia.org/wiki/Bézier_curve#Linear_curves (accessed on 29 June 2022).

56. Shoelace Formula. Available online: https://en.wikipedia.org/wiki/Shoelace_formula (accessed on 29 June 2022).

57. He, K.; Zhang, X.; Ren, S.; Sun, J. Deep residual learning for image recognition. In Proceedings of the IEEE Conference on Computer Vision and Pattern Recognition, Las Vegas, NV, USA, 27–30 June 2016; pp. 770–778.

58. Xie, S.; Girshick, R.; Dollar, P.; Tu, Z.; He, K. Aggregated residual transformations for deep neural networks. In Proceedings of the IEEE Conference on Computer Vision and Pattern Recognition, Honolulu, HI, USA, 21–26 July 2017; pp. 1492–1500.

59. Zagoruyko, S.; Komodakis, N. Wide residual networks. In Proceedings of the BMVC, York, UK, 19–22 September 2016; pp. 1–12. [CrossRef]

60. Ioffe, S.; Szegedy, C. Batch Normalization: Accelerating Deep Network Training by Reducing Internal Covariate Shift. In Proceedings of the 32nd International Conference on Machine Learning, Lille, France; 2015; Volume 37, pp. 448–456.

61. Copernicus Open Access Hub Home Page. Available online: https://scihub.copernicus.eu/ (accessed on 14 December 2021).

62. Sentinel-1 Toolbox. Available online: https://sentinels.copernicus.eu/web/ (accessed on 29 June 2022).

63. Xia, G.S.; Bai, X.; Ding, J.; Zhu, Z.; Belongie, S.; Luo, J.; Datcu, M.; Pelillo, M.; Zhang, L. DOTA: A Large-Scale Dataset for Object Detection in Aerial Images. In Proceedings of the IEEE Conference on Computer Vision and Pattern Recognition, Salt Lake City, UT, USA, 18–23 June 2018; pp. 3974–3983.
64. RoLabelImg. Available online: https://github.com/cgvict/roLabelImg (accessed on 29 June 2022).
65. Lin, T.-Y.; Maire, M.; Belongie, S.; Hays, J.; Perona, P.; Ramanan, D.; Doll'ar, P.; Zitnick, C. Microsoft COCO: Common objects in context. *arXiv* **2014**, arXiv:1405.0312.
66. Cheng, G.; Han, J.; Zhou, P.; Guo, L. Multi-class geospatial object detection and geographic image classification based on collection of part detectors. *ISPRS J. Photogramm. Remote Sens.* **2014**, *98*, 119–132. [CrossRef]
67. Han, J.; Ding, J.; Li, J.; Xia, G.-S. Align Deep Features for Oriented Object Detection. *IEEE Trans. Geosci. Remote Sens.* **2021**, *60*, 5602511. [CrossRef]
68. Han, J.; Ding, J.; Xue, N.; Xia, G.-S. ReDet: A Rotation-equivariant Detector for Aerial Object Detection. In Proceedings of the IEEE/CVF Conference on Computer Vision and Pattern Recognition, Nashville, TN, USA, 20–25 June 2021; pp. 2785–2794. [CrossRef]
69. Law, H.; Deng, J. CornerNet: Detecting Objects as Paired Keypoints. *Int. J. Comput. Vis.* **2018**, *128*, 642–656. [CrossRef]
70. Qian, W.; Yang, X.; Peng, S.; Yan, J.; Zhang, X. RSDet++: Point-based Modulated Loss for More Accurate Rotated Object De-tection. *Proc. IEEE Trans. Circuits Syste. Video Technol.* **2022**, *14*. [CrossRef]
71. Xu, Y.; Fu, M.; Wang, Q.; Wang, Y.; Chen, K.; Xia, G.-S.; Bai, X. Gliding Vertex on the Horizontal Bounding Box for Multi-Oriented Object Detection. *IEEE Trans. Pattern Anal. Mach. Intell.* **2020**, *43*, 1452–1459. [CrossRef] [PubMed]
72. Yang, X.; Yan, J.; Liao, W.; Yang, X.; Tang, J.; He, T. SCRDet++: Detecting Small, Cluttered and Rotated Objects via Instance-Level Feature Denoising and Rotation Loss Smoothing. *IEEE Trans. Pattern Anal. Mach. Intell.* **2022**, 1. [CrossRef]
73. Pan, X.; Ren, Y.; Sheng, K.; Dong, W.; Yuan, H.; Guo, X.; Ma, C.; Xu, C. Dynamic Refinement Network for Oriented and Densely Packed Object Detection. In Proceedings of the IEEE/CVF Conference on Computer Vision and Pattern Recognition, Virtual, 14–19 June 2020; pp. 11207–11216.
74. Yang, X.; Liu, Q.; Yan, J.; Li, A.; Zhang, Z.; Yu, G. R3det: Refined single-stage detector with feature refinement for rotating object. In Proceedings of the AAAI Conference on Artificial Intelligence, Virtual, 2–9 February 2021; pp. 3163–3171. [CrossRef]
75. Simonyan, K.; Zisserman, A. Very Deep Convolutional Networks for Large-Scale Image Recognition. *arXiv* **2014**, arXiv:1409.1556.
76. Bochkovskiy, A.; Wang, C.; Liao, H.M. YOLOv4: Optimal Speed and Accuracy of Object Detection. *arXiv* **2020**, arXiv:2004.10934.
77. An, Q.; Pan, Z.; Liu, L.; You, H. DRBox-v2: An Improved Detector with Rotatable Boxes for Target Detection in SAR Images. *IEEE Trans. Geosci. Remote Sens.* **2019**, *57*, 8333–8349. [CrossRef]
78. Ding, J.; Xue, N.; Long, Y.; Xia, G.; Lu, Q. Learning RoI Transformer for Oriented Object Detection in Aerial Images. In Proceedings of the IEEE/CVF Conference on Computer Vision and Pattern Recognition, Long Beach, CA, USA, 16–20 June 2019; pp. 2849–2858.
79. Wang, J.; Yang, W.; Li, H.-C.; Zhang, H.; Xia, G.-S. Learning Center Probability Map for Detecting Objects in Aerial Images. *IEEE Trans. Geosci. Remote Sens.* **2020**, *59*, 4307–4323. [CrossRef]

Article

Deep Learning Approach for Object Classification on Raw and Reconstructed GBSAR Data

Marin Kačan *, Filip Turčinović, Dario Bojanjac and Marko Bosiljevac

Faculty of Electrical Engineering and Computing, University of Zagreb, Unska 3, 10000 Zagreb, Croatia
* Correspondence: marin.kacan@fer.hr

Abstract: The availability of low-cost microwave components today enables the development of various high-frequency sensors and radars, including Ground-based Synthetic Aperture Radar (GBSAR) systems. Similar to optical images, radar images generated by applying a reconstruction algorithm on raw GBSAR data can also be used in object classification. The reconstruction algorithm provides an interpretable representation of the observed scene, but may also negatively influence the integrity of obtained raw data due to applied approximations. In order to quantify this effect, we compare the results of a conventional computer vision architecture, ResNet18, trained on reconstructed images versus one trained on raw data. In this process, we focus on the task of multi-label classification and describe the crucial architectural modifications that are necessary to process raw data successfully. The experiments are performed on a novel multi-object dataset RealSAR obtained using a newly developed 24 GHz (GBSAR) system where the radar images in the dataset are reconstructed using the Omega-k algorithm applied to raw data. Experimental results show that the model trained on raw data consistently outperforms the image-based model. We provide a thorough analysis of both approaches across hyperparameters related to model pretraining and the size of the training dataset. This, in conclusion, shows how processing raw data provides overall better classification accuracy, it is inherently faster since there is no need for image reconstruction and it is therefore useful tool in industrial GBSAR applications where processing speed is critical.

Keywords: object classification; radar image reconstruction; convolutional neural networks; ResNet18; GBSAR; Omega-K algorithm

Citation: Kačan, M.; Turčinović, F.; Bojanjac, D.; Bosiljevac, M. Deep Learning Approach for Object Classification on Raw and Reconstructed GBSAR Data. *Remote Sens.* **2022**, *14*, 5673. https://doi.org/10.3390/rs14225673

Academic Editors: Tianwen Zhang, Tianjiao Zeng and Xiaoling Zhang

Received: 13 September 2022
Accepted: 2 November 2022
Published: 10 November 2022

Publisher's Note: MDPI stays neutral with regard to jurisdictional claims in published maps and institutional affiliations.

1. Introduction

Synthetic Aperture Radar (SAR) technology is crucial in many modern monitoring applications where optical images are not sufficient or restrictions in terms of light conditions or cloud coverage play a major role. To reach adequate resolution the antenna-based radar system should have a large sensor antenna. The resolution of an optical image obtained using the Sentinel-2 satellite with a mean orbital altitude of 786 km is around 20 m [1]. In order to achieve an equal resolution using a C-band sensor (common in SAR satellites) from that altitude, the sensor antenna would have to be over 2 km long, which is not practical. To virtually extend (synthesize) the length of the antenna (or antenna array), SAR concept utilizes sensor's movement to combine data acquired from multiple positions and reconstruct radar image of the observed area. In the Sentinel-1 SAR satellite launched in 2014, the movement of the 12.3-m-long antenna provides coverage of a 400 km wide area with a spatial resolution of 5 m [1].

The same principle can be applied in a terrestrial remote sensing imaging system—Ground-based SAR (GBSAR). The main concept of GBSAR is based on the sensor antenna which radiates perpendicular to the moving path, but the sensor moves along a ground track, covering the area in front of it. Even though, in many applications, distances between sensors and observed objects can be up to several kilometers, a smaller distance in combination with wider frequency bandwidth can be interesting for sensing small deformations. Satellite frequencies above the X band are rarely used in SAR due to H_20 absorption. However, since that is not a strict limitation for terrestrial GBSAR, higher frequencies and wider bandwidths can be used to achieve higher resolutions in radar images, which is essential for anomaly detection and small object recognition.

Such images, in combination with machine learning algorithms, enable surface deformation monitoring [2–6], snow avalanche identification [7], bridge [8] and dam structure monitoring [9], open pit mine safety management [10], and terrain classification [11]. Moreover, machine learning enables object classification using only radar images. Applications include the classification of military targets from the MSTAR dataset [12–19], ship detection [20–24], and subsurface object classification with an ultra-wideband GBSAR [25]. Radars with 32–36 GHz and 90–95 GHz frequency bands have been shown to accurately locate small metallic targets in the near-field region [26]. To facilitate deep learning research for SAR data, LS-SSDD-v1.0 [27], a dataset for small ship detection from SAR images was released, together with many standardized baselines. This resulted in subsequent advances in deep learning methods for ship classification [28,29], detection [30], and instance segmentation [31,32] in SAR images.

On the other hand, object classification on GBSAR data has not been explored sufficiently since most of the aforementioned work was focused on solving problems encountered in typical GBSAR applications. The potential of GBSAR systems lies in industrial applications, such as monitoring and object detection in harsh environments, and classification of concealed objects. We perform object classification on two modalities of GBSAR data: images reconstructed using the Omega-K algorithm and raw GBSAR data. Both approaches are based on a popular convolutional neural network (CNN) architecture—ResNet18—with certain modifications to accommodate processing raw GBSAR data.

A crucial part of the image reconstruction algorithm is an approximation step, which negatively influences the integrity of the data. We tested and quantified its impact on classification results by comparing multiple models based on raw and reconstructed GBSAR data. The idea of using raw GBSAR data is similar to various end-to-end learning approaches [33], which became more prevalent with the advent of deep learning. In such a paradigm, the model implicitly learns optimal representations of raw data [34], without any explicit transformations during data preprocessing.

The rest of the paper is organized as follows: Section 2 introduces the theory behind GBSAR and the radar image reconstruction algorithm. It also covers the implementation used in generating measurement sets presented in Section 3. Section 4 describes relevant deep learning concepts, such as feature engineering, end-to-end learning, and multi-label classification. It describes two concrete approaches to address GBSAR object classification, which correspond to the two input modalities. The experimental setup and evaluation results are described and analyzed in Section 5. Further discussions and interpretations are given in Section 6. Section 7 provides conclusions and future work.

2. Gbsar Theory and Implementation

2.1. GBSAR

The main idea of GBSAR is, following SAR concept, to virtually extend sensor antenna by utilizing sensor movement along the set track while it emits and receives EM waves. The set of measurements provides extraction of the distances to the observed object in each sensor position and, consequently, radar image reconstruction. Range and azimuth resolution of the radar image are mostly determined by the sensor frequency bandwidth and length of the used track, respectively. Hence, wider bandwidth and longer track (for example some sort of rail) provide better range and azimuth resolution [35].

There are two operational modes regarding sensor movement: continuous and stop-and-go mode. In the first one sensor moves with constant speed from start to the end of the used rail, while in the second one sensor movement is paused at each sampling position to acquire data without impact of motion. In stop-and-go mode step size and total aperture length can be precisely set but it should be emphasized that chosen values affect azimuth resolution. In both modes, the moving sensor has to repeatedly provide information about distance which is commonly obtained by using Frequency Modulated Continuous Wave (FMCW) radar principle as the base radar system due to its relatively simple and well known implementation.

FMCW radar radiates continuous signal whose operating frequency changes during transmission. Operating frequency sweeps through some previously defined frequency band B with a known function which is usually linear (most commonly sawtooth type function is used [36]). The same frequency sweep is observed in the echo or received signal which is delayed in time for the time required by the signal to travel to the object and be reflected back. Emitted and received signals are then mixed in order to eliminate high-frequency content in the received signal and use the low frequency difference to extract the delay in the received signal. To provide the analytical basis we start with a typical sawtooth wave which is made of periodic repetition (period T) of upchirp (frequency increases) or downchirp (frequency decreases) signals. Upchirp's frequency changes according to the linear function with rate of change equal to $\gamma = \frac{B}{T}$. Carrier frequency is denoted by f_c. In the complex form, emitted signal is given by the function

$$S_t(t) = e^{i(2\pi f_c t + \pi \gamma t^2)}. \tag{1}$$

Received signal is delayed in time for the time it takes the signal to travel to the observed object and returns back. Geometry of a standard GBSAR system and position of the object are described in Figure 1. Time delay between the transmission and detection is denoted by t_d,

$$S_r(t) = S_t(t - t_d). \tag{2}$$

After the mixing process we obtain signal S without high-frequency content $2\pi f_c t$

$$S = S_r(t)\overline{S_t}(t) = S_t(t - t_d)\overline{S_t}(t) = e^{-i2\pi t_d(f_c + \gamma t)} e^{i\pi \gamma t_d^2}. \tag{3}$$

All information about the observed object, that can be extracted by radar, is stored in the signal S and reconstruction algorithms operate directly on the signal S. Signal S can be interpreted in various different ways and the algorithm used for the reconstruction depends on the interpretation of S. Using Fourier transform in the spatial domain signal S can be interpreted as the spatial spectrum in azimuthal K_x and distance K_r variables [37].

$$S(K_r, K_x) = e^{-iK_x x_0} e^{-i\sqrt{K_r^2 - K_x^2} y_0} e^{i\frac{vc\Delta K_r K_x}{4\pi \gamma}} A e^{-i\frac{\pi}{4}}. \tag{4}$$

Position of the object is saved in the x_0 and y_0 coordinates in the first two complex exponential functions. Coordinates are multiplied by wave vector variables K_x and $\sqrt{K_r^2 - K_x^2}$. Third exponential function is a residual of a finite recording step size and finite radar's bandwith B. It manifests through the ΔK_r parameter in the Fourier domain. Radar moves with a constant speed v, c is a speed of light and γ represents chirp's rate of change.

We note that this is only one of many ways of interpreting signal S and the choice of reconstruction algorithm depends on this interpretation. This kind of interpretation leads to frequency domain reconstruction algorithms. Time domain algorithms usually have higher algorithm complexity but are more robust to the sensor irregular movement during radar operation [38].

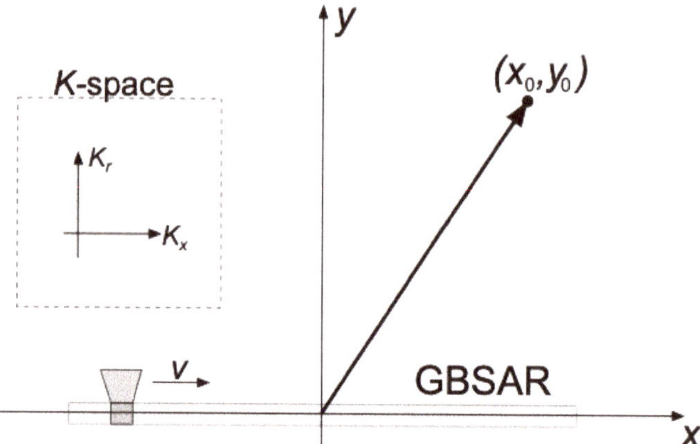

Figure 1. Ground-based SAR geometry. Coordinates (x_0, y_0) represent position of a point object in real space. In the dashed square, coordinates of the Fourier domain space or K-space are presented. K_r is a wave vector coordinate in the range direction while K_x is a wave vector coordinate in the x direction.

2.2. Image Reconstruction

Image reconstruction algorithms are used to generate radar image from the signal S. In this work, signals were recorded using GBSAR radar which operates in the stop-and-go mode. Received signal will be interpreted according to (3) and image reconstruction will be based on Omega-K algorithm [39]. This algorithm has many different implementations for images generated by GBSAR [37,40], but the main idea of these types of algorithms is to apply several processing steps on the spatial spectrum and then use inverse Fourier transform in order to extract position of the object.

Spatial spectrum (4) can be interpreted as a product of several functions carrying information about the object and information about various side effects of SAR radar acquisition. The task of the Omega-K algorithm is to separate those two parts and filter information about side effects out. Second part of the spatial spectrum S,

$$e^{i\frac{vc\Delta K_r K_x}{4\pi\gamma}},\tag{5}$$

represents residual frequency modulation due to finite step between two image acquisition. If we filter out this part then the remaining part of the spatial spectrum contains information about spatial coordinates of the object

$$e^{-iK_x x_0} e^{-i\sqrt{K_r^2 - K_x^2} y_0},\tag{6}$$

which can be extracted using the inverse Fourier transform. It gives $\delta(x - x_0, y - y_0)$, or object position. Constant

$$Ae^{-i\frac{\pi}{4}}\tag{7}$$

does not affect image reconstruction process.

Every Omega-K algorithm is a discrete implementation of previously described steps. They differ by the way they treat spatial spectrum S in order to separate information about position of the object from various side effects of image acquisition. They usually have part for the residual frequency modulation compensation, interpolation from spherical to Cartesian coordinate system and they end with 2D inverse Fourier transform (2D IFFT).

Reconstruction algorithms are used in order to recreate captured image according to predefined signal model, such as (3). Ideally, reconstruction algorithms take signal S and reproduce the image of the observed object without the loss of information in the signal S. In reality, all those steps necessarily introduce an error in the reconstruction process and affect the information in S. Numerical algorithms, such as FFT, have very well known and well-described numerical error and those steps do not affect information in S significantly. On the other side, approximations, such as frequency modulation compensation, residual-phase compensation, and interpolation from spherical to Cartesian coordinate system, are approximation steps associated to the predefined signal model. Algorithms used in those steps are not as researched as the FFT algorithm is, they are not implemented in numerical libraries, and they necessarily degrade the information in S as a result of their approximation character [37]. Although it is easier for human beings to notice useful information and recognize captured object in the reconstructed image than it is in the raw data, it does not mean that there is more information in the reconstructed image. As is described before, just the opposite is true due to all approximation algorithms. Reconstructed representation of the signal S is just easier for human beings to grasp than the raw data. That means that a neural network based classifier, that operates directly on the raw data, has a significant potential for a higher accuracy than the classifier that operates on the reconstructed images.

2.3. Implementation

In order to obtain radar images of objects from the ground while keeping flexibility regarding step size, total aperture length, polarization, and observing angle, GBSAR named GBSAR-Pi was developed. It is based on Raspberry Pi 4B (RPi) microcomputer that controls voltage controlled oscillator (VCO) in FMCW module Innosent IVS-362 [41]. The module operates in 24 GHz band and, besides VCO, has integrated trasmitting and receiving antenna, and mixer. The mechanical platform which contains RPi with AD/DA converter, FMCW module and amplifier is tailor 3D printed to enable precise linear movement and change of polarization. Polarization can be manually set to horizontal (HH) or vertical (VV). Movement of the platform is provided by 5V stepper motor. Figure 2 shows developed GBSAR-Pi: on one side of 1.2 m long rail track is stepper motor which is controlled by the RPi located in the movable white 3D printed platform. The platform also contains FMCW module on the front and touch-screen display used for measurement parameter setup, monitoring, and ultimately representation of reconstructed radar image.

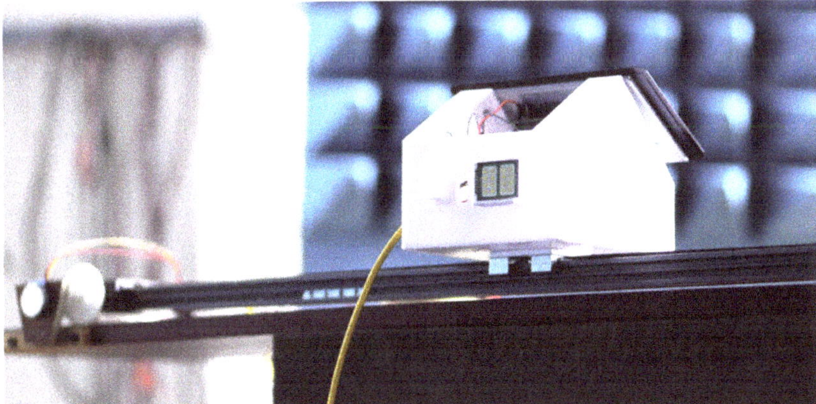

Figure 2. Developed GBSAR-Pi.

Developed GBSAR-Pi works in stop-and-go mode. The process of obtaining measurement for one radar image is following: RPi over DA converter controls VCO on FMCW module to emit upchirp signals using V_{tune} pin, as shown in Figure 3. The module in each step transmits and receives signals, mixes them, and the resultant signal in low frequency band is sent to the microcomputer over In phase pin (IF1 in Figure 3) and AD converter. In order to maximize SNR, in each step the system emits multiple signals and stores the average of received ones. When the result is stored, RPi runs stepper motor via four control pins to move the platform for one step of previously set size. The enable pin in Figure 3 is used to switch off the power supply of the VCO. The process continues until it reaches last step. After that obtained matrix of average signals from each step is stored locally and sent to the server.

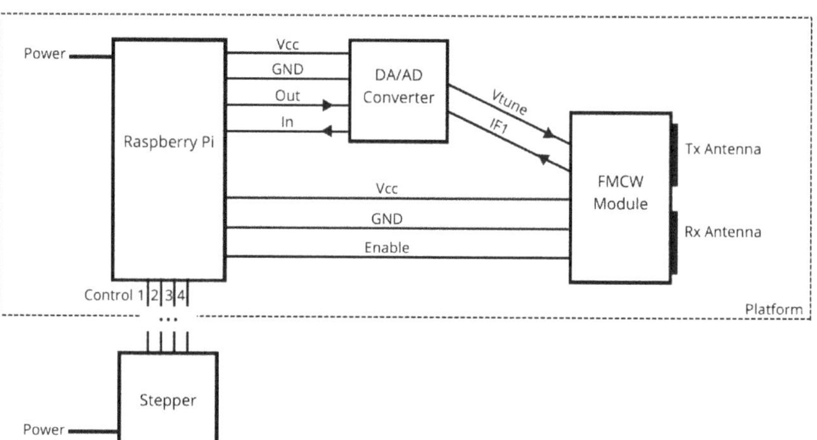

Figure 3. GBSAR-Pi scheme.

FMCW signals transmitted by the module have central frequency of 24 GHz and sweep range is set to 700 MHz bandwidth. Hence, from well known equation for FMCW range resolution $R_r = \frac{c}{2B}$, the used bandwidth provides range resolution of the system $R_r = 21.4$ cm. Each sweep in emitted signal is generated by RPi with 1024 frequency points and has a duration of 166 ms which gives chirp rate change of frequency $\gamma = 4.2 \times 10^9$. Signal stored in each step is average of 10 received signals. FMCW module has output power (EIRP) 12.7 dBm. Number of steps and step size are adjustable and in our measurements there were two cases: case of 0.4 cm step size and 160 steps, and 1 cm step size and 30 steps case. Antennas polarization can be changed manually by rotating case with FMCW module set on the front side of the platform. Since the antennas are integrated within the module, the system is limited to HH and VV polarizations only. GBSAR-Pi use two battery sources: first one charges RPi, AD/DA converter and FMCW module, and second one stepper motor. The whole system is, thus, fully autonomous with the system being controlled by display set on the platform and optionally an external keyboard.

Following Section 2.2, image reconstruction algorithm Omega-K was implemented using python programming language and additional libraries numpy and scipy. Regarding the algorithm, most important methods of the implementation are numpy FFT (Fast Fourier Transform) and IFFT (Inverse FFT), and scipy interpolate which provides one-dimensional array interpolation (interp1d). The visualization of a reconstructed radar image was accomplished using libraries matplotlib (pyplot) and seaborn (heatmap). Implementation consists of following steps: Hilbert transform, RVP (Residual Video Phase), Hanning window, FFT, Reference function multiply (RFM), interpolation, IFFT, and visualization. Complete program code with adjustable central frequency, bandwidth, chirp duration, and step size is given in [42].

3. Data Acquisition

3.1. Measurements

The measurements were taken using GBSAR-Pi. There were two sets of measurements: the first was conducted in laboratory conditions, while the second set of measurements was obtained in a more complex "real world" environment. Thus, we named the collected datasets LabSAR and RealSAR, respectively.

In LabSAR, three test objects set at the same position in an anechoic chamber are recorded. The test objects were a big metalized box, a small metalized box, and a cylindrical plastic bottle filled with water. Hence, the objects were different in size and material reflectance. The distance between GBSAR-Pi and the observed object did not change between measurements and was approximately 1 m. The azimuth position of the object was at the center of the GBSAR-Pi rail track. Only one object was recorded in each measurement for this setup. All measurements were conducted using horizontal polarization. 160 steps of size 4 mm gave total aperture length of 64 cm.

In RealSAR, the conditions were not as ideal. The measurements were intentionally conducted in a room full of various objects in order to produce additional noise. Once again, there were three test objects. However, the objects were empty bottles of similar size and shape. The bottles could be distinguished based on the different reflectances of the material. The three bottles used as test objects were made from aluminium, glass, and plastic. Compared to LabSAR, there was much more variance in recorded scenes since any of the eight (2^3) possible subsets of three objects could appear in a given scene (including an empty scene without any objects). Moreover, the different object positions, which vary in both azimuth and range direction, and different polarizations used for recording also contribute to the complexity of the task. We note that any scene can include at most one object of a certain material. Compared to LabSAR, the step size was increased to 1 cm, while the total aperture length was decreased to 30 cm. RealSAR objects are shown in Figure 4. Comparison between measurement sets is given in Table 1.

Figure 4. RealSAR objects: aluminium, glass and plastic bottle, and GBSAR-Pi.

Table 1. LabSAR and RealSAR measurement set comparison. First five rows describe objects and scenes, while others GBSAR-Pi parameters used in the measurements.

	LabSAR	RealSAR
Objects (# of measurements)	big (37) and small (70) metalized box, and bottle of water (43)	aluminium (172), glass (172) and plastic (179) bottle
# of scene combinations	3	8
Object distance [cm]	100	between 20 and 70
Object position	center	various
# of images	150	338
Azimuth step size [cm]	0.4	1
Azimuth points	160	30
Total aperture length [cm]	64	30
Range resolution [cm]	21.4	21.4
Polarization	HH	HH and VV
FMCW central frequency [GHz]	24	24
Bandwidth [MHz]	700	700
Chirp duration [ms]	166	166
Frequency points	1024	1024
Time per measurement [min]	15	4

3.2. Datasets

Using the measurement sets LabSAR and RealSAR, four datasets were created: LabSAR-RAW, LabSAR-IMG, RealSAR-RAW, and RealSAR-IMG. LabSAR-RAW and RealSAR-RAW consist of raw data from measurement sets mentioned in their names. At the same time, LabSAR-IMG and RealSAR-IMG contain radar images generated using the reconstruction algorithm on that raw data. Since the number of frequency points in GBSAR-Pi did not change throughout the measurements, the dimensions of matrices in RAW datasets depend only on the number of steps: in LabSAR-RAW, each matrix is 160×1024, while in RealSAR-RAW, it is 30×1024. Dimensions of both IMG datasets are the same: 496×369 px. The LabSAR datasets contain 150 and the RealSAR datasets 337 matrices (RAW) and images (IMG). Specifically, in LabSAR datasets, there are 37 data points containing a big metalized box, 70 data points with a small metalized box, and 43 data points containing a bottle of water. It is important to emphasize that measurement conditions and test objects of LabSAR proved not to be adequately challenging for object classification. Models trained on LabSAR-RAW and those trained on LabSAR-IMG all achieved extremely high accuracy and could, thus, not be meaningfully compared. This is why we primarily focused on classification and comparisons based on RealSAR datasets, while the LabSAR datasets were used to pretrain neural network models.

Along with single object measurements, RealSAR datasets also include all possible combinations of multiple objects and scenes with no objects, which affects the number of measurements per object. Therefore, 337 raw matrices of RealSAR measurements include 172 matrices of scenes with an aluminium bottle, 172 matrices with a glass bottle, and 179 matrices with a plastic bottle. RealSAR datasets also contain 29 measurements of a scene without objects.

Four examples of two RealSAR datasets are shown in Figure 5. Left image of each example is a heatmap of raw data (an example from RealSAR-RAW), while right one is an image reconstructed using that data (an example from RealSAR-IMG). Aforementioned dimensions of matrices in RealSAR-RAW dataset can be seen in the raw data examples. Horizontal axis represents number of steps in the GBSAR measurement (in our case 30 steps), while vertical axis number of frequency points in one FMCW frequency sweep (in our case 1024). Four depicted examples stand for four scenes: (a) empty scene, (b) scene with aluminium bottle, (c) with aluminium and glass bottles, and (d) with all three bottles. The examples highlight three possible problems for image-based classification:

- Scaling: pixel colors in each reconstructed image represent different intensity. The pattern visible in reconstructed empty scene (Figure 5a) is actually crosstalk generated by the FMCW module and it is present in all measurements but is only noticeable in that image due to set scale. The scale was not fixed in order to prevent some objects (such as plastic bottle) to fade out from the reconstructed image.
- Angle of measurement: limited field of view due to finite GBSAR track size influences the representation of edge objects although virtual extension of azimuth axis is used in reconstruction algorithm to avoid compression of multiple images in the main field of view. In Figure 5b, aluminium bottle is set in the middle of the scene while in example Figure 5c it is set slightly to the right and more distant. It can be seen that the same object is differently represented in two reconstructed images.
- Difference in reflectance: in scenes where reflectance of one bottle is much higher than other bottle, after reconstruction process, one with lower reflectance can be hard to see. This problem can be spotted in reconstructed image of Figure 5d in which plastic bottle is not visible.

On the other hand, since input data in raw-based classification model is not preprocessed, the integrity of the data is preserved and, hence, such model has an advantage over image-based one.

Both variants of the RealSAR dataset were split into training, validation, and testing splits in the ratio of approximately 60:20:20. The splits consist of 204, 67, and 67 examples, respectively. A single class, in our setting, is any of the 2^n subsets of n different objects that can be present in the scene. Thus, in our case, there are 8 classes. The split was performed in a stratified fashion [43], meaning that the distribution of examples over different classes in each split corresponded to the distribution in the whole dataset. Since RealSAR-IMG was created by applying the Omega-K algorithm on examples in RealSAR-RAW, both variants contain the same examples. The training, validation, and testing splits, as well as any subset of the training set used in experiments, also contain the same examples in both RAW and IMG variants.

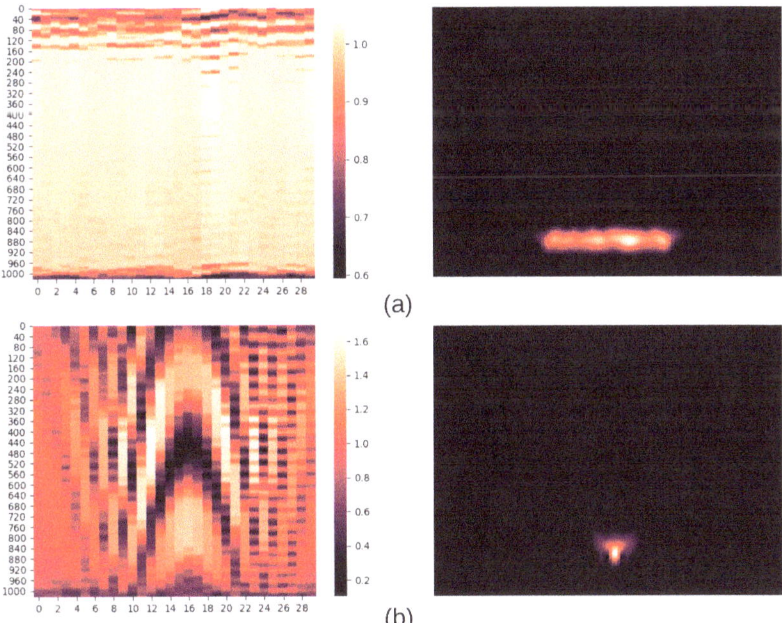

Figure 5. *Cont.*

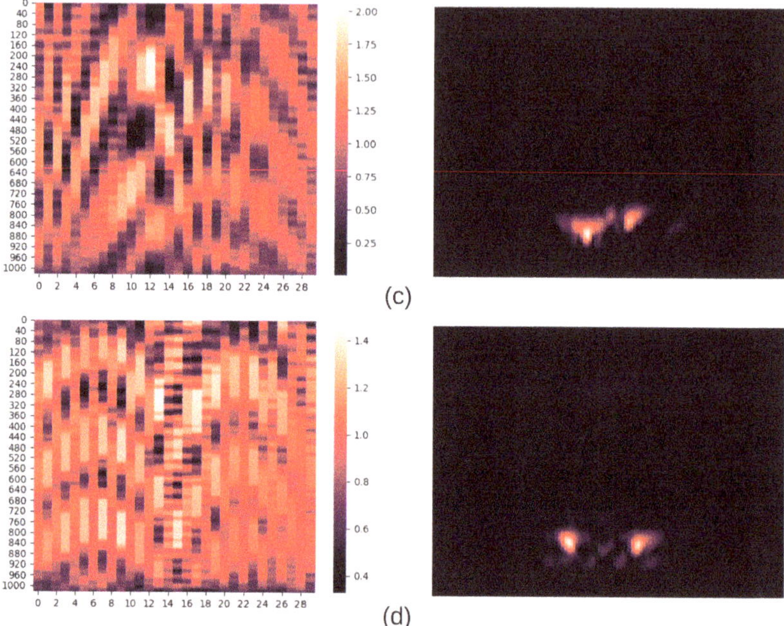

Figure 5. Pairs of RealSAR-RAW (left) and RealSAR-IMG (right) examples. Example pair (**a**) represents an empty scene, (**b**) scene with an aluminium bottle, (**c**) scene with an aluminium and a glass bottle, while (**d**) contains all three bottles.

4. Deep Learning

4.1. End-to-End Learning vs. Feature Engineering

In classical machine learning, we usually need to transform our raw data to make it suitable as input to our machine learning model. We extract a set of features from each raw data point using classical algorithms or hand-crafted rules. The process of choosing an appropriate set of features with corresponding procedure for their extraction is called feature engineering [44].

With the advent of deep learning, approaches with fewer preprocessing steps have become more popular. Instead of manually transforming raw data into representations appropriate for the model, new model architectures were developed that were able to consume raw data [33]. Such models would implicitly learn their own optimal representations [34] of raw data, often outperforming model which learned on fixed, manually extracted features [45]. This paradigm is also referred to as end-to-end learning.

We designed our models and experiments to compare and contrast these two paradigms. The first approach is to use an existing image classification architecture ResNet18, which has been shown to be able to tackle a diverse set of computer vision problems [46,47]. This model is trained on RGB images reconstructed from radar data using Omega-K algorithm [39]. The second is an end-to-end learning approach, where the input is raw data collected with our GBSAR-Pi. The model architecture is based on ResNet18, with a few modifications to accommodate learning from raw data.

4.2. Multi-Task Learning and Multi-Label Classification

Our task consists of detecting the presence of multiple different objects in a scene. Instead of training a separate model for recognizing each individual object, we adopt the multi-task learning paradigm [48] and train only one model which recognizes all objects at once. The model consists of a single convolutional backbone which learns shared feature representations [49] that are then fed to multiple independent fully-connected binary classification output layers (heads). Each classification head is designated for classifying the presence of one of the objects which can appear in a scene. We calculate the binary cross-entropy loss for each head individually. The final loss which is optimized during training is the arithmetic mean of all individual binary cross entropy losses. The described formulation is also known as multi-label classification [50]. An alternative formulation would be to do multi-class classification. In multi-class classification, each possible subset of objects which can occur in a scene would be treated as a separate class. There would only be one head which would classify examples into only one of the classes. Although we do care that our model learns to recognize all possible object subsets well, we chose not to address the problem in this way because it does not scale. Namely, for n objects, there are 2^n different subsets. This means the number of outputs of a multi-class model grows exponentially with the number of objects that can appear in a scene. In contrast, in the multi-label formulation, there are only n outputs for n different objects.

4.3. Models

We train and evaluate multiple approaches for raw data and image-based classification. For image data, we test two popular efficient convolutional architectures for low-powered devices: MobileNetV3 [51] and ResNet18 [46], and compare their results. For raw radar data classification we consider two baseline approaches to compare against. The first is a single-layer fully connected classifier, while the second one uses an LSTM [52] network to process data before classifying it. We also apply the two convolutional architectures-MobileNetV3 and ResNet18-to classifiy raw radar data because of multiple reasons. Firstly, these networks have been shown to work well across a wide array of classification tasks. Secondly, the inductive biases and assumptions of convolutional neural networks regarding spatial locality [53] make sense for raw data as well. This is because raw radar data points are 2D matrices of values. The horizontal dimension corresponds to the lateral axis on which the radar moves as it records the scene (shown in Figure 5a). The values in the vertical dimension represent the signal after the mixer in FMCW system whose frequency correlates with the distance of objects. Finally, our main contribution in this regard is a modification to the ResNet18 network (RAW-RN18) when applied raw radar data, which prevents subsampling of the input matrix in the horizontal dimension.

Each of our image-based classification models takes a batch of RGB images with dimensions (496, 369, 3) as input. The pixels of RGB images are preprocessed by subtracting the mean and dividing by the standard deviation of pixel values, with statistics calculated on the train set. This is common in all computer vision approaches. For each image in the batch, the model outputs three numbers, each of which is then fed to the sigmoid activation function. The output of each sigmoid function is a posterior probability distribution $P(Y_i = 1|x)$, where $i \in \{aluminium, glass, plastic\}$. Each of the three distributions is the probability that the corresponding object (aluminium, glass, or plastic) is present in the scene. To summarize: $P(Y_i = 1|x) = \sigma(model(x)_i)$.

Raw data classification models take a batch of 2D matrices of real numbers with dimensions (1024, 30, 1) as input. We preprocess raw data matrices in the same way that we do for images, with the raw data statistics also calculated on the train set. Raw data models produce the outputs in the same format as the image-based models. Figure 6 describes the machine learning setup for both input modalities.

All models are trained in the multi-label classification fashion. For each of the three outputs, we separately calculate a binary cross-entropy loss: $L = -y \log P(Y = 1|X) - (1 - y) \log(1 - P(Y = 1|X))$, where y is the ground truth label and $P(Y = 1|X)$ is the predicted probability of the corresponding object. The total loss is calculated as the arithmetic mean of the three individual losses.

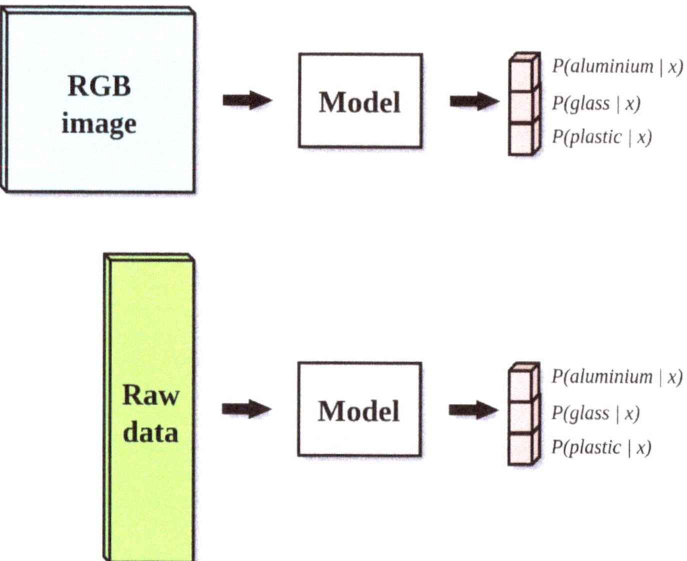

Figure 6. The machine learning setup for all image and raw data models. For image-based classification, the dimensions of input images are $(496, 369, 3)$. For raw data classification, the dimensions of the input matrix are $(1024, 30, 1)$. All models produce three posterior probability distributions $P(Y_i = 1|x)$, where $i \in \{aluminium, glass, plastic\}$.

During training, we use random color jittering and horizontal flipping as data augmentation procedures. These are common ways of artificially increasing dataset size and improving model robustness in computer vision. To extend color jittering to raw data classification, we add Gaussian noise with variance $\sigma^2 = 0.01$ to each element of the input matrix.

4.3.1. Baselines

We develop two baseline approaches for raw data classification. The first is a single-layer fully connected classification neural network. It flattens the input matrix of dimensions $(1024, 30)$ into a vector of 30,720 elements. The vector is then fed to a fully-connected layer which produces the output vector with three elements. The network is shown in Figure 7. The second approach uses a single-layer bidirectional LSTM network for classification [54], which processes the inputs sequentially in the horizontal dimensions. It treats the input matrix with dimensions $(1024, 30)$, as a sequence of 30 tokens with size 1024. Each 1024-dimensional input vector is embedded into a 256-dim representation by a learned embedding matrix [55], which is a standard way of transforming LSTM inputs. Because LSTM network is bidirectional, it aggregates the input sequence by processing it in both directions, using two separate unidirectional LSTM networks. The dimension of the hidden state is 256. The final hidden states of both directions are concatenated and then fed to a fully-connected classification layer. The network is shown in Figure 8.

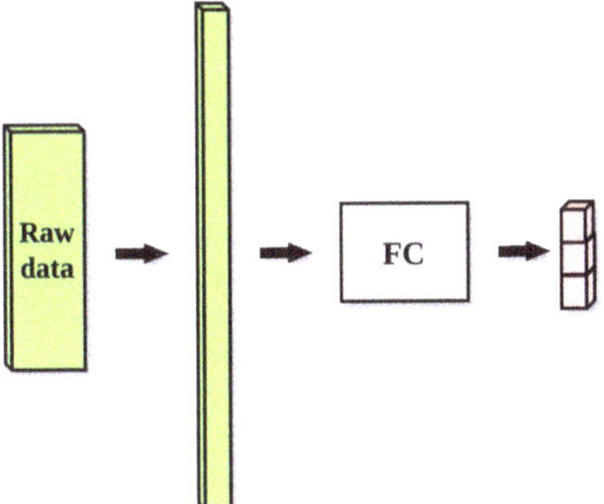

Figure 7. Single-layer fully connected neural network classifier for raw data. The input matrix of dimensions (1024, 30, 1) is flattened into a vector of 30,720 elements. The vector is then fed to a fully connected classifier with the sigmoid activation function, which produces an output vector size 3 which represents three posterior probability distributions $P(Y_i = 1|x)$, where $i \in \{aluminium, glass, plastic\}$.

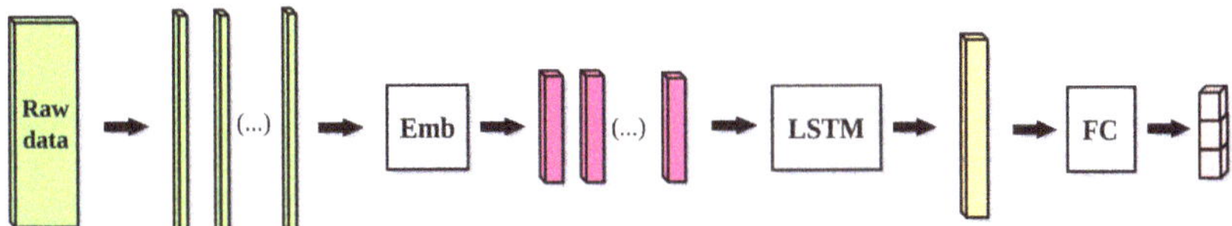

Figure 8. Raw data classifier based on the long short term memory network. Since the LSTM is a sequential model, the input matrix with dimensions (1024, 30) is processed as a sequence of 30 vectors with size 1024. Each 1024-dimensional input vector is embedded into a 256-dim representation by a learned embedding layer (Emb). The dimension of the hidden state is 256. The bidirectional LSTM network aggregates the input sequence by processing it in two directions and concatenating the final hidden states for both directions. As with the fully connected classifier, the resulting vector is given to a fully connected classifier with the sigmoid activation function, which the three posterior probability distributions.

4.3.2. Computer Vision Models

For image-based classification, we employ two popular efficient convolutional neural network classification architectures: MobileNetV3 [51] and ResNet18 [46]. These lightweight, efficient, networks have been shown to work well across a wide array of classification tasks. We chose them because they are tailored to work on low-powered devices, such as Raspberry Pi, which we want to use to run model inference in real-time. Furthermore, the dataset which we developed is small so we opt for smaller models and forgo using larger classification architectures. We use both networks without any modifications for image and raw data classification.

Our main contribution came from analysing the ResNet18 architecture and devising an architectural modification to make the network more suitable for raw data classification. The ResNet18 network is composed of 17 convolutional layers and one fully-connected output layer. After the initial convolutional layer (conv1) which is followed by a max pooling layer, the remaining 16 convolutional layers are divided into 4 groups (conv2, conv3, conv4, conv5). Each group consists of 2 residual blocks, each of which consists of 2 convolutional layers. Residual blocks have skip (residual) connections which perform identity mapping and add the input of the residual block to the output of the two convolutional layers. In the vanilla ResNet18 architecture, the convolutional layers transform the input by gradually reducing the spatial dimensions and increasing the depth (number of feature maps). The spatial dimensions are halved in 5 places during the forward pass of the network, by using a stride of 2 in the following layers:

- The first convolutional layer (conv1);
- The max pooling layer before the first group of convolutional layers;
- The first convolutional layer of each of the remaining groups (conv3_1, conv4_1, conv5_1).

Figure 9 shows all of the ResNet18 layers and their groups, with the four convolutional layers which perform subsampling emphasized. Because the input tensor is halved in both spatial dimensions 5 times, output of the last convolutional layer is a tensor whose both spatial dimensions are 32 times smaller than that of the input. The depth of the tensor is 512. Global average pooling (GAP) is used to pool this tensor across the spatial dimensions into a vector of size 512. This shared representation is given as input to the fully-connected output layer which outputs a posterior probability $P(y_i = 1|x)$ for each object i.

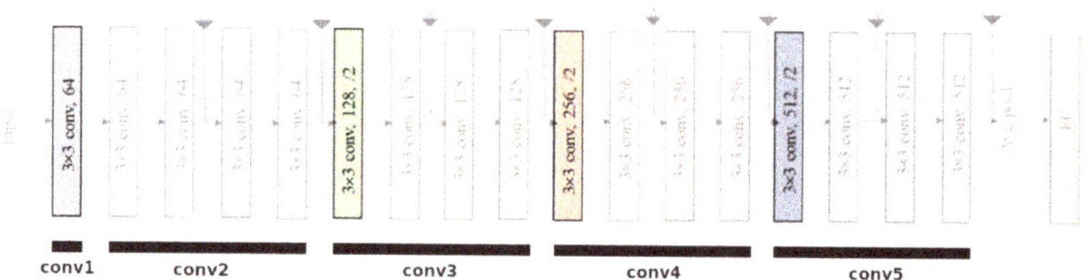

Figure 9. All 17 convolutional layers and one fully-connected layer of the unmodified ResNet18 network. The convolutional layers are grouped into 5 groups: conv1, conv2, conv3, conv4, and conv5. There are four convolutional layers that downsample the input tensor by using a stride of 2. They are the first convolutional layers in groups conv1, conv3, conv4, and conv5. These layers are emphasized in the image. There is also a max pooling layer between groups conv1 and conv2, which also downsamples the image by using a stride of 2. Since the input image is downsampled five times, the resulting tensor which is output by the final convolutional layer is 32 times smaller in both spatial dimensions than the input image.

We considered the dimensionality of our input data: the vertical dimension is 1024, while the horizontal dimension is 30. Furthermore, while the two spatial dimensions of reconstructed images are similar (496 and 369), with an aspect ratio of 1.34, we can see this is not the case with raw radar data, where the horizontal dimension is much smaller than the vertical. In general, since we are focused on GBSAR with a limited aperture size, the horizontal dimension—which corresponds to the number of steps—will usually be low.

With that in mind, we see that the resulting tensor after the convolutional layers in ResNet18 is downsampled to have spatial dimensions (32, 1). Our modification changes the network so that it does not downsample the input data in the horizontal dimension at all. As we described, subsampling is performed in four of the convolutional layers and one max pooling layer of the network. In all cases, the subsampling is performed by using a stride of 2 in both spatial dimensions for the sliding window of either a convolutional or a max pooling layer. We change the horizontal stride of these downsampling layers from 2 to 1, while we keep the vertical stride unchanged. Figure 10 shows how the raw data input matrix is gradually subsampled in ResNet18 after the first convolutional layer (conv1), and each convolutional group (conv2, conv3, conv4, conv5), with and without our modification. The dimensions in the figure are not to scale since it would be impractical to display. Rather, it shows symbolically that without our modification, the horizontal dimension is subsampled from 30 to 1, while with the modification it remains 30.

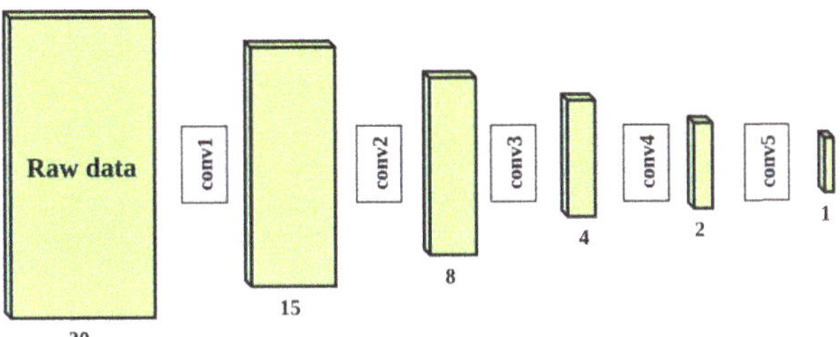

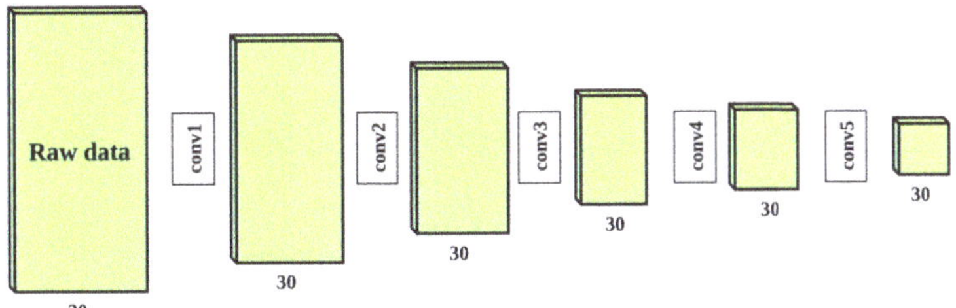

Figure 10. The spatial dimensions of the input raw data matrix and subsequent subsampled intermediate representations after each group of convolutional layers. For groups conv1, conv3, conv4, and conv5, the subsampling is completed in the first convolutional layer of the group. For group conv2, the subsampling is completed in the max pooling layer immediately before the first convolutional layer of conv2. The first diagram show how, without any modification to the ResNet18 architecture, both the vertical and horizontal dimensions are halved five times. The vertical dimension is downsampled from 1024 to 32, while the horizontal dimension is downsampled from 30 to 1. The second diagram shows how only the vertical dimension is subsampled after our modification to the ResNet18 architecture, while the horizontal dimension remains constant. This is because our modification sets the horizontal stride of subsampling layers to 1. Note that the dimensions in the figure are not to scale due to impracticality of displaying very tall and narrow matrices.

This modification does not change the dimensionality of any kernels of the convolutional layers, so it does not add any parameters to the model. Thus, it does not prevent us from using any pretrained set of parameters of the original ResNet18 architecture. Even though the horizontal stride of the modified network is different, which can change the horizontal scale and appearance of features in deeper layers of the network, using existing ImageNet-pretrained weights is still a sensible initialization procedure. ImageNet pretraining has been shown to be consistently beneficial in a wide array of image classification tasks, some of which have different image dimensions, scales of objects appearing in the images, and even cover an entirely different domain of images than the ImageNet-1k dataset [56,57]. We empirically validate the contribution of pretraining for weight initialization in our experiments.

The modification results in the dimension of the tensor output by the last convolutional layer in the network being (32, 30), since the input is not subsampled in the horizontal dimension. The depth of that tensor remains 512, and the global average pooling layers aggregates this tensor across the spatial dimensions into a vector of size 512. Thus, the vector representation which is fed to the fully-connected output layer is of the same dimensionality as before the modification, the output layer also does not require any changes.

5. Experiments

5.1. Experiment Setup

To evaluate our models we perform experiments on the RealSAR dataset. We evaluate and compare two main approaches, each with a number of modifications regarding data augmentation procedures, weight initialization, the size of the training set and the classification architecture used. The first approach consists of models which work on reconstructed images. We train and evaluate them on the RealSAR-IMG dataset. In the second approach, we evaluate models which work with raw radar data. Experiments for the second approach are performed on the RealSAR-RAW dataset. All experiments were completed in a multi-label classification setting, where each label denotes the absence or presence of one of the three different objects that we detect. The training procedure minimized the mean of binary cross-entropy losses of all labels.

We chose all hyperparameters manually, by testing and comparing different combinations of values on the validation set. In all experiments for both models, we use the Adam optimizer, with a learning rate of 2×10^{-4}, and the weight decay parameter set to 3×10^{-4}.

Experiments on the full training set (156 examples) are performed over 30 epochs, with a batch size of 16, which results in 300 parameters updates. To ensure comparisons are fair, when training on subsets of the training set, the number of epochs is chosen, such that the total number of parameter updates stays the same.

To prevent our models from overfitting and increase robustness to noise, we use two stochastic data augmentation procedures during training: random horizontal flipping and color jittering. Since the radar moves laterally when recording a scene the raw recorded data are equivariant to changes in the lateral positions of objects in the scene. The Omega-K algorithm preserves the equivariance as well. Thus, for both the raw and the vision representation of any given scene, the horizontally flipped representation corresponds to rearranging the scene to be symmetric with respect to the middle of the lateral axis. Thus, the horizontal flip transformations does not change the semantics, i.e., the labels of that representation, and can be used as an augmentation procedure. Color jittering is used in computer vision to stochastically apply small photometric transformations on images. This artificially increases the dataset, since in every epoch, each image is transformed with different parameters, sampled randomly. One novelty we introduce is to extend this technique to raw data as well, by adding small Gaussian noise to each example. We validate this contribution empirically.

We trained both the raw and the image-based models with different modifications. To explore the contribution of transfer learning [34], we used three different weight initialization procedures: random initialization [58], initializing with ImageNet-1k pretrained weights, and initializing by pretraining on the LabSAR dataset. To test how the model improves as more data are collected, we performed experiments on 25%, 50%, 75%, and 100% of examples in the training set. The validation and test sets remained fixed throughout all experiments to ensure fair comparisons. Through these experiments we converged to the two best models: one based on raw radar data, and the other based on reconstructed images. We used these two models to performed further analyses.

To test whether combining models trained on different input modalities yields an increase in performance, we evaluated an ensemble model. The ensemble consists of the best RAW and the best IMG model. Its output vector is the mean of the output vectors of the two models.

5.2. Metrics

All of our models are trained in the multi-label classification setting. The model output P is a vector of n numbers, where n is the number of different objects which can appear in a scene. The element i of the output vector ($P[i]$) is the probability given by the model that the corresponding object i is present in the scene. The probability of its absence is $1 - P[i]$. Thus, the detection of each individual object can be seen as a separate binary classification problem.

In binary classification evaluation, we have a set of binary ground truth labels for all examples, and the corresponding model predictions, either as probabilities or as binary values (0 and 1). When the predictions are binary values, the pair consisting of the ground truth label and the prediction for any example in the dataset can be classified into one of four sets:

- True negative (TN): the object is not present, and the model predicts it correctly;
- False negative (FN): the object is present, but the model incorrectly predicts it is not;
- False positive (FP): the object is not present, but the model incorrectly predicts it is;
- True positive (TP): the object is present, and the model predicts it correctly.

Various metrics can be defined using the described quantities. Accuracy measures the percentage of correct predictions:

$$\text{Acc} = \frac{\text{TN} + \text{TP}}{\text{TP} + \text{TN} + \text{FP} + \text{FN}} \tag{8}$$

Precision (P) measures how many positive predictions were correct:

$$P = \frac{\text{TP}}{\text{TP} + \text{FP}} \tag{9}$$

It decreases as the number of false positives increases. Recall (R) measures how many positive examples were captured:

$$R = \frac{\text{TP}}{\text{TP} + \text{FN}} \tag{10}$$

It decreases as the number of false negatives increases. If we want a strict metric which will penalize both excessive false positives and false negatives, we can use the $F1$-score, which is the harmonic mean of precision and recall:

$$F1 = \frac{2 \cdot P \cdot R}{P + R} = \frac{2 \cdot \text{TP}}{2 \cdot \text{TP} + \text{FP} + \text{FN}} \tag{11}$$

Since the output $P[i]$ of the model is the probability of the positive class, we classify an example as a positive if the probability is higher than a certain threshold $T[i]$. Otherwise, we classify it as a negative. The natural threshold to choose is 0.5. However, the actual optimal threshold varies depending on the metric. For example, recall increases as the threshold decreases. The minimum threshold (0) will maximize recall (1.0), since it will capture all positive examples. A threshold that is higher than the highest prediction probability will minimize recall (0.0), since it will not capture any positive examples. Conversely, a higher threshold tends to result in higher precision, since only very certain predictions will be classified as positive.

To remove the variability of the choice of the threshold from model evaluation and comparison, the average precision (AP) metric is often used. It is calculated as the average over precision values for every possible recall value. To extend it to the multi-label setting (multiple binary classification problems), we use the mean average precision (mAP) which is the arithmetic mean of the AP values of all individual tasks.

Once the best model has been chosen according to mAP, in order to use it to make predictions, we have to find the optimal threshold parameter for each individual output. We want our model to perform well on all classes, i.e., all 8 possible subsets of objects that can appear in the scene. Thus, we choose a vector of thresholds such that it maximizes the macro-$F1$ score on the validation set [59].

The macro-$F1$ score is a multi-class extension of the $F1$ score. It is calculated as the arithmetic mean over the $F1$-score of each class.

$$\text{macro-}F1 = \frac{1}{n} \sum_{k=1}^{n} F1_k \tag{12}$$

To calculate the $F1$-score of a given class c_k in a multi-class setting, the results are reformulated as a binary classification problem. Class c_k is treated as the positive class, while all other classes are grouped into one, negative, class. The multi-class confusion matrix is accordingly transformed into a binary confusion matrix. The quantities necessary for calculating the binary precision and recall values are then obtained as shown on Figure 11.

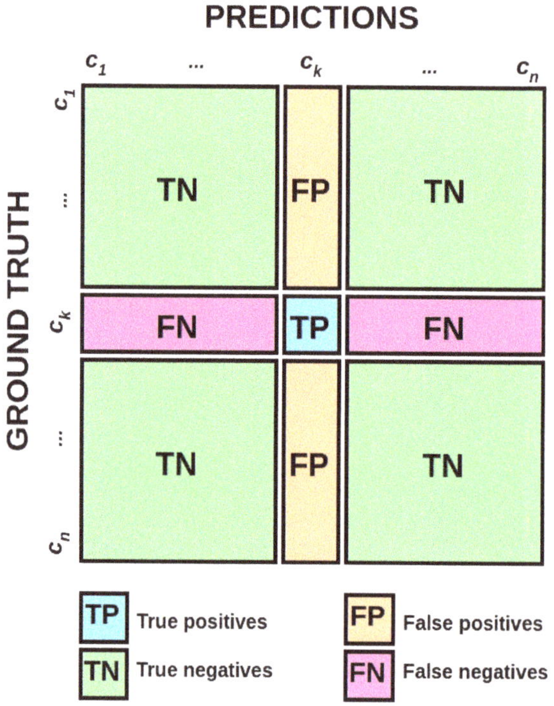

Figure 11. The transformation of a multi-class confusion matrix into a binary confusion matrix for class c_k. The macro-$F1$ score is calculated as the arithmetic mean of $F1$ scores of all classes [60].

5.3. Results

We consider two main approaches for GBSAR object classification, which correspond to the two possible input modalities: raw radar data (RAW) and reconstructed images (IMG). For raw data approaches, we compare five different classification architectures described in Section 4: the fully-connected baseline model (FC), the LSTM-based baseline model (LSTM), the ResNet18 (RN18) and MobileNetV3 (MNv3) convolutional architectures applied without any modifications, and, finally, our ResNet18 modified to handle raw radar data (RAW-RN18). For image-based approaches, we compare classifiers based on two convolutional architectures: ResNet18 and MobileNetV3.

Firstly, we validate the contribution of extending jittering as a data augmentation procedure to the raw input data modality. Table 2 shows the mAP results for all raw data approaches with and without jittering. For a given class, the AP metric averages the precision score over all possible recall scores, which correspond to different threshold values for classification. This makes our comparison of different variants of our approaches invariant to the threshold value. The mean AP score (mAP) is obtained by averaging the AP over all classes. It can be seen that jittering improves performance across the board. We also see that popular computer vision architectures ResNet18 and MobileNetV3 offer considerable improvements compared to the two baseline approaches, even when applied to raw data without any modification.

Table 2. Comparison of performance of all raw data models with and without jittering. The metric used is mean average precision (mAP).

Augmentation	Model				
	FC	LSTM	MNv3	RN18	RAW-RN18
No jittering	90.89	92.51	96.07	96.95	99.51
With Jittering	90.97	92.87	6.23	97.12	99.73

We compare all considered raw and image-based classification models, combined with different weight initialization procedures. The weight initialization procedures that we consider are the following: random initialization (random), pretraining on the LabSAR dataset (LabSAR), pretraining on the ImageNet-1k dataset (ImageNet), and pretraining on ImageNet-1k followed by LabSAR (ImageNet + LabSAR). Table 3 shows the mAP results for all combinations for the raw and image-based classification models. We see that ResNet18 and MobileNetV3 improve upon the baseline for both input modalities. However, for raw data, our modified ResNet18 (RAW-RN18), which ensures that the input tensor is not subsampled in the spatial dimension, significantly outperforms those models. As expected, we see that the vanilla, unmodified ResNet18 and MobileNetV3 architectures are more apt for image-based input, as they achieve significantly better results there compared to raw data. We also observe that out of the two convolutional architectures that we considered, the ResNet18 network consistently outperforms MobileNetV3 for this task. Finally, we can see that pretraining on ImageNet generally improves performance, while LabSAR pretraining is only beneficial in some cases, with a smaller impact. The knowledge learned from pretraining on LabSAR measurements with limited variance due to the controlled laboratory conditions in which they were captured proved not to transfer as significantly to the more complex RealSAR dataset.

Table 3. Comparison of performance of all considered raw and image-based classification models in combination with all different weight initialization procedures on the validation set. The metric used is mean average precision (mAP).

Input Type	Model	Weight Initialization			
		Random	ImageNet	LabSAR	ImageNet + LabSAR
RAW	MNv3	96.23	96.59	96.31	96.39
	RN18	97.12	97.42	97.27	97.38
	RAW-RN18	99.73	99.71	99.67	**99.95**
IMG	MNv3	98.29	98.71	98.34	98.66
	RN18	99.35	**99.75**	99.04	99.64

Based on the results of these experiments, we converged to two main approaches for subsequent experiments. For raw data, we use our modified ResNet18 (RAW-RN18) model, while for image data we use the standard ResNet18. To observe how the size of the training dataset impacts performance, we train our models on subsets that contain 25%, 50%, and 75% of all training set examples. Thus, we perform training with all of the following combinations:

- Model type: IMG, RAW;
- Weight initialization: random, LabSAR pretrain, ImageNet-1k pretrain;
- Training set size: 1/4, 1/2, 3/4, Full;

The results on the validation set are displayed in Table 4. They show that models based on raw data experience a smaller drop in performance when trained on a smaller training set.

Table 4. Comparison of performance of the two best model configurations for image and raw data classification. The chosen image model was an unmodified ResNet18 (RN18), while the chosen raw data model was a ResNet18 with our modification which prevents horizontal subsampling (RAW-RN18). We compare the two models across different weight initialization procedures and training set sizes on the validation set. The metric used is mean average precision (mAP).

Model	Training Set Size			
	0.25	0.5	0.75	Full
IMG, RN18, random	96.56	98.72	99.33	99.35
IMG, RN18, ImageNet	96.25	98.63	99.33	**99.75**
IMG, RN18, LabSAR	98.07	99.65	99.03	99.04
IMG, RN18, ImageNet + LabSAR	96.56	98.94	99.31	99.64
RAW, RAW-RN18, random	98.25	99.13	99.66	99.73
RAW, RAW-RN18, ImageNet	98.66	99.51	99.60	99.71
RAW, RAW-RN18, LabSAR	98.74	98.89	99.23	99.67
RAW, RAW-RN18, ImageNet + LabSAR	98.45	99.63	99.82	**99.95**

We chose one RAW model and one IMG model with the best performance on the validation set to perform further analyses. The best IMG model was pretrained on ImageNet, while the best RAW model was pretrained first on ImageNet, then on LabSAR. Per-class AP scores on the test set for the two chosen models are displayed in Table 5. The mAP scores on the test set of both models are significantly lower than results on the validation set. This is expected since all results in Table 4 are chosen from the best epoch, as measured on the validation set. The validation set is used to choose the best hyperparameters for the model, as is standard practice in machine learning [43,61]. On the other hand, the test set results of chosen models are realistic mAP performance estimates. The RAW model has the higher AP score for classes aluminium and plastic, while the IMG model is slightly better for the glass bottles. The RAW model also has the higher mean AP score. This suggests that classification based on raw radar data, which circumvents lossy reconstruction steps, coupled with architectural modifications of neural networks can yield better performance than traditional computer vision approaches on reconstructed images. Figure 12 shows the average test mAP of the RAW and IMG pairs of models for each combination of the weight initialization scheme and training set size. We notice the general trend of increasing performance as the training set grows. The plots in the graph do not seem to be in a saturation regime, so additional data would be expected to increase the performance further. ImageNet pretraining is shown to be beneficial both in regards to the total performance and to the stability across different training set sizes.

Table 5. Per-class AP performance of the best IMG and RAW models on the test set.

	Aluminium	Plastic	Glass	Mean
IMG, ImageNet	98.07	97.42	98.07	97.85
RAW, ImageNet + LabSAR	98.71	98.12	97.85	98.23

To obtain a thorough evaluation of the performance of our two chosen models, we also test them in a multi-class setting. There are eight classes that cover all possible combinations of objects in the scene. In this way, we can get more insight into how the model behaves for objects when they appear individually in the scene versus how different groups of objects interact. This is especially interesting since the considered objects have varying reflectances and, thus, certain pairs of objects might be more difficult to discern among than others. We average the F1 score of each class to obtain the macro-F1 score. Although the comparison of different approaches in Table 4 was performed over all thresholds (mAP), for a multi-class comparison of the two chosen models, we need to choose concrete threshold parameters. For each of the two models, we find a threshold vector which maximizes the macro F1 score over all classes on the validation set. Table 6 shows the validation F1 score of each class, along with the macro F1 score, for both models. Once the threshold has been chosen using the validation set, we evaluate the models on the test set. The test set results are shown in Table 7. The RAW model outperforms the IMG model in all classes except for the glass–plastic combinations. The test set results also include the results of the ensemble model created by averaging the predictions of the two models.

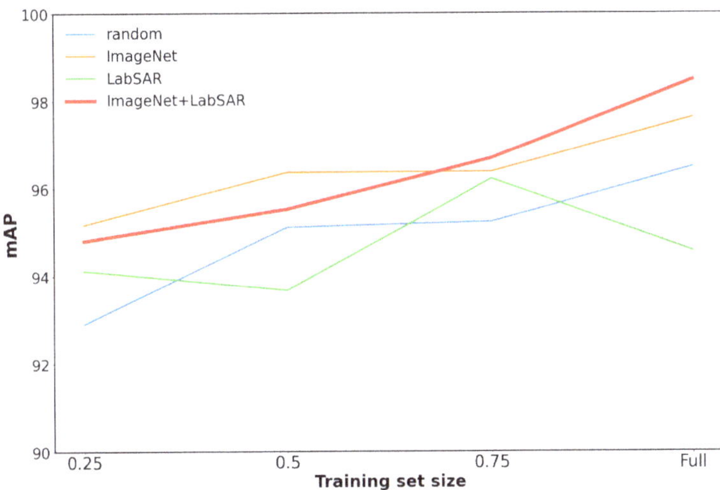

Figure 12. Average test mAP of the RAW and IMG pair of models for each combination of weight initialization scheme and training set size.

Table 6. Per-class F1 scores and the macro F1 score for the best threshold values on the validation set. E—Empty, A—Aluminium, P—Plastic, and G—Glass.

	E	A	P	G	A, P	A, G	G, P	A, G, P	Mean
IMG, ImageNet	90.91	100	83.33	87.50	94.12	100	100	100	94.48
RAW, ImageNet + LabSAR	90.91	93.33	95.24	94.74	100	100	100	100	96.78

Table 7. Per-class F1 scores and the macro F1 score of the two best models and their ensemble on the test set. Classes: E—Empty, A—Aluminium, P—Plastic, and G—Glass.

	E	A	P	G	A, P	A, G	G, P	A, G, P	Mean
IMG, ImageNet	80.00	66.67	78.26	93.33	88.89	75.00	**94.74**	94.12	83.88
RAW, ImageNet + LabSAR	**83.33**	**75.00**	**90.00**	**94.12**	**94.12**	**87.50**	81.82	**100.00**	**88.24**
Ensemble	81.12	71.12	82.55	93.33	91.25	79.73	86.74	97.24	85.39

Out of the 67 test examples, the IMG model misclassifies 11 examples, while for the RAW model, the number of incorrect classifications is 8. This yields accuracy scores of 83.6% and 88.1%, respectively. Looking at the error sets, we found that all of the examples misclassified by the RAW model are also misclassified by the IMG model. In other words, there are no examples where the IMG model outperformed the RAW model. This explains why the ensemble model did not improve upon the performance of the single RAW model. The general idea of ensemble models is to aggregate predictions of different base estimators, each of which is more specialized in a certain sub-region of the input space than other estimators. In our case, the image-based model is not better than the raw model in any sub-region of the input space.

The confusion matrices of both models are shown in Figure 13a,b. The numbers of misclassified examples for different combinations of ground truth and prediction values are contained in elements off the diagonal.

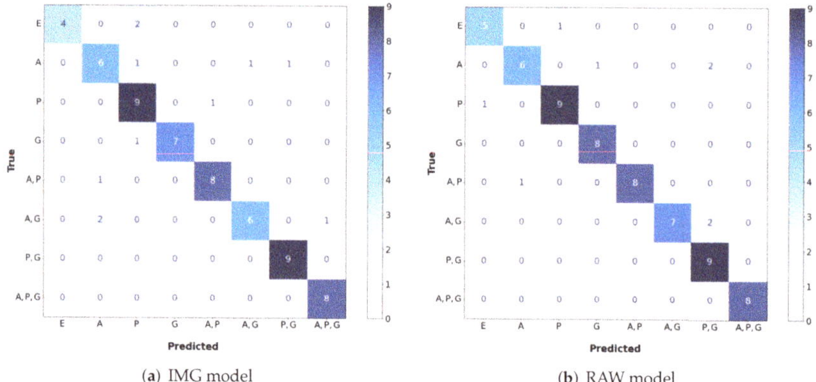

(a) IMG model   (b) RAW model

Figure 13. Confusion matrices for the best IMG and RAW models. E—Empty, A—Aluminium, P—Plastic, and G—Glass.

6. Discussion

The main goal of this work was to compare two deep learning approaches for object classification on GBSAR data. Experimental results presented in the paper show that the model trained on raw data (RAW) consistently outperforms the image-based model (IMG), which uses the same data preprocessed with an image reconstruction algorithm. Specifically, out of 67 test examples, the IMG model misclassified 11, while the RAW model misclassified 8. The overlap of the two misclassification sets is peculiar. All 8 examples misclassified by the RAW model are also misclassified by the IMG model. This means there is no example in the test set where using the IMG model was beneficial compared to the RAW model. The three test examples where the IMG model made a mistake while the RAW model was correct (one of which is shown in Figure 14a) indicate that reconstruction algorithms degrade the information in signals obtained by GBSAR due to approximations, as described in Section 2.2.

Similar behavior is shown in Table 7: the F1 score is higher in the RAW model for every class except for the combination of glass and plastic (G, P). Half of the 8 examples incorrectly classified by the RAW model were predicted to be in the (G, P) class, as seen in the confusion matrix in Figure 13b. This significantly decreased the precision score, and, consequently, the F1 score, of the RAW model on that class. The table also shows that both models have difficulties classifying aluminium and combinations with aluminium. Aluminium has a much higher reflectance than glass and especially plastic. Consequently, it is hard for models to differentiate between scenes where the aluminium object appears alone versus ones where it appears together with another object of significantly lower reflectance. One such example is shown in Figure 14b, in which a scene containing both an aluminium and a plastic bottle is mistaken by both models for a scene containing only an aluminium bottle. The confusion matrices in Figure 13a,b also capture this phenomenon. Most of the misclassifications of both models are found in rows and columns that represent the aluminium class. The confusion matrices also highlight the difficulty of distinguishing an empty scene from one containing only a plastic bottle due to the very low reflectance of plastic.

In addition to reflectance and the aforementioned approximations, there is another factor impacting the results of the IMG model—the heatmap scale of reconstructed images. Total GBSAR signal intensities of scenes containing aluminium are much larger than signal intensities of empty scenes and scenes containing plastic bottles. Consequently, using a fixed scale in the visualization would result in objects fading out from reconstructed images due to their very high or very low intensity responses. By choosing not to set a fixed scale, we reduced the likelihood of such a scenario. However, this means that the resulting pixel intensities in reconstructed images are not absolute. For example, two pixels with equal color intensities in a scene with aluminium and an empty scene correspond to different raw signal intensities from which they were reconstructed. Hence, similar reconstructed images of two scenes with different objects can lead to misclassifications, such as the one presented in Figure 14a. RAW model classified that test example correctly as glass bottle, but IMG model mistook it with plastic one. In problems containing materials with less variance in terms of reflectance, the scale might be fixed.

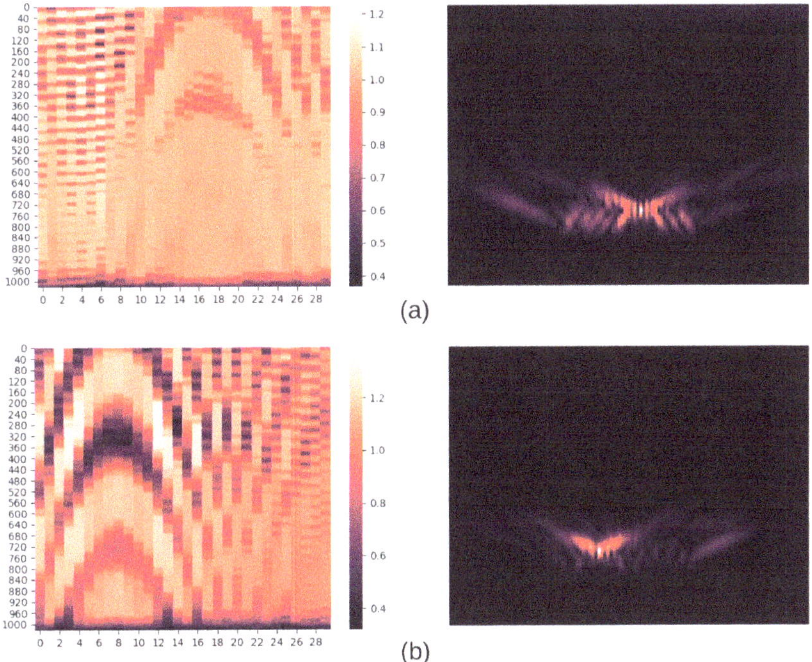

Figure 14. RealSAR-RAW (left) and RealSAR-IMG (right) misclassified examples. Example (**a**) is misclassified by the IMG model while example (**b**) is misclassified by both RAW and IMG models. In (**a**) the recorded scene included a glass bottle, but IMG model classified it as plastic bottle. In (**b**) the recorded scene included an aluminium and a plastic bottle but both models classified it as scene with an aluminium bottle only.

Even though the comparisons between the two input modalities suggest the raw-data approach is superior, the usage of reconstructed images has certain advantages. Reconstructed images are more interpretable to humans, especially regarding the locations of the objects in the scene. It is extremely challenging for humans to discern objects and their locations in raw GBSAR data. Reconstructed images also allow for meaningful error analyses, making it easier to deduce why particular examples are misclassified. A valid paradigm might be to use raw data for the actual classification and only generate reconstructed images when we need to analyze specific examples or localize objects.

Regarding weight initialization, both models reached the highest classification accuracy when they were pretrained on the ImageNet dataset. ImageNet pretraining also decreased the variance of the results with respect to the size of the training set, as seen in the plot in Figure 12. Table 4 shows that the ImageNet-pretrained variant of the model based on reconstructed images achieved the highest mean average precision. In the case of the raw data model, the highest mAP was achieved by pretraining on both ImageNet and LabSAR datasets. This highlights the importance of pretraining and suggests that the RealSAR dataset might also be useful in future radar-related deep learning research.

7. Conclusions

The presented paper investigates the potential and benefits of processing raw GBSAR data in automated radar classification which can be particularly useful for industrial applications focused on monitoring and object detection in environments with limited visibility where this approach can save considerable resources and time. This differs from, and in this paper is contrasted with, the standard practice of applying classification algorithms on reconstructed images. The testing setup was developed around an FMCW based GBSAR system which was designed and constructed using low-cost components. The developed GBSAR was used in a series of measurements that were performed on several objects made of different materials with the final intention to realize a complete GBSAR radar system with embedded computer control capable of classifying these objects.

For classification purposes, a detailed analysis and comparison of two deep learning approaches to GBSAR object recognition was performed. The first approach was a more conventional SAR classification approach based on reconstructed GBSAR images, while in the second approach the classification was performed essentially on raw (unprocessed) data. Multiple different deep learning architectures for both approaches were trained, tested, and compared. These included two baselines based on fully-connected layers and LSTM networks, two standard convolutional neural network architectures-ResNet18 and MobileNetV3-popular in computer vision, and a modified variant of the ResNet18 network which makes it suitable for processing raw GBSAR data. The modification consists of preventing subsampling of the input in the horizontal dimension which is generally small due to the nature of GBSAR data. This modification is our main contribution regarding deep learning model design. The contribution was validated through experiments which showed that the best results on raw data are achieved by the modified version of ResNet18. Additionally, the practice of color jittering as a common data augmentation procedure in computer vision was extended to classification of raw GBSAR data and was shown to provide consistent, though small, improvements. Furthermore, it was shown that classification on raw data outperforms the classification based on reconstructed images. This was partially expected since the SAR image reconstruction algorithms necessarily introduce certain approximations, and with this negatively influence the integrity of the recorded data. In addition to better classification performance, raw data classification is inherently faster since it avoids the need for image reconstruction, and with this is more suitable for embedded computer implementation which opens possibilities for various application scenarios. On the other hand, it limits human visual confirmation and disables approaches in which radar images are combined with optical ones. However, keeping in mind that the primary focus is on applications in embedded systems, this is not a serious hindrance.

Even though this study has shown the applicability of this concept, it was tested using a relatively small dataset in order to focus on the comparison between approaches. Using larger datasets, more classes, and more general problem formulations could lead to more powerful and useful models. The other directions is to seek improvements in terms of more efficient networks and implementation in embedded computers with limited resources. With this in mind, for potential use in future research we generated and publicly released these datasets of raw GBSAR data and reconstructed radar images which we intend to update with time.

Author Contributions: Conceptualization, M.K., F.T., D.B. and M.B.; methodology, M.K., F.T., D.B. and M.B.; software, M.K. and F.T.; validation, M.K., F.T., D.B. and M.B.; formal analysis, M.K. and F.T.; investigation, M.K. and F.T.; resources, M.B.; data curation, M.K. and F.T.; writing—original draft preparation, M.K. and F.T.; writing—review and editing, D.B. and M.B.; visualization, M.K. and F.T.; supervision, D.B. and M.B.; project administration, M.B.; funding acquisition, M.B. All authors have read and agreed to the published version of the manuscript.

Funding: This work was supported in part by Croatian Science Foundation (HRZZ) under the project number IP-2019-04-1064.

Data Availability Statement: RealSAR datasets (RAW and IMG) used in this research are publicly available: https://data.mendeley.com/datasets/m458grc688/draft?a=ff342b09-dd03-4d09-a169-560af2f87773 (accessed on 1 November 2022).

Conflicts of Interest: The authors declare no conflict of interest.

Abbreviations

The following abbreviations are used in this manuscript:

SAR	Synthetic Aperture Radar
GBSAR	Ground Based SAR
CNN	Convolutional Neural Network
FMCW	Frequency Modulated Continuous Wave
FFT	Fast Fourier Transform
IFFT	Inverse FFT
RPi	Microcomputer Raspberry Pi 4B
VCO	Voltage-Controlled Oscillator
SNR	Signal to Noise Ratio
AP	Average Precision
mAP	Mean Average Precision

References

1. Copernicus Space Component Mission Management Team. Sentinel High Level Operations Plan (HLOP). Available online: https://sentinels.copernicus.eu/documents/247904/685154/Sentinel+HLOP+-+Issue+3.1+-+16+Dec+2021.pdf (accessed on 23 July 2022).
2. Liu, B.; He, K.; Han, M.; Hu, X.; Ma, G.; Wu, M. Application of UAV and GB-SAR in Mechanism Research and Monitoring of Zhonghaicun Landslide in Southwest China. *Remote Sens.* **2021**, *13*, 1653. [CrossRef]
3. Izumi, Y.; Zou, L.; Kikuta, K.; Sato, M. Anomalous Atmospheric Phase Screen Compensation in Ground-Based SAR over Mountainous Area. In Proceedings of the IGARSS 2019—2019 IEEE International Geoscience and Remote Sensing Symposium, Yokohama, Japan, 28 July–2 August 2019; pp. 2030–2033. [CrossRef]
4. Chet, K.V.; Siong, L.C.; Hsin, W.H.H.; Wei, L.L.; Guey, C.W.; Yam, C.M.; Sze, L.T.; Kit, C.Y. Ku-band ground-based SAR experiments for surface deformation monitoring. In Proceedings of the 2015 IEEE 5th Asia-Pacific Conference on Synthetic Aperture Radar (APSAR), Singapore, 1–4 September 2015; pp. 641–644. [CrossRef]
5. Ma, Z.; Mei, G.; Prezioso, E.; Zhang, Z.; Xu, N. A deep learning approach using graph convolutional networks for slope deformation prediction based on time-series displacement data. *Neural Comput. Appl.* **2021**, *33*, 14441–14457. [CrossRef]
6. Tarchi, D.; Antonello, G.; Casagli, N.; Farina, P.; Fortuny-Guasch, J.; Guerri, L.; Leva, D. On the Use of Ground-Based SAR Interferometry for Slope Failure Early Warning: The Cortenova Rock Slide (Italy). In *Landslides: Risk Analysis and Sustainable Disaster Management*; Sassa, K., Fukuoka, H., Wang, F., Wang, G., Eds.; Springer: Berlin/Heidelberg, Germany, 2005; pp. 337–342. [CrossRef]
7. Martinez-Vazquez, A.; Fortuny-Guasch, J. A GB-SAR Processor for Snow Avalanche Identification. *IEEE Trans. Geosci. Remote Sens.* **2008**, *46*, 3948–3956. [CrossRef]
8. Miccinesi, L.; Consumi, T.; Beni, A.; Pieraccini, M. W-band MIMO GB-SAR for Bridge Testing/Monitoring. *Electronics* **2021**, *10*, 2261. [CrossRef]
9. Qiu, Z.; Jiao, M.; Jiang, T.; Zhou, L. Dam Structure Deformation Monitoring by GB-InSAR Approach. *IEEE Access* **2020**, *8*, 123287–123296. [CrossRef]
10. Du, S.; Feng, G.; Wang, J.; Feng, S.; Malekian, R.; Li, Z. A New Machine-Learning Prediction Model for Slope Deformation of an Open-Pit Mine: An Evaluation of Field Data. *Energies* **2019**, *12*, 1288. [CrossRef]
11. Kang, M.K.; Kim, K.E.; Cho, S.J.; Lee, H.; Lee, J.H. Wishart supervised classification results of C-band polarimetric GB-SAR image data. In Proceedings of the 2011 IEEE International Geoscience and Remote Sensing Symposium, Vancouver, BC, Canada, 24–29 July 2011; pp. 459–462. [CrossRef]
12. Feng, S.; Ji, K.; Zhang, L.; Ma, X.; Kuang, G. SAR Target Classification Based on Integration of ASC Parts Model and Deep Learning Algorithm. *IEEE J. Sel. Top. Appl. Earth Obs. Remote Sens.* **2021**, *14*, 10213–10225. [CrossRef]
13. Liu, H.; Li, S. Decision fusion of sparse representation and support vector machine for SAR image target recognition. *Neurocomputing* **2013**, *113*, 97–104. [CrossRef]
14. Huang, Z.; Pan, Z.; Lei, B. Transfer Learning with Deep Convolutional Neural Network for SAR Target Classification with Limited Labeled Data. *Remote Sens.* **2017**, *9*, 907. [CrossRef]
15. Chen, S.; Wang, H.; Xu, F.; Jin, Y.Q. Target Classification Using the Deep Convolutional Networks for SAR Images. *IEEE Trans. Geosci. Remote Sens.* **2016**, *54*, 4806–4817. [CrossRef]

16. Liu, J.; Xing, M.; Yu, H.; Sun, G. EFTL: Complex Convolutional Networks With Electromagnetic Feature Transfer Learning for SAR Target Recognition. *IEEE Trans. Geosci. Remote Sens.* **2022**, *60*, 1–11. [CrossRef]
17. Zhang, J.; Xing, M.; Xie, Y. FEC: A Feature Fusion Framework for SAR Target Recognition Based on Electromagnetic Scattering Features and Deep CNN Features. *IEEE Trans. Geosci. Remote Sens.* **2021**, *59*, 2174–2187. [CrossRef]
18. Pei, J.; Huang, Y.; Huo, W.; Zhang, Y.; Yang, J.; Yeo, T.S. SAR Automatic Target Recognition Based on Multiview Deep Learning Framework. *IEEE Trans. Geosci. Remote Sens.* **2018**, *56*, 2196–2210. [CrossRef]
19. Wagner, S.A. SAR ATR by a combination of convolutional neural network and support vector machines. *IEEE Trans. Aerosp. Electron. Syst.* **2016**, *52*, 2861–2872. [CrossRef]
20. Li, J.; Qu, C.; Peng, S.; Jiang, Y. Ship Detection in SAR images Based on Generative Adversarial Network and Online Hard Examples Mining. *Dianzi Yu Xinxi Xuebao/J. Electron. Inf. Technol.* **2019**, *41*, 143–149. [CrossRef]
21. Li, J.; Qu, C.; Peng, S.; Deng, B. Ship detection in SAR images based on convolutional neural network. *Xi Tong Gong Cheng Yu Dian Zi Ji Shu/Syst. Eng. Electron.* **2018**, *40*, 1953–1959. [CrossRef]
22. Zhang, T.; Zhang, X. High-Speed Ship Detection in SAR Images Based on a Grid Convolutional Neural Network. *Remote Sens.* **2019**, *11*, 1206. [CrossRef]
23. Li, J.; Qu, C.; Shao, J. Ship detection in SAR images based on an improved faster R-CNN. In Proceedings of the 2017 SAR in Big Data Era: Models, Methods and Applications (BIGSARDATA), Beijing, China, 13–14 November 2017; pp. 1–6. [CrossRef]
24. Xia, R.; Chen, J.; Huang, Z.; Wan, H.; Wu, B.; Sun, L.; Yao, B.; Xiang, H.; Xing, M. CRTransSar: A Visual Transformer Based on Contextual Joint Representation Learning for SAR Ship Detection. *Remote Sens.* **2022**, *14*, 1488. [CrossRef]
25. Koyama, C.N.; Sato, M. Detection and classfication of subsurface objects by polarimetric radar imaging. In Proceedings of the 2015 IEEE Radar Conference, Johannesburg, South Africa, 27–30 October 2015; pp. 440–445. [CrossRef]
26. Yigit, E.; Demirci, S.; Unal, A.; Ozdemir, C.; Vertiy, A. Millimeter-wave Ground-based Synthetic Aperture Radar Imaging for Foreign Object Debris Detection: Experimental Studies at Short Ranges. *J. Infrared Millim. Terahertz Waves* **2012**, *33*, 1227–1238. [CrossRef]
27. Zhang, T.; Zhang, X.; Ke, X.; Zhan, X.; Shi, J.; Wei, S.; Pan, D.; Li, J.; Su, H.; Zhou, Y.; et al. LS-SSDD-v1.0: A Deep Learning Dataset Dedicated to Small Ship Detection from Large-Scale Sentinel-1 SAR Images. *Remote. Sens.* **2020**, *12*, 2997. [CrossRef]
28. Zhang, T.; Zhang, X. A Dual-Polarization Information Guided Network for SAR Ship Classification. *arXiv* **2022**, arXiv:2207.04639.
29. Zhang, T.; Zhang, X. A polarization fusion network with geometric feature embedding for SAR ship classification. *Pattern Recognit.* **2022**, *123*, 108365. [CrossRef]
30. Zhang, T.; Zhang, X.; Liu, C.; Shi, J.; Wei, S.; Ahmad, I.; Zhan, X.; Zhou, Y.; Pan, D.; Li, J.; et al. Balance learning for ship detection from synthetic aperture radar remote sensing imagery. *ISPRS J. Photogramm. Remote Sens.* **2021**, *182*, 190–207. [CrossRef]
31. Zhang, T.; Zhang, X. A Mask Attention Interaction and Scale Enhancement Network for SAR Ship Instance Segmentation. *arXiv* **2022**, arXiv:2207.03912. https://doi.org/10.48550/arXiv.2207.03912.
32. Zhang, T.; Zhang, X. HTC+ for SAR Ship Instance Segmentation. *Remote Sens.* **2022**, *14*, 2395. [CrossRef]
33. Schmidhuber, J. Deep learning in neural networks: An overview. *Neural Netw.* **2015**, *61*, 85–117. [CrossRef]
34. Bengio, Y. Deep Learning of Representations: Looking Forward. In *SLSP 2013: Statistical Language and Speech Processing, Proceedings of the First International Conference, SLSP 2013, Tarragona, Spain, 29–31 July 2013*; Lecture Notes in Computer Science; Dediu, A., Martín-Vide, C., Mitkov, R., Truthe, B., Eds.; Springer: Berlin/Heidelberg, Germany, 2013; Volume 7978, pp. 1–37. [CrossRef]
35. Amézaga, A.; López-Martínez, C.; Jové, R. A Multi-Frequency FMCW GBSAR: System Description and First Results. In Proceedings of the 2021 IEEE International Geoscience and Remote Sensing Symposium IGARSS, Brussels, Belgium, 11–16 July 2021; pp. 1943–1946. [CrossRef]
36. Jankiraman, M. *FMCW Radar Design*; Artech House: London, UK, 2018.
37. Guo, S.; Dong, X. Modified Omega-K algorithm for ground-based FMCW SAR imaging. In Proceedings of the 2016 IEEE 13th International Conference on Signal Processing (ICSP), Chengdu, China, 6–10 November 2016; pp. 1647–1650. [CrossRef]
38. Cruz, H.; Véstias, M.; Monteiro, J.; Neto, H.; Duarte, R.P. A Review of Synthetic-Aperture Radar Image Formation Algorithms and Implementations: A Computational Perspective. *Remote Sens.* **2022**, *14*, 1258. [CrossRef]
39. Giroux, V.; Cantalloube, H.; Daout, F. An Omega-K algorithm for SAR bistatic systems. In Proceedings of the 2005 IEEE International Geoscience and Remote Sensing Symposium, IGARSS'05, Seoul, Korea, 29 July 2005; IEEE: New York, NY, USA, 2005; Volume 2, pp. 1060–1063.
40. Hamasaki, T.; Ferro-Famil, L.; Pottier, E.; Sato, M. Applications of polarimetric interferometric ground-based SAR (GB-SAR) system to environment monitoring and disaster prevention. In Proceedings of the European Radar Conference 2005—EURAD 2005, Paris, France, 3–4 October 2005; pp. 29–32. [CrossRef]
41. Innosent. Radar Sensor IVS-362. Available online: https://www.innosent.de/en/radarsensoren/ivs-series/ivs-362/ (accessed on 19 July 2022).
42. RSHub FER. Omega-K Algorithm in Python. Available online: https://github.com/filt27/OmegaK/blob/main/OmegaK.py (accessed on 24 July 2022).
43. Ojala, M.; Garriga, G.C. Permutation Tests for Studying Classifier Performance. *J. Mach. Learn. Res.* **2010**, *11*, 1833–1863. [CrossRef]

44. Zheng, A.; Casari, A. *Feature Engineering for Machine Learning: Principles and Techniques for Data Scientists*, 1st ed.; O'Reilly Media, Inc.: Sebastopol, CA, USA, 2018.

45. Collobert, R.; Weston, J.; Bottou, L.; Karlen, M.; Kavukcuoglu, K.; Kuksa, P.P. Natural Language Processing (Almost) from Scratch. *J. Mach. Learn. Res.* **2011**, *12*, 2493–2537. [CrossRef]

46. He, K.; Zhang, X.; Ren, S.; Sun, J. Deep Residual Learning for Image Recognition. In Proceedings of the 2016 IEEE Conference on Computer Vision and Pattern Recognition, CVPR 2016, Las Vegas, NV, USA, 27–30 June 2016; IEEE Computer Society: New York, NY, USA, 2016; pp. 770–778. [CrossRef]

47. He, K.; Zhang, X.; Ren, S.; Sun, J. Identity Mappings in Deep Residual Networks. In *ECCV 2016: Computer Vision—ECCV 2016, Proceedings of the 14th European Conference, Amsterdam, The Netherlands, 11–14 October 2016*; Leibe, B., Matas, J., Sebe, N., Welling, M., Eds.; Part IV; Lecture Notes in Computer Science; Springer: Berlin/Heidelberg, Germany, 2016; Volume 9908, pp. 630–645._38. [CrossRef]

48. Collobert, R.; Weston, J. A unified architecture for natural language processing: Deep neural networks with multitask learning. In *Machine Learning, Proceedings of the Twenty-Fifth International Conference (ICML 2008), Helsinki, Finland, 5–9 June 2008*; Cohen, W.W., McCallum, A., Roweis, S.T., Eds.; ACM International Conference Proceeding Series; ACM: Rochester, NY, USA, 2008; Volume 307, pp. 160–167. [CrossRef]

49. Bengio, Y.; Courville, A.C.; Vincent, P. Representation Learning: A Review and New Perspectives. *IEEE Trans. Pattern Anal. Mach. Intell.* **2013**, *35*, 1798–1828. [CrossRef]

50. Boutell, M.R.; Luo, J.; Shen, X.; Brown, C.M. Learning multi-label scene classification. *Pattern Recognit.* **2004**, *37*, 1757–1771. [CrossRef]

51. Howard, A.; Pang, R.; Adam, H.; Le, Q.V.; Sandler, M.; Chen, B.; Wang, W.; Chen, L.; Tan, M.; Chu, G.; et al. Searching for MobileNetV3. In Proceedings of the 2019 IEEE/CVF International Conference on Computer Vision, ICCV 2019, Seoul, Korea, 27 October–2 November 2019; IEEE: New York, NY, USA, 2019; pp. 1314–1324. [CrossRef]

52. Hochreiter, S.; Schmidhuber, J. Long Short-Term Memory. *Neural Comput.* **1997**, *9*, 1735–1780. neco.1997.9.8.1735. [CrossRef]

53. Krizhevsky, A.; Sutskever, I.; Hinton, G.E. ImageNet classification with deep convolutional neural networks. *Commun. ACM* **2017**, *60*, 84–90. [CrossRef]

54. Yildirim, Ö. A novel wavelet sequence based on deep bidirectional LSTM network model for ECG signal classification. *Comput. Biol. Med.* **2018**, *96*, 189–202. [CrossRef] [PubMed]

55. Mikolov, T.; Sutskever, I.; Chen, K.; Corrado, G.S.; Dean, J. Distributed Representations of Words and Phrases and their Compositionality. In *Advances in Neural Information Processing Systems 26, Proceedings of the 27th Annual Conference on Neural Information Processing Systems 2013, Lake Tahoe, NV, USA, 5–8 December 2013*; Burges, C.J.C., Bottou, L., Ghahramani, Z., Weinberger, K.Q., Eds.; Curran Associates, Inc.: Red Hook, NY, USA, 2013; pp. 3111–3119.

56. Studer, L.; Alberti, M.; Pondenkandath, V.; Goktepe, P.; Kolonko, T.; Fischer, A.; Liwicki, M.; Ingold, R. A Comprehensive Study of ImageNet Pre-Training for Historical Document Image Analysis. In Proceedings of the 2019 International Conference on Document Analysis and Recognition, ICDAR 2019, Sydney, Australia, 20–25 September 2019; IEEE: New York, NY, USA, 2019; pp. 720–725. [CrossRef]

57. Cherti, M.; Jitsev, J. Effect of pre-training scale on intra- and inter-domain, full and few-shot transfer learning for natural and X-ray chest images. In Proceedings of the International Joint Conference on Neural Networks, IJCNN 2022, Padua, Italy, 18–23 July 2022; IEEE: New York, NY, USA, 2022; pp. 1–9. [CrossRef]

58. Glorot, X.; Bengio, Y. Understanding the difficulty of training deep feedforward neural networks. In Proceedings of the Thirteenth International Conference on Artificial Intelligence and Statistics, AISTATS 2010, Chia Laguna Resort, Sardinia, Italy, 13–15 May 2010; Volume 9, pp. 249–256.

59. Fujino, A.; Isozaki, H.; Suzuki, J. Multi-label Text Categorization with Model Combination based on F1-score Maximization. In Proceedings of the Third International Joint Conference on Natural Language Processing, IJCNLP 2008, Hyderabad, India, 7–12 January 2008; The Association for Computer Linguistics: Stroudsburg, PA, USA, 2008; pp. 823–828.

60. Krüger, F. Activity, Context, and Plan Recognition with Computational Causal Behaviour Models. Ph.D. Thesis, University of Rostock, Rostock, Germany, 2016.

61. Yao, Y.; Rosasco, L.; Caponnetto, A. On Early Stopping in Gradient Descent Learning. *Constr. Approx.* **2007**, *26*, 289–315. [CrossRef]

MDPI

St. Alban-Anlage 66

4052 Basel

Switzerland

Tel. +41 61 683 77 34

Fax +41 61 302 89 18

www.mdpi.com

Remote Sensing Editorial Office

E-mail: remotesensing@mdpi.com

www.mdpi.com/journal/remotesensing